The Properties of Fresh Concrete

The Properties of Fresh Concrete

TREVAL C. POWERS

John Wiley & Sons, Inc.
New York • London • Sydney • Toronto

Library of Congress Catalog Card Number: 68-28504
SBN 471 69590 4
Printed in the United States of America

Preface

Many years of experience with concrete and countless laboratory research projects are the background for present-day specifications, pamphlets, and manuals pertaining to the making of concrete mixtures. These writings tell us how to make concrete, but they do not give or require more than a superficial understanding of it; indeed, they have made it so easy to learn the simple steps that the value of delving more deeply may be quite overlooked.

If concrete were invariably satisfactory in every respect, there would be little basis for contending that the making of it should be based on a better understanding of certain empirical relationships as well as fundamental matters than is generally evident. But concrete is not always satisfactory, and perhaps one of the evidences of that is that research continues in every industrialized nation.

In current research there is a trend toward the kind of study that leads to reliable generalizations, which means, of course, better comprehension of the fundamental nature of the material. Sometimes it is questionable, I think, to assume that more laboratory experimentation is the best next step toward that goal. Perhaps the time has come to give priority to analysis of information already available.

In writing this book I tried to analyze existing data and hypotheses about the nature of concrete in its fresh state, interpreting properties in terms of structure insofar as I was able. One result is that I, the author, came out with some old concepts modified or discarded, some new ones added, and generally a better comprehension of the subject than I had before I started. There is at least that much basis for believing that some readers will have a similar experience.

I do not consider this work an exercise in analysis but an essential step toward understanding the whole of concrete technology, which means the technology of mature concrete as well as that of concrete in its fresh state. It is a fund of basic knowledge for those responsible for the making of concrete, and is essential for those undertaking research on almost any of its aspects.

To see more clearly why this is true, consider first the nature of mature concrete. It is composed of particles of rock dispersed in a matrix of hardened cement paste and air bubbles. We are now beginning to understand its properties in terms of its internal structure. Particularly, its properties can be described and to a degree predicted in terms of the structure of the matrix, the structure or structures of the particles making up the mineral aggregate, and the forces of adhesion developed at the interface between the matrix and the mineral matter dispersed in it.

Fresh concrete, too, is composed of particles of rock dispersed in a matrix material composed of paste and air bubbles. The paste is not hard in the sense that mature paste is hard, but it has the properties of a solid to one degree or another. At the same time it is capable of taking on the properties of a fluid.

Such a material has structure, and it is this structure from which the structure of hardened paste grows. It is, so to speak, the pattern for the structure of mature paste, even though the original cement grains may become replaced completely by the physically and chemically different material called cement gel. The pattern just mentioned begins on a microscopic scale and develops into a structure composed of submicroscopic structural units.

Fresh concrete has a structure that comprises the structural features of fresh paste mentioned above, the pattern of fissures and other flaws that develops while the concrete is settling, and the pattern of density variation that develops during the same period. If freshly mixed concrete is ever isotropic, it becomes anisotropic before it sets.

Thus, whatever the structure of a given specimen of concrete may be, it is one that grew out of the structure of the original mixture and one that incorporates the gross flaws and anisotropisms that developed before the mixture became too firm to settle. As already said, an understanding of these matters is prerequisite to understanding mature concrete.

The first four chapters of this book give terminology, data on materials, and the development of a method of analysis of mixtures. Based on parameters already in use by soils technologists and physical chemists, the method is not original. It was adopted because it combines effectiveness with simplicity. The analysis revealed three distinct categories of concrete mixtures and established guidelines for limiting generalizations based on data from only one category.

The fifth chapter is a review of the principal schemes for formulating concrete mixtures (mix design), and the sixth is an analysis of the theoretical aspects of mix design, the analysis being made in terms of material developed in the first four chapters. The Weymouth theory of particle interference, and later modifications of it, are given special attention.

Entrained air is discussed in Chapters 7 and 8. The mechanisms of entrainment and entrapment are described, and the determinants of air-void characteristics are discussed, especially void-size distribution.

At various points in the book the prominent role played by interparticle forces in fresh concrete is described. In Chapter 9 the nature of these forces is discussed. I attempt to present this subject in terms that do not require previous knowledge of physical chemistry, yet without avoiding all quantitative aspects of it.

Workability of fresh concrete is dealt with incidentally—enough to show that workability is a term that implies much more than the term rheology and that we can deal quantitatively only with rheology. Chapter 10 is a discussion of the rheology of fresh concrete in two conditions: the plastic state and the fluidized state, the latter being maintained by continual vibration.

The settlement, bleeding, and shrinking of fresh concrete which occurs in the quiescent period following mixing and placing is discussed in Chapter 11. Two aspects are emphasized: what the phenomena tells us about structure and the practical consequences of settlement and bleeding, particularly the development of zones of sedimentation and the development of structural flaws or zones of weakness.

In Chapter 12 the theoretical aspects of the bleeding rate are discussed. Bleeding involves fluid through a granular body and is thus basically the same kind of phenomenon dealt with by Darcy, Kozeny, Carman, and various others. It is shown that use of the Kozeny-Carman analysis and equation has led to some erroneous conclusions. A different analysis is presented.

Cyril Stanley Smith* wrote recently, "Of all the approaches to nature, a science based on structure seems most able to unite the microscopic and macroscopic, theory and practice, intuition and logic, beauty and utility." As far as concrete is concerned, I think we are only beginning to understand it in terms of structure, and the area covered by this book is not the one in which greatest progress in that direction has been made. Nevertheless, a beginning is something of value, and I hope

* *Scientific American,* **271,** No. 3, **76** (September **1967**).

that it will not only prove to be a stimulus to students and researchers but will also help those in the field to observe concrete with knowing and critical eyes.

This work was begun before the time of my retirement from the research staff of the Portland Cement Association. Throughout 35 years with the Association I enjoyed freedom for study that I doubt could have been surpassed anywhere and for that I am grateful. I thank Mr. Joseph Walker, Vice President for Research and Development, who arranged some financial support and made available to me material needed for continuing the work after retirement.

I am indebted to many former colleagues, especially Dr. Stephen Brunauer, Dr. L. E. Copeland, and Dr. H. H. Steinour; they served as counselors in many matters of science. My special thanks go to Dr. A. J. Gaskin, Dr. Paul Seligman, and the late Professor Gerald Pickett, who each read portions of the manuscript.

Indebtedness to many authors and to copyright owners is acknowledged at appropriate points throughout the book.

With much more than gratitude I acknowledge the contributions of Trista W. Powers who not only endured with forbearance and good humor the long ordeal of her husband's authorship but also typed the entire manuscript.

T. C. Powers

Barrington, Illinois
1967

Contents

The Properties of Fresh Concrete

1

Particles and Aggregates

1. INTRODUCTION

Fresh concrete is composed of particles of solid matter held in a matrix of cement paste and air-filled spaces; the air-filled spaces are such that most of them can be called bubbles. The paste part of the matrix is composed of an aqueous solution and coated particles of cement. The solution derives its solutes from the constituents of cement. The cement grains acquire their coating during their first few minutes of contact with water, the coating being a gelatinous material. The gelatinous material is itself particulate, the particles being colloidal hydrates of some or all the chemical constituents of the cement grains.

Given enough time and proper environmental conditions, all or nearly all of the cement becomes converted by physicochemical processes to a rocklike mass of particles of hydrated material, mostly in the form of colloidal material having a specific surface area about 500 times the specific surface area of the original grains of cement. This particulate material, which replaces the original cement grains, is called *cement gel;* it occupies the space in the mixture originally occupied by cement grains and the aqueous solution, commonly referred to as water.

When we study the properties of mature, hardened concrete, the coherent, particulate material, called the cement gel, and the adsorbed water associated with it are the main focal point. However because we are studying the properties of concrete in the fresh state, the smallest particles to be considered are the gel-coated cement particles or the cement particles in their original state. Even when it is present, the gel coating is not mentioned unless there is a special reason to do so.

Indeed, in this chapter our attention will be confined to aggregates in the dry state.

We are not concerned with questions pertaining to the chemistry and mineralogy of rocks, nor with the many facets of the problem of selecting materials suitable for specific uses and situations. We assume that the aggregate complies with ASTM Designation C33, at least with respect to deleterious substances if not grading. We discuss in this and following chapters the characteristics of particles and aggregates in relation to the art of making concrete, especially concrete mixtures that are plastic before they become rigid and strong.

1.1 Aggregates

In concrete technology any aggregation of solid particles except portland cement or other mineral or rock powders is called an *aggregate.* The noun may be modified by adjectives such as *fine, medium,* or *coarse* to indicate the relative mean particle diameter. The term "combined aggregate" is used to designate an aggregate obtained by mixing two or more aggregates, one usually being finer than the other. The term "fine aggregate" generally indicates the material in a natural deposit, or in the product of a crusher, most of which can pass through a standard No. 4 sieve, the clear opening of which is 0.187 in., or 4.76 mm, but cannot pass through a No. 200 sieve, with a clear opening of 0.0029 in. or 74 μ (ASTM Designation C125).

The term "sand" is defined in the same way as fine aggregate with the restriction that it refers to the material "resulting from natural disintegration and abrasion of rock, or of processing completely friable sandstone" (ASTM Designation C125.). However in practice the term, with modifiers, often is applied to manufactured material; for example, stone sand or manufactured sand. In general a concrete aggregate is an aggregation of solid particles, whether the particles are derived from rock by natural or artificial means.

Common usage has established a less-literal meaning of the term "aggregate"; we may speak of a rock fragment or a pebble as an aggregate particle or as a piece of aggregate, thus making the word designate the particulate form of rock used for concrete and for road ballast.

1.2 Solid Volume

A stone is a solid in the sense that it is a rigid body bounded by visible, tangible surfaces, but it is usually not solid in the sense that the space within its visible boundaries is filled completely with mineral matter. Some kinds of stones contain isolated voids; most contain interconnected pores permeable to liquid and gases but not to a fine powder such as cement.

In this book the amount of space in a concrete mixture from which cement is excluded by a particle is called the *solid volume* of that particle; it usually (but not always, with vesicular material) corresponds to the volume bounded by the visible, tangible surface of the particle. When a dry particle is added to a certain volume of a plastic concrete mixture, it displaces a quantity of the mixture equal to its solid volume as just defined. At the same time it may absorb some water from the mixture, thereby tending to reduce the overall volume of the mixture. The amount of the reduction in volume is equal to the volume of water absorbed, provided that the particle does not at the same time release occluded air, which may become trapped in the mixture. In any case, the solid volume of the particle includes any internal pore spaces that might be permeable to liquid as well as any that might be impermeable.

2. SPECIFIC PHYSICAL PROPERTIES

In the following sections we discuss some of the principal properties of aggregate particles. Not all the properties enter explicitly into the considerations of later chapters, but their consideration helps us to understand the nature of the principal constituents of concrete.

2.1 *Specific Gravity and Apparent Specific Gravity*

As applied to solids, specific gravity is defined in ASTM Designation E-12 as follows:

"The ratio of the mass of a unit volume of a material at a stated temperature, to the mass of the same volume of gas-free distilled water at a stated temperature. If the material is a solid, the volume shall be that of the impermeable portion."

The term "apparent specific gravity" has the same definition, except that it is in terms of weight rather than mass.

2.2 *Bulk Specific Gravity*

As applied to concrete aggregate, the bulk specific gravity is a figure used for computing approximately the solid volume of a material as defined in Section 1.2 from the oven-dry weight of the material. The ASTM definition (slightly paraphrased) is as follows:

The ratio of the oven-dry weight in air of a given volume of permeable material (the volume including both permeable and impermeable voids normal to the material) at a stated temperature, to the weight in air of an equal volume of gas-free distilled water at a stated temperature.

The ASTM methods (Designations C127 and C128) are such that the bulk specific gravity is the ratio of the dry weight of a particle to the weight of water displaced by the particle in its saturated, surface-dry state. Because the method stipulates that the water used for the displacement measurement should be at 68 to 77°F (20 to 25°C), the figure obtained is not one by which the solid volume, as defined above, can be computed exactly. For example, if 2.5 g of rock displaces 1 g of water at 21°C under the conditions stipulated by the method, the bulk specific gravity is 2.5; but because 1 g of water at 21°C is 1.002 cm^3, the bulk density is $2.5/1.002 = 2.495$ g/cm^3.

If we use the figure of bulk specific gravity as if it were bulk density, it does not ordinarily produce significant error, except if we are calculating the air content of a mixture from the difference between total volume and the sum of the volumes of water and solids.

2.2.1 Bulk specific gravity, SSD basis. The ASTM method also provides a factor for calculating the solid volume, approximately, of a material (as defined in Section 1.2) from the weight in air of the material in the saturated, surface-dry state (SSD) to the weight of the water displaced by the SSD material. Such a figure is useful in the laboratory when aggregates are brought to the saturated, surface-dry state before we weigh the quantities to be used.

2.2.2 The SSD state. The ASTM methods for bringing a sample of aggregate to the saturated, surface-dry state, one for fine and another for coarse aggregate, is difficult to apply to some materials, especially lightweight, coarse aggregates. For coarse aggregate, Landgren [1] has proposed a modification of the standard method. The difficulty can be partially circumvented by establishing a "specific gravity factor," as we discuss in Chapter 5, Section 7.9.

2.2.3 Range of values. The bulk specific gravities of rocks commonly used in concrete generally are upwards of 2.4 but seldom higher than 2.8. For special purposes, particularly concrete for radiation shielding, aggregates having relatively high specific gravities are sometimes produced. This may be done by using metallic products such as steel punchings or iron shot that have specific gravities of about 7.6; various heavy ores and minerals having high densities also may be used. Also, for shielding, certain hydrous ores having specific gravities as low as 1.8 may be used [2,3].

The bulk specific gravities of lightweight aggregates generally cannot be determined exactly, but the specific gravity factors already mentioned, based on the apparent displacement of a lightweight aggregate in the matrix of lightweight concrete, indicate values for coarse aggregate as low as 1.4.

2.3 Specific Solid Volume

In this book we usually derive the volume of a solid material from its weight by means of a factor having dimensions in the cgs system of volume per unit mass. This factor is called *specific solid volume.* In exact terms, it should be the reciprocal of bulk density as defined in ASTM Designation E12; in practice, it is usually taken as the reciprocal of ASTM bulk specific gravity, although conversion of bulk specific gravity to bulk density is preferred practice.

Adopting the convention above, instead of the customary density or specific gravity, has some worthwhile advantages. In algebraic expressions involving weights to be converted to volumes, the symbols can be written or typed on the same line. For example, if a is the weight in air of a quantity of fine aggregate in grams, and if its specific solid volume is v_a cm^3/g, the expression for solid volume is written av_a, the two symbols always appearing together when the quantity in solid-volume units is intended. It soon becomes habitual to read the combination as a single symbol of solid volume; yet the two may be separated according to the rules of algebraic expression whenever weights constituting a given volume are wanted. Of course, if weight is expressed in pounds, v_a must be expressed in cubic feet per pound.

2.4 Absorption

Most kinds of particles that have been exposed to dry air for a while are capable of absorbing water. The process by which the liquid penetrates the rock and the increase in weight resulting from the penetration are called *absorption.* In the latter sense the absorption of a given material is usually given as the increase over the oven-dry weight during 24 hr of immersion (although longer soaking periods may be used), exclusive of the weight of surface moisture, expressed as the weight increase per unit dry weight (ASTM C127 and C128). Absorption by various aggregates, including lightweight materials, range from virtually zero to over 30 per cent of the dry weight [1].

When dry, absorptive material is used for making concrete, it may be necessary to estimate the amount of water absorbed by the aggregate particles at a particular moment. For example, if we want to make a consistency test 10 minutes after the mixing water is added to a batch of concrete, we must estimate the quantity of water that has been absorbed up to that moment, or we cannot correlate consistency with the effective water content. If we want to correlate the strength of hardened concrete with the volume composition of the matrix of the aggregate after the period of settlement and bleeding is over (Chapter 11), we must estimate the amount of water that has been displaced from the matrix by settlement of solids, and the amount of water that has been absorbed from the matrix by the aggregate particles up to the moment when settlement ceases. Because of the inaccuracy of such estimates, it is preferable in laboratory experiments to use presaturated aggregates.

2.5 Absorptivity

A principal difference between different kinds of rock particles is the rate at which they approach the saturated state after being given access to free water [4]. (This is of importance in the study of certain aspects of hardened concrete, particularly the effects of frost action on it.) The rate may be expressed in terms of the *coefficient of absorptivity* [5], which may be defined as the square of the quantity of water absorbed per unit exposed area, divided by the elapsed time; that is to say,

$$\left(\frac{q}{A}\right)^2 = K_a t$$

where q is the volume absorbed through area A, t is the elapsed time, and K_a is the coefficient of absorptivity. This relationship between absorption and time is applicable during the initial stage of the process of absorption through a plane surface.

For four samples of limestone, Dolch [6] reported values of K_a ranging from 4.59×10^{-9} to 3.24×10^{-5} cm/sec, the lower value corresponding to the denser specimen.

2.6 Permeability

When specimens of rock are subjected to fluid pressure on one of their sides, most of them prove to be permeable, as is of course indicated by their absorptivity (Section 2.5). Coefficients of permeability of some rocks to water for several samples are given in Table 1.1. As for other properties, rocks bearing the same name comprise a wide range of permeability. For example, Dolch [6] reported coefficients of permeability for four samples of limestone having bulk densities ranging from 2.20 g/cm^3 to 2.73, the most porous having a permeability two thousand times as high as the least porous. Similar comparisons can be made in Table 1.1.

Table 1.1 Coefficients of Permeability of Various Samples of Rock

Density (g/cm³)	w_e/V*	K† (cm/sec)	Description of Rock
2.99	0.0057	3.45×10^{-13}	Trap rock; dense; some crystal-boundary pores.
2.70	0.0082	9.20×10^{-13}	Marble; fine grained; dense.
2.94	0.0065	1.15×10^{-12}	Quartz diorite; coarsely crystalline; crystal-boundary pores.
2.65	0.0018	1.26×10^{-12}	Quartz-feldspar; felsite; very dense.
2.71	0.0046	1.72×10^{-12}	Limestone, crystalline.
2.78	0.0180	3.34×10^{-12}	Limestone, crystalline; fine-grained marble.
2.75	0.0310	8.05×10^{-11}	Limestone, crystalline; fine-grained marble.
2.60	0.0140	1.15×10^{-10}	Quartzite, imperfectly cemented; sandstone.
2.72	0.0510	2.30×10^{-10}	Limestone; uniform; fairly dense; pure.
2.69	0.0073	7.48×10^{-10}	Granite, gray.
2.58	0.0430	1.72×10^{-9}	Sandstone, porous.
2.60	0.0052	2.18×10^{-9}	Granite

* w_e/V is the weight of water in grams per unit volume of specimen found in the specimen after the permeability test; it is approximately the porosity so far as permeability to water is concerned.

† K is the coefficient of permeability to water at 27°C.

2.7 *Elastic Properties*

2.7.1 Compressibility. The volume of a particle of rock in concrete is subject to variation if there are variations in temperature or if there are changes in the ambient pressure. Ordinarily, a particle of rock at the surface of the earth is under atmospheric pressure, its volume being determined by that pressure and thus varying with atmospheric pressure. However, atmospheric pressure is negligible in relation to the internal forces of cohesion that restrain the tendency of the atoms toward spontaneous dispersion; hence the ordinary variations of atmospheric pressure have negligible effects. However, if a rock is subject to high hydrostatic pressure, the effects are noticeable and sometimes technologically important. Fluid pressure (either from a gas or a liquid) compresses whatever material there is within the surfaces with which the fluid may make contact, and the characteristic diminution of volume per unit increase in pressure, that is, the compressibility coefficient, depends on the chemical composition and physical structure of the material within those boundaries. Thus, if the specimen is porous, but the pores are completely accessible to the fluid, the compressibility depends only on the nature of the mineral substance, and the fractional change in the overall volume of the specimen, including the pores, is the same as the change in any part of the solid substance of which the specimen is made.

On the other hand, if the specimen contains pores that are completely inaccessible to the fluid, so that the compressive force is developed only at the visible boundary of the specimen, the coefficient of compressibility of the specimen is greater than that of the solid substance of which the specimen is composed. If the fluid partly penetrates the pore system of a specimen, the observed coefficient of compressibility is greater than that of the solid substance of which the specimen is composed but less than the compressibility of the whole specimen under compressive force developed only at the visible surfaces.

As situated in hardened concrete, a particle of rock is subjected to compressive force when, at constant temperature, the concrete matrix loses water and shrinks, for the matrix by itself is able to shrink much more than the rock particle under the same conditions; thus in concrete the rock particle mechanically opposes the tendency of the matrix and restrains the movement, the restraint being larger the smaller the coefficient of compressibility of the rock. That is a principal reason for being concerned over the compressibility coefficient of aggregate particles, there being marked differences in compressibility among different rocks.

Table 1.2 gives data on the compressibility of various rocks. It should be understood that wide variations among rocks of the same name are to be expected and that the values given are not necessarily typical of the named material. Coefficients of compressibility are given for specimens that were loaded hydraulically, but covered to prevent penetration of the liquid into the specimens. During the first loading, most specimens appear to be more compressible, at the beginning, under small pressures than they are later when under higher pressure. To show this characteristic, the coefficient of compressibility (reciprocal of the bulk modulus) is reported for the initial tangent of the stress-strain curve (pressure indicated to be zero) and for a pressure of 720 kg/cm^2 (10,238 psi).

The relatively high compressibility at the beginning of compression is probably due at least in part to a consolidation of the structure of the rock; that is to say, there seems to be a closing up of structural flaws. The specimen of marble represented in Table 1.2 is probably an extreme example. Used as aggregate, it would probably give concrete of unusually high drying shrinkage. Its initial compressibility is higher than that of any of the other rocks represented, and it is about 12 times as high as the compressibility at the higher pressure level. Obsidian, a glass, showed the same compressibility at both pressure levels.

Generally, the densest rocks are the least compressible, but there are significant exceptions.

Table 1.2 Ziesman's Data on Compressibility of Rock[a]

Name of Rock	Pressure[b] (kg/cm²)	Density (g/cm³)	Specific Volume[b] (cm³/g)	Compressibility[c] cm²/dyne $\times 10^{13}$	in.²/lb $\times 10^8$
Quincy granite, 100 ft down	0	2.59	0.386	76.6	52.8
	720	2.59	0.386	24.6	17.0
Quincy granite, 235 ft down	0	2.64	0.379	89.8	61.9
	600	2.64	0.379	23.6	16.3
Rockport granite	0	2.63	0.380	91.7	63.2
	720	2.63	0.380	26.6	18.3
French Creek norite	0	3.05	0.328	59.0	40.7
	720	3.05	0.328	15.2	10.5
Olivine diabase	0	2.96	0.338	17.1	11.8
	720	2.96	0.338	12.6	8.7
Obsidian	0	2.34	0.427	29.5	17.9
	720	2.34	0.427	29.5	17.9
Quartzitic sandstone	0	2.64	0.379	58.7	40.5
	720	2.64	0.379	29.9	20.6
Marble	0	2.71	0.369	180.0	124.0
	720	2.71	0.369	15.0	10.3
Limestone	0	2.69	0.372	29.2	20.1
	720	2.69	0.372	23.0	15.9
Dolomite	0	2.82	0.355	37.1	25.6
	720	2.82	0.355	14.2	9.8
Portlandite [$Ca(OH)_2$]	700	2.24	0.446	36.5	25.2

[a] Data are from *Proc. National Academy of Sciences,* **19,** 666–679, 1933.
[b] Hydraulic pressure on a specimen covered to prevent penetration of the pores in the material.
[c] Compressibility $= (1/V)(dV/dP)$ where V is the volume of the specimen and P the pressure.

Table 1.2 is not representative of the full range of densities and compressibilities found among rocks used for concrete. Particularly the natural and artificial lightweight aggregates have a relatively low density and high compressibility. On the basis of density we might expect the compressibility of some lightweight material to be as high as 100×10^{-6} in.²/lb, which is to say, five to ten times the compressibility of ordinary aggregates.

2.7.2 Young's modulus. When a rock is loaded in uniaxial compression, the stress-strain curve may or may not appear to be straight. If it is straight, the ratio of stress to strain may be taken as the value of *Young's modulus* over the range of stress for which the relationship

Table 1.3 Values of Young's Modulus for Materials Used as Aggregates

Name of Aggregate	Young's Modulus kg/cm² $\times 10^{-5}$	psi $\times 10^{-6}$	Author[a]
Diorite	10.41	14.8	Dantu (1957)
Glass	7.42	10.55	Dantu (1957)
Steel	22.00	31.3	Dantu (1957)
Limestone (Plattin)	7.95	11.3	La Rue (1946)
Limestone (Burlington) HM Carthage	6.32	9.0	La Rue (1946)
Limestone (Burlington) LM Carthage	4.73	6.7	La Rue (1946)
Limestone (Bowling Green)	1.31	1.9	La Rue (1946)
Calcareous-siliceous gravel		8.9	Hirsch (1962)
Limestone		4.6	Hirsch (1962)

[a] Data from the first two authors was collected by Hansen, *Proc. Swedish Cement and Concrete Research Institute of Technology,* **31,** 1960. For Hirsch, see *Proc. ACI,* **59,** 427–447, 1962.

is linear. If the stress-strain relationship is curved, the ratio of stress to strain at an arbitrarily selected stress is sometimes reported, such a ratio being called the *secant modulus of elasticity* [7]. An alternative is to report the slope of the initial tangent to the curve, which is called the *initial tangent modulus of elasticity.* Some data on rocks have been reported on the basis of the resonance-frequency of vibration of a prismatic specimen (ASTM Designation C215) or on the basis of the velocity of sound conduction [8]. The value derived from the results of such measurements is called the *dynamic modulus of elasticity,* or sometimes the "sonic modulus." Generally the dynamic methods give values higher than the values derived from static loading, particularly if the stress-strain curve is not linear.

Table 1.3 gives values of Young's modulus for some aggregate materials, glass and steel being included mainly for comparison, although glass aggregate has been used in experimental work, and steel for special purposes (Section 2.2.2). The various values for limestone are indicative of the range of physical properties that may be found among rocks bearing the same name.

2.7.3 Poisson's ratio. Values for Poisson's ratio of various rocks used in concrete have been obtained by the static method, and sometimes by the sonic method (ASTM Designation C215). For the latter method the torsional mode of vibration as well as that for lateral vibration must be determined. Kaplan [9] reported values determined by the latter method from specimens of seven different kinds of rock. The lowest value was 0.15, the highest 0.36, and the average was 0.27. Without the rock giving the lowest value (a quartzite gravel) the average was 0.29.

2.8 *Thermal Properties*

Like all building materials, concrete undergoes changes of volume in response to changes of temperature. When building massive structures such as dams, thermal volume change is a matter of concern, principally because of stresses arising from temperature gradients between the surface and interior regions. There is reason also to believe that thermal volume change, together with that associated with drying and wetting, influences the durability of a concrete structure [10] because of the difference between the thermal properties of the aggregate and those of its matrix.

2.8.1 Coefficient of expansion (thermal coefficient). Among the common rocks found in concrete aggregates wide differences in the thermal coefficient may be encountered, as is shown in Table 1.4. However, as for other physical properties, it is hardly possible to say whether

Table 1.4 Thermal Coefficients of Expansion of Various Rocks[a]

Name of Rock	Number of Specimens	Average Linear Expansion (millionths) per °F	Average Linear Expansion (millionths) per °C
Granites and rhyolites	27	1.0 to 6.6	1.8 to 11.9
Diorites and andesites	17	2.3 to 5.7	4.1 to 10.3
Gabbros, basalts, diabases	15	2.0 to 5.4	3.6 to 9.7
Sandstones	24	2.4 to 7.7	4.3 to 13.9
Quartzites	20	3.9 to 7.3	7.0 to 13.1
Dolomites	7	3.7 to 4.8	6.7 to 8.6
Limestones	65	0.5 to 6.8	0.9 to 12.2
Siliceous limestones	6	2.0 to 5.5	3.6 to 9.9
Cherts	49	4.1 to 7.3	7.4 to 13.1
Marbles	29	0.6 to 8.9	1.1 to 16.0
Slates and argillites	5	4.5 to 4.8	8.1 to 8.6

[a] Data are from Rhodes and Meilenz, *Proc. ACI*, **42**, 590, 1946. Reproduced by permission American Concrete Institute.

Table 1.5 Thermal Conductivity and Specific Heat for Certain Aggregates[a]

Material	70°F		90°F		110°F		130°F	
	C	H	C	H	C	H	C	H
Quartz sand	1.784	0.167	1.779	0.178	1.772	0.190	1.769	0.207
Basalt gravel	1.103	0.183	1.100	0.181	1.099	0.187	1.096	0.200
Dolomite gravel	2.490	0.192	2.446	0.196	2.407	0.204	2.298	0.212
Granite gravel	1.684	0.171	1.678	0.169	1.674	0.175	1.664	0.185
Limestone gravel	2.329	0.179	2.275	0.181	2.231	0.187	2.189	0.196
Quartzite gravel	2.711	0.165	2.689	0.173	2.667	0.181	2.641	0.189
Rhyolite gravel	1.085	0.183	1.092	0.185	1.100	0.191	1.104	0.193

[a] Conductivity (C) and specific heat (H) are for the mean temperature indicated. Conductivity is in Btu/(ft hr °F). Data are from Final Report, Boulder Canyon Project, Part VII, *Cement and Concrete Investigations*, Bulletin 1: "Thermal Properties of Concrete," Bureau of Reclamation, Denver, Colorado, 1940.

the thermal coefficient of a given specimen will be relatively high or low on the basis of the name given it.

The thermal coefficient of mature, hardened, cement paste is about 6.1×10^{-6} °F^{-1} (11×10^{-6} °C^{-1}) if the paste is either dry or saturated. Looking at Table 1.4 with these figures in mind, we see that in many if not most cases the thermal coefficient of the aggregate material in concrete will not be the same as that of its cement-paste matrix. As a matter of fact, when the paste is not fully saturated, its apparent thermal coefficient may be as much as three times the figure given above, and because this is the most common state for concrete in service, we may conclude that the thermal properties of aggregate and paste are almost never alike.

2.8.2 Thermal conductivity and specific heat. In Table 1.5 values for thermal conductivity and specific heat are given for several aggregates at different temperature levels, each of these properties being influenced considerably by temperature. Each of the different gravels were composed of particles of the kind indicated, hand picked from one or two sources of lithologically heterogeneous material. However we cannot assume that rocks of the same name from other sources have the same thermal coefficients as those given here.

Among the materials represented, we see that at a given temperature specific heat does not range widely from the average, but thermal conductivity does. For specific heat, the lowest value is 93 per cent of the average, and the highest 108 per cent, whereas for thermal conductivity the corresponding figures are 58 per cent and 132 per cent.

The thermal conductivity of lightweight aggregate tends to be less than that of material of ordinary density because of the relatively high porosity of the individual particles; the specific heat would be relatively low for the same reason if it were based on the apparent volume of the material instead of weight.

2.9 *Compressive Strength*

Aggregates composed of rocks having normal density are usually considerably stronger than the concrete of which they are a part; whereas the specified compressive strength of concrete is usually not more than 6000 psi (420 kg/cm^2), dense rocks may be expected to have compressive strengths upwards of 13,000 psi (915 kg/cm^2), as shown in Table 1.6, due to Woolf [11].

Of course lightweight material may be much weaker than dense rocks, and indeed may have a strength less than that of the concrete of which it is a part.

Table 1.6 Compressive Strength of Types of Rock Commonly Used as Concrete Aggregates[a]

Type of Rock	Number of Samples[b]	Compressive Strength (psi)		
			After Deletion of Extremes[d]	
		Average[c]	Max	Min
Granite	278	26,200	37,300	16,600
Felsite	12	47,000	76,300	17,400
Trap	59	41,100	54,700	29,200
Limestone	241	23,000	34,900	13,500
Sandstone	79	19,000	34,800	6,400
Marble	34	16,900	35,400	7,400
Quartzite	26	36,500	61,300	18,000
Gniess	36	21,300	34,100	13,600
Schist	31	24,600	43,100	13,200

[a] Data are from Woolf, ASTM Special Publication No. 169A, 1966, p. 470.

[b] For most samples, the compressive strength is an average of three to 15 specimens.

[c] Average of all samples.

[d] Ten per cent of all samples tested, with highest or lowest values, have been deleted, for they are not considered to be typical of the material.

2.10 *Surface Texture*

According to Mather, "Surface texture is the property the measure of which depends upon the relative degree to which particle surfaces are polished or dull, smooth or rough, and upon the type of roughness" [12]. Mather suggests that "roughness . . . should probably be expressed in terms of arithmetic average deviation of the actual surface from the mean surface." Usually the particles are simply described as rough, fairly rough, smooth, or very smooth.

It has been shown by certain tests that particle roughness tends to enhance the tensile strength of concrete while at the same time tending to increase the water requirement. It seems reasonable to assume that for a given particle shape and grading, rough particles give less dense aggregates than smooth particles do.

3. PARTICLE SIZE AND SHAPE

3.1 *Size of an Irregular Particle*

The term "particle size" is used variously, and sometimes without appreciation of its ambiguity when it is applied to irregular particles. Definition of particle size, together with the problem of size measurement, is a complex subject (see, for example, the review by Hawksley [13] and especially that of Mather) [12]. It is not necessary here to explore the subject fully, but some of the main points must be considered. Hawksley stated the problem as follows:

"The size of a spherical particle is defined completely by its diameter, but the "size" of an irregular particle . . . depends on what property of the particle is being considered. Different methods of particle size measurement assess different size-dependent properties, such as volume, or surface, or resistance to motion in a fluid. . . . There are many size dependent processes, and for each there is a corresponding 'size.' "

3.1.1 Shape classification. A method of shape classification based on the lengths of three axes was devised by Zingg [14,15]. These axes are called *greatest length, a, intermediate length, b,* and *shortest length, c.* Shape is described in terms of ratios as follows:

Elongation ratio $= q = b/a.$

Flatness ratio $= p = c/b.$

Shape factor $= F = p/q = ac/b^2.$

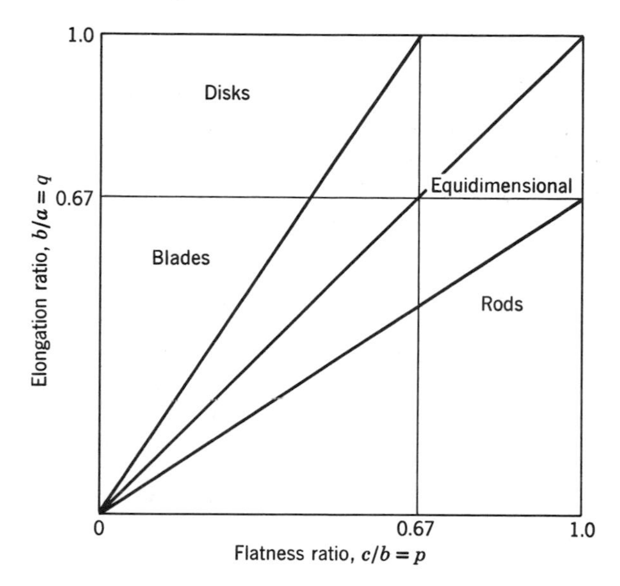

Fig. 1.1. Shape categories bounded by the arbitrary limits of the elongation ratio and the flatness ratio. (After Lees [15]).

If the shape factor is greater than 1.0, the intermediate length is closest to the shortest in length, and prolateness is indicated. If the shape factor is less than 1.0, the intermediate length is closest to the longest length, and oblateness is indicated.

Zingg classified shapes according to the scheme given in Figure 1.1. Such a classification or a related one is used in some specifications for aggregates; but it has obvious limitations.

A method for measuring the elongation and flatness of particles has been standardized by the U.S. Army, Corps of Engineers [16]; Mather [12] refers to several other methods.

3.2 Equivalent Sphere Diameters and Sphericity Factor

We are concerned mainly with the functions of size that give volume and surface area. "Since theories and methods," wrote Hawksley, "are derived for spherical particles, the 'size' of an irregular particle is conveniently expressed as the diameter of a sphere having the same volume, surface, etc." [13]. This approach to the question leads to the following definitions.

Volume diameter d_o^v, is the diameter of a sphere of the same volume as the irregular particle.

Surface diameter d_o^s, is the diameter of a sphere of the same surface area as the irregular particle.

Specific surface diameter d_o^{sp}, is the diameter of a sphere of the same surface area per unit solid volume as the irregular particle.

Stokes' diameter d_o^{st}, is the diameter of a sphere, in a given fluid, of the same free-falling terminal velocity as the irregular particle. It is evaluated by measuring terminal velocity and employing Stokes's law. This method is not usually applied to particles of the sizes found in concrete aggregates, but it is applied to cement and other powdered materials.

Sphericity factor. The sphericity of an irregular particle can be expressed in terms of a sphericity factor ψ defined as follows:

$$\psi = \frac{\text{specific surface diameter}}{\text{volume diameter}} = \frac{d_o^{sp}}{d_o^v}$$

From the definitions already given, we have also

$$\psi = \frac{(d_o^v)^2}{(d_o^{st})^2}$$

In terms of Stokes' diameter d^s, for particles that are not too elongated, Hawksley gives, as a good approximation,

$$\psi^{1/4} = \frac{d_o^s}{d_o^v}$$

and

$$\psi^{3/4} = \frac{d_o^{sp}}{d_o^s}$$

An earlier and perhaps more widely known definition of sphericity is that attributed to Wadell [17,12]. It is the ratio of the diameter of a sphere, with the same volume as the particle, to the diameter of a circumscribing sphere. On the basis of the assumption that the volume of the particle is approximately $(\pi/6)\ abc$, the sphericity factor is

$$\psi_W = \left(\frac{abc}{a^3}\right)^{1/3} = \left(\frac{bc}{a^2}\right)^{1/3}$$

where ψ_W is Wadell's sphericity factor, approximately.

3.3 Roundness

Roundness is a term used to give verbal descriptions, it being considered to have various degrees. They are indicated by the following sequence of terms [12]:

1. Angular: little evidence of wear.
2. Subangular: evidence of some wear, faces untouched.
3. Subrounded: considerable wear, faces reduced in area.
4. Rounded: faces almost gone.
5. Well rounded: no original faces.

Roundness can be defined more quantitatively as the ratio of the average radius of curvature of the corners and edges of the particle to the radius of the maximum inscribed circle [12].

3.4 *Angularity*

Angularity is used in at least two different ways. One is that indicated in the expression of degrees of roundness above. In another usage it is defined as the reciprocal of the sphericity factor. An indication of how the angularity factor varies with shape is given by the following figures: sphere, 1.0; cube, 1.24; tetrahedron, 1.98, fiber, 3+.

3.5 *Sieve Size*

3.5.1 Sieve-size quotients. In concrete practice information as to size of aggregate particles is usually limited to that which can be obtained with sieves. An analysis is carried out by means of a series of sieves having conveniently related sizes of openings. Such a series is described in terms of quotient of the linear dimension of the opening of a given sieve to that of the next larger sieve. The size quotient of a series may be such as $\sqrt[4]{2}$, or $\sqrt{2}$, or 2, the quotient 2 being the one most commonly used for the sieve analysis of concrete aggregates. This series is based on a sieve having 200 square openings per inch ($40{,}000/\text{in.}^2$) the width of an opening being 0.0029 in. or 74 μ.

Because aggregates are usually composed of a continuous series of sizes, the sieve analysis is reported in terms of the amount retained or passed by each of the sieves in the series or by the amounts caught between sieves. (See Fig. 1.2.) Also, size-frequency curves may be constructed [18].

The size of an irregular particle that just can be passed through a given sieve cannot be identified with any of the sizes we discuss in Section 3.2. We designate it as having size l_i, where l_i is the width of opening of the ith sieve of the series. l_i would approximate the volume diameter if the particle were spherical, but the sieve size of any irregular particle is dependent on the shape of the particle, on the method of sieving, and on the length of time sieving is continued. For this reason, sieving for purposes of product control should be done strictly according to standardized procedure.

Sieve Size	Per Cent Fine and Coarse Separated			Per Cent Fine and Coarse Combined		
	Individual	Retained	Passing	Individual	Retained	Passing
3 in.						
1½ in.	1	1	99	1	1	99
¾ in.	43	44	56	27	28	72
⅜ in.	33	77	23	21	49	51
No. 4	23	100	0	14	63	37
No. 4	1	1	99	0	—	—
No. 8	14	15	85	5	68	32
No. 16	15	30	70	6	74	26
No. 30	25	55	45	9	83	17
No. 50	25	80	20	9	92	8
No. 100	18	98	2	7	99	1
Pan	2	100	0	1	100	0
FM		2.79				
Per cent sand				37		
Sieve sizes are based on square opening						

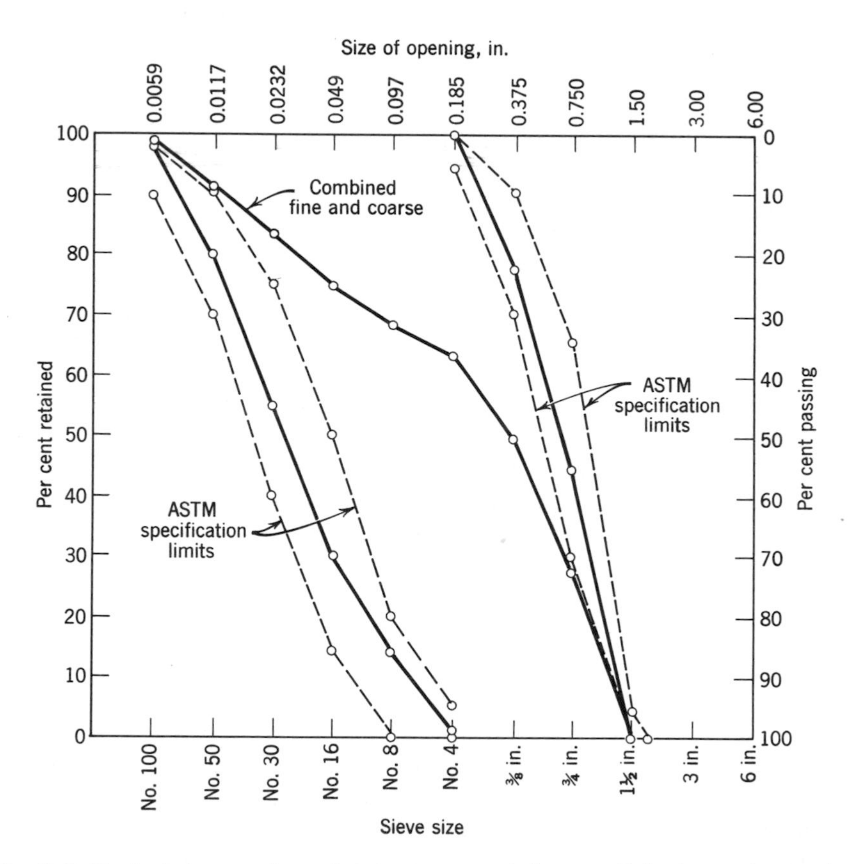

Fig. 1.2. Typical sieve analyses for aggregate graded up to 1½ in. in maximum size. (After W. H. Price, ASTM Special Publication STP 169A, 406, 1966, by permission.)

3.5.2 Size groups. Sieve sizes are generally dealt with in groups whenever it is desirable to deal with one of the functions of size that are not given directly by integral grading curves. A size group is defined by its nominal size limits, l_i and l_{i+1}, i and $i+1$ indicating two consecutive sieves in a standard series. For example $\frac{3}{4}$ in. — $1\frac{1}{2}$ in. is one of the size groups obtained in the series having a size quotient of 2 in ascending order. The size limits indicated are $\frac{3}{4}$ in. and $1\frac{1}{2}$ in.

3.5.3 Average volume diameter. A satisfactory estimate of the mean volume diameter of a size group may be obtained by counting a known weight of particles and computing the equivalent-sphere diameter from the relationship

$$d_o^v = \left(\frac{6\bar{V}}{\pi}\right)^{1/3}$$

where $\bar{V}$ is the average solid-volume per particle. Generally, this is not done.

3.5.4 Average specific surface diameter. The average specific surface diameter can be obtained from the following relationship:

$$d_o^{sp} = \frac{\sigma_a}{6}$$

where σ_a is the specific surface area of the aggregate in area per unit solid volume. Methods of estimating specific surface area are given in Section 5 of this chapter.

3.5.5 Average size of a size group. If aggregate particles were spherical, the maximum and minimum particle diameters for a given size group could be taken as practically the same as the corresponding sieve openings, and the average size would be at some intermediate value depending *only* on the grading within the group. However, if the particles are irregular, the average size for a given group depends not only on grading, but also on particular shape. Lees [15] measured the weights and volumes of individual particles in certain size groups, using four different particle shapes, described as equidimensional, rods, discs, and blades (Figure 1.1), obtaining the results shown in Table 1.7. For these size groups, which are defined by upper and lower size limits that are not far apart, the average equivalent-sphere diameters for the equidimensional and blade categories proved to be near the arithmetical mean of the limiting sieve openings, but for the other categories the results were different: The rodlike particles were larger and the disclike particles were smaller than the arithmetic mean.

Table 1.7 Influence of Shape on Average Particle Size in a Given Size Group[a]

Size Group	Mean of Sieve Openings	Equivalent Sphere Diameter (inches for Shape Indicated)[b]			
		Equi.	Rod	Disc	Blade
$\frac{1}{4}$–$\frac{3}{4}$ in.	0.31	0.32	0.37	0.27	0.31
$\frac{3}{8}$–$\frac{1}{2}$ in.	0.44	0.44	0.52	0.37	0.44
$\frac{1}{2}$–$\frac{3}{4}$ in.	0.63	0.62	0.72	0.53	0.63
$\frac{3}{4}$–1 in.	0.88	0.86	1.00	0.77	0.88
1–$1\frac{1}{2}$ in.	1.25	1.22	1.45	1.07	1.27

[a] Data are by Lees, *J. Brit. Granite and Whinstone Federation*, **4** (2), 1964.
[b] See Figure 1-1.

Because data of the kind presented above have generally not been obtained for the various aggregates reported in the literature, a nominal average size is usually defined in one way or another. The arithmetic mean of the sieve sizes is often taken to be equal to the average particle size, but for groups based on sieve-size quotients as large as 2 it is preferable to use the geometric mean. Such a preference is based on the observation that the cumulative plot of a sieve analysis to semilog scales tends to be linear. Butcher and Hopkins [19] showed that in this case the average size of the group is two-thirds its upper limit; that is,

$$d_i = 0.67 l_i^u \tag{1.1}$$

where l_i^u is the sieve opening (square) of the upper sieve of the size-defining pair.

3.5.6 Nominal average size of an aggregate. The nominal average size of an aggregate can be calculated in terms of the relative proportions of the size groups of which it is composed according to the following relationship:

$$d = \left[\frac{1}{\Sigma\, p_i d_i^{-3}}\right]^{1/3} \tag{1.2}$$

where d is the nominal average size of the aggregate, p_i is the volume fraction of the whole aggregate of the ith size group, and d_i is the nominal average size of that group, which may be defined as in Eq. 1.1.

3.5.7 Range of sizes. When an aggregate comprises more than one size group, the range of sizes is given in terms of the limiting sieve sizes, just as is done for a single group. For example, #4–$1\frac{1}{2}$ indicates

Table 1.8 Size-Group Numbers and Average Sizes

Size Group	Group Number	Relative Size	Average Size[a] mm	in.
#200–#100	1	1	0.099	0.0025
#100–#50	2	2	0.198	0.0050
#50–#30	3	4	0.397	0.0101
#30–#16	4	8	0.794	0.0202
#16–#8	5	16	1.59	0.0404
#8–#4	6	32	3.17	0.0805
#4–$\frac{3}{8}$ in.	7	64	6.34	0.161
$\frac{3}{8}$–$\frac{3}{4}$ in.	8	128	12.7	0.323
$\frac{3}{4}$–$1\frac{1}{2}$ in.	9	256	25.4	1.00

[a] See Eq. 1.1.

an aggregate the particles of which are retained on a No. 4 sieve, but which pass a sieve having a $1\frac{1}{2}$ in. square opening. Similarly, a sand might be designated as 0–#4. In this case the cipher is nominal because the specifications under which commercial aggregates are produced generally require washing out practically all material that would pass a No. 200 sieve, and because the smallest possible size is finite anyhow. However No. 200 is not usually given as the minimum size unless the material has actually been screened at that minimum size.

Another way of designating size range, used by Plum (1950) [20], is to indicate the number of size groups in the aggregate. It is convenient to assign number 1 to the #200–#100 group, obtaining the designations given in Table 1.8.

4. SINGLE NUMBER SIZE AND GRADING INDICES

4.1 *Fineness Modulus*

Instead of dealing explicitly with particle-size distribution, single-number indices that are functions of size distribution have been devised. The one most widely used in the United States is called the *fineness modulus,* and was introduced by Abrams in 1918 [21]. The fineness modulus for a given size group is given by the following expression:

$$FM = 7.94 + 3.32 \log d_i$$

where FM stands for fineness modulus and d_i is the nominal average diameter of the group in inches, it being calculated from the average of the logarithms (base 10) of the limiting sieve sizes. The constants are so selected that for the group (#200–#100), $FM = 0$. Because the average diameters of the size groups increase in the ratio of 2:1, it follows that the fineness modulus of Group 2 is 1.0, that of Group 3 is 2, etc. Thus the fineness modulus of an aggregate composed of several groups can be calculated as follows:

$$FM = 0p_1 + 1p_2 + 2p_3 + 3p_4 \cdots (i - 1)p_i$$

where p_i is the weight fraction of the material in the ith group. In practice, the fineness modulus is usually obtained by adding the cumulative percentage of the amounts *retained* on the respective sieves in the series beginning with No. 100, and dividing the sum by 100.

Abrams pointed out that when the per cent passing the successive sieves is plotted against the logarithms of sieve size, the fineness modulus may be seen as a number proportional to the area above the sieve-analysis curve. Defined in this way, the fineness modulus can be obtained from a sieve analysis regardless of the system of sieves used. Popovics [22] derived the following general relationship:

$$FM = 0.0332 \left[100 \log (10D) - 0.4343 \int_{0.1}^{D} \frac{f(d)}{d}\, dd \right]$$

where d is the particle size in millimeters; D is the maximum size; $f(d)$ is that cumulative distribution of particle size which gives the relation between the particle size d and the quantity of those particles smaller than d, in per cent. Popovics gave the following alternative form of the equation above:

$$FM = 0.0332 \int_{0.1}^{D} \log (10d)\, f'(d)\, dd$$

In this form we see the fineness modulus is proportional to the logarithm of an average particle size, as defined by the right-hand side of the equation. For example, if d_{av} stands for average size, and d_1 and d_2 are lower and upper limits of a size fraction, and the semilog sieve analysis is linear, the logarithmic average size of a size fraction is

$$\log d_{av} = \frac{\log d_1 + \log d_2}{2}$$

The fineness modulus of any grading is related to the weighted logarithmic average size as follows:

$$FM = \frac{\log (10d_{av})}{\log 2} = \frac{\log (10d_{av})}{0.301}$$

Popovics showed that when particle size is taken as the average size of a size group, the value of the fineness modulus computed from the equation above will depend on the assumption made about size distribution within the group. For a size quotient no greater than 2, the sizes based on the arithmetic mean of the limiting sieve sizes, a linear distribution of sizes, or a logarithmic distribution, give small differences in FM. However, any of the values based on average size are slightly larger than that obtained by summing the percentages retained on the sieves as originally proposed by Abrams.

The fineness modulus is widely used as a means of designating the relative fineness of sand. The ASTM specifications for fine aggregate requires the fineness modulus to be not less than 2.3 nor more than 3.1. Fineness modulus is used also, although less commonly, for coarse aggregates; an aggregate graded (#4–1 in.) for example, usually has a fineness modulus of about 7.0. The fineness modulus of a combined aggregate may be computed readily from the weight proportions of fine and coarse and their respective fineness moduli.

Because, as we point out above, the volume diameter is not uniquely defined by the limiting sieve sizes, a given fineness modulus does not correspond to the same average size unless all aggregates compared are composed of particles having the same shape. An aggregate composed of crushed material will have a larger mean volume diameter than that of one composed of rounded particles having the same fineness modulus.

It was pointed out by Young [23] that among various natural aggregates there is a roughly linear inverse empirical relationship between Abrams' fineness modulus and the specific surface area of the aggregate. However, the fineness modulus is not mathematically related to specific surface area of aggregates, except for certain artificial gradings, and therefore is not a dependable means of estimating relative specific surface under all conditions.

4.2 Standard Deviation

Zietsman [24] pointed out that the fineness modulus is an average size and that grading can be expressed as a standard deviation σ from that mean. Fulton [25] gives the following expression for standard deviation from that weighted mean:

$$\sigma = p_1(0 - FM)^2 + p_2(1 - FM)^2 + p_3(2 - FM)^2 + p_4(3 - FM)^2 + p_5(4 - FM)^2 \cdots$$

The lower the figure for standard deviation, the larger the fraction near the weighted mean size. For example, if all the particles were in group

Table 1.9 Examples of Gradings Having the Same Fineness Modulus *FM* but Different Standard Deviation[a]

Sand	Grading[b]						FM	σ
	p_1	p_2	p_3	p_4	p_5	p_6		
A	0.055	0.115	0.270	0.193	0.275	0.092	2.79	1.35
B	0.180	0.077	0.198	0.090	0.224	0.231	2.79	1.80
C	0.029	0.075	0.224	0.468	0.153	0.051	2.79	1.05

[a] Data are from F. S. Fulton, "Concrete Technology, A South African Handbook," Portland Cement Institute, Johannesburg, 1961.
[b] p_1 is the weight fraction of Group 1, (#200–#100), p_2 is the weight fraction of Group 2, (#100–#50), and so on for the whole aggregate.

4, that is, $p_4 = 1.0$, the fineness modulus would be 3.0, and the standard deviation would be zero. In Table 1.9 Fulton gives examples for three different aggregates having the same fineness modulus.

4.3 *Average Diameter*

An aggregate may be characterized in part by its average particle size. The nominal average diameter may be calculated from the sieve analysis by means of Eq. 1.2. As discussed before, accuracy is influenced by particle shape.

4.4 *Surface Modulus*

Talbot [26] defined a number proportional to the specific surface area called the surface modulus. It was determined from the sieve analysis as follows:

$$SM = \text{surface modulus} = 100(p_1 + \tfrac{1}{2}p_2 + \tfrac{1}{4}p_4 \cdots \tfrac{1}{128}p_6 \cdots)$$

where the p's have the same significance as above. If the particles in one group have the same shape as those in all the rest, if the grading of sizes within each group is the same, if the specific gravity is the same for all sizes, and if the specific surface area of one of the groups is known, a figure for specific surface area can be obtained from the surface modulus. (See Section 5.2.)

4.5 *Murdock's Index*

By what he described as a process of successive approximation, followed by use of an electronic computer, Murdock [27] developed the following grading index:

$$f_s = (2p_1 + 7p_2 + 9p_3 + 9p_4 + 7p_5 + 4p_6 + 1p_7 - 1p_8 - 2p_9 - 2.5p_{10}) + 0.33$$

Murdock called f_s a *surface* index, but that name apparently reflects the evolution of the factor, for its development was the result of finding the specific surface area, or surface modulus, to be inadequate for general application. It can be described as a weighted average size, with an added constant.

The expression above gives coefficients for each of the size groups including the tenth, a 3-in. maximum. For lesser maximum sizes, or for gap gradings, the corresponding coefficient is simply omitted. The coefficients are based on data from mixtures having consistencies expressed in terms of the Glanville compacting factor; the values are probably influenced by that circumstance.

Murdock's explanation of the constant was the following: "The computer analysis has suggested the use of negative values for individual surface indices for sizes larger than about $\frac{3}{8}$ in., and this is counterbalanced by the inclusion of a constant, 0.33."

5. SPECIFIC SURFACE AREA

Various attempts have been made to determine actual specific surface area. Edwards [28] separated aggregates into size groups and then counted the number of particles per unit weight in each group. From these numbers and the specific gravity of the material the average volume per particle in each group was obtained. The average volume per particle gave the mean volume diameter d_o^v. Edwards then assumed that the specific surface area $= 6/d_o^v$, which is correct for spheres but incorrect for irregular particles. (See Section 3 of this chapter.) The values obtained were ψ times the correct value, ψ being the sphericity factor.

5.1 Surface Area Estimated from Coefficient of Permeability

The specific surface areas of aggregates have been measured directly by Loudon [29] using a modification of the Carman [30] permeability method. Shacklock and Walker [31] contributed additional data; a diagram of their apparatus is shown in Fig. 1.3. The accuracy of this method depends on the reliability of the Carman equation when it is applied to aggregates having various shapes. There is good reason to question the theoretical basis of the equation (see Chapter 12), but a considerable amount of experimental data indicate that for an application such as

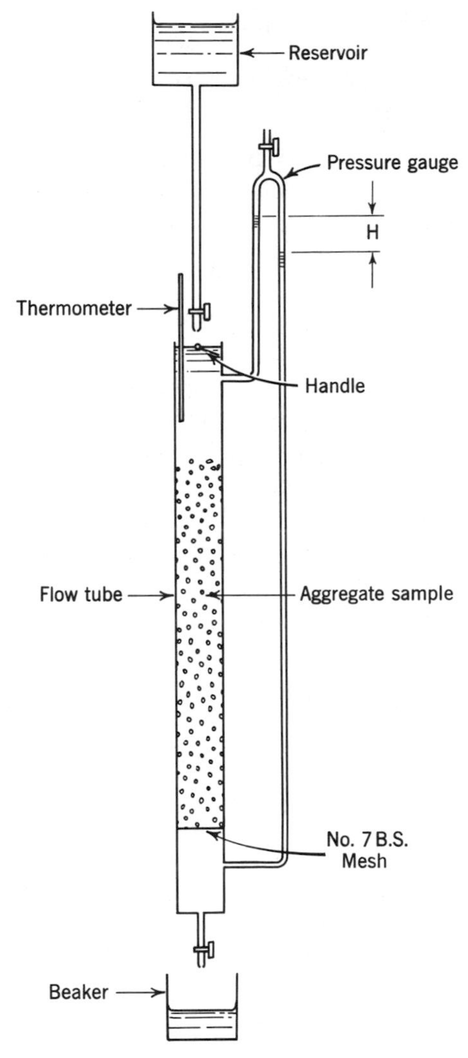

Fig. 1.3. The apparatus used by Shacklock and Walker to measure the specific surface area of sand.

that mentioned above, it is fairly accurate: nevertheless, its possible limitations should be kept in mind.

5.2 Surface Area Estimated from Sieve Analysis

Loudon [29] calculated the specific surface area of each size group for *spherical* particles by assuming that the diameter of the smallest sphere is equal to the clear opening of the sieve; then by assuming the mean size of a size group to be the geometric mean of the limiting sizes, he obtained

$$\begin{pmatrix}\text{specific surface area of}\\ \text{a size group of spheres}\end{pmatrix} = \frac{6}{(l_i l_{i-1})^{1/2}}$$

This formula is sufficiently accurate if the size quotient l_i/l_{i-1} does not exceed 2; for a quotient of 2 it amounts to assuming the mean sphere diameter d to be $0.71 l_i$. Loudon's calculated results, together with some added by the author, are given in Table 1.10 for two series of sieve sizes. The figures in this table make it possible to compute the specific surface area of an aggregate composed of spheres from the sieve analysis. If the sieving were perfect, and if the grading within each group conformed to the assumption stated above, the result would be exactly correct.

Loudon used the figures on Table 1.10 as a basis for estimating the sphericity factor of irregular particles. Using the permeability method he determined the specific surface area of size groups of the various kinds of sand, obtaining results given in Table 1.11 along with corresponding calculated figures for spheres. In Table 1.12 are presented simi-

Table 1.10 Calculated Specific Surface Areas of Size Groups of Spheres[a]

Size Group	Size Limits (cm or μ)	Specific Surface (cm²/cm³)	BSS[d] Sieve Size	Size Limits (μ)	Specific Surface (cm/cm³)
$\frac{3}{4}$–$1\frac{1}{2}$ in.[b]	19.1 to 38.1	2.22	#7–#10	2410 to 1676	29.6
$\frac{3}{8}$–$\frac{3}{4}$ in.[b]	9.52 to 19.1	4.44	#10–#14	1676 to 1204	41.9
#4–$\frac{3}{8}$ in.[b]	4.76 to 9.52	8.93	#14–#18	1204 to 853	59.2
#8–#4[b]	2.38 to 4.76	17.9	#18–#25	853 to 599	83.8
#7–#14[c]	2410 to 1204	35.2	#25–#36	599 to 422	119.0
#14–#25[c]	1204 to 599	70.6	#36–#52	422 to 295	170.00
#25–#52[c]	599 to 295	143.0	#52–#72	295 to 211	242.0
#52–#100[c]	295 to 152	284.0	#72–#100	211 to 152	336.0
#100–#200[c]	152 to 76	558.0	#100–#150	152 to 104	476.0
			#150–#280	104 to 76	675.0

[a] Data are from A. G. Loudon, "Computation of Permeability from Simple Soil Tests," *Geotechnique,* **3,** 165–183, 1952–1953.

[b] ASTM Sieves, size quotient 2 (these items inserted by author).

[c] British Standard Sieve Series, size quotient 2.

[d] British Standard Sieve Series, size quotient $\sqrt{2}$.

lar data by Shacklock and Walker, but covering a wider range of sizes. (The sphericity factor is the ratio of the specific surface of the size group of spheres to the specific surface of the same size group of irregular particles, as measured by Loudon's method.)

Going back to Section 3.2 and the definition of the sphericity factor, we see that ψ is defined in terms of a sphere having the same volume diameter as the irregular particle. However, by Loudon's method, the measured specific surface area of a size group of irregular particles is compared with the specific surface area of spheres having the nominal volume diameter, that is, a volume diameter equal to the mean of the sieve openings. As discussed in Section 3.5.5, the volume diameter of irregular particles may be considerably larger than that of spheres, and therefore the specific surface areas for the spheres corresponding to a given size group of irregular particles should generally be smaller the larger the mean volume diameter of the particles. Thus to obtain the correct sphericity factor from Loudon's data it would be necessary to multiply each reported value by the ratio of the measured mean volume diameter of the particles to the nominal volume diameter. The effect would be to reduce the reported sphericity factors, the reduction generally being greater the smaller the flatness ratio of the particles [14].

Table 1.11 Measured Specific Surface Areas of Size Groups from Various Fine Aggregates[a]

Material	Size Group BSS[b]	Specific Surface of Sand (cm^2/cm^3)	Specific Surface of Spheres (cm^2/cm^3)	Angularity Factor[c] ($1/\psi_L$)
Leighton Buzzard Std. sand	#18–#25	91	83.8	1.1
Leighton Buzzard sand	#25–#52	150	143.0	1.1
Ham River sand	#25–#52	164	143.0	1.1
Ham River sand	#52–#100	285	284.0	1.1
Ham River sand	#14–#25	94	70.6	1.2
Crushed marble	#25–#52	220	143.0	1.5
Crushed quartzite	#25–#52	233	143.0	1.6
Crushed sandstone	#25–#52	241	143.0	1.7
Crushed basalt	#25–#52	242	143.0	1.7
Crushed pyrex glass	#25–#52	275	143.0	1.9

[a] Data are from A. G. Loudon, "Computation of Permeability from Simple Soil Tests," *Geotechnique*, **3**, 165–183, 1952–1953.

[b] British standard sieves.

[c] ψ_L is the sphericity factor by Loudon's method.

Table 1.12 Specific Surface Area of Concrete Aggregates*

	Specific Surface Area (cm^2/cm^3) and Angularity Factor for Materials Indicated						
	Rounded		Irregular		Crushed		
Size Group†	Sand and Gravel	$1/\psi$‡	Sand and Gravel	$1/\psi$‡	Granite	$1/\psi$‡	Spheres§
$\frac{1}{2}$–$\frac{3}{4}$ in. (1.5)	4.12	1.1	5.47	1.4	6.63	1.7	3.86
$\frac{3}{8}$–$\frac{1}{2}$ in. (1.5)	5.67	1.0	7.39	1.4	7.98	1.5	5.45
$\frac{1}{4}$–$\frac{3}{8}$ in. (1.5)	7.49	1.0	11.68	1.6	10.31	1.4	7.50
$\frac{3}{16}$–$\frac{1}{4}$ in. (1.5)	10.45	1.0	15.6	1.4	14.6	1.3	10.9
$\frac{1}{8}$–$\frac{3}{16}$ in. (1.5)	16.6	1.1	25.4	1.6	28.8	1.9	15.4
#7–$\frac{1}{8}$ in. (1.3)	24.8	1.2	36.9	1.7	48.8	2.3	21.6
#14–#7 (2.0)	56.5	1.6	50.0	1.4	98.5	2.8	35.2
#25–#14 (2.0)	105.4	1.5	103.5	1.5	149.0	2.1	70.6
#52–#25 (2.0)	159.0	1.1	164.0	1.2	260.0	1.8	143.0

* Data are from B. W. Shacklock and W. R. Walker, "The Specific Surface of Concrete Aggregates and Its Relation to the Workability of Concrete," Cement and Concrete Association Research Report No. 4, 1958.
† British standard sieves; the size quotient as given in parentheses.
‡ These figures are for angularity.
§ The last column was calculated by the author.

This means, of course, that the correction is such as to increase the angularity factor. The subscript L is used to designate a sphericity factor subject to the error just discussed.

To estimate the specific surface area of an aggregate, Loudon recommended calculating it from the sieve analysis on the basis of spherical particles, and then applying the sphericity factor, the reciprocal of which he called *angularity factor*. On this basis, we find the following relationship for a sieve series having a size quotient of 2:

$$\sigma_{ag} = \frac{5.58}{\psi_L} SM = \frac{558}{\psi_L}\left(p + \frac{1}{2}P_1 + \frac{1}{4}p_2 + \frac{1}{8}p_3 \cdots\right) \qquad (1.3)$$

where σ_{ag} is the specific surface area of the aggregate area per unit solid volume. The number 558 is the calculated specific surface area, in cm^2/cm^3, of the (#200–#100) group of spheres. (See Table 1.10.)

The value of $1/\psi_L$ for any aggregate may be estimated on the basis of the descriptions and values given in Table 1.11. Loudon recommended the following factors for natural sands:

$$\text{Rounded particles:} \quad \frac{1}{\psi_L} = 1.1.$$

$$\text{Medium angularity:} \quad \frac{1}{\psi_L} = 1.25.$$

$$\text{Angular:} \quad \frac{1}{\psi_L} = 1.4.$$

For crushed materials, the angularity factor may be as high as 1.9, as may be seen in Table 1.11.

The surface modulus of a concrete sand might be 21, for example, and that of a coarse aggregate, $1\frac{1}{2}$ in. maximum size, might be 0.78. Their calculated specific surface areas would be

$$\sigma_a = \frac{5.58}{\psi_L} \times 21 = \frac{117}{\psi_L}$$

and

$$\sigma_b = \frac{5.58}{\psi_L} \times 0.78 = \frac{4.35}{\psi_L} \text{ cm}^2/\text{cm}^3$$

where σ_a and σ_b are the specific surface areas of fine and coarse aggregate, respectively.

Multiplying by 2.54 we obtain

$$\sigma_a = \frac{297}{\psi_L} \quad \text{and} \quad \sigma_b = \frac{11}{\psi_L} \text{ in.}^2/\text{in.}^3$$

For reasons already discussed, the figures for specific surface area arrived at above are liable to be in error if the particles have an appreciable degree of flatness. To obtain better accuracy, it would be necessary to correct the sphericity factors on the basis of the ratio of real volume diameter to nominal volume diameter, and if the shapes differed from group to group, it would be necessary to apply a different corrected sphericity factor to each term of Eq. 1.3.

Popovics [22] found that the specific surface area is related to grading as follows:

$$\sigma_{ag} = b \int_{d_{\min}}^{d_{\max}} \frac{f'(d)}{d} \, dd$$

where $f'(d)/d \, dd$ is the differential function of the grading curve and b is a constant depending on particle shape and on the units of measure.

If the particle-size distribution curve [the differential curve $f'(d)$] is plotted on a semilogarithmic system of coordinates, the specific surface area of the aggregate is proportional to the area under the curve. For spherical particles, Popovics gave this formula:

$$\sigma_{ag} = 0.1382S$$

where the specific surface area is in cm^2/cm^3 and S is the area under the differential curve with sizes expressed in millimeters.

6. AGGREGATE STRUCTURE

6.1 General

An aggregation of randomly arranged solid particles requires more space than the sum of the solid volumes of the particles, for obvious reasons. In the dry or inundated state each particle is in contact with many of its immediate neighbors, and therefore the aggregate exhibits some degree of mechanical stability. Because of this, it is appropriate to use the term "aggregate structure." The characteristics of aggregate structure depend on particle size, particle shape, and grading, and to a lesser degree on the size and shape of the container, if any, that confines the aggregate. Also, it depends on the circumstances under which the aggregation takes place; if the particles are subjected only to the force of gravity, the structure is relatively loose and unstable, but if subjected to vibration or tamping, the structure becomes denser and more stable.

In the following chapters we see that in concrete the aggregate particles become slightly dispersed, and thus the aggregate structure loses its mechanical stability; nevertheless, the idea of aggregate structure finds application when accounting for the water requirement of certain classes of concrete mixtures—the most common classes—as we show later.

An attribute of aggregate structure is the fraction of a unit volume of aggregate occupied by solids, that is, its *density,* or the complement of density, *voids content.* The term "voids ratio" designates the volume of void spaces per *unit solid volume* of material. *Aggregate density,* or *solidity ratio* pertains to the solid content of a unit compacted volume.

6.2 Standard Laboratory Methods of Compacting Aggregate

The specific voids content, or aggregate density, or solidity ratio, may vary for the same material according to the method used for compacting the aggregate in the measuring container. ASTM Designation C29-60 describes three standard methods of compacting aggregate as a means of determining the weight of unit aggregate volume. A commonly used method is that of filling a standard cylindrical container in three successive layers, poking each layer 25 times with a round-pointed, $\frac{5}{8}$-in. steel rod. The condition produced is called "dry-rodded volume."

7. VOIDS CONTENTS OF AGGREGATES

7.1 *Pertinence to Concrete Technology*

During earlier periods of concrete technology it was often assumed that, among comparable aggregates, the one which had the smallest percentage of voids was the most suitable for concrete. At the same time it was specified that the minimum size must be limited and that the percentage of the smallest particles must be held to a low value, a stipulation that limited the aggregate density. The assumption actually was that the voids content should be the lowest possible with particles larger than approximately the opening of a No. 200 sieve, 74 μ. It is now known that even on this restricted basis, a minimum volume of voids is not the best target for the mix designer, except for the leanest class of mixtures; usually, the best grading contains less fine aggregate than that giving a minimum content of voids. Nevertheless, production of satisfactory concrete requires aggregates with a low content of voids even if not the lowest possible, and this requires finding proper combinations of sizes within the allowable size range.

Various schemes for selecting the proportions of concrete mixtures, called methods of mix design, have been devised, and some of them are based on the packing characteristics of dry mixtures. Because of this, we emphasize certain characteristics of dry mixtures, and later, compare them with the characteristics of wet, plastic mixtures to show why mix design based on packing characteristics of dry materials is not wholly satisfactory.

7.2 *Effect of Particle Shape*

Particle shape has already been considered in connection with particle size and the surface area of particles. It is now to be considered further in connection with aggregate structure.

From some points of view, perfect sphericity is the ideal particle shape, for spheres have the smallest specific surface area for a given nominal size, and a random aggregation of spheres has the least percentage of voids for a given grading. In practice, however, spherical particles are not generally available. Among natural and manufactured aggregates

Table 1.13 Relation between Angularity Factor and Voids Content of Size Groups[a]

Material	Size Group[b]	Voids Content[c] (per cent)	Angularity Factor ($1/\psi_L$)
Crushed Pyrex-brand glass	#25– #52	57.8	1.9
Crushed basalt	#25– #52	58.5	1.7
Crushed sandstone	#25– #52	56.5	1.7
Crushed quartzite	#25– #52	56.0	1.6
Crushed marble	#25– #52	53.0	1.5
Ham River sand	#25– #52	46.0	1.1
Leighton Buzzard sand	#25– #52	44.0	1.1
Glass beads No. 1	1100 to 600 μ	41.0	1.0
Glass beads No. 4	350 to 120 μ	43.0	1.0

[a] Data are from A. G. Loudon, "Computation of Permeability from Simple Soil Tests," *Geotechnique*, **3**, 165–183, 1952–1953.
[b] Sizes are given in terms of British standard sieves except for the last two items, which are given in microns.
[c] The content of voids is that of aggregate formed by pouring dry particles into water.

particle shapes vary widely. In Section 5.1 we have seen how an angularity factor $1/\psi_L$ may be obtained by means of a permeability test, or how it may be estimated on the basis of data given in Tables 1.12 and 1.13. Another basis for estimating the relative angularity of aggregate particles is the relative volume voids in given size groups. From Table 1.13 and Fig. 1.4, derived from Loudon's data, it appears that

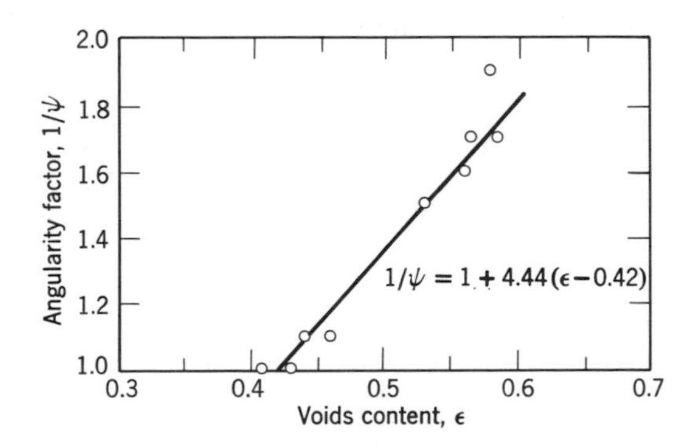

Fig. 1.4. An empirical relationship between angularity factor and voids content in a size group in the loose state. (Data are from Table 1.13.)

the angularity factor is related to the voids content of the aggregate as follows:

$$\frac{1}{\psi_L} = 1 + 4.44(\epsilon_1 - 0.42) \tag{1.4}$$

where ϵ_1 is the void fraction in a one-size-group aggregate. It should not be overlooked that this relationship pertains to groups based on a size quotient of 2, and for *loose* aggregates formed by pouring the dry material into a water-filled container.

Shergold [32] made a systematic investigation of voids content as an index to sphericity, using narrow size limits such as $\frac{1}{2}$ to $\frac{3}{8}$ in. The procedure included compacting the samples by rodding. Results were expressed in terms of an "angularity number," which is of the same kind as the quantity in parenthesis in the equation above. Shergold's numerical factor was 0.33 instead of 0.42. He found a range of about 9 points in angularity number, whereas the range is about 16 points in Fig. 1.4.

Goldbeck [33] used loose aggregate volume as a basis for estimating particle shape. Table 1.14 is derived from his data, and it includes angularity factors calculated on the basis of Fig. 1.4 and Eq. 1.5.

On comparing the results reviewed above, we see that voids measurement is more sensitive to differences in shape if it is based on the loose aggregate volume.

Gray [34] prepared 20 manufactured sands comprising different gradings and angularity factors, as shown in Table 1.15 (Angularity factors based on Eq. 1.4 were added to the group data by the present author.)

Table 1.14 Angularity as Indicated by Relative Voids Content[a]

	#4–#8		#8–#16		#16–#30		#30–#50	
Material	v/V	$1/\psi_L$	v/V	$1/\psi_L$	v/V	$1/\psi_L$	v/V	$1/\psi_L$
Stone sand:								
Crushed	0.539	1.53	0.579	1.70	0.591	1.76	0.590	1.75
Ground	0.490	1.22	0.519	1.44	0.569	1.66	0.591	1.76
Rounded	0.457	1.16	0.464	1.20	0.473	1.23	0.495	1.33
Natural sand	0.417	1.00	0.460	1.02	0.480	1.03	0.493	1.03

[a] Data are from A. T. Goldbeck, "Stone Sand for Use as Fine Aggregate," *Concrete*, **59,** 28–32, 1951. v is the volume of voids, V is the volume of aggregate. The angularity factor $1/\psi_L$ was calculated by the author from Eq. 1.4.

Table 1.15 Relation between Angularity Factor and Content of Voids in the Dry-rodded State[a]

Voids Content of Size-Group[b] (per cent)	Angularity Factor[c] ($1/\psi_L$)	Voids Content, Per Cent, of (0–#4) Sands Dry-Rodded State, for the Fineness Moduli Indicated				
		2.41	2.58	2.73	2.88	3.06
49	1.31	33.9	34.1	34.0	34.0	34.7
51	1.40	35.7	35.7	35.7	35.4	36.3
53	1.49	37.1	37.3	37.1	36.7	37.3
55	1.58	38.1	38.2	37.7	37.8	38.7

[a] Data are from J. E. Gray, "The Effect of Shape of a Particle on Properties of Air-entrained Sand, Stone, Mortar," *Crushed Stone J.*, **36**(3), 3–8, 1961.
[b] Not rodded or tamped.
[c] Calculated from Eq. 1.4 by the author.

For particles of a given shape, the greatest difference in voids content due to grading is about 0.8 percentage points, whereas the average difference due to the angularity factor at a given grading was 4 percentage points, and the difference was nearly the same for each of the gradings. In these aggregates the largest three size groups constituted from 30 to 50 per cent of the total weight of the six size groups. Thus we may conclude that for gradings within a given size range, and having ratios of fine to coarse within the usual range, particle shape has a much larger effect on voids content than does grading. It was pointed out by Dunagan [35] and by Neville [36] that when angular particles of a given size group are mixed with various proportions of rounded particles of the same size group, the voids content of the mixture is reduced about in proportion to the proportion of rounded material.

In Section 5.2 we have seen that the sphericity factors obtained by Loudon are subject to a considerable amount of error, because he uses nominal instead of actual volume diameters when computing the specific surface area of the equivalent spheres. Because Loudon's data are used in Fig. 1.4, it follows that the numerical coefficient in Eq. 1.4 is subject to the same error.

7.3 *Effect of Grading*

When two different size groups are mixed together, the voids content of the combined aggregate is less than that of either group alone. Of course this is true also of two different aggregates, each comprising any number of size groups, provided that the average size of one aggregate differs from that of the other. The factors that determine the content of voids in combined aggregates in terms of properties of the components are analyzed in Section 8.

7.3.1 Grading diagram. When the various sizes in an aggregate are expressed as fractions of the nominal diameter of the largest particle, and plotted against the weight or volume fractions, the result is a grading diagram (Andreasen and Anderson) [37]. The property illustrated by such a diagram is called the grading. Aggregates may have different average sizes but identical gradings. In the following discussion the distinction between grading and size or size range should be kept in mind.

7.3.2 The similarity principle. From considerations of geometry, we may understand that the voids content in regularly arranged uniform particles, such as a pyramid of spheres, is independent of the size of the particles. It is true also that if two aggregates of randomly arranged particles have the same grading and particle shape, they also should have an identical voids content. For example, if an aggregate is composed of particles ranging from 0.001 to 0.1 in., and another is composed of particles from 0.01 to 1.0 in., and if the two have the same grading, they have an equal voids content. Why this is so becomes intuitively apparent when we consider that if the smaller aggregate were examined with a ten-power glass, particles and voids should appear exactly as in the unmagnified view of the larger material; in other words, two such aggregates are *geometrically similar*.

The similarity principle and exceptions to it are illustrated in Table 1.16. The exceptions are shown by the first three items, which pertain to particle sizes under 16 μ. These sizes show a substantially higher content of voids than the rest. The relatively high voids content of an aggregate composed of very small particles is due to forces of interparticle attraction which, although present in all aggregates are ineffective in those composed of large particles.

The effect of differences in particle shape on the voids content is also apparent in Table 1.16; see also Table 1.13. The voids content in any size group of cement particles larger than 16 μ and that of a size group of crushed limestone are similar; it is higher than that of sand or gravel size groups.

Geometric similarity always indicates an equality in the voids content, but geometric dissimilarity does not necessarily indicate unequal voids contents.

Table 1.16 Data Illustrating the Principle of Geometric Similarity, and Departures from It, with Respect to Voids Content

Size Group	Approximate Mean Volume Diameter: Microns (when not mm)	Approximate Mean Volume Diameter: Inches	Voids Content[a] (per cent)
	Cement Particles (Angular)[b]		
—	81.6	—	45
—	49.2	—	45
—	30.2	—	44
—	15.3	—	46
—	8.3	—	58
—	2.4	—	80
	Crushed Limestone (angular)[c]		
#4–$\frac{3}{8}$ in.	6.7 mm	0.265	44
$\frac{3}{8}$–$\frac{3}{4}$ in.	13.5 mm	0.53	45
$\frac{3}{4}$–$1\frac{1}{2}$ in.	26.9 mm	1.06	45
	Sand (rounded)[d]		
#100– #50	210	0.0082	38
#50– #30	420	0.0165	38
#30– #16	840	0.0331	37
#16– #8	1680	0.0661	36
#8– #4	3360	0.132	36
	Gravel (rounded)		
#4– $\frac{3}{8}$in.	6.7 mm	0.265	35
$\frac{3}{8}$–$\frac{3}{4}$ in.	13.5 mm	0.53	36
$\frac{3}{4}$–$1\frac{1}{2}$ in.	26.9 mm	1.06	38

[a] Compacted by rodding or tamping.
[b] Data are from P. S. Roller, U. S. Bureau of Mines Technical Paper No. 490, 1931.
[c] Data are from W. M. Dunagan, *Proc. ACI*, **36,** 649–684, 1940.
[d] Data are from the Portland Cement Association.

8. ANALYSIS OF BINARY MIXTURES

Because the production of concrete involves, among other things, control of the voids content in the aggregate, the factors controlling voids content should be considered analytically. This can be done well enough by considering any aggregate to be a binary mixture, one component being fine aggregate and the other coarse. Relationships between voids content and composition tend to be nonlinear, but by choosing certain functions of voids content and composition these relationships can be handled with the least difficulty. As to the function of voids content, there are two to choose from. One is the volume of mixture, overall, per unit quantity of solid material. This is called the *specific aggregate volume*. The other is volume of voids per unit of solid material; it is called the *voids ratio*. Either of these parameters is dealt with as a function of the ratio of weight or solid volume of the coarse component to the total weight or solid volume of the material in the mixture.

8.1 Specific Aggregate Volume

Specific aggregate volume is defined as the volume of space required to accommodate a unit weight of the aggregate. Thus, designating the weight of fine aggregate by a and that of the coarse by b, we have

$$v_{ag} = \frac{V}{a + b} \tag{1.5}$$

where V is the overall volume of the aggregate (the aggregate volume) and v_{ag} is the specific aggregate volume of a combined aggregate comprising components a and b. In a series of mixtures of given fine and coarse aggregates the specific aggregate volume is a function of the relative proportions of a and b. Thus

$$v_{ag} = f\left(\frac{b}{a + b}\right) = f(\text{N}) \tag{1.6}$$

Figure 1.5 is a specific aggregate volume diagram. It represents mixtures of 0–#4 sand and #4–1½ in. gravel in the compacted state (dry-rodded). At N = 0 the mixture contains no coarse aggregate and thus v_{ag} at that point is the specific aggregate volume of the fine aggregate; in this case $(v_{ag})_a = A = 0.541$ cm³/g. At the other side, N = 1.0, the specific aggregate volume of the coarse aggregate is plotted, the value being $(v_{ag})_b = C = 0.569$ cm³/g. Intermediate mixtures have specific aggregate volumes smaller than that of either of the two components tested separately, as is indicated by the locus of the curve through experimentally determined points.

The point $v_b = 0.377$ cm³/g represents the specific *solid volume* of the coarse aggregate. The straight line from A to D' is the locus for

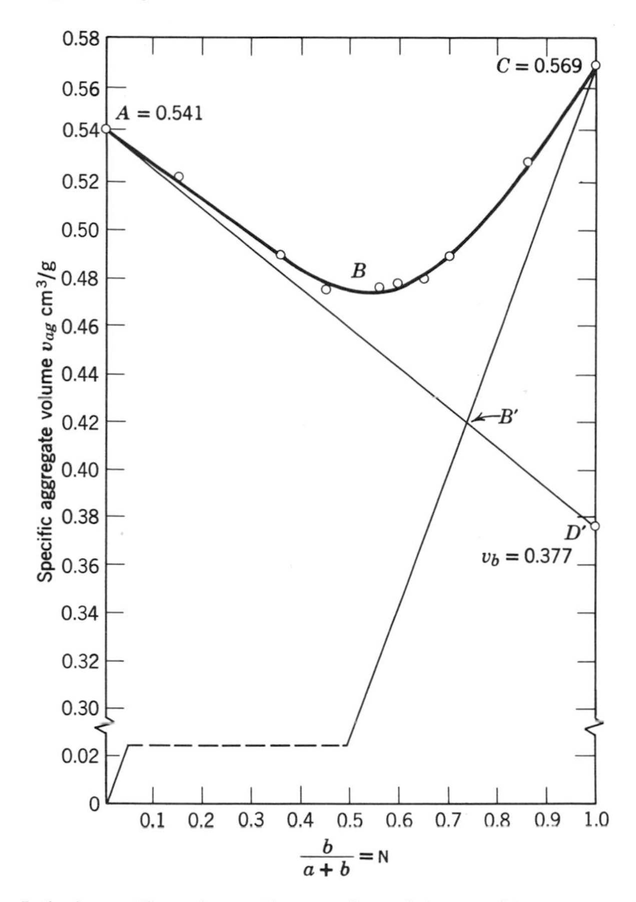

Fig. 1.5. A specific volume diagram for mixtures of two aggregates.

hypothetical mixtures in which increments of fine aggregate are replaced by equal volume increments of solid rock, the properties of the fine aggregate remaining unaffected by the presence of the coarse particles.

The straight line from C to the origin represents for any given value of N the smallest volume that could accommodate the coarse aggregate in that mixture. Thus at N = 0.5, the smallest possible space into which the amount of coarse aggregate in the mixture could be placed is one-half of 0.569 cm^3/g of coarse aggregate.

The factors determining the locus of the experimental curve are discussed in following sections, but in terms of voids ratio rather than specific aggregate volume.

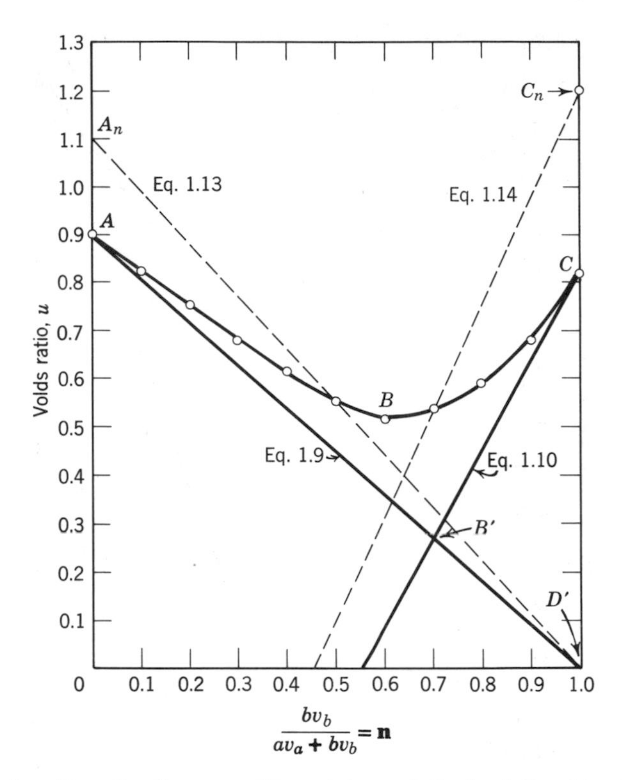

Fig. 1.6. A voids-ratio diagram with certain analytical relationships illustrated.

*8.2 The Voids Ratio**

For reasons that will become clearer as the discussion proceeds, we find it advantageous to use voids ratio rather than specific aggregate volume, and to abandon weight as the specified reference quantity in favor of solid volume. Data for the same series of mixtures shown in Fig. 1.5 are plotted that way in Fig. 1.6, a voids-ratio diagram.

Voids ratio, u, is defined as follows:

$$u = \frac{\epsilon}{1 - \epsilon} = \frac{V}{av_a + bv_b} - 1 \tag{1.7}$$

in which $\epsilon/(1 - \epsilon)$ is the volume of voids in the aggregate per unit solid volume of the aggregate, V is the overall volume aggregate, a and b are

* In previous publications I used the term *specific void content* instead of *voids ratio*. Because the latter is well established in soil technology and because it is desirable to use as few special terms as possible, the change was made.

the weights of the fine and coarse aggregates, as before, and v_a and v_b are the respective values of *specific solid volume,* that is, solid volume per unit weight of material.

For the scale of abscissas we have*

$$\mathbf{n} = \frac{bv_b}{av_a + bv_b} \tag{1.8}$$

It should be noted that $\mathbf{n}$ is the solid volume of coarse material per unit total solid volume in the aggregate and is *not* the solid volume per unit overall volume of the mixture. (Other features of Fig. 1.6 are discussed further on.)

8.3 Hypothetical Mixtures

In preparation for further consideration of actual mixtures of granular materials it is useful to consider first the voids-ratio diagram for a pair of materials that can be mixed together without one of them altering the structure of the other. Consider first a fine aggregate of this kind, to which we shall add small increments of coarse aggregate, mixing the two together so that the coarse particles become dispersed in the matrix of fine material. In such a state the coarse material has no aggregate structure, each particle of coarse material being independent of other coarse particles. Thus, as we add coarse particles, we add solid substance but no voids. Hence we should expect the voids content of the mixtures to be less than that of the fine aggregate alone by an amount proportional to the volume fraction of coarse material in the mixture; that is to say,

$$u = A(1 - \mathbf{n}) \qquad \text{(hypothetical)} \tag{1.9}$$

where u is the voids ratio, $\mathbf{n}$ is the volume fraction of the coarse material, and A is the voids ratio of the fine component alone. In Fig. 1.6 Eq. 1.9 is represented by the line from A to D'.

If the process of adding solid material and replacing material containing voids could be continued without limit, the final result would be a unit volume of voidless material, represented at Point D' in Fig. 1.7. However, the added material is particulate, and when the particles be-

* In this chapter and others expressions such as Eq. 1.8 appear frequently. The reader will become accustomed to reading pairs such as av_a and bv_b as representing solid volumes ("absolute" volumes). By using this device it becomes easy to convert any given relationship from the volume basis to the weight basis when that is desirable.

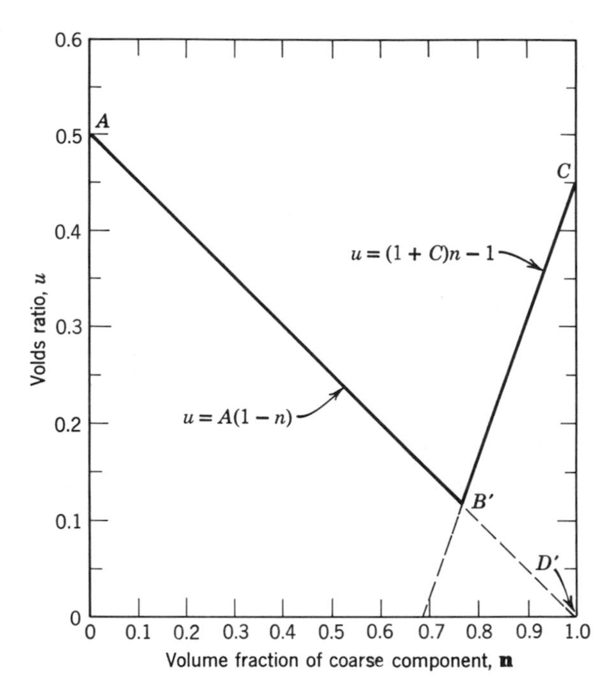

Fig. 1.7. A voids-ratio diagram for mixtures of two hypothetical aggregates.

come so concentrated in the matrix that they come into contact with each other, no more can be added without increasing the voids content in the system. At this point aggregate structure becomes dominant. The limiting concentration for this hypothetical series of mixtures is marked B' in Fig. 1.7. For values of $\mathbf{n}$ beyond that point, the locus is a straight line from B' to the point corresponding to the voids ratio of the coarse component, marked C on the diagram.

For mixtures to follow the line $B'C$, the fine material must not alter the structure of the coarse. That is, if we start with a quantity of coarse aggregate and add small increments of the fine component, the fine particles must be so small and mobile as to occupy the voids of the coarse aggregate structure. The voids ratio of any such mixture is given by the following equation:

$$u = (1 + C)\mathbf{n} - 1 \qquad \text{(hypothetical)} \qquad (1.10)$$

Because Eqs. 1.9 and 1.10 have the same value at Point B', we can easily obtain the coordinates of that point in terms of parameters A

and C. The result is the following:

$$\mathbf{n}_{B'} = \frac{1 + A}{1 + A + C} \tag{1.11}$$

$$u_{B'} = \frac{AC}{1 + A + C} \tag{1.12}$$

About the only fine component that could work with pebbles as illustrated in Fig. 1.7, would be a liquid such as water, in which case the values of A would correspond to the volume of "holes" characteristic of the structure of water. Actual mixtures of ordinary particulate materials do not conform to the triangle $AB'C$ because two aggregates cannot ordinarily be mixed without their structures becoming altered; however, the triangle does serve as a frame of reference, showing the degree to which various combinations approach the greatest possible density. It serves not only for dry aggregates but also for concrete mixtures, as we see later.

8.4 Mixtures in which the Fine Aggregate is Dominant

Looking at the left-hand side of Fig. 1.6, we see that the points for actual mixtures diverge from the straight line from A to D'. Figure 1.8

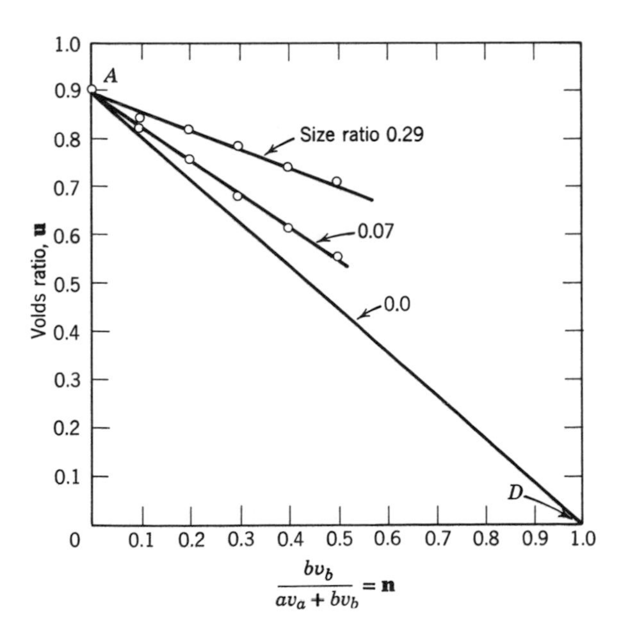

Fig. 1.8. A diagram illustrating the effect of adding coarse particles to fine aggregate.

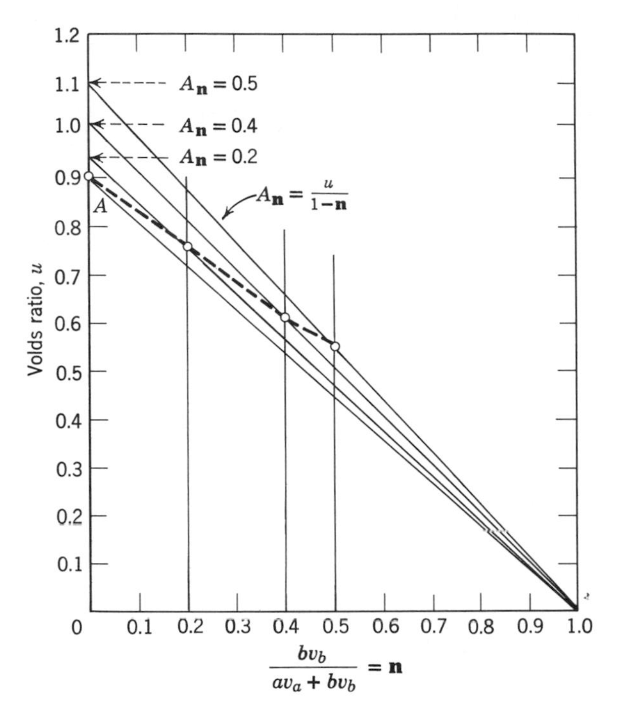

Fig. 1.9. The graphical method to evaluate the effect of coarse particles on the voids ratio of the fine aggregate in a mixture of fine and coarse aggregate.

shows other data for two series of mixtures reported by Andreasen and Anderson. These data show that the degree of divergence is greater the larger the ratio of the average size of the fine aggregate to that of the coarse aggregate.

8.4.1 Voids content of the fine component in a mixture. The effect of the coarse aggregate in a mixture on the structure of the fine aggregate can be expressed in terms of the increase in the voids content of the fine aggregate. As shown in Fig. 1.9, which shows one of the curves of Fig. 1.8, each point on the curve can be regarded as being on a straight line from $A_\mathbf{n}$ to D', where $A_\mathbf{n}$ is the apparent voids ratio of the fine aggregate in the state of packing it has in mixture **n**. The apparent voids ratio can be calculated from the coordinates of the point by the following relationship:

$$A_\mathbf{n} = \frac{u}{1 - \mathbf{n}} \tag{1.13}$$

For example, the voids ratio of the sand alone is $A = 0.9$, but at $\mathbf{n} = 0.4$ $u = 0.61$ and $A_{0.4} = 0.61/0.60 = 1.017$, which is an increase of about 13 per cent from the voids ratio of the sand alone.

Weymouth [38] offered an explanation of the increases in voids content based on the observation that the void space between a flat surface and particles packed against it is greater, per unit solid volume of particulate material in that region, than is the void space per unit volume of material in the interior of an aggregate where the particles nest together. He reasoned that similar "boundary voids" exist wherever small particles make contact with large ones. Such an explanation seems reasonable when we are thinking of the relationship between two widely different sizes, but if there is a continuous gradation of sizes in the combined aggregate, the explanation seems less clear.

A better explanation might be based on the observation that when an isolated large particle is in a matrix of fine aggregate it is the center of a local disturbance or deformation of the structure of the fine aggregate that reaches beyond the region of direct contact. The effect is analogous to a mechanical strain such as that produced by an expanding solid body in an elastic matrix. In this case the "expanding body" causes no stress and suffers no compression, and thus the effect of the deformation is confined entirely to the matrix.

In this connection we may note that the measuring container must have some effect on the percentage of voids found in a given aggregate. Just as a large rock particle creates a disturbance in a matrix of fine aggregate, the container has a similar peripheral effect, that effect extending for some distance inward; it raises the voids content above the value that would be found with a very large sample. Thus the values of A and C in Fig. 1.6 are determined not only by the characteristics of the aggregate but to some degree, generally minor, by the size and shape of the measuring container. The effect is greater the larger the mean particle diameter relative to the diameter of the container. Usually, the effect of the container on aggregate structure is ignored.

Correction factors to offset the effect of the size of the container on voids content of a given aggregate have been derived from the geometry of packing [35]. From the foregoing discussion it is apparent that results obtained this way are questionable.

8.5 *Coarse Aggregate Dominant*

Let us now consider the opposite extreme, where the coarse aggregate is the dominant component. As discussed in Section 8.3, the locus of the curve would be the straight line from C to B' if the fine aggregate could be contained by the voids in the coarse aggregate. Under special conditions a close approach to this kind of packing can be achieved, as was shown by McGeary [39]. For the coarse component he used equal spheres having a mean diameter of 0.124 in. and for the fine material spheres having a diameter of 0.0016 in., the size ratio being 0.013. He filled the container with the larger spheres, and then by means of vibration caused the fine components to sift down through the structure of the coarse. McGeary pointed out that such a procedure is a process of packing, and the result is not indicative of what can be achieved by producing a mixture before the compaction occurs.

In Chapter 2 we find that when the fine components is portland cement, and the coarse is composed of relatively large particles, a small amount of the powder can be contained by the voids of the coarse material. But in general this is not the case.

In Fig. 1.6 the plotted points on the curve *ABC* represent actual mixtures of fine and coarse aggregate compacted as much as possible by jigging. Points to the right of *B* represent mixtures in which the coarse aggregate is dominant. In this range the divergence of the curve *CB* from the line *CB'* indicates that an actual mixture contains a larger fraction of voids than that which corresponds to the assumption that the fine aggregate is accomodated by the voids of the coarse aggregate. Although a mixture of fine and coarse aggregates has a smaller fraction of voids than that of either of the separate components, the reduction is not accomplished by the fine material simply filling the voids of the coarse. Usually, as in the present case, even a small amount of fine aggregate causes dispersion of the coarse particles, the voids in the compacted coarse aggregate usually being too small to contain all sizes in the fine aggregate.

Conversely, when we separate an aggregate into fine and coarse portions by means of a sieve, the process is not that of removing fine particles from the voids of the coarse aggregate; instead, we create two aggregates, each having a structure that did not exist in the combined aggregate. If we use a sieve of such a size as to divide a given aggregate into two equal parts, each aggregate so obtained is in a "shrunken" state as compared with its state in the combined aggregate; this is especially so for the coarse aggregate.

The degree to which the coarse aggregate is dispersed by the fine aggregate is discussed in Section 8.6 in terms of the "clearance" between coarse particles; here we evaluate it in terms of the parameters of the voids-ratio diagram. Let $C_{\mathbf{n}}$ stand for the voids ratio of the coarse aggregate as it exists in mixture **n**, all the space surrounding the particles being considered void even though the aggregate is actually partly filled with fine particles. Then

$$C_{\mathbf{n}} = \frac{1 + u}{\mathbf{n}} - 1 \tag{1.14}$$

For example, in Fig. 1.6, the value of u at $\mathbf{n} = 0.5$ is 0.54; hence, $C_{\mathbf{n}} = (1.54/0.7) - 1 = 1.2$, which is to be compared with $C = 0.82$, the voids ratio of the coarse aggregate in the compacted state.

8.6 *The Theory of Particle Interference*

8.6.1 Definition of particle interference. As we have already seen, mixing a small proportion of pebbles with sand causes an increase of the voids content of the sand, the increase apparently being due to the disturbance of the normal structure of the sand by the larger particles. Such an effect could be said to be due to the interference of the larger particles with the normal packing of the sand; that is, it could be ascribed to particle interference. However, the evolution of ideas in this area was such that the term particle interference is usually used with the connotations given it by Weymouth [40] who, as mentioned above, ascribed the interference effect just described to "boundary voids." Weymouth applied the term particle interference to the effect that causes the upward deviation of the voids-ratio curve from the initial linear portion after the proportion of the coarse component exceeds a certain value. That is to say, the initial point of divergence, which can be seen in Fig. 1.6 at $\mathbf{n} = 0.5$ marks the beginning of particle interference. The term therefore pertains to an effect that is due not only to the coarse particles individually, but also to the smallness of the distance between coarse particles.

8.6.2 The Weymouth criterion for particle interference. Weymouth developed his theory for application to plastic concrete mixtures (this application is discussed in Chapter 5), but it is applicable with a minimum of complications to mixtures of fine and coarse aggregates and is most easily comprehended when discussed in terms of such mixtures. Weymouth proposed the following criterion for particle interference: If the average clear distance between the particles of any given size is smaller than the diameter of the next smaller size, particle interference is present and has influenced the packing of the mixture.

If an aggregate is composed of particles of only two sizes, the meaning of the statement above is clear, but, for an aggregation of a continuous series of sizes, the meaning is not clear. In practice Weymouth treated arbitrarily defined size groups as if they were each composed of one size only; indeed, he contended that the principle of particle interference is valid regardless of the range of sizes within the size group, although he actually defined a size group as we do in Section 3.5.2. He applied his criterion to successive pairs of size groups in an aggregate regardless of the number of size groups smaller than the smaller of the pair, and of the number larger than the larger of the pair, as will be discussed farther on.

8.6.3 Weymouth's formula. The derivation of the Weymouth formula for clearance, as condensed by Butcher and Hopkins, is as follows:*

$$d_b = NgD_b^3 \tag{1.15}$$

where d_b is the solid volume of coarse particles per unit volume of the mixture, the coarse particles being dispersed to some degree; N is the number of particles; D_b is the average nominal diameter of the coarse particles as computed from Eq. 1.1;† g is a proportionality constant that depends on particle shape, it being $\pi/6$ for spheres.

If the same particles are imagined to become enlarged without change of shape to diameter $D_b + t$, the enlarged diameter being such that the enlarged particles would be in contact with each other when in a compact array, we would find

$$d_o = Ng(D_b + t)^3 \tag{1.16}$$

where d_o is the volume of the enlarged particles in a unit volume of compacted coarse aggregate; that is, it is the density of the imaginary aggregate. However, because the real and imaginary aggregates are geometrically similar, it is also the density of the real coarse aggregate in its compacted state. Now we may write

$$\frac{d_o}{D_b} = \frac{(D_b + t)^3}{D_b^3}$$

and

$$t = \left[\left(\frac{d_o}{d_b}\right)^{1/3} - 1\right] D_b \tag{1.17}$$

where t is the clearance.

When data are given in terms of the voids-ratio diagram, the values needed for Eq. 1.17 are not given directly; in that case the following

* Weymouth's nomenclature (modified) is used here.

† Weymouth actually used the arithmetical mean of the limiting sieve sizes of a group.

equivalent may be used.

$$t = \left[\left(\frac{1+u}{\mathbf{n}(1+C)}\right)^{1/3} - 1\right] D_b \tag{1.18}$$

8.6.4 The criterion of Butcher and Hopkins. Butcher and Hopkins pointed out [19] that if an aggregate is composed of more than two size groups, interference with a given group can hardly be ascribed to the next smaller particles only, because a given particle is likely to be in contact with particles of all sizes. They suggested a criterion fundamentally like Weymouth's but applied it to two components composing the whole mixture. That is to say, they proposed that if a mixture is free from particle interference, the average clear distance between the particles larger than any arbitrarily selected intermediate size is at least equal to the average diameter of all the smaller particles. Butcher and Hopkins considered concrete rather than aggregate alone, and therefore in their examples the cement was included when computing the average diameter of the fine component of the mixture.

8.6.5 Hughes' modification. Hughes [41] offered a further modification of the Weymouth theory. Accepting the Butcher and Hopkins modification in principle, Hughes suggested that it is predominantly the coarser fine particles that tend to produce "interference with the finer coarse particles." Accordingly, he defined an "average-volume mean diameter" for the fine component as follows:

$$D_F = [\Sigma\, p_i d_i^3]^{1/3} \tag{1.19}$$

Equation 1.19 gives a value for the nominal average diameter which is heavily weighted by the coarser particles. For example a certain sand having an average diameter of 0.246 mm by Eq. 1.2 shows a value of 2.1 mm by Eq. 1.19. Thus, where a clearance of 0.246 mm would be required by the Butcher and Hopkins criterion, 2.1 mm would be required by the Hughes criterion. For mixtures containing cement, Hughes would exclude the cement when calculating D_F because the value of D_F would not be changed significantly by including the cement.

8.6.6 Experimental data. None of the three criteria described above prove to be generally reliable when applied to various experimental observations. However each appears to be valid for certain restricted conditions. The Hughes criterion seems to be approximately correct when applied to fine and coarse aggregates having ordinary gradings; the others are accurate when applied to mixtures of two size groups only and presumably would be accurate for mixtures of any pair of geometrically similar aggregates (see Section 7.3.2) regardless of the size range

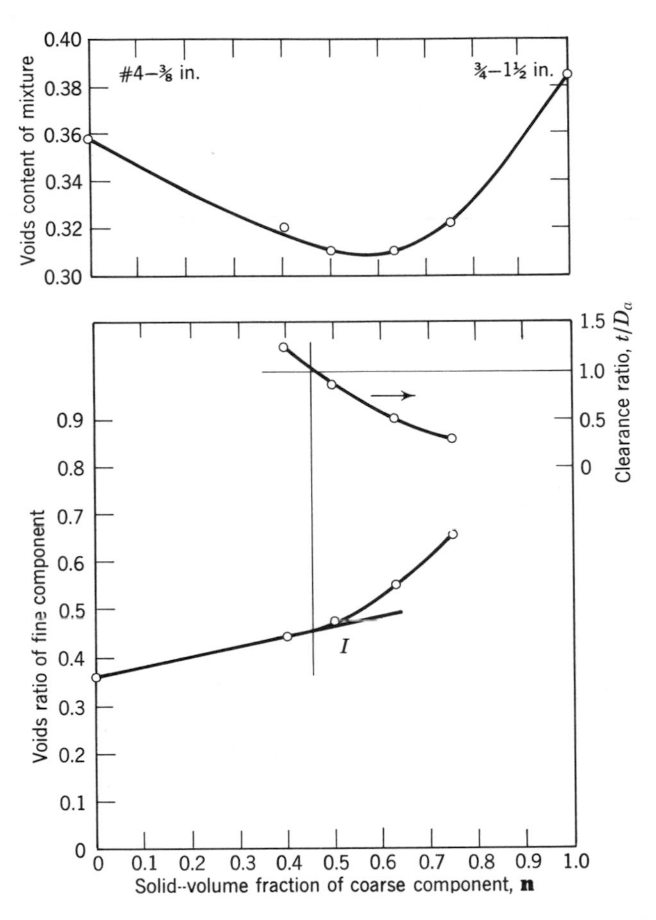

Fig. 1.10. The graph demonstrating Weymouth's method to detect the point of the beginning of particle interference, point *I*.

of the aggregates. Some of the evidence on which these conclusions are based is discussed below.

Consider the data shown in Fig. 1.10. The top curve is a voids-content diagram for mixtures of two size groups (#4–$\frac{3}{8}$ in.) and ($\frac{3}{4}$–$1\frac{1}{2}$ in.). The bottom curve is the voids-ratio curve of the fine component; each point is the voids content of a given mixture divided by the solid-volume fraction of the fine component in that mixture. The intermediate curve gives the clearance ratio, which is to say, the computed clear distance between the particles of the coarse component divided by the nominal average diameter of the particles of the fine component D_a.

The voids-ratio curve begins as a straight line with an upward slope, the slope being due to the increase in voids content in the fine component, which is caused by the presence of individual particles of the coarse component, as we discussed above. The upward departure of the plotted points from this straight line at the point marked I is, according to the Weymouth theory, due to particle interference. It is apparent that this point of departure (I) occurs at or near the point where the computed clear distance between the coarse particles is equal to the average diameter of the particles of the fine component, that is, where the clearance ratio is unity.

Thus the Weymouth criterion seems to be accurate, and the Butcher and Hopkins as well, for in this case the two values for critical clearance are identical. However, we may note that the mixtures represented in Fig. 1.10 are composed of aggregates that are geometrically similar, or substantially so, and we shall see that such similarity is apparently necessary to the success of the original theory.

In connection with the discussion above, consider the data of Fig. 1.11, representing mixtures of sand comprising six size groups, with a gravel comprising three, the aggregates thus being dissimilar. Weymouth's original criterion cannot be applied, for the two components

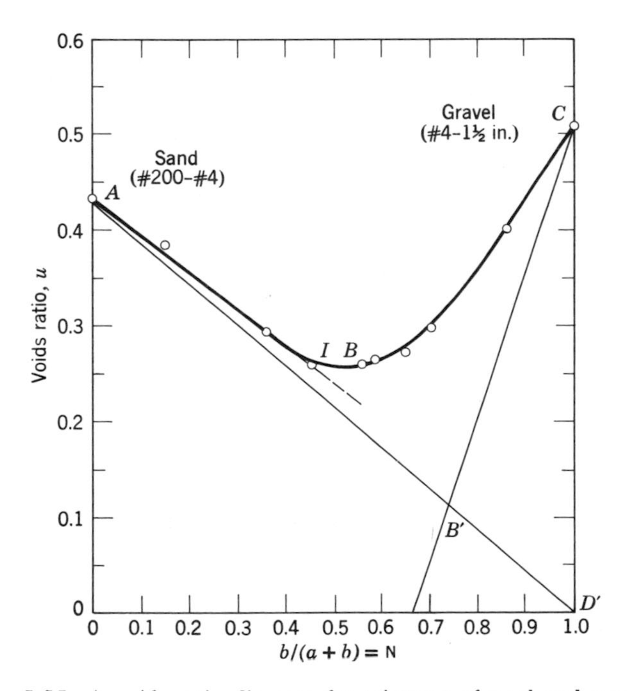

Fig. 1.11. A voids-ratio diagram for mixtures of sand and gravel.

cannot be treated as comparable size groups. However, the other two criteria are applicable.

Interference appears at the point $\mathbf{n} = 0.45$ and $u = 0.26$; $C = 0.51$. The average sizes for the two components, computed by Eq. 1.1, are, for the fine component, $D_a = 0.246$ mm, and for the coarse, $D_b = 9.59$ mm (0.097 in. and 0.38 in.). Using the data above for the observed point I, we find by Eq. 1.18 that $t_{cr} = 2.2$ mm.

Thus we find that at the point where particle interference begins, the clearance between coarse particles is $2.2/0.246 = 9$ times the average diameter of the sand particles. This is very far from confirmation of the Butcher-Hopkins criterion. Actually, the clearance is almost the same as the average nominal diameter of the two largest size groups of the sand, which is 1.97 mm. Also it is nearly the same as the Hughes diameter, which is 2.1 mm. However, as we see farther on, this case does not establish general validity for the Hughes criterion.

We might apply the Weymouth criterion to the case above by considering the coarse aggregate as one size group (comprising three regular groups) and on that basis comparing the clearance with the average size of the three largest size groups in the sand. That average size turned out to be 1.03 mm, which is about one-half the observed critical clearance.

Figure 1.12 is an example in which the sand comprises the three smallest size groups, and the gravel comprises two: (#4–$\frac{3}{8}$ in.) and ($\frac{3}{8}$–$\frac{3}{4}$ in.). The beginning of particle interference appears to be somewhere between $\mathrm{N} = 0.60$ and 0.65. Using the latter value, we estimate the other coordinate of the critical mixture to be $u = 0.255$. Then, with $C = 0.505$, and $D_b = 7$ mm,* Eq. 1.18 gives $t_{cr} = 0.6$ mm. This is about the smallest critical clearance that could be reconciled with the data; it is to be compared with 0.19 mm, the estimated average size of the sand (its sieve analysis is not available), and with the average size of the largest size group in the sand, which is about 0.40 mm. We see that the critical clearance is not only much larger than the average size, it is larger also than the average size of the largest group in the sand; it is in fact comparable with the upper limit of sand sizes, about 0.6 mm. The Hughes diameter for this sand cannot be computed exactly for lack of sieve analysis data, but it cannot be much if any greater than 0.4 mm, which is two-thirds the observed critical diameter.

To apply the original Weymouth criterion in this case, we consider the coarse aggregate as one size group (two regular size groups) and the two coarsest groups of the sand as the interfering group. The esti-

* The average size of the coarse aggregate cannot be estimated exactly, mainly because the amount of undersize material in it is not known.

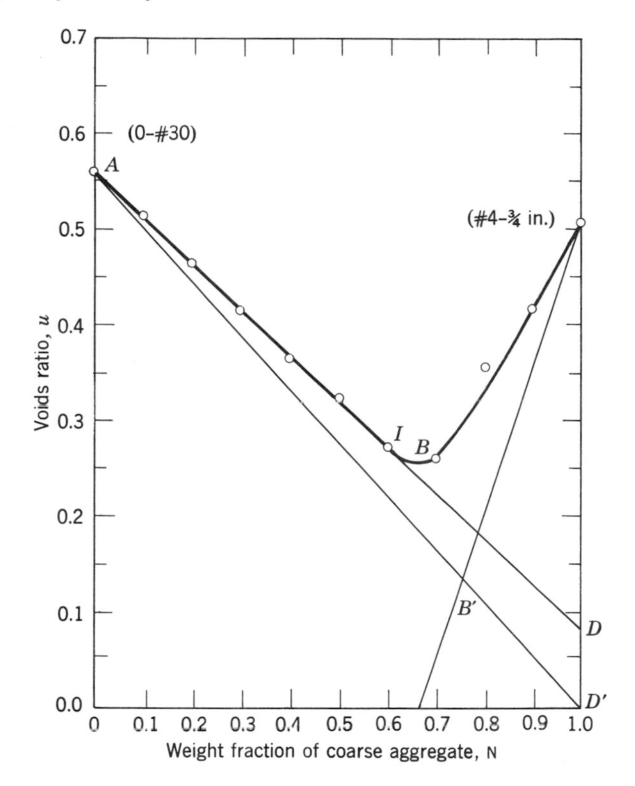

Fig. 1.12. A voids-ratio diagram for mixtures of fine sand normal gravel. (Data are from National Sand and Gravel Association, Circular No. 8, 1958.)

mated average size of the latter is 0.3 mm. By the Weymouth criterion, this value should be the critical clearance (or dispersion) of the coarse component, but, as already noted, the actual critical clearance was found to be two times that value.

Thus, for the data of Fig. 1.12, all three criteria fail by a wide margin.

In the examples considered so far we have found both conformity and nonconformity with theory. Other data show that none of the three criteria can be considered a reliable basis for proportioning aggregates or as a dependable diagnostic method. Discrepancies between the underlying assumptions and the experimental results depend on factors not considered in the assumptions. One such factor, probably the principal one, is grading.

The data in Table 1.17 pertain to various mixtures of sand and gravel. Voids-ratio diagrams for several series of mixtures were examined and the points I were noted. From the coordinates of these points the values for critical clearance was noted, with the results shown in the table.

Table 1.17 Clearance between Coarse Aggregate Particles at the Critical Concentration, for Mixtures of Sand and Gravel

Grading of Coarse Aggregate (CA) (Weight Per Cent)			Proportion of CA (Per Cent)	t_{cr}* (mm)	D_b† (mm)
#4–$\frac{3}{8}$ in.	$\frac{3}{8}$–$\frac{3}{4}$ in.	$\frac{3}{4}$–$1\frac{1}{2}$ in.			
0.0	40.0	60.0	46.0	3.18	16.5
15.0	34.0	51.0	46.0	2.30	11.0
30.0	28.0	42.0	45.0	2.08	9.0
50.0	20.0	30.0	47.0	1.81	8.4
100.0	0.0	0.0	50.0	1.14	6.3
0.0	100.0	0.0	50.0	2.67	12.7
30.0	70.0	0.0	50.0	2.02	8.8
0.0	25.0	75.0	50.0	2.87	18.5
30.0	17.5	52.5	50.0	1.75	9.2
30.0	0.0	70.0	54.0	1.46	9.4

* t_{cr} = critical clearance.
† D_b = average nominal diameter of coarse aggregate.

The upper block comprises data for mixtures in which the grading of the coarse aggregate was varied systematically, as shown in the first three columns. The lower block gives other combinations. The sand was the same as that represented in Fig. 1.11; its average size (Eq. 1.1) was 0.246 mm, and its Hughes diameter 2.1 mm. By the Butcher-Hopkins criterion 0.246 should be the critical clearance in each case; by the Hughes criterion, 2.1.

In all cases the observed critical clearance is much larger than the average size of the sand as computed from Eq. 1.2; the Butcher-Hopkins criterion fails completely. In three of the 10 cases the observed clearance is within ±10 per cent of the Hughes diameter, but the other seven generally differ from that value by an amount exceeding the estimated experimental uncertainty. Moreover, the upper block of data show that the variations are systematic rather than random.

A point that has not been dealt with by the several authors is the sensitivity of the figure for critical clearance; that is, how much should two values differ for the differences to be considered practically significant?

For the upper block of data the grading of the coarse aggregate was varied by keeping the ratio of the intermediate to the largest size constant while introducing the smallest size (pea gravel) progressively. It is evident that when the average size of the coarse aggregate was thus reduced by the introduction of pea gravel, the figure for critical clearance was correspondingly reduced, from 3.18 to 1.14 mm. It thus seems that the critical clearance might depend on the average size of the voids of the coarse aggregate as well as on the size and grading of the fine aggregate. If the size range and grading of the sand were also introduced as a variable, we might reasonably expect even further departures from any single criterion for critical clearance.

A limitation of the Hughes criterion is seen clearly by comparing the last item of the first block in Table 1.17 with the first item of the second. Each of the coarse aggregates was a single size group of similar material, and therefore of nearly equal density, that is, equal d_o. We should therefore expect them to show about the same value for critical clearance when mixed with the same kind of sand, but actually the larger aggregate required over twice the clearance required by the smaller for mixtures to be without particle interference. Nevertheless, of the three, the Hughes criterion seems more successful than the others, and it could be useful when applied to materials having ordinary grading.

The examples of success of the Weymouth criterion we noted above are those involving only two regular size groups. We infer but have not proved that the size range of each group might be widened without changing the criterion, provided that geometric similarity is preserved. It seems on an intuitive basis that when interference occurs between two sizes of geometrically similar aggregates, all the particles are involved equally, and the average size is a significant figure. However, when the aggregates being mixed together are dissimilar, different sizes play unequal roles, and the significance of average size varies with the nature or degree of dissimilarity. (Further discussion of this subject in connection with theories of design of plastic concrete mixtures will be found in Chapter 6.)

8.7 Size Ratio and Minimum Void Content

In mixtures of two sizes of particles the minimum voids content is smaller the smaller the ratio of the small size to the large. Figure 1.13, a somewhat idealized voids-ratio diagram prepared by Furnas [42], shows that as the size ratio approaches zero, the minimum voids content approaches the value of point B', as discussed before. Presumably the same conclusion applies to two geometrically similar graded aggregates.

However, it is not generally true that the minimum voids content is a function of size ratio only. This is shown in Table 1.18. Because the same sand was used in all mixtures, the size ratio decreased as

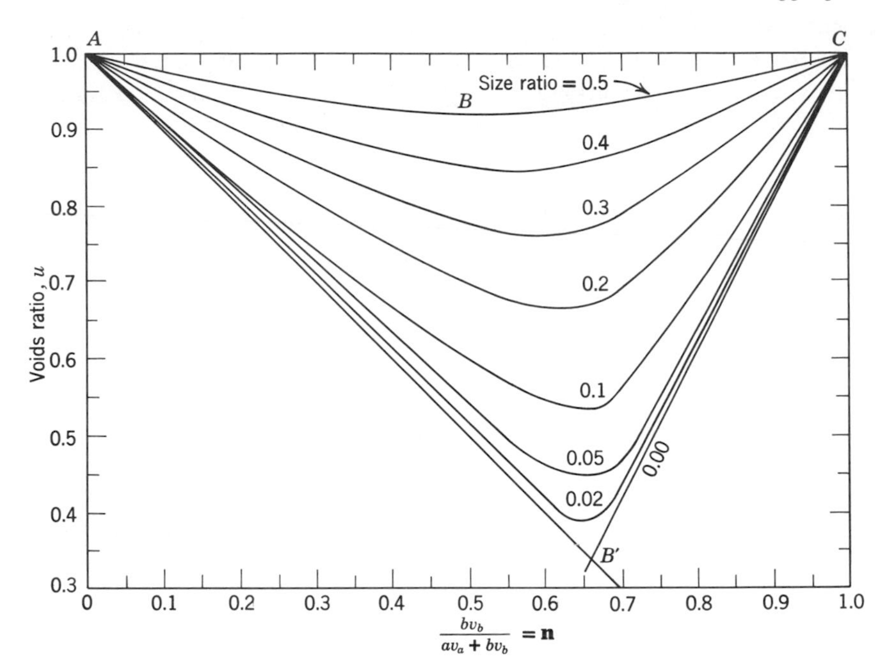

Fig. 1.13. An illustration of the voids ratio as a function of the size ratio. (After Furnas [42].)

the average size of the coarse aggregate increased. The data show that as long as the maximum size remains the same, a change of grading of the coarse aggregate has practically no effect on the minimum voids content. For example, in the third block of data, reducing the average size of the coarse aggregate from 16.5 mm to 9.0 mm had no effect on the voids content. However, if the size ratio is decreased by increasing the maximum size of the coarse aggregate, the effect on the voids content is significant, as is shown in Section 8.8.

8.8 Range of Sizes versus Voids Content

Andreasen and Anderson [37], relying on the similarity principle, showed that if an aggregate contains all possible sizes from the maximum down to zero, voids content is a function of grading only, and for a given grading it is independent of the maximum size. The grading for which this conclusion is true can be represented by the following equation.

$$p = \left(\frac{d}{D}\right)^q$$

Table 1.18 Voids Contents for Various Gradings of Concrete Aggregate[a]

Composition of Coarse Aggregate (Weight Per Cent)			Voids in Coarse Aggregate (Per Cent)	Proportion of Sand[b] (Per Cent)	Voids in Combined Aggregate (Per Cent)	Average Size of Coarse Aggregate (mm)
#4–$\frac{3}{8}$ in.	$\frac{3}{8}$–$\frac{3}{4}$ in.	$\frac{3}{4}$–$1\frac{1}{2}$ in.				
($\frac{3}{8}$–$\frac{3}{4}$ in.)/($\frac{3}{4}$–$1\frac{1}{2}$ in.) = 0/100						
0.0	0.0	100.0	38.5	50.0	17.6	25.4
30.0	0.0	70.0	33.1	40.0	17.1	9.4
50.0	0.0	50.0	31.0	40.0	18.0	7.9
100.0	0.0	0.0	35.9	45.0	21.3	6.3
($\frac{3}{8}$–$\frac{3}{4}$ in.)/($\frac{3}{4}$–$1\frac{1}{2}$ in.) = 25/75						
0.0	25.0	75.0	38.5	45.0	19.5	18.3
15.0	21.3	63.7	34.4	45.0	20.1	11.1
30.0	17.5	52.5	33.3	40.0	20.0	9.2
50.0	12.5	37.5	33.7	45.0	20.8	7.9
($\frac{3}{8}$–$\frac{3}{4}$ in.)/($\frac{3}{4}$–$1\frac{1}{2}$ in.) = 40/60						
0.0	40.0	60.0	37.4	45.0	20.2	16.5
15.0	34.0	51.0	35.2	50.0	20.0	11.0
30.0	28.0	42.0	33.1	45.0	20.0	9.0
50.0	20.0	30.0	33.0	47.0	19.8	8.4
($\frac{3}{8}$–$\frac{3}{4}$ in.)/($\frac{3}{4}$–$1\frac{1}{2}$ in.) = 60/40						
0.0	60.0	40.0	37.4	45.0	20.0	14.8
15.0	51.0	34.0	34.7	47.0	20.0	10.6
30.0	42.0	28.0	33.0	47.0	20.3	8.9
50.0	30.0	20.0	33.6	45.0	20.1	7.8
($\frac{3}{8}$–$\frac{3}{4}$ in.)/($\frac{3}{4}$–$1\frac{1}{2}$ in.) = 100/0						
0.0	100.0	0.0	37.4	45.0	20.9	12.7
15.0	85.0	0.0	35.3	45.0	20.4	10.0
30.0	70.0	0.0	34.0	45.0	21.3	8.8
50.0	50.0	0.0	34.0	45.0	21.8	7.7

[a] Data are from T. C. Powers, "Studies of Workability of Concrete" *Proc. ACI*, **28,** 419–451, 1932.

[b] This is the percentage of sand giving a minimum voids content. The sand was graded (0– #4); its voids content was 0.313; its *FM* was 2.90.

where p is the fraction of the total aggregate that is finer than nominal size d and D is the maximum size in the aggregate; q is a constant that determines the voids content in the aggregate. In practice voids content is not fixed by q alone because the minimum particle size in a real aggregate is not infinitesimal, and because even if it were infinitesimal the finest particles would not pack as presumed by the theory.

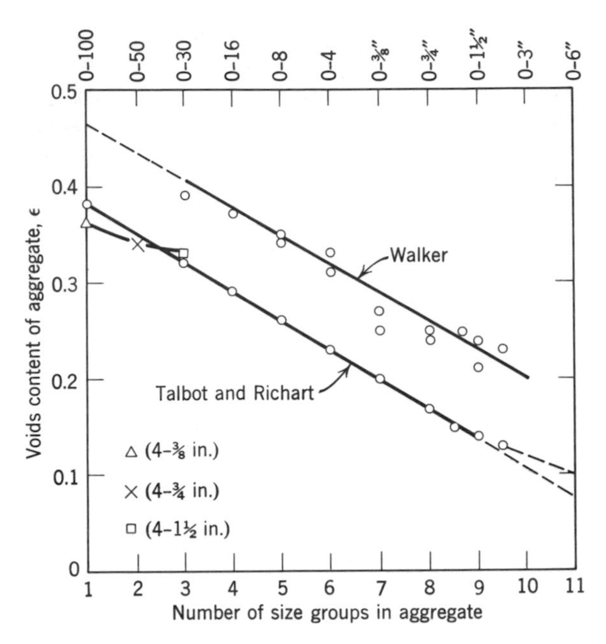

Fig. 1.14. An illustration of the voids content as a function of the number of size groups.

Within the range of sizes normally found in concrete aggregates, it is generally true that the greater the range from minimum to maximum size, the lower the minimum possible voids content.

Talbot and Richart [43] reported data from which the lower line of points in Fig. 1.14 was derived. Each point represents the densest mixture obtained with aggregates graded according to Andreasen's theory, except that there was a finite limit (not given) on minimum size. Although the plot appears to be linear, it cannot actually be so because the line as drawn would indicate a zero voids content at a maximum size of 48 inches, whereas zero voids content would have to be approached asymptotically.

The points near the upper line in Fig. 1.14 represent data reported by Walker [44]. They can be approximated by a line parallel to the Talbot and Richart line; the higher position may be attributed to a higher voids content in individual groups, or perhaps to a difference in the method of compacting. The scatter of points is probably due to a lack of geometric similarity among groups, some having relatively high angularity factors, and possibly different grading.

The three points that differ in shape from the rest (triangle, cross, and square) represent coarse aggregates made up of 1, 2, and 3 size groups, the minimum size being Sieve No. 4. The voids contents of these three groups were as follows: for #4–$\frac{3}{8}$ in., 0.358; for $\frac{3}{8}$–$\frac{3}{4}$ in., 0.374; for $\frac{3}{4}$–$1\frac{1}{2}$ in., 0.385. These numbers should be identical for similar groups; the differences account for the locus of the three points relative to the slope of the Talbot and Richart line.

The lines in Fig. 1.14 can be represented by the following equation:

$$\epsilon = \epsilon_1 - m(n - 1) \tag{1.19}$$

where ϵ is the voids content in a continuously graded mixture of n geometrically similar size groups; ϵ_1 is the porosity of a single group; m is a constant, 0.030 for the data in Fig. 1.14.

8.9 Continuous versus Gap Grading

Feret [45] separated continuously graded natural aggregate into three aggregates and found by systematic experiments that the smallest possible voids content for the given size range was reached when the intermediate size was omitted. Furnas [42] concluded that the densest possible mixtures would result from mixing two widely different sizes. On the other hand, Andreason and Anderson [37] found theoretical and experimental grounds for doubting the fundamental correctness of such a conclusion. They found experimentally, for example, that a binary mixture of 0.153 and 2.12 mm particles of crushed flint (the figures are average sizes of size groups; the sieve-size quotient is about 1.5) gave a voids content of 34 per cent. When this mixture was combined with an intermediate size, 0.53 mm, thus reducing the size gap, the voids content was reduced slightly, to 33 per cent. They concluded that ". . . one can hardly expect greater density from products consisting of a few fitted together sizes than from products in which all sizes are present within certain limits." In general, the data show that if the size groups are similar, size *range* rather than grading governs voids content, provided that the ratio of fine aggregate to coarse aggregate is at the optimum value. However if size groups are geometrically dissimilar, another degree of variation is introduced. If it happens that an intermediate size has a relatively high angularity factor, it may be advantageous to omit it. Also, if the angularity factor increases as the size range is increased by introducing larger size groups, the normal advantage of increasing size range may not be realized.

REFERENCES

[1] Landgren, R., "Determining the Water Absorption of Coarse Lightweight Aggregates for Concrete," *Proc. ASTM*, **64**, 846–865, 1964.

[2] Polivka, M., "Hardened Concrete: Radiation Effects and Shielding," ASTM Special Technical Publication No. 169-A, 298–308, 1966.

[3] Brink, R. H., and A. G. Timms, "Concrete Aggregates: Weight, Density, Absorption, and Surface Moisture," ASTM Special Technical Pubilcation No. 169-A, 432–442, 1966.
[4] Powers, T. C., "Basic Considerations Pertaining to Freezing and Thawing Tests," *Proc. ASTM,* **55,** 1132–1155, 1955.
[5] Powers, T. C., and T. L. Brownyard, "Studies of the Physical Properties of Hardened Portland Cement Paste, Part 7, Permeability and Absorbtivity," *Proc. ACI,* **43,** 865–880, 1947.
[6] Dolch, W. L., "Studies of Limestone Aggregates by Fluid-Flow Methods," *Proc. ASTM,* **59,** 1204–1215, 1959.
[7] Walker, S., "Modulus of Elasticity of Concrete," Bulletin 5, Structural Materials Research Laboratory, Lewis Institute, Chicago, 1923.
[8] Whitehurst, E. A., "Evaluation of Concrete Properties from Sonic Tests," ACI Monograph No. 2, 1966.
[9] Kaplan, M. F., "Flexural and Compressive Strength of Concrete as Affected by the Properties of Coarse Aggregates," *Proc. ACI,* **55,** 1193–1208, 1959.
[10] Cook, H. K., "Concrete Aggregates: Thermal Properties," ASTM Special Technical Publication No. 169-A, 476–486, 1966.
[11] Woolf, D. O., "Concrete Aggregate: Toughness, Hardness, Abrasion, Strength and Elastic Properties," ASTM Special Technical Publication No. 169-A, 462–475, 1966.
[12] Mather, B., "Concrete Aggregates: Shape, Surface Texture, and Coatings," ASTM Special Technical Publication No. 169-A, 415–431, 1966.
[13] Hawksley, P. G. W., "The Physics of Particle Size Measurement, Part I, Fluid Dynamics and the Stokes' Diameter," *Brit. Coal Util. Res. Assoc., Monthly Bull.* **XV**(4), 1951.
[14] Zingg, T. "Beitrag zur Schotteranalyse," *Schweiz. Mineral. Petrog. Mitt.,* **15,** 39–140, 1935.
[15] Lees, G., "The Measurement of Particle Shape and Its Influence in Engineering Materials," *J. Brit. Granite and Whinstone Federation,* **4**(2), 1964.
[16] Waterways Experiment Station, Concrete Research Division, Corps of Engineers, Jackson, Mississippi, *Method of Test for Flat and Elongated Particles in the Coarse Aggregate,* CRD-C 119-53.
[17] Wadell, H., "Shape Determinations of Large Sedimental Rock Fragments," *Pan-Am. Geologist,* **61,** 187, 1935.
[18] Dallavalle, J. M., *Micromeretics,* 2nd ed., Pitman Publishing Corp., New York, 1948.
[19] Butcher, B. J., and H. J. Hopkins, "Particle Interference in Concrete Mixes," *Proc. ACI,* **53,** 545–556, 1956.
[20] Plum, N. M., "The Predetermination of Water Requirement and Optimum Grading of Concrete," Building Research Studies No. 3, The Danish National Institute of Building Research, Copenhagen, 1950.
[21] Abrams, D. A., "Design of Concrete Mixtures," Bulletin 1, Structural Materials Research Laboratory, Lewis Institute, Chicago, 1918.
[22] Popovics, S., "Formulas on Fineness Modulus and Specific Surface Area," RILEM Bulletin No. 16, 19–28, 1962.
[23] Young, R. B., "Some Theoretical Studies on Proportioning Concrete by the Method of Surface Area of Aggregates," *Proc. ASTM,* **91,** Part II, 445–457, 1919.
[24] Zietsman, C. F., "Mortar- and Concrete-Making Properties of Natural Sands Related to Their Physical Attributes," *Proc. ACI,* **28,** 1041–1056, 1957.

[25] Fulton, F. S., *Concrete Technology,* The Portland Cement Institute, Kew Road, Richmond, Johannesburg, South Africa, 1961, pp. 87–91.

[26] Talbot, A. N., discussion of paper by R. B. Young, "Some Theoretical Studies on Proportioning Concrete by the Method of Surface Area of Aggregates," *Proc. ASTM,* **19,** Part II, 483, 1919.

[27] Murdock, L. J., "The Workability of Concrete," *Research and Application,* No. 11, 1960. Publication of Wimpey Central Laboratory, London. Also, *Magazine of Concrete Research,* **12,** 135–144, 1960.

[28] Edwards, L. N., "Proportioning the Materials of Mortars and Concretes by Surface Areas of Aggregates," *Proc. ASTM,* **18,** Part II, 235–283, 1918.

[29] Loudon, A. G., "Computation of Permeability from Simple Soil Tests," *Geotechnique,* **3,** 165–183, 1952–1953.

[30] Carman, P. C., "The Determination of the Specific Surface of Powders," *J. Soc. Chem. Ind.,* **57**(7), 225–234, 1938.

[31] Shacklock, B. W., and W. R. Walker, "The Specific Surface of Concrete Aggregates and its Relation to the Workability of Concrete," Cement and Concrete Association Research Report No. 4, 1958.

[32] Shergold, F. A., "The Percentage Voids in Compacted Gravel as a Measure of its Angularity," *Magazine of Concrete Research,* **5,** 3–10, 1953. Note: This article gives 27 references to other works pertaining to this subject.

[33] Goldbeck, A. T., "Stone Sand for Use as Fine Aggregate," *Concrete,* **59,** 28–32, 1951.

[34] Gray, J. E., "The Effect of Shape of a Particle on Properties of Air-entrained Stone Sand Mortar," *Crushed Stone J.,* **36**(3), 3–8, 1961.

[35] Dunagan, W. M., "The Application of Some of the Newer Concepts to the Design of Concrete Mixes," *Proc. ACI,* **36,** 649–684, 1940.

[36] Neville, A., *Properties of Concrete,* Wiley, New York, 1963, p. 101 and Fig. 3.1.

[37] Andreasen, A. H. M., and J. Anderson, "The Relation of Grading to Interstitial Voids in Loosely Granular Products," *Kolloid-Z.,* **49,** 217–228, 1929.

[38] Weymouth, C. A. G., "A Study of Fine Aggregate in Freshly Mixed Mortars and Concretes," *Proc. ASTM,* **38,** 354–393, 1938.

[39] McGeary, R. K., "Mechanical Packing of Spherical Particles," *J. Am. Ceram. Soc.,* **44**(10), 513–522, 1961.

[40] Weymouth, C. A. G., "Effect of Particle Interference in Mortars and Concrete," *Rock Products,* **26,** February 25, 1933.

[41] Hughes, B. P., "Particle Interference and the Workability of Concrete," Digest supplement to the paper that appeared in *Proc. ACI,* **63,** 369–372, 1966.

[42] Furnas, C. C., "The Relations Between Specific Volume, Voids, and Size Composition in Systems of Broken Solids of Mixed Sizes," Department of Commerce, Bureau of Mines, Serial No. 2894, *Reports of Investigations,* 1–10, 1928.

[43] Talbot, A. N., and F. E. Richart, "The Strength of Concrete: Its Relation to the Cement, Aggregates, and Water," Bulletin 137, Engineering Experiment Station, University of Illinois, Urbana, Illinois, 1923.

[44] Walker, S., "Modulus of Elasticity of Concrete," Bulletin 5, revised ed., Structural Materials Research Laboratory, Lewis Institute, Chicago, 1923.

[45] Feret, R., "On the Compactness of Hydraulic Mortars," *Annales Ponts Chaussees,* 7th Series, **4,** 5–164, 1892.

2

Physical Properties of Cement

1. INTRODUCTION

In this chapter we consider the physical properties of cement, first with respect to the particles of which it is composed, and then with respect to the properties of cement as a powder. The word "cement" as used here may pertain to any one of the classes of cement described in *ASTM Standards,* Part 9. Physically, cement is not unlike other mineral powders, and therefore much of what is said is applicable, at least qualitatively, to other mineral powders and to mixtures of cement with other powders.

The final step of manufacturing cement is that of pulverizing cement clinker together with a small amount (3 to 4 per cent) of calcium sulfate as gypsum or the equivalent amount of some other form of calcium sulfate. Clinker is the granular product of the rotary cement kiln, the name being indicative of its appearance. The manufacturer regulates the fineness of grinding according to the requirements of standard specifications for the type of cement being made.

2. PROPERTIES OF CEMENT PARTICLES

2.1 *Specific Gravity and Specific Volume*

2.1.1 Specific gravity. As determined by the standard ASTM method (ASTM Designation C188-44) the specific gravities of various portland cements range from about 3.10 to 3.23. When the value for a given cement is not known, it is usual to assume 3.15.

The ASTM method for determining specific gravity is based on the displacement of kerosene by a weighed amount of dry cement. Provided that the cement is weighed on an equal-arm balance (rather than on a spring balance), the figure obtained meets the definition of *apparent* specific gravity, which is "The ratio of weight in air of a unit volume of material at a stated temperature to the weight in air of equal density of an equal volume of gas-free distilled water at a stated temperature. If the material is a solid, the volume shall be that of the impermeable portion." (ASTM Designation C12-61T) The value differs from *bulk specific gravity* (see Chapter 1, Section 2.1) only to the extent that individual cement particles have pores permeable to the liquid used in the test.

As actually performed, the ASTM method gives the volume in milliliters of a given weight of material at 20°C. Because the volume of 1 g of water is 1 ml only at 4°C, the figure for specific gravity should be understood to indicate the ratio of the mass of 1 ml of the substance at a temperature of 20°C to the mass of 1 ml of water at 4°C. This means that the ASTM figure for specific gravity is practically the same as the density in g/cm^3.

Because the specific surface area of cement as measured by gas adsorption is usually considerably higher than that measured by the permeability method or sedimentation analysis, it has been concluded by some that individual cement grains contain appreciable fractions of void space, either as natural pores or as cracks remaining from the crushing process; these are considered to be penetrable by gas (nitrogen) but not by kerosene. However, experimental evidence such as that in Table 2.1 is contrary to this view.

Table 2.1 gives observed and calculated densities of three compounds

Table 2.1[a]

Compound	Observed Density (g/cm^3)	Calculated Density (g/cm^3)
β-Dicalcium silicate	3.28	3.31
γ-Dicalcium silicate	2.97	2.96
Tricalcium silicate	3.12–3.25	3.227[b]

[a] Data are from L. Heller and H. F. W. Taylor, *Crystallographic Data for the Calcium Silicates*, Department of Scientific Industrial Research, Building Research Station, Her Majesty's Stationary Office, London, 1956.
[b] Alite.

related to portland cement. The calculated values are based on crystallographic data. Because most of the grains of the sample contain more than one crystal, the observed specific gravity of the grain should be less than that of individual crystals if there are intercrystalline voids inaccessible to fluid. The data given in the table indicate that if there is any difference between the two values it is very small.

The electron microscope reveals that a sample of cement may contain a small amount of hydration product adhering to the grain surfaces. The surface area of such material is measured by gas adsorption, but it is not measured by either of the two methods normally used for portland cement. Because the hydration product has a specific surface area about 500 times that of cement, a very small amount of hydration product can account for a considerable difference between the apparent specific surface areas measured by the different methods.

It is significant that the ASTM method gives about the same specific gravity for the products of various degrees of grinding of the same material. For example, the results from an ordinary commercial clinker are presented in the accompanying table.

Specific Surface by Turbidimeter (cm^2/g)	Specific Gravity
1085	3.154
1540	3.150
2045	3.155
2550	3.156

If the coarse particles have pores impenetrable by kerosene, the apparent specific gravity should increase as fineness increases. It is true that the average value for the two coarsest materials is smaller than the average for the other two, but if the difference is statistically significant, it indicates only a slight degree of porosity of the coarsest particles.

Thus there is reason to believe that the specific gravity of cement as measured by the ASTM method meets the definition of bulk specific gravity as well as that of apparent specific gravity.

Even though the ASTM method may give the apparent specific gravity (*true* specific gravity, by a looser definition) the ASTM value is not suitable for calculating the space occupied by cement in water. For that calculation the *apparent specific solid volume,* or *displacement factor,* as defined below should be used.

2.2 Apparent Specific Solid Volume in Water—The Displacement Factor

When cement and water are mixed together, some constituents of cement dissolve rapidly, and chemical reactions occur. As a result, the cement and water are altered; cement particles lose some substance and at the same time acquire a small amount of water by chemical combination at the surface. The water becomes a solution having a higher density than that of pure water.

The effects of reaction and solution show up immediately if we attempt to evaluate the specific gravity of cement from its displacement in water. For example, a certain cement having a specific gravity of 3.133 seemed to have a specific gravity of about 3.24 when it was measured in water; the value could not be established exactly because it changed with time and depended on the amount of water used with the cement. All this makes it impossible to compute accurately the volume of a mixture of cement and water from the specific gravities and weights of dry cement and pure water. However it is possible to establish a basis for computation sufficiently accurate for most purposes.

Calculation of the solid volume of cement in fresh paste or concrete can be based on the apparent specific gravity or apparent specific solid volume of cement immersed in water, which is a displacement factor. Typical values are given in Table 2.2 together with ASTM values for four types of cement. The average difference between the apparent specific solid volume and the ASTM value for all these cements was

Table 2.2 Typical Values for Apparent Specific Solid Volume of Cement in Water[a]

ASTM Cements		ASTM Specific		Apparent Density in Water[b] (g/cm³)			Apparent Specific Solid Volume[b] (cm³/g)		
Type	Number Averaged	Gravity	Volume	min	max	avg	max	min	avg
I	(17)	3.153	0.3172	3.153	3.256	3.210	0.3171	0.3071	0.3115
II	(6)	3.208	0.3118	3.229	3.304	3.259	0.3097	0.3027	0.3069
III	(4)	3.126	0.3199	3.164	3.210	3.188	0.3160	0.3115	0.3137
IV	(4)	3.223	0.3103	3.261	3.297	3.282	0.3066	0.3033	0.3047

[a] Data are from C. L. Ford, "Determination of the Apparent Density of Hydraulic Cement in Water Using a Vacuum Pycnometer," *ASTM Bulletin* No. 231, 1958.
[b] After 30 min of contact at room temperature.

0.0107 cm^3/g, and the corresponding difference in density was 0.056 g/cm^3. The differences varied considerably among the different cements.

Steinour [3] found that the difference between the ASTM apparent specific gravity of cement and the apparent density in water shows a good correlation with the sum of the amount of gypsum that dissolves and a quantity that is a function of the tricalcium aluminate content of the cement. However, rather than attempt a calculation of the difference, it is desirable to make a determination for each different cement. For many purposes it would suffice to use an average value. For example, where the actual specific gravity is 3.15, a value of 3.21 might be used, or its reciprocal, $v_c^o = 0.312$.

While a mixture remains plastic, its volume decreases proportionately to the apparent decrease in specific solid volume of the cement. Therefore, to obtain the highest degree of accuracy possible, the determination of the apparent specific solid volume of cement should be synchronized with other operations. For example, if the volume of a concrete mixture is measured 10 min after the first contact is made between cement and water, the apparent specific solid volume of cement should be determined after the same interval. Such a precaution, together with others, makes it possible to calculate the fractional volumes of solid, water, and air in a sample of concrete with a reasonable degree of accuracy. Because the change of specific solid volume after the first five minutes is relatively slow, a value established at any time between 15 and 30 min after mixing is usually accurate enough.

2.2.1 Methods of determination. To determine the apparent specific solid volume of cement in water a special vacuum pycnometer such as that described by Ford [2] may be used. Because the specific volume of water as well as that of cement may be changed by the interaction, and because a solution is involved, the ratio of water to cement in the pycnometer should not be much different from the ratio generally found in concrete. However Ford found it was not feasible to use less than about 1 ml of water per gram of cement, an amount somewhat above the range of water-cement ratios used in normal practice. The method as described by Ford is given below:

APPARATUS

1. (a) Vacuum pycnometer made by joining a 50-ml Erlenmeyer flask to a stopcock (Note 1) of 6-mm bore. To the other end of the stopcock is joined the inner part of a $\frac{14}{35}$ ground glass joint as shown in [Fig. 2.1]. For efficient work at least three of these pycnometers should be available.

(b) Connecting assembly consisting of the outer portion of a $\frac{14}{35}$

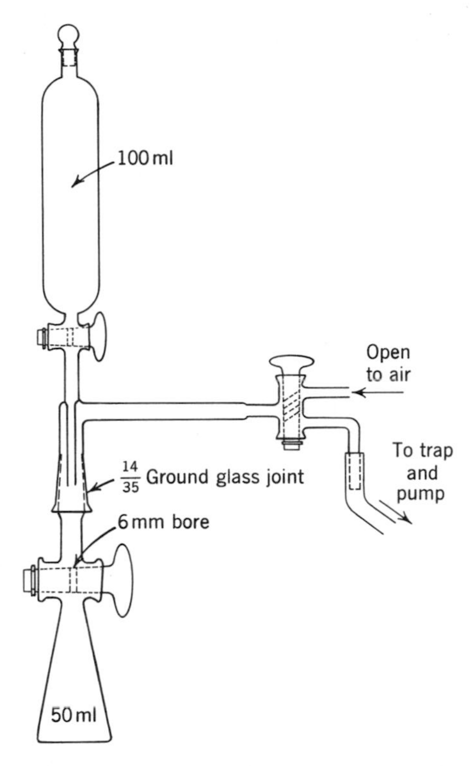

Fig. 2.1. Apparatus used to determine the apparent specific volume of cement. (By permission of American Society for Testing and Materials.)

ground glass joint connected below a 100-ml separatory funnel with a small-bore inner tube to supply water to the pycnometer, and a side arm and three-way stopcock (Note 1) for connection to either a vacuum pump or to the atmosphere, as shown in [Fig. 2.1].

(c) Water trap in a Dewar flask filled about $\frac{1}{4}$ full with methanol or ethanol, then with dry ice to about $\frac{3}{4}$ full. The water trap is inserted in the vacuum line between the three-way stopcock and the vacuum pump.

(d) Funnel with 5 mm outside diameter, stem about 15 cm long.

(e) Vacuum pump (0.3 μ of mercury or less).

Note 1: High-vacuum stopcocks and high-vacuum lubricant should be used throughout.

CALIBRATION OF PYCNOMETERS

2. (a) Be sure that the pycnometers are clean and dry (see section 2.(d) and 3.(f) for methods). Weigh the first one to the nearest milli-

gram and record as pycnometer, empty. Connect to vacuum apparatus. Close separatory funnel stopcock, and open three-way stopcock to pump, start pump and evacuate 5 min. During this time weigh the second pycnometer.

(b) Fill the separatory funnel with freshly boiled and cooled distilled water. At the end of the 5-min evacuation period, close the stopcock to the pump and carefully open the separatory funnel stopcock until the pycnometer is filled to a little above its stopcock. When effervescence of the water has nearly ceased, open the three-way stopcock to the air, remove the pycnometer and with the aid of a clamp immerse it up to the stopcock for 30 min in a constant temperature (preferably several degrees above room temperature) bath, with the stopcock still open. Record the water temperature.

(c) During the 30-min period, proceed as above with the second and third pycnometers. At the end of the 30-min period, close the pycnometer stopcock, remove from the bath and dry the outside with a towel. Empty the excess water from the tube above the stopcock, then carefully dry the interior of the tube with absorbent tissue such as Kleenex. Weigh and record as pycnometer plus water. Follow the same procedure with the second and third pycnometers.

(d) After weighing, shake the water from the pycnometers, insert a small glass tube connected to the laboratory vacuum line, and draw air through until dry. Drying may be accelerated by rinsing the pycnometers once or twice with acetone before applying suction.

PROCEDURE

3. (a) Weigh the first clean, dry pycnometer and record as pycnometer, empty (separate recording from that used in calibration). Weigh roughly in a weighing pan approximately 40 g of cement. Clamp the pycnometer to a support stand and insert the funnel. Transfer the cement to the pycnometer using a No. 16 bare copper wire to help force the cement through the funnel stem. Weigh and record as pycnometer plus cement.

(b) Be sure that the separatory funnel stopcock is closed and that the connecting tube beneath it is free of water. Connect the pycnometer to the vacuum apparatus. Close the three-way stopcock very cautiously to prevent drawing the cement out of the pycnometer flask. It is sometimes helpful also to tap the flask very lightly. Evacuate for 15 to 20 min for ordinary cements or 30 min for air-entraining cements.

(c) During the evacuation period weigh the second pycnometer, add cement and reweigh.

(d) At the end of the evacuation period, close the three-way stopcock to the pump and fill the pycnometer with water. (If any air bubbles appear, evacuation was not complete and the test must be repeated.) Place in the constant temperature bath for 30 min. Record the water temperature. Dry and weigh as described under "Calibration of Pycnometers," section 3.(c). Record the weight as pycnometer plus cement plus water.

(e) During the 30-min period, carry on the procedure as described above, successively with the second and third pycnometers.

(f) Clean the pycnometers as soon as possible after the weights with cement and water are obtained. If the cement is allowed to set, the pycnometers will be ruined. The cleaning may be done as follows:

"Invert (stopcock open), shake out some of the water, close the stopcock, shake until all the cement is loosened from the bottom, open the stopcock and shake out as much of the slurry as possible. Insert a small glass tube connected to a water line (same tube that was used for drying), turn on the water and flush well. Shake out the water, then with the aid of the funnel, partially fill with dilute hydrochloric acid (1-1), shake and empty. Rinse first with city water injected through the glass tube then with distilled water. Dry as described under "Calibration of Pycnometers," section 3.(d).

CALCULATIONS

4. Calculate the density of the cement according to the method illustrated by the same calculations shown below:

Determination of Volume of Pycnometer

1. Weight of pycnometer filled with water	170.045 g
2. Weight of pycnometer empty	115.205 g
3. Weight of water to fill pycnometer (item 1 − item 2)	54.840 g
4. Temperature of water	28.5°C
5. Specific volume of water (handbook)	1.00390 ml per g
6. Volume of pycnometer (item 3 × item 5)	55.054 ml

Determination of Density of Cement in Water

7. Weight of pycnometer empty	115.204 g
8. Weight of pycnometer + cement	155.217 g
9. Weight of pycnometer + cement + water	197.526 g
10. Temperature of water	28.5°C
11. Weight of cement (item 8 − item 7)	40.013 g
12. Weight of water (item 9 − item 8)	42.309 g
13. Specific volume of water (handbook)	1.00390 ml per g
14. Volume of water (item 12 × item 13)	42.474 ml
15. Volume of pycnometer, empty (item 6)	55.054 ml
16. Volume of cement (item 15 − item 14)	12.580 ml
17. Density of cement (item 11 ÷ item 16)	3.181 g per ml

Steinour [3] described another method that is relatively simple and in some ways superior to that described by Ford. The only equipment needed is a balance and an Erlenmeyer flask with the rim ground so that it may be closed tightly with a glass plate. A cement paste is made using a water-cement ratio the same as, or similar to, that of the mixture in which the cement is to be used; the ratio is usually high enough to eliminate practically all air entrainment. The flask is filled exactly with the paste, making sure the paste is approximately at room temperature. Then the weight of the paste in the flask is determined. From the quantities used in the mixture, the ratio of cement to water is computed: c/wv_w, g/cm³ of water. Also, from the capacity of the flask and the weight of the paste in the flask, the density of the paste and its apparent water content are computed as follows:

Let V = volume of flask.
ρ_p = density of paste.
v_p = specific volume of paste $= 1/\rho_p$.
ρ_w = density of water at existing temperature.
ρ_c^o = apparent density of cement as measured in water $= 1/v_c^o$.

Other symbols are used as before. Then

$$\rho_p = \frac{w}{V} = \frac{c}{w}\frac{w}{\rho_w}\frac{1}{V} \tag{2.1}$$

or

$$\frac{w}{V\rho_w} = \frac{\rho_p}{\rho_w + \dfrac{c}{w}\rho_w} = \frac{\rho_p}{\rho_w\left(1 + \dfrac{c}{w}\right)} \tag{2.2}$$

and

$$\rho_c^o = \frac{(c/w)\rho_w\rho_p}{\dfrac{c}{w}\rho_w + \rho_w - \rho_p} = \frac{\rho_p}{\left(1 + \dfrac{w}{c}\right)\left(1 - \dfrac{\rho_p}{\rho_w}\right)} \tag{2.3}$$

or

$$v_c^o = v_p + \frac{w}{c}(v_p - v_w) \tag{2.4}$$

Obviously Steinour's method cannot be used for an air-entraining cement unless air entrainment is prevented. This has been done by using an electrical mixer under an evacuated bell jar.

It is important to take notice of the fact that by the above methods of taking into account the interaction between cement and water on the volume of a mixture, all the volume change due to interaction is attributed to the cement, even though some of it actually occurs in the water. The consequence of adopting this procedure is that we must *always use the specific gravity of pure water in calculations of paste or concrete volume.*

2.3 *Particle Size*

As discussed in Chapter 1, the size of an irregular particle cannot be given exactly by one number. The nominal sizes of cement particles are usually given in terms of sieve size, or Stokes' diameter. (See Chapter 1, Section 3.) The mean specific surface diameter can be estimated from figures for specific surface area (a typical value is 6 μ).

2.3.1 Maximum size. Owing to the practical limits on how small the openings can be made, sieves are useful mainly for estimating the maximum size of cement particles. Modern cements are pulverized so finely that practically all the particles pass a No. 100 sieve (150 μ, square opening), and 95 per cent or more will pass a No. 200 sieve (74 μ). ASTM Type III cement (high early strength) is so fine that nearly all the particles pass the No. 200 sieve, and about 99 per cent pass the No. 325 (44 μ). Thus, if we define a practical maximum size of cement grains as that sieve size which will pass about 99 per cent of the product, we find that, among the various types of cement now being manufactured in the United States, the maximum particle size ranges from about 44 to about 100 μ.

2.3.2 Minimum size. The pulverizing process produces a wide range of sizes extending into the range of submicroscopic dimensions. It seems likely that there is a lower limit on particle size, but the limit has not been established exactly. Haegermann [4] reported sizes as small as 0.03 μ, observed by means of the electron microscope, but whether such particles constitute a significant fraction of the whole remains in doubt. Plum [5] estimated that the fraction smaller than about 0.3 μ is negligible and considered that size the practical minimum.

2.4 *Particle Shape*

Cement powder is the product of crushing and grinding brittle clinker, and therefore cement particles are angular. As shown in Table 1.16, any size group larger than about 16 μ may have the same voids content, about 45 per cent, compacted, which corresponds to a voids content of about 50 per cent before it has been compacted. By Eq. 1.4, we find the angularity factor to be about 1.36, which is comparable with the angularity factor of some crushed limestones.

3. PROPERTIES OF CEMENT POWDER

3.1 *Particle-Size Distribution (Grading)*

The particle-size distribution of cements has been estimated from sedimentation analysis of samples dispersed in fluid. It has also been estimated by procedures based on elutriation. The most convenient method based on sedimentation analysis employs the Wagner turbidimeter (ASTM Designation C115-58). Lea [6] reported that it gives the same results as the Andreason pipette method for particles larger than a nominal 7.5 μ the usual lower limit for the turbidimeter analysis. Because relative size is indicated by the velocity of the fall in a medium of known viscosity, the reported values are in terms of Stokes' diameters as defined in Chapter 1, Section 3.2. An up-to-date method is described by Hime and LaBonde [7].

Size-distribution curves for three cements differing in fineness are shown in Fig. 2.2. The lower part of each curve is based partly on Plum's estimate of the minimum size and is drawn so that each is compatible with other information concerning the specific surface of this fraction of the cement (see later). Obviously this section of each graph is tentative.

Although the three curves in Fig. 2.2 represent different average sizes, they do not necessarily represent different gradings. Indeed, the gradings are practically the same, as is shown in Fig. 2.3. In this diagram particle size is expressed as a percentage of the maximum size, the maximum size in each case being estimated from the trend of the curve, or for the two coarser cements from a combination of curve trend and sieve analysis.

The cements represented in Figs. 2.2 and 2.3 were produced by a one-pass grinding method called open circuit ball-mill grinding. Some cements are produced by a continuous process that involves regrinding a coarse fraction by means of a feedback circuit (closed-circuit grinding), and such cements show gradings somewhat different from that shown in Fig. 2.2. There is some evidence also that cement clinkers differing widely in chemical composition (for example ASTM Type IV and ASTM Type III) produce different gradings for the same grinding process. However, all such differences are not of much consequence as far as the physical properties of cement in fresh concrete are concerned,

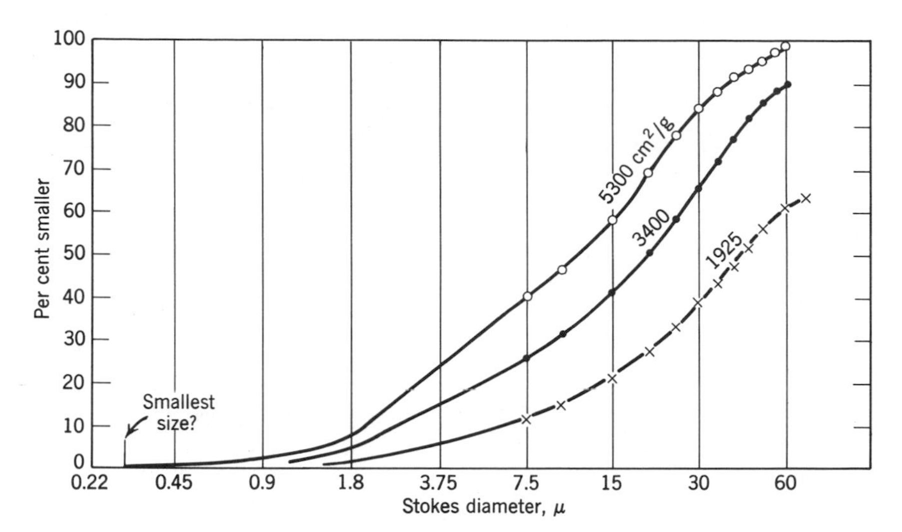

Fig. 2.2. Size-analysis curves. The top curve represents six Type-III cements with an estimated maximum size of 65 μ. The middle curve represents twelve Type-I cements with an estimated maximum size of 110 μ. The bottom curve represents one Type-I cement with an estimated maximum size of 240 μ. (Data are from Lerch and Ford [11] and from PCA Series 290.)

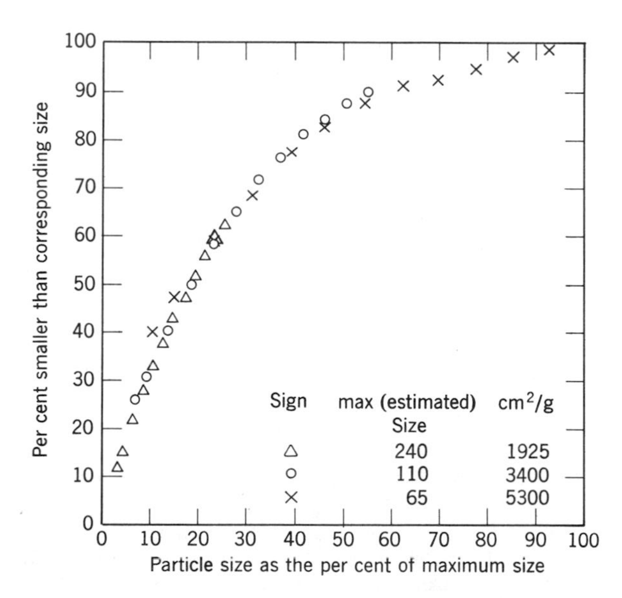

Fig. 2.3. Grading curves derived from the curves of Fig. 2.2.

the specific surface area being the important function of particle size in most considerations.*

3.2 Specific Surface Area

Presently accepted as the best method for evaluating specific surface area of cement is the one called the *air permeability method,* first introduced to cement technology by Lea and Nurse [8]. The same method is applied by means of the Blaine apparatus (ASTM Designation C204-55). Results by the permeability method are similar to those obtained from the sedimentation process as carried out by means of the Andreasen apparatus when the analysis is carried down to a minimum size of about 0.6 μ. Actually, as discussed in Chapter 12, Sections 2.5 and 3.3, the area given by the permeability method is, on the average, 1.22 times that computed from the sedimentation analysis on the assumption that the particles are spheres.

Sedimentation analysis is also carried out by means of the Wagner turbidimeter, but when we carry it out according to the ASTM method (ASTM Designation C115-58) the values for the specific surface area obtained are only a little more than half those obtained by the permeability method or by the Andreasen sedimentation method. The fault with the ASTM turbidimeter method is not in the apparatus, but in the assumption made about the specific surface area of the fraction remaining in suspension at the end of the period of observation, the maximum Stokes' diameter of this fraction being about 7.5 microns.

Table 2.3 gives data for four different groups of cement. The three cements of each group were manufactured from the same clinker at a commercial mill. The coarsest cement of each group is comparable in fineness with cements produced before 1900; the medium fineness is similar to much of present-day ASTM Type I cement and the finest is typical of high early strength cements (ASTM Type III).

Within the range of finenesses shown in Table 2.3, the specific surface area of cement is approximately directly proportional to the fraction of the total weight that is smaller than 7.5 μ. This is shown in Fig. 2.4 for the data of Table 2.3. Each one of the four diagrams represents three degrees of fineness from the same clinker. To the degree that the experimental points approximate the same straight line, the specific sur-

* With respect to drying shrinkage, grading may be of some importance. At a given specific surface area a cement made by closed-circuit grinding will have a fraction of +60 μ particles greater than that found in cement made by closed-circuit grinding; for that reason, a partially mature paste made with the cement produced by closed-circuit grinding will have the smaller residue of anhydrous cement and thus somewhat greater shrinkage.

Table 2.3 Particle Size and Specific Surface Area of Commercial Cements

	Sieve Analysis				Specific Surface Area			
					cm²/g		cm²/cm³	
0–7½[a] (per cent)	(325–200) (44 μ)	(200–100) (74 μ)	(100–50) (150 μ)	>44 μ	Turb.[b]	Perm.[c]	Turb.[b]	Perm.[c]
Cement Mill No. 1								
11.8	9.5	21.9	7.1	38.5	1040	1925	3260	6035
21.0	6.6	5.3	0.0	11.9	1665	3145	5220	9860
34.3	4.1	2.1	0.0	6.3	2280	4490	7160	14095
Cement Mill No. 2								
17.0	6.2	10.4	3.8	20.4	1470	2855	4560	8855
21.2	6.6	4.5	0.0	11.1	1740	3405	3465	10580
36.2	1.3	0.5	0.0	1.8	2500	5260	7770	16355
Cement Mill No. 3								
16.3	6.8	18.7	2.7	28.3	1375	2330	4365	7395
23.9	4.7	9.5	0.5	14.7	1820	3100	5780	9840
31.8	3.0	2.7	0.5	6.2	2200	4010	6995	12750
Cement Mill No. 4								
14.5	12.1	13.2	0.4	25.7	1250	2060	4040	6655
21.9	2.0	0.7	0.0	2.7	1830	2715	5885	8730
30.1	0.5	0.1	0.0	0.6	2290	3995	7360	12845

[a] 0 to 7.5 μ fraction. The percentage is given by a Wagner turbidimeter analysis.
[b] Determined by using the Wagner turbidimeter. (The surface area includes that of +60 μ particles.)
[c] Determined by using the permeability method.

face area of the zero to 7.5 μ fraction can be considered a constant the value of which is given by the slope of the straight line. The intercept of the ordinate at zero of an extension of this line gives the corresponding specific surface area of the coarse fraction.

From the point of view presented above, cement can be regarded as a mixture of two aggregates, fine and coarse, each aggregate having a characteristic specific surface area. In these cases the indicated specific surface area of the fine part ranged from about 10,000 to 12,750 cm²/g,

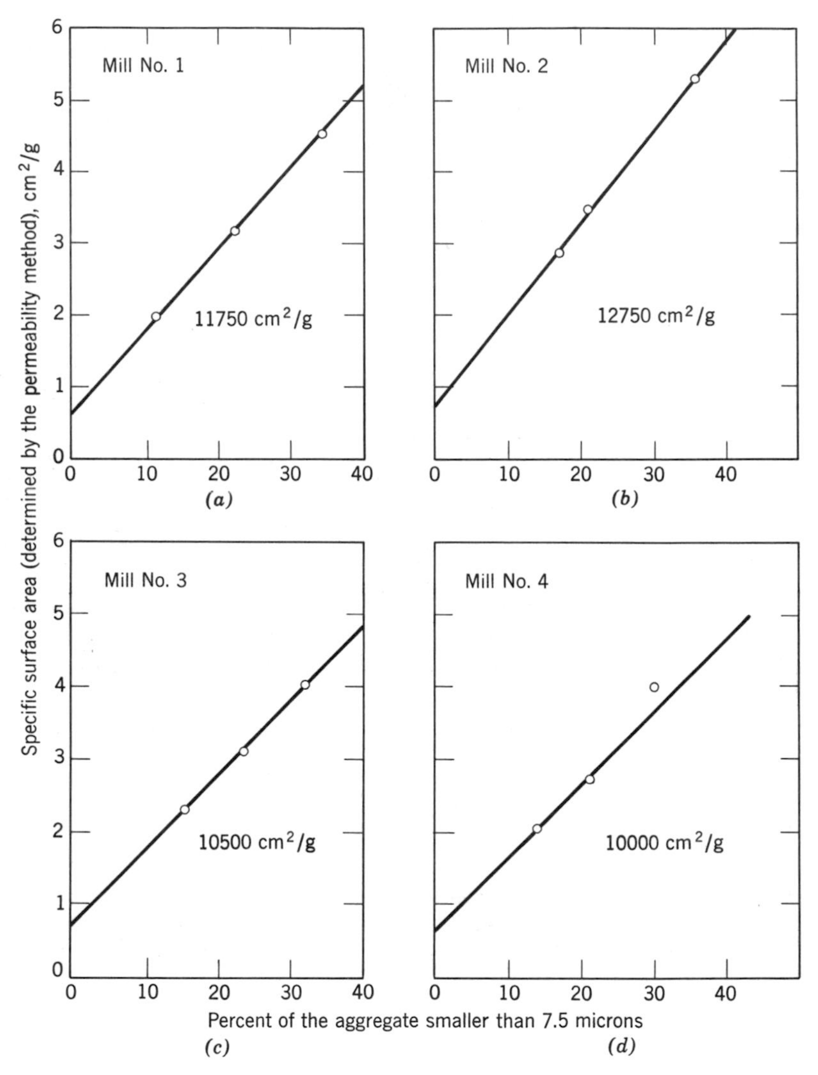

Fig. 2.4. Graphical method for estimating the specific surface area of the 7.5 μ fraction in portland cement. The number in each graph gives the slope of the line, and thus the specific area of the fine fraction. The fraction finer than 7.5 μ was that indicated by the Wagner turbidimeter. The intercept at ordinate zero is the average specific surface area of the coarse fraction as given by Wagner turbidimeter data. (Data are from PCA Series 290.)

and that of the coarse part from 600 to 700 cm^2/g. Thus, of the total surface area in a cement having a specific surface area of, for example, 3400 cm^2/g, the zero to 7.5 μ fraction constitutes about 26 per cent of the total, and about 3100 cm^2 of the 3400 will be accounted for by that fraction.

The specific surface areas for the fine parts just given are from the areas as computed from the permeability data by the Kozeny-Carman equation. The corresponding equivalent-sphere values are about 80 per cent of the values indicated, that is, 8000 to 10,600 cm^2/g, or about 25,200 to 33,400 cm^2/cm^3. The latter figures are to be compared with $6/(3.75 \times 10^{-4}) = 16{,}000\ cm^2/cm^3$, which is the value assumed for any cement when computing the specific surface area from data obtained with the Wagner turbidimeter.

It should be understood that graphs such as those in Fig. 2.4 cannot be linear over an unlimited range. The different points represent the products of a grinding process by which the maximum size becomes smaller the larger the specific surface area of the product; probably the minimum size is unaffected. The linearity of the plot indicates that within a limited range of increase of fineness, the concomitant reduction in the size range of the coarse fraction has no effect on the specific surface area of that fraction large enough to stand out above the random factors influencing the positions of the data points, whereas the specific surface area of the fine fraction remains virtually independent of the degree of grinding. If the grinding process were carried much further, the locus could not remain linear.

With reference to Fig. 2.2, it was pointed out that the lower part of each of the particle size distribution curves was drawn to be compatible with the estimated specific surface area of the zero to 7.5 μ fraction. The curves as drawn are compatible with a specific surface area of about 12,000 cm^2/g.

3.3 Voids Content

When we lay off size-analysis curves in size groups having a size ratio of 2, as in Fig. 2.2, we find that cements of ordinary fineness comprise about seven size groups. This range corresponds to that of a mineral aggregate continuously graded from No. 200 to $\frac{3}{8}$ in. The voids content in such an aggregate can be estimated by means of Eq. 1.19, which is

$$\epsilon_n = \epsilon_1 - 0.031(n - 1)$$

On the basis of data given in Table 1.16, we may estimate the voids content of a single size group, ϵ_1, to be 0.45. Using this figure in the equation above, we obtain

$$\epsilon_n = 0.45 - 0.031(7 - 1) = 0.26$$

for the compacted state. We therefore might expect compacted cement to have a voids content of about that value.

The actual voids content in dry cement can take on one or another of many different values according to the method by which the cement is handled. When a few hundred grams are poured into a laboratory liter graduate and then alternately jolted and vibrated, the voids content can be reduced to about 47 per cent, as compared with the 26 per cent just estimated for a geometrically similar aggregation of larger particles. Some of the difference may be attributed to difference in grading, and possibly the smallest size groups have a higher angularity factor than that of the coarser groups on which the estimate was based; however, there is no reasonable doubt that the difference is mostly due to the effects of interparticle cohesion, discussed in the next section.

As we mention above, the voids content in any of the three or four largest size groups obtained from cement is about 45 per cent. Normally when such groups are combined with others to form a graded aggregate, the voids content in the aggregate becomes considerably less than that of any single size group, as is stated in the equation above. With cement, however, if we combine the sizes smaller than 16 μ with the coarser sizes, the voids content may actually be increased: in the present example we found a voids content of 47 per cent, as compared with 45 per cent for the larger individual size groups.

Feldman, Sereda, and Ramachandran [9] found that the voids content of a dry cement, when pressed under conditions favorable to compaction, was related to the applied pressure according to the following equation, which was derived from their published graph:

$$u = 1.212 - 0.188 \log P$$

where u is the voids ratio, cm^3 per cm^3 of solid material, and P is the applied pressure in pounds per square inch. The equation represents data from applied pressures ranging from 8150 to 118,000 psi. For a porosity of 0.26, or $u = 0.26/0.74 = 0.35$, the calculated pressure is 38,500 psi; that is to say, the voids content of cement powder can be reduced to that of a comparable aggregation free from the effects of interparticle cohesion by applying a pressure of about 40,000 lb/in.^{2}.

3.4 Interparticle Cohesion

When a heap of ordinary dry aggregate is produced, as for example by dropping particles into a container, the particles normally slide and roll over each other into positions giving mechanical stability. Presumably each particle is able to find multiple points of support from underlying particles and perhaps from the container wall. On the other hand, when dry cement is poured into a container, it does not immediately achieve stability at all, owing to the fluidized state associated with its air content (see below). As excess air bleeds out and the particles settle, mechanical stability is eventually attained, but there is relatively little sliding or rolling of particles such as would maximize the number of contacts per particle. Consequently, the powder structure tends to become stabilized at a relatively high voids content, as discussed above.

In Chapter 3 we see how sand particles can be caused to stick together by forces arising from the surface tension of water. In that case it is necessary for the sand to contain a little more than enough water to moisten the surfaces and for the water content to be considerably less than that which would inundate the aggregate. In either the dry or inundated state, sand particles do not cohere. In the present case, however, cement particles exhibit cohesion in either the dry or inundated states, as well as any intermediate moisture state. Thus when dealing with cement or other equally fine powders we find evidence of interparticle cohesive forces that do not depend on the surface tension of water.

Experiments carried out by Roller [10] indicate how small the particles must be for forces of cohesion to be appreciable in relation to the weight of the particles. Roller separated cement powder and gypsum powder into various size fractions and determined the minimum porosity that could be produced in each fraction by an arbitrary procedure of tapping the container. As we have already seen (Table 1.16) the porosity of groups having average sizes of 30, 49, and 82 μ was about 45 per cent, a value that seems normal for a size group of angular particles. However, for the very small cement particles, 8.3 and 2.4 μ, the porosities were 59 per cent and 79 per cent. Because such porosities are too high to be reasonably accounted for in terms of differences in particle shape and because there is no reason to assume the particle shapes in the finest fractions are much different from those in the coarser fractions, it is concluded that the observed effect is due to interparticle cohesion. Roller found a rather definite size limit of 16 μ, above which packing of cement particles was normal. It seems therefore that the cohesiveness of cement powder is due mainly to the presence of the ultrafine particles; we should therefore expect any aggregate containing an appreciable amount of flour-fine material to fail to pack as would be expected from its grading alone.

Interparticle cohesion is manifested not only by the relatively high voids content in cement but also in various other ways. For example, settled dry cement shows itself to be a weak solid; it may be cut with a knife so as to leave a vertical or even overhanging surface, and if the dimensions are such that the maximum stresses do not become greater than the strength of the mass, the structure does not collapse. Also the effect of small interparticle forces manifested at a large number of points per unit volume is seen in the magnitude of pressure required to reduce the void content to a normal value, as described above.

Under some circumstances, stored cement absorbs moisture from the atmosphere in a sufficient amount to become more or less cemented and therefore difficult or impossible to handle. The cohesiveness discussed above is not due to such chemical action.

3.5 Specific Powder Volume or Specific Weight

There being no already established, unambiguous term, we shall here speak of the space occupied by a unit weight of cement powder as *specific powder volume*. It is essentially the same as *specific aggregate volume*, as defined in Chapter 1. Its value is very sensitive to the method of manipulation, and under some circumstances it changes with time, owing to settlement. A typical figure for cement packed by vibration or jigging is about 600 cm^3/kg or 0.92 ft^3 per hundred pounds. Cement finer or coarser than the average has a higher or lower than average specific powder volume.

The reciprocal of specific volume is the more familiar specific weight. Thus 600 cm^3/kg corresponds to 1670 g/liter or 109 lb/ft^3. In earlier days of cement technology, when cement was proportioned by powder volume for concrete mixtures, its specific weight was taken to be 94 lb/ft^3. For the relatively coarse product manufactured at that time this was a good average figure for loose cement that had had time to settle but which had not been subjected to vibration or jigging. At present the standard bag contains 94 lb of cement, but the bag capacity is not necessarily 1 ft^3. (The quantity called a barrel is 376 lb.)

3.6 The Fluidized State

Dry cement becomes mobile when its particles become slightly separated by a fluid. The fluidized state of cement powder can be produced by stirring; it can be maintained by causing air to flow upward through the powder. At a suitable air velocity, particles are kept slightly separated, yet the different sizes do not become appreciably segregated, owing to the cohesive forces that tend to hold fine and coarse particles together. In such a state cement can be pumped through pipe lines or caused to flow by gravity (it is in this state when it is first packed into bags). Not until after the excess air has had time to bleed from the powder does the cement in the bag lose fluidity. As cement settles, expelling air, interparticle forces become more and more manifest, and, as already pointed out, settlement ceases while the powder still has a relatively high voids content.

The fluidized state can be produced simply by stirring or otherwise agitating cement in the presence of air, after which the fluidized state persists for a limited time. The length of time it remains fluidized is believed to be dependent on the same physical factors that limit the duration of bleeding of cement paste (see Chapter 11).

4. MIXTURES OF DRY CEMENT AND AGGREGATE

4.1 The Voids-Ratio Diagram

In terms of present-day concrete technology, or of its probable form in the future, there is no practical reason to study dry mixtures of cement and aggregate. However, the relationships found in such mixtures are of interest from the analytical point of view; they are related on the one hand to mixtures of aggregates discussed in Chapter 1, and on the other to mixtures of cement, aggregate, and water discussed in the following chapters.

A voids-ratio diagram (Chapter 1, Section 8.2) for mixtures of dry cement and standard Ottawa sand is shown in Fig. 2.5. The average volume diameter of the sand was about 706 μ, and that of the cement, as calculated from sedimentation analysis, was about 29 μ, the size ratio thus being 0.04. The envelope *AB′C*, and the curve *ABC* have the same features as for sand and gravel mixtures; adding small amounts of sand to dry cement, the sand being the coarse aggregate in this case, increases the content of voids in the cement, or adding small amounts of cement to the sand increases the voids content of the sand, thus giving minimum voids content at the point *B* rather than at *B′*.

The mixtures shown in Fig. 2.5 were consolidated in a liter graduate by means of a vibrating platform on which the graduate rested, and by subsequent jigging by hand until the smallest volume attainable by this method was reached. Although the powder structure of the dry cement produced in this way had a much higher voids content than is normal for a geometrically similar but coarser aggregate, as explained above, the powder seemed to have a definite characteristic structure that became dilated when sand particles were introduced, just as the structure of dry sand seems to become dilated by the introduction of small amounts of coarse particles.

Similarly, although most of the cement particles are small enough

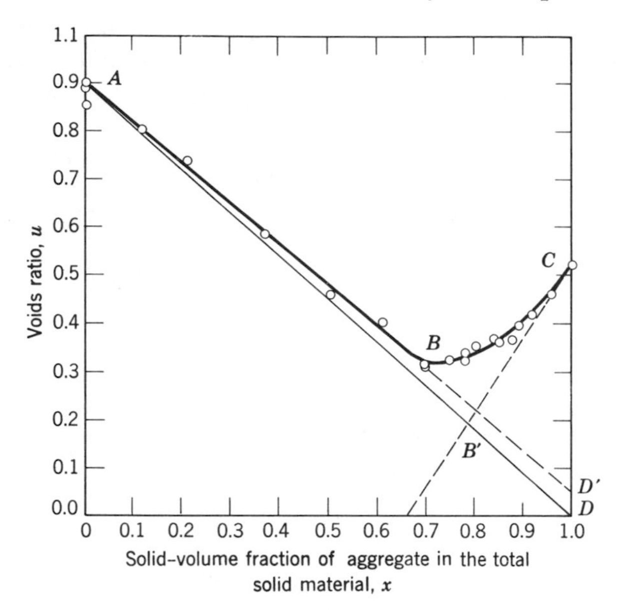

Fig. 2.5. A voids-ratio diagram for dry mixtures of cement and standard Ottawa sand.

to fit into the voids of the sand, the curve from B to C, in relation to the line from B' to C, shows that if the cement content of the mixture exceeded, say, 5 per cent, the voids content of the aggregate became appreciably increased.

When preparing Fig. 2.5, we assumed that the path from B to C should be curved, and, although the points scattered considerably from the curve as it was drawn, it seemed likely that such scatter was due to inaccuracy of the data. However, if we examine the same data from a different point of view, we are led to a different interpretation. Figure 2.6 shows the result of plotting the ratio of void space to cement volume (v/cv_c) against the ratio of aggregate volume to cement volume $av_a/cv_c = M$. The points A and B are the same as those appearing in Fig. 2.5; point F will be explained.

We see that the points describe two, perhaps three, straight lines. The first is the line from A to B; the second is the line from the origin through B to F, and the third is from F to C. Plotting the data in this manner revealed a characteristic of the relationship not apparent in Fig. 2.5, specifically that beyond point B the voids-cement ratio is directly proportional to M, for a limited range.

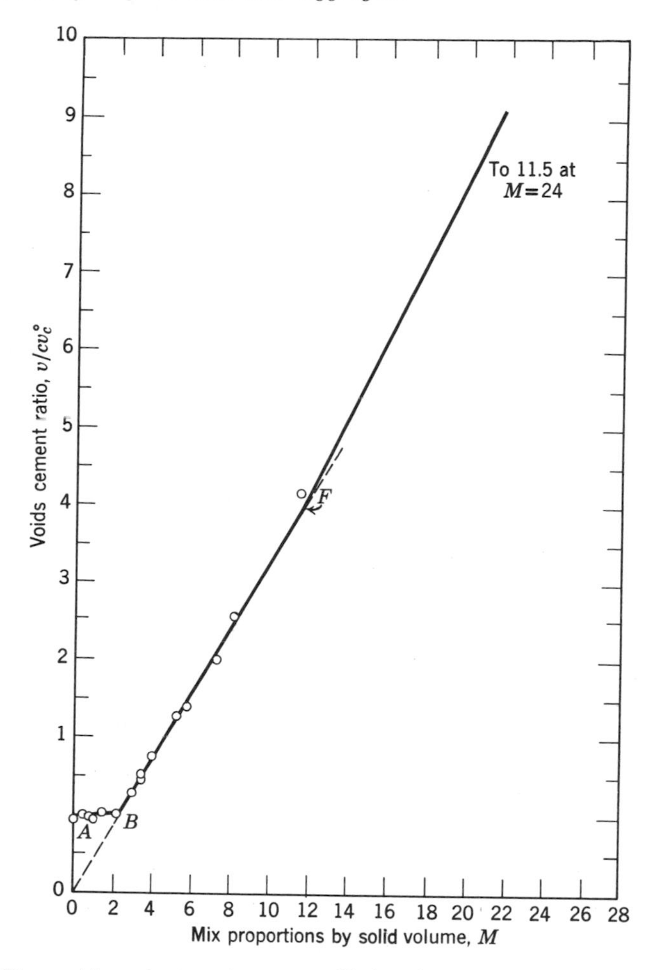

Fig. 2.6. The voids-cement ratio versus M for the same dry mixtures of cement and standard Ottawa sand used in Fig. 2.5.

When Fig. 2.5 was redrawn to conform to this relationship, it consisted of three straight lines, as is shown in Fig. 2.7. Now we see that point B seems to lie on a straight line from the origin to point E and that point F is the intersection of that line with line $B'C$.

The value of u at point E is the slope of the straight line from B to E, and it thus gives the rate at which the voids ratio of the mixture increases as x is increased from x_B to x_F, in this case from 0.7 to 0.92. Beyond point F the rate must increase because points on the line from B' to C correspond to the smallest volumes of space that can accommodate

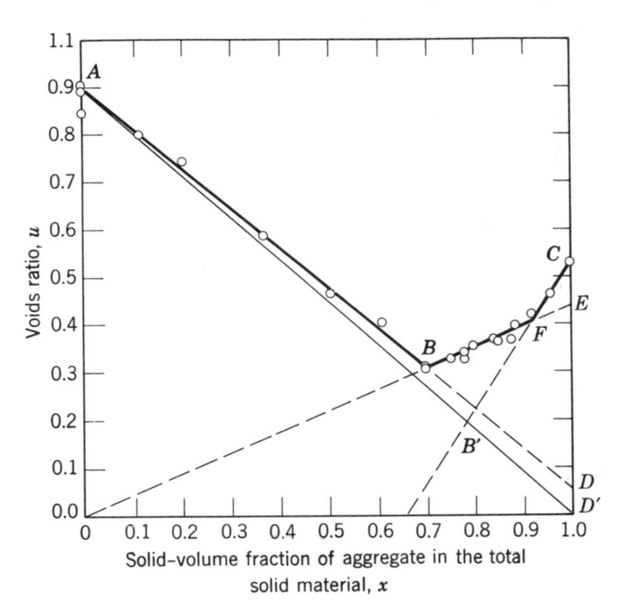

Fig. 2.7. A revised voids-ratio diagram for dry cement and Standard Ottawa sand.

the sand; in other words, points on that line correspond to the aggregate volume of the coarse component, and in that range any cement present is entirely within the voids of the undispersed coarse-aggregate structure.

At point B the sand particles are at a critical degree of dispersion, as is discussed in Chapter 1, Section 8.6. Point I, where particle interference begins, is a little to the left of point B, and at point B the particle-interference effect has just become large enough to offset the reduction in voids content due to the powder-displacement effect of the individual particles. To the right of this point, the voids ratio is proportional to x and is apparently independent of the voids content of the fine component.

A question naturally arises as to why point B occurred at the particular composition it did. Let us begin to develop an answer by considering the composition at point F.

As shown in Fig. 2.6, there are about 12 volumes of sand to one of cement at point F. The sand has a voids ratio of 0.52, and thus at point F the volume of the aggregate voids is $12 \times 0.52 = 6.24$ cm³/cm³ of aggregate, solid volume. The voids ratio of the cement as it exists at point B, which is about the densest packing we might reasonably assume at point F, is $0.305/0.3 = 1.02$ and thus the powder volume of the cement may be assumed to be $1 + (1.02 \times 1) = 2.02$ cm³. That volume is contained in 6.24 cm^3 of interstitial space, which means that the space is about 32 per cent full of powder. Evidently, this is the fraction of total interstitial space that is composed of spaces large enough to accommodate all the cement particles, including the largest ones with a diameter of about 100 μ. (The process of jigging and vibration described above was apparently able to establish the same frequency of mutual sand-grain contact as is characteristic of this sand alone in the compacted state.)

The grains of the sand discussed above are nearly spherical, and the average grain diameter is about 706 μ. This means that where the sand grains are packed in a face-centered cubic array the voids can accommodate spheres up to about 300 μ; where they are hexagonally close packed, they can accommodate 50 μ spheres; randomly packed, there would be void sizes ranging between these limits. It thus appears that many voids were large enough to hold the largest particles of cement. Thus the position of point F on the line $B'C$ is easy to account for.

However, point F marks a limit: when more cement was added, dispersion of the aggregate began, and the line FB, in Fig. 2.6, indicates that the amount of dispersion was directly proportional to the amount of cement added in excess of that in the mixture at point F. The dispersion may be thought of as a process of creating additional spaces large enough to accommodate the largest cement particles; at the same time, spaces originally too small for cement powder become enlarged and able to admit it. When point B is reached this process is complete; all the interstitial space is now able to hold cement powder. This explanation seems to account for a reduction of interstitial void space as cement powder replaces sand, or an increase as sand replaces cement powder.

4.2 *Particle Interference*

The beginning of particle interference in the mixtures of dry cement and sand discussed above seems to be at $u = 0.32$ and $x = 0.7$; $C = 0.525$. Calculation of the clearance between sand particles by Eq. 1.18 gives 63 μ. The average diameter of this cement is about 30 μ, and the average size of the largest size group is roughly 70 μ. Thus we find that particle interference begins when the clearance is about equal to the diameter of the largest size group in the cement, a result in line with the critical clearances found for some of the mixtures of sand and gravel discussed in Chapter 1.

It seems clear that the pattern seen in Fig. 2.7 is characteristic of a coarse component having voids large enough to accommodate the largest particles of the fine component. However, if the size range of the coarse component overlaps that of the fine, there may be no sector corresponding to FC, the coarse component being subject to dispersion by the first increment of the fine.

REFERENCES

[1] Heller, L., and H. F. W. Taylor, *Crystallographic Data for the Calcium Silicates,* Department of Scientific Industrial Research, Building Research Station, Her Majesty's Stationery Office, London, 1956, pp. 15–19.

[2] Ford, C. L., "Determination of the Apparent Density of Hydraulic Cement in Water Using a Vacuum Pycnometer," ASTM Bulletin No. 231, 1958; Research and Development Laboratories of the Portland Cement Association, Research Department Bulletin No. 101, November 1958.

[3] Steinour, H. H., "Further Studies of the Bleeding of Portland Cement Paste," Research Department Bulletin No. 4, Research and Development Laboratories of the Portland Cement Association, December 1945.

[4] Haegermann, G., "On the Order of Size of the Smallest Grains in Ground Portland Cement Clinker," *Zement,* **31,** 441, 1942.

[5] Plum, N. M., "The Predetermination of Water Requirement and Optimum Grading of Concrete," Building Research Studies No. 3, The Danish National Institute of Building Research, Copenhagen, 1950.

[6] Lea, F. M., *The Chemistry of Cement and Concrete,* St. Martins, New York, 1956, p. 332.

[7] Hime, W. G., and E. G. Labonde, "Particle Size Distribution of Portland Cement from Wagner Turbidimeter Data," *J. PCA Research and Development Laboratories,* **7**(2), 66–75, May 1965.

[8] Lea, F. M., and R. W. Nurse, "The Specific Surface of Fine Powders," *J. Soc. Chem. Ind.,* **58,** 277–282, 1939.

[9] Feldman, R. F., P. J. Sereda, and V. S. Ramachandram, "A Study of the Length Changes of Compacts of Portland Cement on Exposure to H_2O," Highway Research Record No. 62, 106–118, Highway Research Board, Washington, D.C. Research Paper No. 248 of the Division of Building Research, National Research Council, Ottawa, Canada.

[10] Roller, Paul S., "The Bulking Properties of Microscopic Particles," *Ind. Eng. Chem.,* **22,** 1206–1208, 1930.

[11] Lerch, W., and C. L. Ford, "Long-time Study of Cement Performance in Concrete: Chapter 3; Chemical and Physical Tests of the Cements." *Proc. ACI,* **44,** 744–795, 1948. Data are from Table 3.6.

3

Mixtures of Aggregate, Cement, and Water

1. INTRODUCTION

In Chapter 1 we have dealt with various dry mixtures of aggregate. Now we turn to mixtures containing water and cement and thus begin to deal with concrete in its freshly mixed state.

Properties of mature concrete such as strength and permeability to fluids may take on any of a wide range of values, depending on variable factors that are subject to control. With given materials, strength and permeability depend mainly on the characteristics of the matrix of the aggregate particles. In the fresh state consistency and the various attributes collectively called "workability" also depend on the characteristics of the matrix of the aggregate and on the concentration of aggregate material in that matrix.

The matrix is always composed of cement, water, and air; sometimes it contains supplementary material as well. Cement and water exist as a paste, and the air may be enclosed in the paste in the form of more-or-less spherical bubbles; otherwise, most of the air may be present in interconnected spaces. If the fresh mixture is a plastic one, its air content exists as bubbles, but if it is a nonplastic mixture, such as is used by the block industry, much of the air is in relatively large, interconnected spaces.

When poorly designed mixtures, intended to be plastic, are ineptly handled, the coarse aggregate often segregates from a batch during handling and later appears as "rock pockets" that exhibit interstitial spaces containing little but air. Voids formed in this way are not considered here because they signify nothing but the result of carelessness or inadequate knowledge of the properties of mixtures. We are concerned with water-filled space and with air voids that are structural components of concrete and that can generally be called bubbles even though some of them may have irregular form.

The air content of the matrix of plastic mixtures usually amounts to 5 per cent or more of the matrix volume, but when the mixture contains purposely entrained air (Chapters 7 and 8), it may constitute 20 per cent or more of the matrix.

The values of strength and other desirable qualities that may ultimately be reached by the physicochemical process called "curing" are limited by the composition of the matrix of the aggregate, particularly its cement content. The composition of the matrix is usually expressed as the ratio of its cement content to its content of air and water, this ratio being called the *cement-voids ratio;* or the inverse is used, the *voids-cement ratio*. The sum of the water and air contents is called the *content of total voids*. Under most conditions the total-voids space is filled mostly, but not entirely, with water, which makes it possible to state a fairly good approximation of the composition of the matrix in terms of the water-cement ratio or its reciprocal; however, this practice is questionable, especially when the mixture contains purposely entrained air.

We are mainly concerned with the factors that determine the amounts of water and air in mixtures containing given amounts of cement. Following the same approach we use in Chapter 1, we see that cement may be regarded as the fine component of a two component mixture, the aggregate, of course, being the coarse component. Water and air together are regarded as composing the void space in such binary mixtures. In general, we find a close parallel between the factors that govern the voids contents of wet mixtures and those of dry ones. At the same time we observe important differences, among them being those phenomena arising from the surface tension of water, particularly the phenomenon known as bulking. We begin by discussing mixtures of sand and water.

2. MIXTURES OF SAND AND WATER

2.1 Bulking

The voids content in a given sand, after the sand has been mixed with enough water to moisten it and handled in a certain manner, is a function of the water content of the mixture. A typical relationship is shown in Fig. 3.1 in which the upper curve represents the loose condition that results from pouring the dry or moistened sand into the measuring container; the other curve represents the volumes resulting from "shaking and packing to refusal" [1]. In Fig. 3.1 the voids indicated at the left terminus of the curve are entirely air voids, and at the other extreme, the inundation point, the voids are nearly water filled. The increase in the voids content at intermediate water contents is called *bulking*, and the amount of bulking relative to the aggregate volume in the dry state is customarily expressed as a *bulking factor*. It should be understood that the bulking of sand is not caused by adding water; bulking appears when the dampened sand is handled, as when it is stirred or is transferred from one container to another.

The magnitude of the bulking factor at the moisture content giving the greatest amount of bulking is smaller the larger the average size of the sand particles. The bulking effect is negligible with aggregates that do not contain particles smaller than about $\frac{1}{4}$ in. or 6 mm.

Prescribed proportions for concrete mixtures are given in terms of solid volumes or, much more commonly, the corresponding weights of the solid ingredients. However measurement is often made in terms of

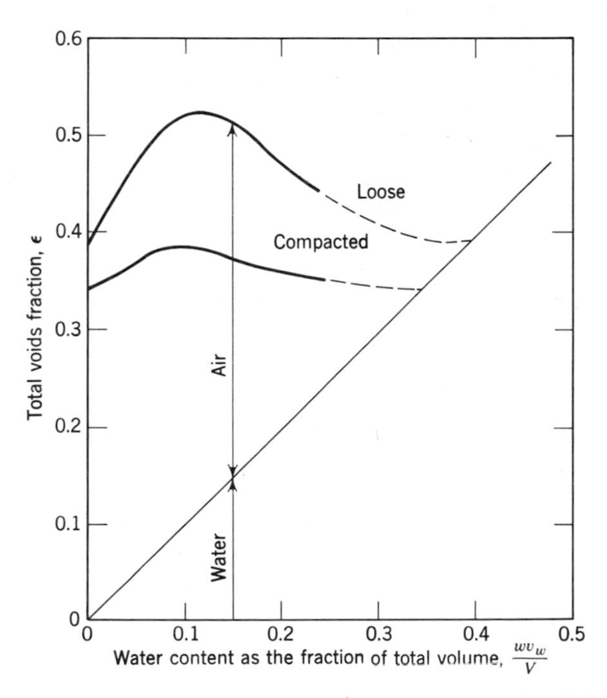

Fig. 3.1. The bulking of moist sand. (Data are from H. N. Walsh [1].)

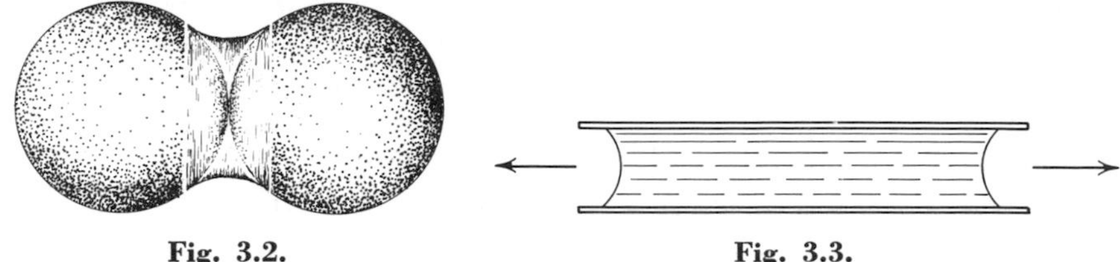

Fig. 3.2. **Fig. 3.3.**

aggregate volume, also called the *bulk volume*. When this method of measurement is used it is necessary to use a bulking factor to compute the volume of sand that will contain the prescribed amount of solid material. For example, if the moist sand as measured has a bulking factor of 1.3 and if the sand in the dry, loose state contains 100 lb/ft^3 aggregate volume, we must take 1.3 ft^3 of loose moist sand to obtain 100 lb of solid material. When establishing the bulking curve the points should be based on the degree of compaction that results from the actual measuring process. Usually compaction is effected only by the forces corresponding to the momentum at which the moist sand enters the measuring container.

At an earlier time, a device called an inundator was used to avoid the problem of bulking when sand was to be measured. The inundator was essentially a cylindrical container mounted on trunnions so that it could be filled in the upright position and emptied by inverting it. An excess of water was put in the container before the moist sand, and the excess water overflowed through a spillway. This apparatus has been supplanted by gravimetric equipment, with which it is necessary only to correct the wanted weight of solid material for the moisture content.

When an aggregate is used in plastic concrete, it becomes inundated or nearly so, and its bulking partly or completely disappears, depending on factors yet to be considered.

2.2 *Cause of Bulking*

Bulking of moist fine aggregate is caused principally by the surface tension of water and forces that arise from it.* The mechanics of the effect can be understood by examining the simple model shown in Fig. 3.2, which represents two spherical particles moistened with a limited quantity of water. The water exists as a film over the surface of the particles and as a pendular body between the particles. The particular distribution of water illustrated in the figure is produced spontaneously

* For a good description of the nature of surface tension see Sears [2].

because that is the configuration representing the least extent of water surface possible under the conditions depicted; it represents minimum free surface energy. The pendular body of water presents two principal curvatures of surface, one being concave toward the exterior and the other convex.

Because of surface curvature and surface tension there is a force that holds the spheres together, and the magnitude of the force and its relation to surface curvature and surface tension can be derived as follows. Consider a spherical droplet of water. Hydrostatic pressure in the droplet is given in Eq. 3.1:

$$P = P_A + \frac{2\gamma}{r} \tag{3.1}$$

where P is the pressure in the droplet, P_A is the ambient (atmospheric) pressure, γ is the surface-tension coefficient of water (at room temperature the surface tension of water is about 72 dyn/cm), and r is the radius of the droplet, $1/r$ being the curvature of the surface. If the radius of the droplet is, for example, 10^{-3} cm (10 μ) then $(2\gamma/r) = 1.44 \times 10^5$ dyn/cm^2, which is 0.144 bars, or 0.142 atm. This is the amount by which the pressure in the droplet is caused by surface tension to exceed atmospheric pressure.

Now consider a cylindrical glass tube containing some water, the water presenting a concave surface toward the air (menisci), as is indicated in Fig. 3.3. Equation 3.1 applies, but because the convex curvature of the droplet has been called positive, the concave curvature of the meniscus must be negative, and thus $2\gamma/(-r)$ stands for a pressure deficiency relative to atmospheric pressure. Because the water in the pendular body (Fig. 3.2) has both positive and negative curvatures, hydrostatic pressure in the ring of water is given by

$$P = P_A + \gamma\left[\frac{1}{r_1} + \frac{1}{r_2}\right] \tag{3.2}$$

The numerical values of r_1 and r_2 being of opposite sign, the second term may be positive or negative depending on whether the convex (+) or the concave (−) curvature is the larger. When the interstitial space is considerably less than full of water, the negative curvature is the larger, and thus the second term is normally negative. Pressure deficiency, or hydrostatic tension, is made possible by the fastening of the curved water surfaces to the solid bodies by forces of adhesion (adsorption). Stress in the water is the algebraic sum of the compressive stress due to the atmosphere and the tensile stress due to surface curvature. In the present case the net stress may be positive (compressive), but it is the negative component that holds the particles together.*

When there are forces tending to hold particles together normal packing is prevented, for the particles are not free to slide or roll; bulking is the observed result. When the water content of interstitial space is low, hydrostatic tension is relatively high, but the cohesive force per particle is relatively low because the fraction of particle cross-section area over which negative pressure is effective is small. As the water content is increased, hydrostatic tension diminishes, but the effective area increases. Thus the effective cohesive force between particles rises to a maximum and then diminishes as the water content increases, the corresponding effects on bulking being such as those shown in Fig. 3.1.

When water fills a certain fraction of interstitial space, curvatures of water surface are smaller the larger the particles, and therefore hydrostatic tension is lower the larger the particles. This, together with the relative particle weights, accounts for the smaller bulking factors found for coarse sands and the absence of bulking in clean coarse aggregate.

The bulking of sand has been explained by assuming that water holds the particles apart. Water probably does prevent actual contact between solid surfaces (see Chapter 9), but sand particles are so large relative to the separation that can be maintained by water under static conditions that the effect is undetectable. We should note in passing, however, that in aggregations of submicroscopic particles this is not true; separation due to water films as well as cohesive forces must be considered to account for the volume, and the volume changes, of such aggregations, hardened cement paste being an important example.

3. CEMENT, AGGREGATE AND WATER—NONPLASTIC MIXTURES

When mixed with relatively small amounts of water, cement or mixtures of cement and aggregate show the same kind of bulking as does moist sand. Examples are shown in Fig. 3.4, where neat cement and two mixtures of cement and standard Ottawa sand are represented. The degree of bulking is indicated by the arching of the curve from the point representing the voids content of the dry material to the point of minimum voids content, indicated by an arrow. Considering the uppermost curve, we see that bulking of neat cement reaches its maximum when the

* It is well to carry in mind the fact that liquid can support tensile stress, which is to say that negative pressure can exceed 1 atm, if the liquid body has a small enough volume or cross-section, or if in a large body conditions for nucleation of a bubble are unfavorable.

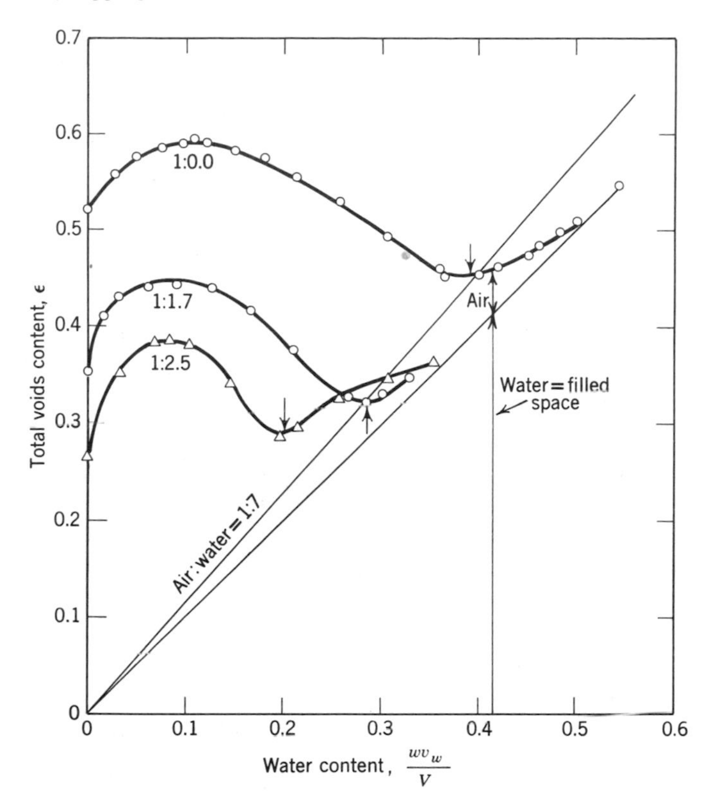

Fig. 3.4. Voids content versus water content for neat cement and for mixtures of cement and standard sand. (Data are from Richart and Bauer [6].)

total volume of voids constitutes 59 per cent, the water content 10 per cent, and the air content 49 per cent of overall volume. (Those figures would be somewhat different for different cements, particularly if the cements differed in fineness.) Where the volume of air voids greatly exceeds that of water-filled space in a mixture, most of the air is present as a continuous phase; that is, it is simply air at atmospheric pressure permeating the porous granular structure, the structure being held together partly by cohesive forces due to tension in the interstitial water and partly by interparticle forces. (See Chapter 9.)

Mixtures with a relatively high proportion of aggregate to cement and a water content that produces marked bulking emerge from a mixer in the form of a loose aggregation of more or less clustered particles. Except for the clusters, there is little evidence of cohesiveness. However, when the mixture is tamped or vibrated into a suitable mold, it acquires a continuous cohesive structure that has sufficient strength to support its own weight even when the mold is removed immediately. (See Fig. 3.5.) The concrete block industry is based on this characteristic of nonplastic but cohesive mixtures.

The degree of stability shown by a given bulky structure depends partly on the intensity of hydrostatic tension and on the fraction of cross-section area occupied by the tensed component, water, as is the case for moist sand. The internal geometry of the mixture is such that hydrostatic tension decreases and the area subject to tensile force increases as water content increases; hence, maximum strength and stability of the structure occurs at a certain intermediate water content, as is shown by the points of maximum bulking in Fig. 3.4.

Fig. 3.5. Cross section of a nonplastic mixture. The macroscopic voids comprise 15.5 per cent of the overall volume (× 5.0). (Photo by Bernard Erlin, Portland Cement Association.)

4. PLASTIC MIXTURES OF CEMENT, AGGREGATE, AND WATER

4.1 Transition from the Bulky to Plastic State

A bulky mass with a low water content, at least up to the point of maximum bulking, is permeable to air, thus showing continuity of the air-filled spaces in the aggregation. When the water content is progressively increased beyond the point of maximum bulking, the total voids content diminishes, and the proportion of total voids occupied by water

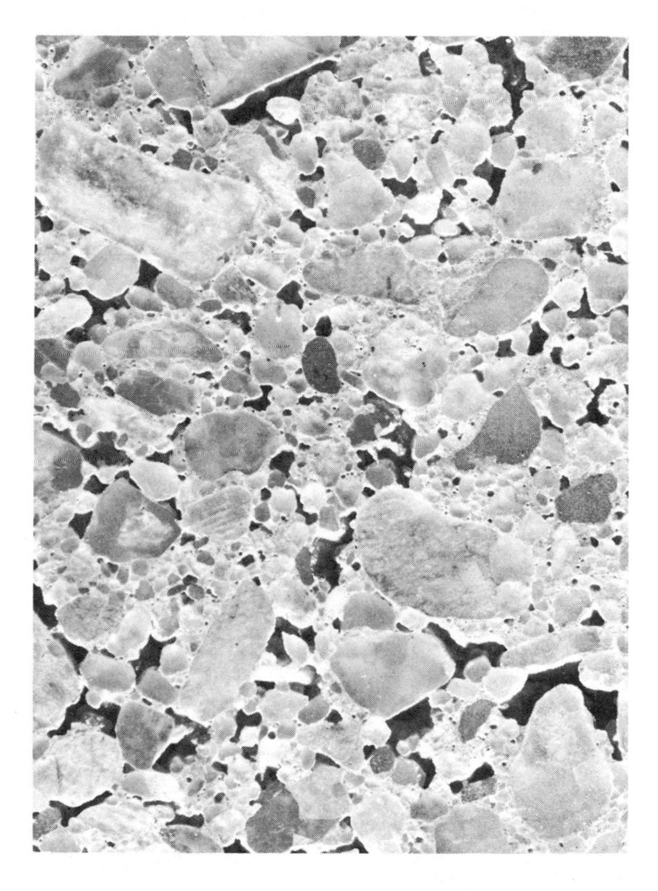

Fig. 3.6. Cross section of a nonplastic mixture containing enough water to produce localized bodies of consolidated material in which air voids are isolated. The macroscopic voids comprise 12.8 per cent of the overall volume (× 5.0). (Photo by Bernard Erlin, Portland Cement Association.)

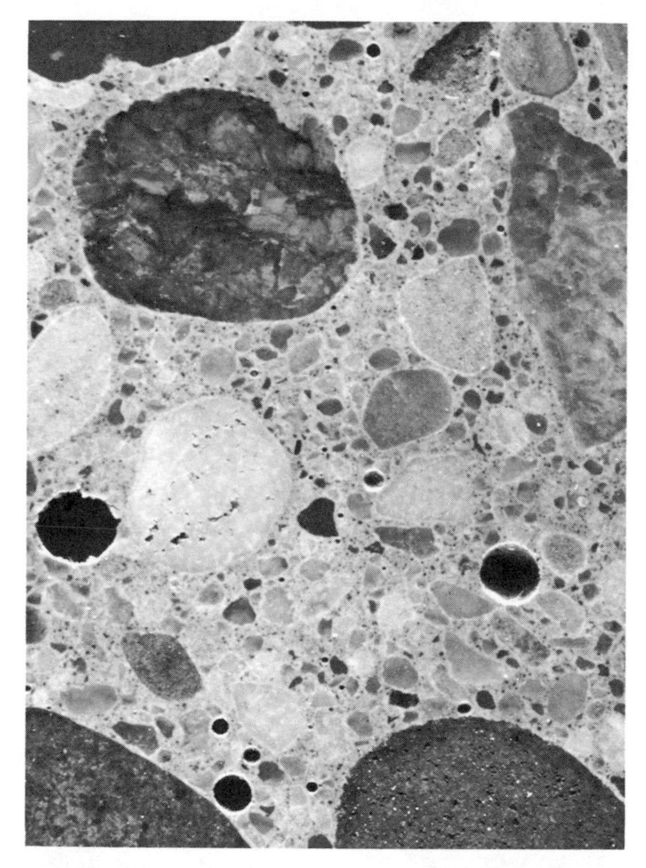

Fig. 3.7. Cross section of concrete made from a plastic mixture showing characteristic shapes and situations of air voids (× 5.0). (Photo by Bernard Erlin, Portland Cement Association.)

increases, some of the air-filled spaces becoming completely surrounded by water. (See Fig. 3.6.) The water surface in such a cavity at the stage indicated may consist partly of meniscuses and partly of irregular contours determined by the shapes of the particles; at this stage the surrounding water is in a state of tension, owing to the tendency of surface tension to draw water into the cavity and owing to the fact that the same water is being similarly drawn toward other cavities. Air in the cavity becomes compressed to the degree that water is able to move into the cavity, and thus the forces due to surface tension are balanced, partly by hydrostatic tension and partly by air pressure in the space.

As water is made more abundant in successive mixtures, more water can be drawn into the cavities, and the air-filled part of the space becomes more nearly spherical; at the same time air pressure becomes a larger fraction of the total force opposing the force due to surface tension. Finally the force is completely balanced by air pressure, and the air-filled part of the cavity becomes quite spherical, hydrostatic tension having vanished. (See Fig. 3.7.) At this stage the mixture is, of course, impermeable to air.

It seems that the smallest water content of which hydrostatic tension is absent is very nearly the same as that at which the minimum voids content in the mixture is observed. Hydrostatic tension being eliminated, that part of the stability of aggregate structure that is due to hydrostatic tension also vanishes. The remaining cohesive force is due to interparticle attraction among small particles. Water has changed its role; when there is a smaller amount of water present, water provides the principal force to hold solid particles together. When there is a greater amount it holds them apart.

Bulky mixtures in which hydrostatic tension is the predominant coehesive force are characteristically dilatant, which is to say, a deformation producing shear strain is accompanied by an increase in overall volume; this is one of the reasons why such materials appear crumbly while being mixed. When a sufficient amount of water is present to cause hydrostatic tension to vanish, true plastic deformation is possible, but the amount of shear strain that can occur plastically is normally limited, owing to the smalless of the initial degrees of separation of the particles; as shear strain increases beyond some small value, a mixture becomes progressively more dilatant. Thus the concept of *capacity for plastic deformation* as an attribute of a concrete mixture appears. (See Chapter 10.)

The points indicated by arrows in Fig. 3.4 may be considered transition points between mixtures that are plastic and those that are not for lack of water. The transition in some cases is from the nonplastic to only a quasiplastic state, owing to shortcomings in the make up of the aggregation. In any case the water content giving a minimum total voids content is called the *basic water content*, and the corresponding consistency, *basic consistency*. At the basic water content, any granular mixture, including neat cement, has a small capacity for plastic deformation.

4.2 Normal Compactness

We should not expect ordinary compacting methods to eliminate all air voids from plastic mixtures. When we use ordinary low-energy methods of compacting, the void space in a mixture containing just enough water to give a minimum voids content comprises about 88 per cent water and 12 per cent air, provided that the cement content of the mixture is not below a certain limit. That is to say, any mixture having a cement content above that limit, shows about the same ratio of air to water, 1:7, at its minimum-void content, and such mixtures may be considered to be in the normally compact state, or may be said to have *normal compactness*.

Examples of the ratios of air to water in mixtures are shown in Fig. 3.4: the 1:0 and 1:1.7 mixtures made with standard sand are normally compact, having reached a minimum voids content when the voids were 88 per cent full of water. In Fig. 3.8 we see two concrete mixtures also showing normal compactness at the point of minimum voids content. On the other hand, we see that the 1:2:5 mixture shown in Fig. 3.4, reached a minimum voids content when the voids were only about 70 per cent full of water (air:water = 1:2.33). This is the result of hindrance caused by an excess of the coarse component of the mixture.

Lack of normal compactness may not be immediately obvious in some series of mixtures. An example is shown in Fig. 3.9, based on data by

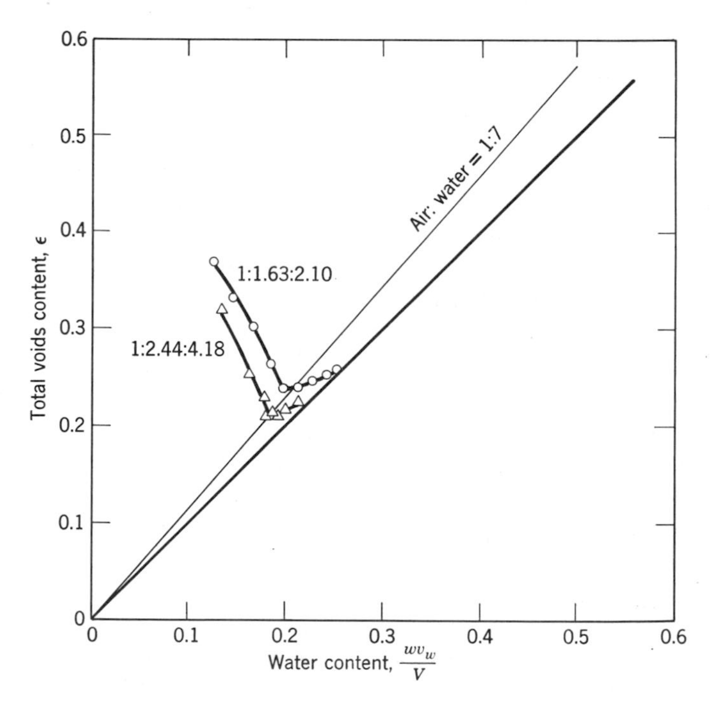

Fig. 3.8. A diagram showing voids content versus water content for two concrete mixes.

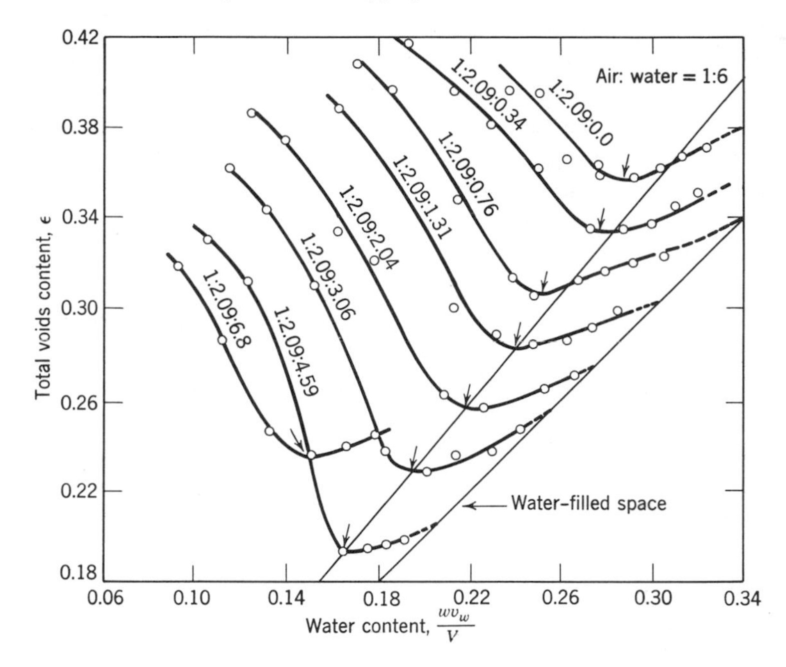

Fig. 3.9. A diagram showing voids content versus water content for mixtures of mortar and crushed limestone coarse aggregate. (Data are from Blanchette [3].)

Blanchette [3]. This series was produced by starting with a 1:2.09 mortar [sand (#200–#14); four size groups] and adding successive increments of crushed limestone. The mortar alone gave maximum density when the voids were about 80 per cent full of water as compared with 86 per cent for normal compactness of this material. The characteristics of the aggregate were improved by adding coarse material, as shown by the increase in the water content of the voids from curve to curve. Normal compactness was observed for the mixes ranging from 1:2.09:2.04 to 1:2.09:4.59. Finally, the percentage of water in the void space dropped sharply when the fraction of coarse aggregate became too high.

Thus we see that in a series of mixtures such as the one described above, lack of normal compactness after low-energy compacting can result either from an excess or a deficiency of coarse aggregate in the combined aggregate. When, however, the ratio of cement to sand is such as to give normal compactness in the mortar alone, all mixtures formed by adding coarse aggregate up to a limit fixed by characteristics of the combined aggregate and by the method of compacting will have normal compactness.

The air-water ratios for normal compactness and the minimum-voids content found in a series of mixtures of given materials, as shown in Figs. 3.4, 3.8, and 3.9, are characteristic of a low-energy compacting procedure, employing small forces. The Richart and Bauer [16] data (Fig. 3.4) involved light rodding or tamping and an arbitrary number of strokes per specimen. Although Blanchette [3] used a power-driven device, his method also is in the low-energy class; he deliberately adjusted his mechanical method to give the same compactness found in fresh concrete being placed for pavement by a method involving spreading and screeding but very little compacting. To accomplish this he caused the apparatus to impart to the mold (beam or cylinder) a total of 340 vertical drops (jigs), each of $\frac{1}{64}$th in., in 1 min. Evidently the effective forces and the amount of work usefully expended were about the same as those expended by the Richart and Bauer manual method.

That the results described above might have been altered if a more effective compacting method had been used is indicated by the data by Plum [4] shown in Fig. 3.10. All plotted points represent the same mixture. It is evident that the minimum voids content occurred at a lower level and

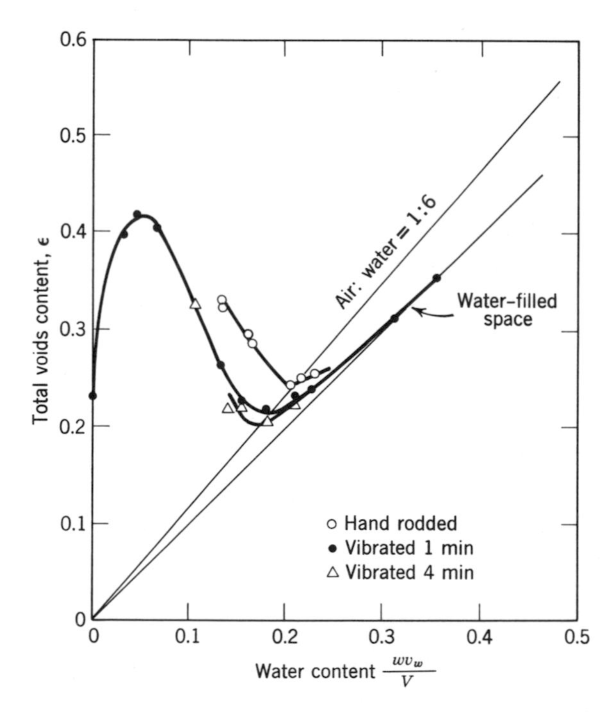

Fig. 3.10. A diagram showing the effect of the compacting procedure on voids content. (Data are from Plum [4].)

at a lower water content as the energy of placement was increased; an expected result. It would seem also that the ratio of air to water at the point when the voids content was at a minimum might also be smaller the greater the compacting energy. However these data indicate that, at least within certain limits of compacting energy, the ratio of air to water is influenced little, if at all, and therefore it might tentatively be concluded that for given materials and different compacting energies, the same ratio of air to water can be used as a criterion for normal compactness.

If Richart and Bauer had used a moderately high-energy compacting method, the 1:2.5 mix shown in Fig. 3.4 could easily have been given normal compactness. At some higher proportion of aggregate to cement, lack of normal compactness (for the given energy of compaction) would be found.

The term "normally compact state" is used here for want of one more apt. It should *not* be construed as a *usually* compact state because, in practice, the majority of mixtures would not be normally compact, as here defined, when they contain the amount of water that gives a minimum voids content.

We shall see that it should be possible generally to produce mixtures capable of reaching normal compactness under low-energy compacting procedures and that it might be advisable to modify any mixture that lacks this capability. However, we must not assume that any mixture containing more air than that corresponding to the definition of normal compactness would be benefited by modifying it so as to reduce the air content. Judgment on this point should be reserved until many other aspects have been considered, particularly the factors influencing volume changes of hardened concrete, and damage by freezing. With respect to the latter it has become standard practice to entrain definite amounts of air in concrete. (See Chapter 7.)

4.2.1 Basic water content. Talbot [5] called the amount of water giving a minimum voids content the *basic water content.* He applied this term whether or not the basic water content was associated with normal compactness, as the term is defined above. In mixtures sufficiently rich in cement or other material of similar fineness normal compactness may be assumed at the basic water content.

4.2.2 Normal consistency of paste. Richart and Bauer [6] found that the consistency of cement paste at its basic water content is nearly the same as the normal consistency as it is defined by ASTM standards. It seems therefore that among mixtures that become normally compact under low-energy compaction there is a basic consistency corresponding to the basic water content. Blanchette found that at the point of the basic water content of mortar or concrete mixtures made with different aggregates, the standard slump tests gave values between 1.5 and 2.0 in., the average being about 1.7 in. Other tests show that neat cement at normal consistency slumps about the same amount. This consistency is not much less stiff than the stiffest that can be handled by low-energy methods; in fact, it is much stiffer than plastic mixtures ordinarily made in American field practice. Because water contents higher than the basic are commonly used, the water content of the void space is usually greater than 88 per cent, unless the mixture is grossly deficient in fine material or unless the air content has been raised by means of an air-entraining agent.

5. ANALYSIS OF PLASTIC MIXTURES

In this section we begin laying the foundation for an understanding of the factors that enter into the formulation of concrete mixtures, discussed at length in Chapter 5.

5.1 The Role of Water and Interparticle Forces

When sand and a sufficient volume of water are placed in a mechanical mixer, the sand exhibits a mobility that it does not show when it is handled dry. Such mobility is the result of dispersing the particles throughout the fluid. However sand particles remain dispersed for only a matter of seconds after agitation has stopped and settle rapidly, tending to form a stable aggregation. On the other hand, when we mix portland cement with water, producing a dispersion of solids throughout the water-filled space, the rate of settlement following the period of agitation is relatively low; whereas a suspension of clean concrete sand settles at a rate that might be expressed in inches of fall per minute, cement particles settle only a few microns per minute. Not only does settlement of cement occur slowly, the condition of the sediment is much different from that formed by sand.

Sand in water forms an aggregate structure having a mechanical stability that arises from particle-to-particle contact; cement forms a loose structure, which, after settlement, has a stability due mainly to interparticle cohesion, such cohesion being evident before settlement as well as after.

The physical state maintained by interparticle cohesion is called the *flocculated state,* or *flocculent state,* and therefore when dealing with cement-water mixtures it is appropriate to use the term *floc-structure* or *paste structure,* rather than aggregate structure. Structure arising from interparticle force among cement particles dispersed in an aqueous solution gives the attributes connoted by the term *paste,* and gives to the whole mixture, to one degree or another, the quality called *plasticity.* (See Chapter 10.)

When cement paste and aggregates are combined in proportions within certain limits, the paste imparts some degree of plasticity to the whole mixture, and the rate of settlement is controlled mainly by the characteristics of the paste. (See Chapter 11.) When we deal with mixtures containing cement, aggregate, and water, we must deal not only with the geometric properties discussed in Chapter 1, but also with the characteristic called *consistency,* and with entrapped or entrained air.*

5.2 The Voids-ratio Diagram

5.2.1 Definition of voids ratio. The voids ratio of a fresh concrete mixture is defined in the same way it is defined for mixtures of fine and coarse aggregate we deal with in Chapter 1; that is, the voids ratio is the total void space (volumes of water and air) per unit solid volume of the solid material in the mixture. Thus, letting u stand for voids ratio, we have

$$u = \frac{\epsilon}{1 - \epsilon} = \frac{V}{cv_c^o + av_a + bv_b} - 1 \tag{3.3}$$

ϵ is the voids content (air plus water) per unit volume of mixture; V is the total volume of the mixture; c, a, and b are weights of cement, fine aggregate, and coarse aggregate, respectively; v_a and v_b are the specific solid volumes of fine and coarse aggregate, respectively, solid volume per unit weight. The symbol v_c^o may be called the *apparent specific solid volume of cement,* or, better, the *displacement factor for cement in water.* It does not correspond to a true specific gravity; it is the apparent space required by 1 g of cement in a paste composed of equal parts by weight of cement and water, one-half hour after the mixture is formed. The factor is based on the contrary-to-fact assumption that the specific gravity of the liquid remains unchanged. (See Chapter 2, Section 2.2.)†

The voids ratio of a given concrete mixture having a certain consistency is, for given materials, a function of the proportion of aggregate in the total solid material. The characteristics of this function can be established empirically by making a series of mixtures having the same consistency, measuring the voids content in each, and then plotting u against the solid-volume fraction of aggregate. A smooth curve drawn through the experimentally established points makes the voids ratio diagram.

* It is customary to speak of entrapped air when no air entraining agent is present and to speak of entrained air as the excess present because of the agent.

† The specific volume of pure water at room temperature is 1.002 cm^3/g or 0.0160 ft^3/lb. Its density is 0.998 g/cm^3; 62.28 lb/ft^3.

The expression for voids ratio can also be written as follows.

$$u = \frac{v}{cv_c^o + av_a + bv_b} \tag{3.4}$$

or, in terms of ratios,

$$u = \frac{v}{cv_c^o} \cdot \frac{1}{1 + M} \tag{3.5}$$

where

$$M = \frac{av_a + bv_b}{cv_c^o} \tag{3.6}$$

and v is the volume of voids, air plus water.

5.2.2 The water ratio. We shall later be concerned with the water content of mixtures as well as with the total voids. The water ratio u_w is defined by the following expression:

$$u_w = \frac{wv_w}{cv_c^o + av_a + bv_b} \tag{3.7}$$

or

$$u_w = \frac{wv_w}{cv_c^o} \cdot \frac{1}{1 + M} \tag{3.8}$$

where w stands for the weight of water in the sample and v_w is the specific volume of *pure* water at the existing temperature. (See Chapter 2, Section 2.2.)

5.2.3 Example of a voids-ratio diagram. Figure 3.11 is a voids-ratio diagram for a series of mixtures of cement and standard Ottawa sand, each mixtures having its basic water content. Each plotted point represents the minimum point on a curve such as one of those shown in Fig. 3.4. The horizontal scale gives the solid-volume fraction of aggregate in the total solids, represented by x; thus

$$x = \frac{av_a + bv_b}{av_a + bv_b + cv_c^o} \tag{3.9}$$

In this case b is zero, the aggregate being sand only.

From the definition of x, it is apparent that it is related to the mix-ratio M as follows:

$$x = \frac{M}{1 + M} \tag{3.10}$$

and

$$M = \frac{x}{1 - x} \tag{3.11}$$

At $x = 1.0$, point C represents the voids ratio of the aggregate in the compact state (dry-rodded). At $x = 0$, point A gives the voids ratio of neat cement, which is commonly called the void-cement ratio based on the solid volume of cement. Thus, at $x = 0$,

$$A = \left(\frac{v}{cv_c^o}\right)_{x=0} = \left(\frac{\epsilon}{1 - \epsilon}\right)_{x=0} \tag{3.12}$$

The void-cement ratio of the matrix in any given mixture is given by the position at point A_x obtained by extending a straight line from D',

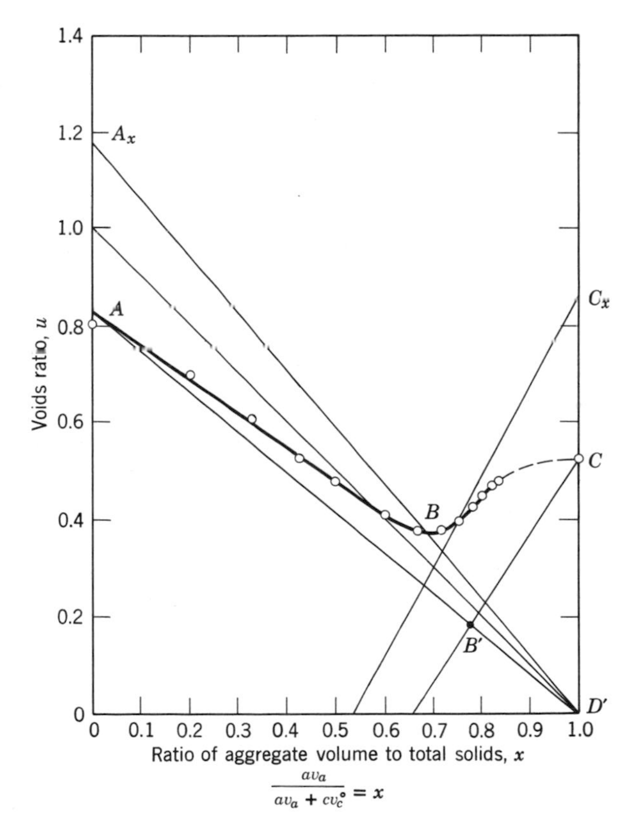

Fig. 3.11. A voids-ratio diagram for mixtures of cement and standard sand, the mixtures having basic water content.

through the point representing the mixture, to the ordinate at $x = 0$. Or it may be obtained analytically by the following relationship:

$$\frac{v}{cv_c^o} = \frac{u_x}{1 - x} = A_x \tag{3.13}$$

where u_x is the voids ratio of the mixture represented at x.

5.2.4 Calculation of mixture composition. All analytical relationships given below, as well as various others, involve nothing more than algebraic manipulations of the following:

$$V = cv_c^o + av_a + bv_b + v \tag{3.14}$$

$$v = wv_w + \mathrm{A} \tag{3.15}$$

where A is the volume of air, and where w and v_w are weight and specific volume of water.

Because the voids content comprises water and air volumes,

$$\epsilon = \frac{wv_w}{V} + \frac{\mathrm{A}}{V} \tag{3.16}$$

Letting Z stand for the ratio of volume of air to volume of water, we have

$$Z = \frac{\mathrm{A}}{wv_w} \tag{3.17}$$

and thus

$$\epsilon = \frac{wv_w}{V}(1 + Z) \tag{3.18}$$

The third expression in each of the following three equations is in terms of the coordinates of the voids-ratio diagram.

Cement content:

$$\frac{cv_c^o}{V} = \frac{1}{1 + M + (v/cv_c^o)} = \frac{1 - x}{1 + u} \tag{3.19}$$

Voids content:

$$\epsilon = \frac{(v/cv_c^o)}{1 + M + (v/cv_c^o)} = \frac{u}{1 + u} \tag{3.20}$$

Water content:

$$\frac{wv_w}{V} = \frac{(wv_w/cv_c^o)}{1 + M + (v/cv_c^o)} = \frac{u_w}{1 + u} \tag{3.21}$$

or

$$\frac{wv_w}{V} = \frac{u_w}{1 + u(1 + Z)} \tag{3.22}$$

where u_w is the water ratio.*

Aggregate content:

$$\frac{av_a + bv_b}{V} = \frac{M}{1 + M + (v/cv_c^o)} = \frac{x}{1 + u} \tag{3.23}$$

The convenience of carrying out calculations in terms of the coordinates of the voids-ratio diagram is indicated by the following example in which we are given the coordinates $x = 0.88$ and $u = 0.18$, these values being typical of an ordinary concrete mixture. The volume composition of the mixture is obtained as follows:

Cement content: $$\frac{cv_c^o}{V} = \frac{0.12}{1.18} = 0.1016$$

Voids content: $$\epsilon = \frac{0.18}{1.18} = 0.1526$$

Aggregate content: $$\frac{av_a + bv_b}{V} = \frac{0.88}{1.18} = 0.7457$$

Quantities can, of course, be converted from volume units to other units by applying suitable conversion factors. For example, to convert from cement content, solid volume per unit concrete volume, to pounds per cubic yard, multiply by 5440; to convert to bags (94 lb) per cubic yard, multiply by 58.3. (Assumption: $v_c^o = 0.31$ cm³/g, or $\rho_c^o = 3.225$ g/cm³.)

5.2.5 Other analytical relationships. The ratio of voids to cement in terms of the coordinates of a mixture is given by the following:

$$\frac{v}{cv_c^o} = \frac{u}{1 - x} = u(1 + M) \tag{3.24}$$

or, in weight units

$$\frac{v}{c} = \frac{uv_c^o}{1 - x} \tag{3.25}$$

Similarly, the water-cement ratio can be calculated:

$$\frac{w}{c} = \frac{v_c^o}{v_w} \cdot \frac{u_x}{1 - x} = \frac{v_c^o}{v_w} u_w(1 + M) \tag{3.26}$$

* The term *water ratio* should not be confused with the water-cement ratio.

or

$$\frac{w}{c} = \frac{v_c^o}{v_w} \cdot \frac{u}{1 - x} \cdot \frac{1}{1 + Z} \tag{3.27}$$

The relationship between the water ratio and the voids ratio is

$$u = u_w(1 + Z) \tag{3.28}$$

The term *cement paste*, or just *paste*, is defined as a mixture composed of cement and water only. Adding Eqs. 3.20 and 3.21 we obtain

$$\frac{p}{V} = \frac{1 - x + u_w}{1 + u} \tag{3.29}$$

where p stands for the volume of cement paste.

The coordinates at point B' are given in Chapter 1 but are repeated here for convenience:

$$u_{B'} = \frac{AC}{1 + A + C} \tag{3.30}$$

$$x_{B'} = \frac{1 + A}{1 + A + C} \tag{3.31}$$

For aggregate mixtures without water, A represented the voids ratio of the fine aggregate, whereas here it may be regarded as the voids ratio of the cement paste in the mixture the water content of which is that giving a minimum voids content. For mixtures without water, C represented the voids ratio of the coarse aggregate, whereas here it may be regarded as the voids ratio of the combined aggregate; or (see Section 5.8.1) it may be considered the voids ratio of the coarse aggregate if the fine component is an invariant mixture of cement and sand and water, that is, mortar.

5.2.6 Apparent voids ratio of aggregate in concrete. As with dry aggregates (Chapter 1) mixtures having their basic water contents and represented in the voids-ratio diagram by points between B and C are those in which the aggregate structure of the coarse component is the dominant factor establishing the characteristics of the mixture. In this range mixtures can be thought of as having been produced by adding small increments of cement to combined aggregate (beginning at point C), each increment of cement being accompanied by enough water to result in the lowest content of voids possible under given test conditions. As we see in Fig. 3.11, the effect of such additions is to disperse the aggregate particles and thus to enlarge, or dilate, the voids in the aggregate, the voids in this case being the space occupied by cement, water, and air.

The degree to which the voids in a given aggregate have apparently become dilated in a given mixture is indicated graphically in the voids-ratio diagram by the position of point C_x with respect to point C. In Fig. 3.11 the value of C_x given shows the apparent voids content of the mixture represented at $x = 0.75$. As in Chapter 1, C_x may readily be evaluated from the coordinates as follows:

$$C_x = \frac{1 + u}{x} - 1 \tag{3.32}$$

Further discussion of aggregate dispersion may be found in Section 5.6.

5.3 *Determinants of the Voids Content in Concrete Mixtures Having Basic Water Contents*

5.3.1 Effect of adding aggregate to paste. The main effect of adding increments of aggregate to paste is to produce a mixture having a lower percentage of voids (water plus air) than that of the neat paste. However, we have already seen that if the same compacting force and energy are applied to the mixture as to the paste, it is necessary to add an increment of water with each increment of aggregate. Because water is added the voids content is not diminished as much as it otherwise would be. Thus each increment of aggregate seems to be accompanied by an increment of water, and it is this voids increment with which we are now concerned.

The effect of adding solid particles to paste is qualitatively like that of adding coarse particles to fine aggregate, discussed in Chapter 1. In mixtures without water the coarse particles apparently distort and dilate normal fine aggregate structure, and thus each coarse particle seems to be accompanied by an increment of voids. In plastic mixtures, in which cement is the fine component, something of the same structure-effect seems to be present, but at the same time the phenomenon seems to be rheological. With its basic water content paste is a plastic body having characteristic stiffness, and it requires a certain force for unit deformation, the required force being a function of the water content of the paste as well as other factors. The consistency is such as to permit molding by a low-energy procedure. Molding involves producing certain degrees of plastic deformation (shear strain) in the mass being molded. When aggregate is introduced, displacing some plastic paste, strains in the paste produced during molding under the same conditions are necessarily greater than they were in the absence of the aggregate. Because the force required is about proportional to the plastic strain produced, the consistency seems stiffer than that of paste alone and, for successful molding, larger forces must be applied. The alternative is to soften the paste by adding water, which is done under practical conditions. (For further discussion see Chapter 10.)

As we shall see, the effect described above is somewhat more complex than it is indicated to be. In very rich mixtures, where the degree of aggregate dispersion is high, the effect seems to depend on the particles individually rather than on properties of the particle aggregation, that is, the aggregate. Theoretically, there is an effect due to volume displacement entirely, and it is thus independent of particle size. However the observed effect of adding aggregate to plastic paste is, as we see farther on, strongly influenced by particle size. Indeed, the effect is indistinguishable, qualitatively at least, from the size-ratio effect in mixtures of dry aggregates we discuss in Chapter 1.

5.3.2 Parameter *D*. The amounts by which the water volume must be increased in very rich mixtures to offset the stiffening effect of added solid particles is shown clearly by a water-ratio diagram, an example of which is given in Fig. 3.12. This diagram represents the same materials presented in Fig. 3.11 (the voids-ratio diagram is reproduced for comparison and for use in later considerations). The first six of the 12 points representing water content fall near a straight line drawn between point A_w and point D. The divergence of this line from the line drawn from A_w to D' is due to the increments of water added along with each increment of aggregate. The divergence per unit increase of x is numerically equal to the distance from point D' to point D. The equation for the line from A_w to D is as follows:

$$u_w = A_w(1 - x) + Dx \tag{3.33}$$

Since, by Eqs. 3.8 and 3.11,

$$\frac{wv_w}{cv_c^o} = u_w(1 + M) = \frac{u_w}{1 - x} \tag{3.34}$$

it follows that the water-cement ratio for any point on the line is given by

$$\frac{wv_w}{cv_c^o} = A_w + D\left(\frac{x}{1 - x}\right) = A_w + DM \tag{3.35}$$

For the range of x over which Eqs. 3.34 and 3.35 apply ($0 < x < x_B$) the ratio of air volume to water volume Z is constant; hence,

$$\frac{wv_w}{cv_c^o}(1 + Z) = \frac{v}{cv_c^o} \tag{3.36}$$

$$A_w(1 + Z) = A \tag{3.37}$$

$$D(1 + Z) = D^* \tag{3.38}$$

and

$$\frac{v}{cv_c^o} = A + D^*M \tag{3.39}$$

D is a factor representing an intrinsic characteristic of a given aggregate that is independent of grading and void content. It probably is the resultant of several factors, among which the mean particle size is the most important; it seems to be proportional to the mean specific-surface

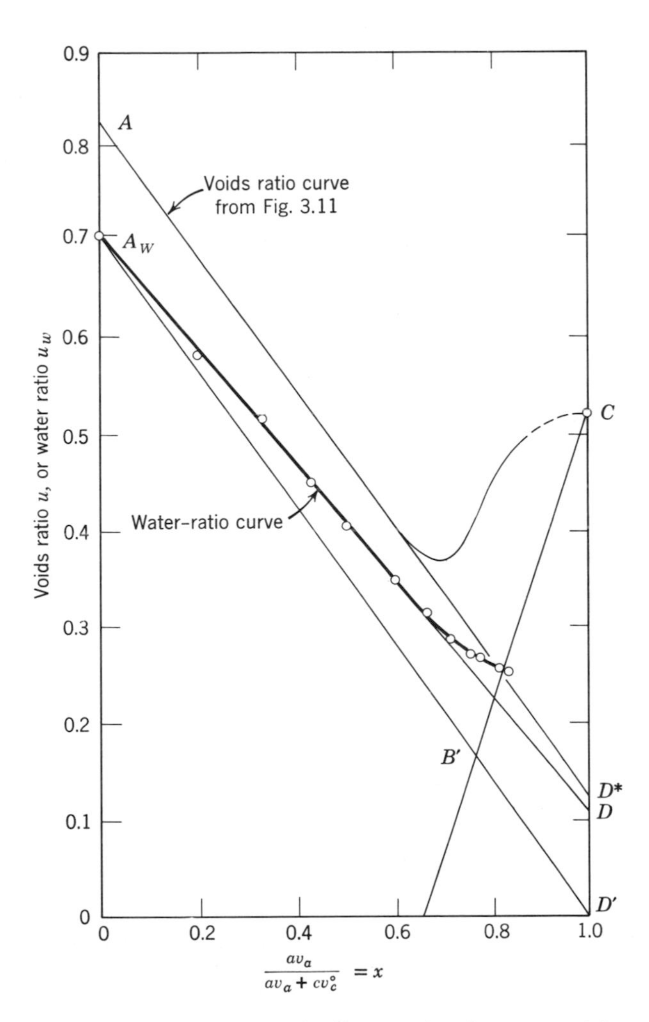

Fig. 3.12. A water-ratio and voids-ratio diagram for the same mixtures of cement and standard sand represented in Fig. 3.11.

diameter. As shown in Chapter 4, D is a function of consistency, increasing with the amount of water in excess of the basic water content. D^* depends on the factors that influence the value of Z as well as on the factors that determine D. In some cases it is convenient to evaluate D^* rather than D.

Data showing the dependence of D^* on mean particle size are given in Tables 3.1 and 3.2. Aggregates 32, 24, and 36 are of special interest because they are closely related to each other. Aggregate 24 is a natural sand comprising six size groups. Aggregate 36 was obtained from the same material by omitting the smallest three of the six groups. Aggregate 32 was obtained by omitting the largest three groups. Because sands 32 and 36 comprise the same number of consecutive groups and are similarly graded, they are almost geometrically alike, as is indicated by the similarity of their voids contents (line 7, Table 3.2), but not quite alike, as is indicated by the small difference in the voids content. Although similar in this respect, the values of D^* for this pair of aggregates are widely different, as we see in line 9. This comparison demonstrates that D^* does not depend on grading.

Aggregates 36 and 308-33 have different voids contents, 0.378 and 0.212, respectively; yet they have similar values for D^*. This comparison demonstrates that D^* is not a function of the voids content.

Table 3.1 Sieve Analyses of Aggregates

Sieve Size	Per Cent Passing Sieve Indicated at Left				
	Agg. 32[a]	Agg. 24[a]	Agg. 36[a]	Agg. 308-33[b]	Agg. 265-11[c]
$1\frac{1}{2}$ in.				100	
$\frac{3}{4}$ in.				66	
$\frac{3}{8}$ in.		100		50	
#4		97	100	33	100
#8		78	88	27	77
#16	100	51	63	22	61
#30	78	25	3	14	31
#50	43	10	0	4.3	7.3
#100	7	4	0	1.6	2.4

[a] Data are from A. N. Talbot and F. E. Richart, "The Strength of Concrete: Its Relation to the Cement, Aggregates, and Water," Bulletin 137, Engineering Experiment Station, University of Illinois, Urbana, Illinois, 1923.

[b] PCA Series 308.

[c] PCA Series 265.

Table 3.2 Characteristics of Aggregates

	Reference Number of Aggregate				
	32[a]	24[a]	36[a]	308-33[b]	265-11[c]
1. Number of size groups	3	6	3	9	6
2. Surface modulus SM	36.4	16.6	10.2	7.5	15.2
3. Angularity Factor $1/\psi$ (estimated)	1.1	1.1	1.1	1.1	1.0
4. Surface area factor $(SM)/\psi$	40	18	11	8.2	15
5. Approximate specific surface, cm^2/cm^3	220	100	61	46	84
6. Fineness modulus FM	1.73	3.33	3.46	5.82	3.21
7. Porosity of aggregate, dry-rodded	0.349	0.332	0.378	0.212	0.296
8. Voids ratio $(\epsilon/(1-\epsilon)_{ag} = C$	0.536	0.497	0.608	0.282	0.42
9. Parameter D^*	0.315	0.11	0.04	0.05	0.07

[a] A. N. Talbot and F. E. Richart, "The Strength of Concrete: Its Relation to the Cement, Aggregate, and Water," Bulletin 137, Engineering Experiment Station, University of Illinois, Urbana, Illinois, 1923.
[b] PCA Series 308.
[c] PCA Series 265.

Perusal of line 6 in Table 3.2 makes it clear that D^* does not correlate with the fineness modulus, although, as we may see in Chapter 5, D or D^* is a function of the water-requirement factor of the aggregate and fineness modulus is supposed to be related to the water requirement. There is a correlation but not among mixtures belonging to the category now under discussion.

There is a fairly close correlation between D^* and the specific surface area of the aggregate. This is shown in Fig. 3.13 for 15 aggregates, including the five represented in Table 3.2. The specific surface area is indicated by the surface modulus SM, derived from the sieve analyses, and the angularity factor $1/\psi$, estimated from descriptions of the aggregates.

The plotted points suggest a smooth curve starting at the origin, but there is reason to believe that the curve is sigmoid as shown. Although some data represent aggregates having negligible specific surface area, no values of D^* smaller than 0.04 have been observed, and, as we mentioned above, there is reason to believe that D^* is a function of the solid volume as well as surface area. The scatter of points is such as to suggest that other variable factors may be involved; however, we cannot draw a conclusion because of probable variations in the relationship between surface modulus and actual specific surface area among different aggregates and

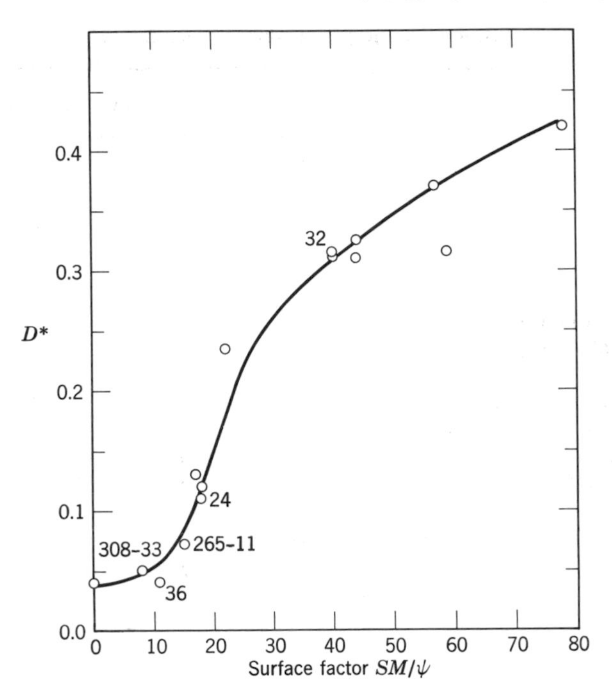

Fig. 3.13. The empirical relationship between parameter D^* and the surface factor of the aggregate.

lack of adequate data on particle shape and the consequent uncertainty in values given ψ.

5.3.3 Parameters *A* and *C*. It is useful to think of parameters A or A_w and C as anchor points, which, as a first approximation, define the merits of a given pair of mixture components (cement and aggregate), merit being here considered in terms of the voids contents in the mixtures. In general the higher the values of A and C, the higher the voids content at any given proportion of the two components.

There may be some exceptions to the conclusions stated above, however, with respect to the value of A_w. At the basic water content, A_w corresponds to standard normal consistency (ASTM Designation C187) or approximately so. When a cement is used that has a relatively high normal consistency value because of its relatively high specific surface area with a given aggregate, any mixture to the left of point B will show a relatively high voids content (water plus air) at a given consistency, but point B is likely to occur at the same or a lower level content, and mixtures to the right of point B will have voids contents no higher, or perhaps somewhat lower than those of mixtures made with cement giving a lower figure for normal consistency. (All this carries a proviso that the aggregate is free from silt and clay.)

The value of A_w for a given cement can be reduced slightly by using one of a certain class of surfactants that reduces interparticle attraction without entraining air. The effect is to produce a softer consistency, but the effect is not as great as might be expected because at basic water content reduction of interparticle attraction is accompanied by a marked increase of dilatancy. (See Chapter 10.)

With the remarks above we encounter for the first time the necessity of thinking of different categories of concrete mixtures, the rules applicable to one being different from those applicable to another.

Parameter C corresponds to the voids content in the aggregate in its dry-rodded state. It has its lowest value for given materials when the size range and grading is such as to give the lowest possible voids content. As shown in Chapter 1, the smallest size is not intended to be less than 74 μ (200 mesh), but the maximum size is usually made the largest that is economically available or usable. For any given size range, and with given gradings of the fine and coarse aggregates, there is a certain ratio of fine to coarse that gives a minimum value of C. However, in a concrete mixture, the *optimum* proportion of fine aggregate is usually smaller than that giving the smallest value of C. (See Chapter 5.) In air-entrained concrete some of the fine aggregate can be replaced by air bubbles, thus further widening the difference between the optimum percentage and that giving a minimum voids content in the aggregate alone.

Although the optimum percentage may differ from that giving minimum value of C, the value of C corresponding to optimum percentage of fine aggregate is not much greater than the lowest. In other words, the best proportions of fine aggregate usually fall in a range where voids content is relatively insensitive to proportions. Thus the principal factor that establishes the value of C is the size range—the number of size groups if the grading is continuous.

5.4 *Minimum-voids Ratio—Point B*

The minimum-voids ratio, u_B, found in a series of mixtures of given materials, generally occurs within the triangle $AB'C$, and point B is always found at a higher level of u than is B'. In dry mixtures, in which the fine component is relatively coarse and its aggregate structure relatively rigid, point B cannot be caused to coincide with point B' without applying compacting forces of such magnitude as to break down the aggregate structure and even crush some of the particles. However, if the fine component were identified as the plastic paste, there would be no voids other than air bubbles; point B would differ from B' only by the amount of air entrapped in the mixture, and that amount could be made very small.

Identifying plastic paste with the fine component would be a reasonable basis for analysis if paste were a singular entity and if our purpose were to study the factors that influence the amount of that paste required per unit volume of aggregate. However, because paste composition is a variable factor of prime importance, practical considerations focus our attention on the amounts of water and air, principally water, per unit quantity of cement. We continue to consider the fine component of the mixture to be cement. Although interstitial spaces among the aggregate and cement particles are mostly full of water, we continue to identify it as void space.

Even though we choose to consider cement alone as a fine component, we should remain aware that cement, the fine component, is rendered plastic by water. Dry cement causes some dispersion of aggregate particles (Chapter 2), but in the plasticized state, that is to say, when the cement is in cement paste, it effects an amount of dispersion that depends primarily on the consistency of the paste. The stiffer the consistency, the greater the dispersion (separation) of aggregate particles. We see later that the consistency of the paste is the dominant factor only in mixtures that are relatively rich in cement. In the leaner mixtures, particularly those to the right of point B, aggregate structure is important. This range comprises much of the concrete made for ordinary use.

The voids content at point B being a natural reference point, we now investigate how the coordinates of that point are related to the parameters of the voids-ratio diagram and to the specific surface area of the aggregate, the latter being given in terms of surface factor SM/ψ.

D^* being related to the surface factor of the aggregate and point B being always found near the line from A to D^*, we might expect point B to be a function of surface factor. Figure 3.14 does indicate such a relationship, but at the same time it shows that other factors are involved; four of the five points belong to the same grading family, but the fifth, grading 36, stands alone.

The basic requirement for conformity to an empirical relationship in terms of surface factor only is either that the voids content of the aggregate has no effect or that the voids content and specific surface area of the aggregate are not independent variables among the series of mixtures under consideration. The plotted data illustrate the falsity of either assumption for the present case. Notice that for aggregates 36 and 24 the values of u_B are nearly the same; also the surface factors for aggregates 36 and 308-33 are similar, but the values of u_B differ greatly. Al-

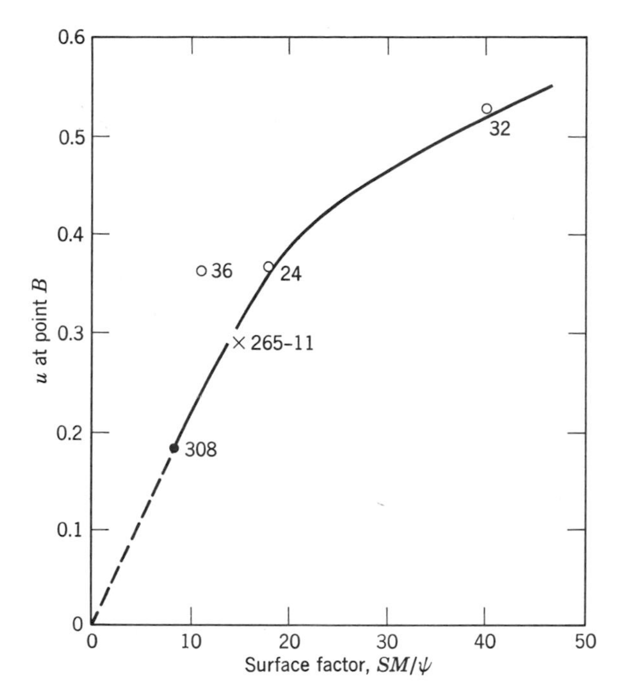

Fig. 3.14. An empirical relationship between the minimum-voids ratio and the surface factor of the aggregate.

though the surface area of 36 was like that of 308-33, the high voids content in 36 (high value of C) pulled the curve away from the AD^* line, so to speak, at a relatively high level of voids content. On the other hand, the other four aggregates apparently share about the same relationship between surface factor and voids content.

A more general relationship, one that takes into account the content of voids as well as specific surface area, can be set up in terms of u_B and $u_{B'}$. The coordinates of point B' can be calculated from Eqs. 3.30 and 3.31, the first one being applicable here. As already mentioned, u_B is always at a higher level than $u_{B'}$, the ratio of the first to the second being usually upwards of 1.5. A linear function to which the data from Fig. 3.14 conform reasonably well is shown in Fig. 3.15. Points representing other data are shown also. The scatter of some of these points may be no greater than uncertainty regarding the values of surface and angularity factors. The point for aggregate 36, which was a standout in the previous plot, conforms to this plot as well as most of the other points.

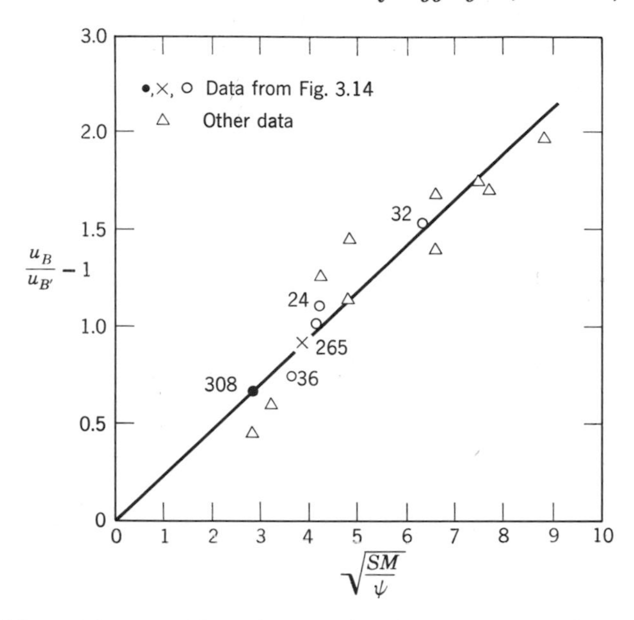

Fig. 3.15. The minimum-voids ratio as a function of $u_{B'}$ and the surface factor of the aggregate.

The relationship illustrated in Fig. 3.15 is given by the following equation:

$$u_B = u_{B'}[1 + 0.238(SM/\psi)^{1/2}]$$

Or, in terms of specific surface area,

$$u_B = u_{B'}(1 + 0.10\sqrt{\sigma_{ab}})$$

$$u_{B'} = \frac{AC}{1 + A + C} \tag{3.40}$$

The value of the numerical constant is questionable because of uncertainties pertaining to the surface area factors for the materials represented in Fig. 3.15. However, it is probably at least of the right order of magnitude.

Equation 3.40 indicates that the minimum voids ratio that we can obtain from a given set of materials by varying the cement content and using the basic water content for each of the mixtures can be expressed as a function of (1) the specific surface area of the aggregate, (2) the voids ratio of the compacted aggregate, and (3) the voids cement ratio of the cement at normal consistency.

In Chapter 5 there is a procedure for estimating which mix of a series will have the lowest water content for a given consistency. It is probably more useful than the relationship given above, but Eq. 3.40 does serve to show that voids content as well as the grading of the aggregate are determinants of the water requirement at point B.

5.5 Bulking of Mixtures at Basic Water Content

The data reported by Talbot and Richart [7] and by Richart and Bauer [6], used for the voids-ratio diagrams of this chapter, are in some important respects the best available. However they were not ideal, mainly because the specimens were compacted by a hand-tamping method that was not standardized with respect to the amount of energy expended. Therefore one cannot say that all the mixtures compared were equal as to consistency: it can only be said that the points plotted represent the

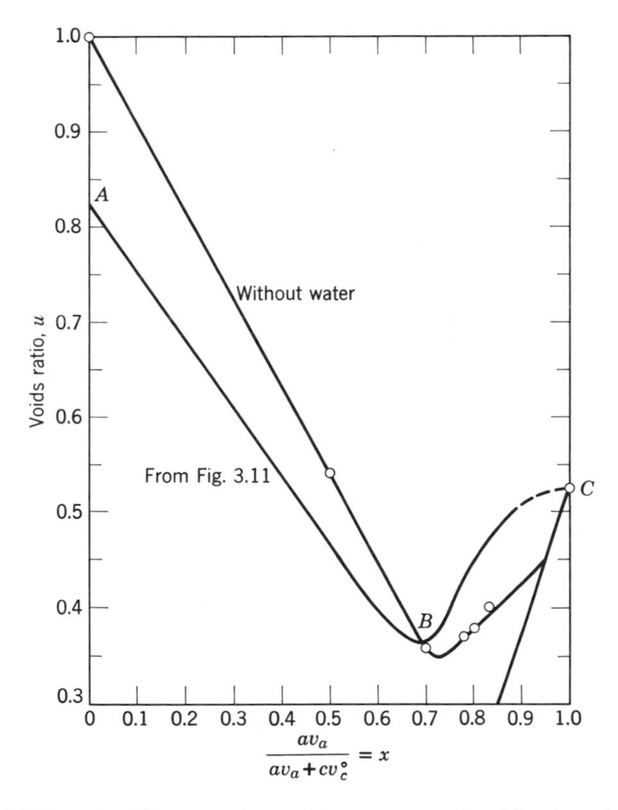

Fig. 3.16. Voids-ratio diagrams for mixtures prepared without water, and with basic water content.

mixtures having the water content giving the least voids content for this method of compacting. The mixtures well to the right of point *B* could have been only quasiplastic, and the diagrams show a bulking effect which was not overcome by the relatively ineffective tamping. Figure 3.16 illustrates the phenomenon: the curve from Fig. 3.11 is reproduced along with the diagram for the same materials without water. The latter is drawn according to considerations brought out in the discussion in Chapter 2 of Fig. 2.6. Between points *B* and *C* bulking is revealed by an upward arching relative to the diagram for dry materials.

Just as with moist sand, the bulking effect at basic water content is larger the smaller the average particle size, or the larger the specific surface area, of the aggregate. The Talbot and Richart data show that with some unusually fine sands the bulking effect may be so great that between *B* and *C* the curve rises above the level of point *C*.

The bulking of mixtures in the range from *B* to *C* is not so much an intrinsic quality of such mixtures as it is the consequence of following certain routine practices and of expending relatively little energy for compaction. Bulking can be eliminated by an appropriate expenditure of mechanical work or by increasing the water content.

An example of the reduction of bulking by increasing the water content enough to maintain constant consistency is found in Fig. 3.17, which gives partial voids-ratio diagrams for mixtures made with the materials represented in Tables 3.1 and 3.2.* For aggregates 32, 24, and 36, the amounts of water in mixtures to the left of point *B* were such as to produce about the same consistency, but to the right of point *B*, the water contents were significantly smaller than the amounts required for the same consistency as that of those on the left side. In these cases we see an arching of the curve denoting bulking due to entrapped air. For the other two aggregates, the water requirement was regulated by consistency as measured by the remolding test. (See Chapter 10.) In these cases there is no evidence of bulking of the mixtures represented between *B* and *C*.

5.6 Dispersion of Aggregate

In Chapter 1 we brought out the fact that if the fine and coarse aggregates are assumed to retain their identities when they are mixed together, the coarse aggregate appears to be dispersed to one degree or another in the fine. It was emphasized that although that way of looking at aggregate is useful, it is nonetheless artificial. Now, when considering mixtures of cement paste and aggregate, there is nothing artificial about assuming that the aggregate retains its identity. If the mixture is plastic, the aggre-

* In this connection it must be kept in mind that each point in the upper three curves of Fig. 3.17 represents the low point on a curve such as those in Fig. 3.4.

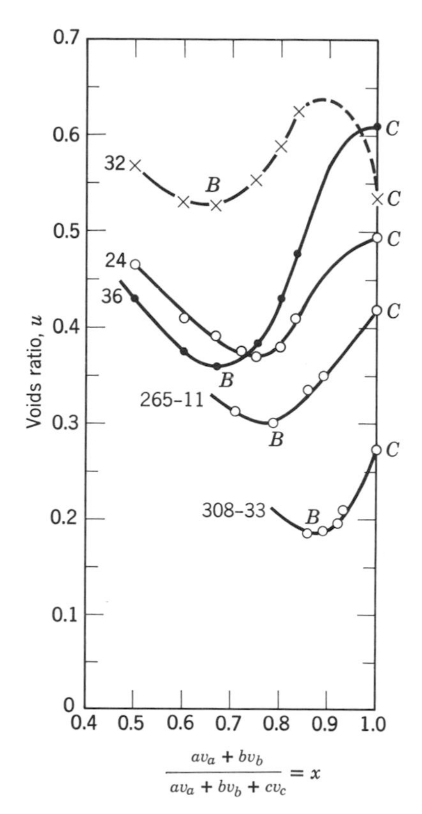

Fig. 3.17. Partial voids-ratio diagrams for mixtures that show bulking and for those that do not.

gate is slightly dispersed by the paste. As we may see in Chapter 5, some authors have proposed methods of formulating concrete mixtures based on the degree of dispersion. In this chapter we are taking a preliminary look at the degree of dispersion.

5.6.1 Computation from "excess" paste. We can imagine a concrete mixture to be composed of aggregate and cement paste, the volume of the cement paste being sufficient to fill the interstitial space in the compacted aggregate, plus an increment that causes a certain dispersion of the aggregate particles; that is to say, there is an "excess" of paste that produces a certain distance of separation between points that would be in contact in the dry, compacted state. The imagined situation is illustrated in Fig. 3.18; drawing *A* represents the dry compacted state, and *B* the state when the aggregate is dispersed in paste. Each particle seems to have acquired a thin coating of paste that maintains the dispersed

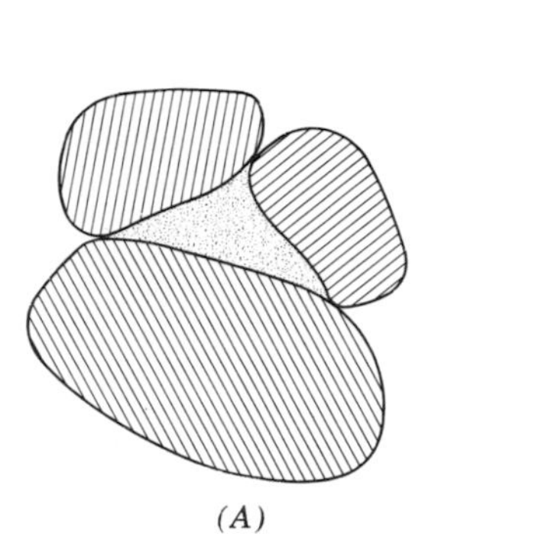

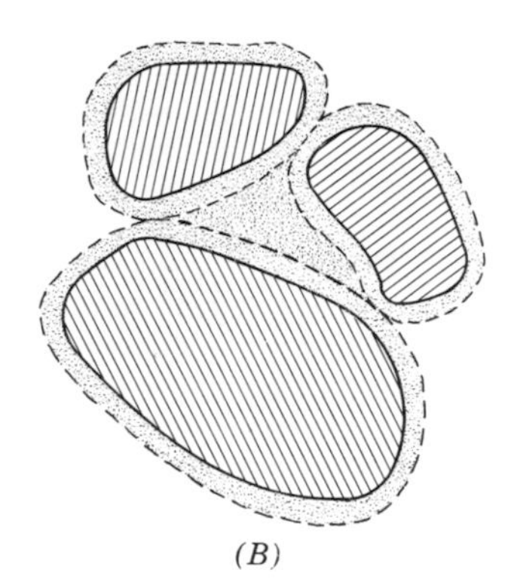

Fig. 3.18.

state, but, if the mixture is plastic, the interstitial space must also be filled. Actually, in plastic mixtures, it is not paste alone that produces dispersion of aggregate particles and plasticity; the volume of the paste is always augmented by a certain amount of air, the amount depending on factors discussed in Chapter 7.

If the distance of dispersion is small relative to the diameter of the particle, the average distance can be estimated by dividing the volume of excess paste by the total area of aggregate surface with which it makes contact; the quotient gives the thickness of the imagined layer, and thus it is approximately one-half the distance of separation, the accuracy being greater the thinner the layer relative to the size of the particle. On this basis, and considering that the matrix of the aggregate is composed of paste and air, we can write the following equation.

$$\frac{t}{2} = \frac{(v_{ag}/V) - (v_0/V)}{[(av_a/V) + (bv_b/V)]\,\sigma_{ab}} \tag{3.41}$$

The distance of dispersion is t. On the right-hand side, the volume of excess paste is divided by the total surface area of the aggregate in concrete volume V; σ_{ab} is the average specific surface area of the combined fine and coarse aggregates, area per unit solid volume; v_0 is what the volume of interstitial space in the aggregate would be if the aggregate were in the compacted state; v_{ag} is the actual interstitial volume and is thus the volume of interstitial material in the mixture.

By dividing the quantities above and below the line in Eq. 3.41 by the volume of aggregate, we obtain voids ratios above the line, and the specific surface area is left below; then, remembering the definitions of C and C_x, we are able to rewrite Eq. 3.41 as follows:

$$t = \frac{2(C_x - C)}{\sigma_{ab}} \tag{3.42}$$

The value of C_x is conveniently computed from Eq. 3.32; the value of the voids ratio u and that of x are known from the composition of the mixture; the value of C and σ_{ab} are determinable characteristics of the aggregate, although the accuracy of the latter is usually questionable.

An example of data from four concrete mixtures and the corresponding computed values of dispersion t are given in Table 3.3. The computed dispersion ranges from 8.6 μ for the leanest mixture to 78 μ for the richest, the latter mixture being close to point B on the voids-ratio diagram. By way of comparison, we may note that about three-fourths of the weight of a typical cement is made up of particles larger than 15 μ, and the largest distance, 78 μ, is about the same as the opening in a No. 200 sieve, which is able to pass most of a modern cement.

Looking at any voids-ratio diagram, we see that C_x increases very rapidly as x becomes smaller than x_B; indeed the method of calculation

Table 3.3 Computed Dispersion of Aggregate Particles on the Basis of Eq. 3.42[a]

1	2[b]	3[c]	4[d]	5[e]	6	7[f]	8[g]
Reference Number	Voids Ratio (u)	x	$\frac{1+u}{x} - 1 = C_x$	C	$C_x - C$	σ_{ab} (cm^2/cm^3)	t (μ)
101	0.208	0.833	0.450	0.266	0.184	47	78
106	0.188	0.877	0.355	0.264	0.091	50	36
111	0.188	0.897	0.324	0.260	0.064	52	12
116	0.199	0.918	0.306	0.258	0.048	56	8.6

[a] Data from PCA Series 303. The consistency, slump = 3 to 4 in.
[b] Ratio of the volume of air and water to total solids.
[c] Volume fraction of aggregate in total solids.
[d] Apparent voids ratio of the aggregate in the mix at x.
[e] Voids ratio of the aggregate in the dry-rodded state.
[f] Specific surface area of the aggregate, computed from $\sigma_a = 5.58(S/\psi)$; the angularity factor $1/\psi$ was estimated at 1.10.
[g] Computed clear distance between aggregate particles, by Eq. 3.42.

becomes inaccurate as the value of t becomes of the same order as the smaller sizes of sand, upwards of 100 μ.

From the nature of the computation it is apparent that the distance t is a minimum distance between particles; the maximum distance may be considerably larger.

5.6.2 Dispersion by Weymouth's formula. In Chapter 1, in connection with the theory of particle interference, we give Weymouth's method of computing the clearance between the particles of coarse aggregate in a combined aggregate. The term *clearance*, apt with respect to considerations of particle interference as defined by Weymouth, is actually the same as that which we have here and elsewhere called "dispersion." In the terminology for voids-ratio diagrams Weymouth's equation is

$$t = \left[\left(\frac{1 + C_x}{1 + C}\right)^{1/3} - 1\right] d \tag{3.43}$$

or

$$t = \left[\left(\frac{1 + u}{x(1 + C)}\right)^{1/3} - 1\right] d \tag{3.44}$$

where d is the average size of the aggregate as computed from Eq. 1.2.

Calculations by the Weymouth equation for the same data of Table 3.3 are given in Table 3.4. The values of t range from 43 to 164 μ, compared with 8.6 to 78 μ for Eq. 3.42. The degree of agreement cannot be considered good, and it is difficult to say which is likely to be the more

Table 3.4 Computed Dispersion of Aggregate Particles on the Basis of Eq. 3.43

1	2[a]	3[b]	4[c]	5[c]	6	7[d]	8[e]
Reference Number	$\frac{100a}{a+b}$	d (mm)	$1 + C_x$	$1 + C$	$\frac{1 + C_x}{1 + C}$	$\frac{t}{d}$	t (μ)
101	36	3.49	1.450	1.266	1.145	0.047	164
106	38	3.44	1.355	1.264	1.072	0.024	83
111	40	3.37	1.324	1.260	1.051	0.017	57
116	43	3.29	1.306	1.258	1.038	0.013	43

[a] Percentage of sand in the aggregate.
[b] Average nominal diameter of the aggregate particles.
[c] Data from Table 3.3.
[d] Ratio of computed dispersion to average diameter.
[e] Computed dispersion (clearance) by Eq. 3.43.

accurate. However, considering the nature of the underlying assumptions, and the uncertainties of estimated values, it would seem that the Weymouth equation is to be preferred.

Examination of the voids-ratio diagram for the data of Tables 3.3 and 3.4 shows that the point of departure from the linear part of the diagram, that is from the line from A to D, occurs at about $u = 0.195$, $x = 0.855$. The calculated clearance at this point is 120 μ. Remembering the discussion in Chapter 2, Section 4, we see that this amount of clearance is more than sufficient to clear the largest size group of the cement. It seems that the point of departure from the line AD is the point where particle interference between the aggregate and the coarsest particles of the cement would occur if the water were absent; the extra dispersion produced by the water is that amount required to *prevent* actual particle interference and give the particles enough freedom to permit a certain amount of plastic deformation by application of relatively small forces. (In this connection see also Chapter 6, Section 3.)

5.7 Mixtures Made Plastic by Vibration

5.7.1 Use of vibrators. In Section 4 we define the range of water contents that give plastic mixtures and discuss the limits of proportions within which normal compactness, as we define it in Section 4, may be produced with low-energy compacting methods. For such methods of placement the mixture must be sufficiently plastic to slump at least 1 in. in the standard test, and in usual practice enough water is used to make it slump considerably more than the minimum required. If the consistency used corresponds to a slump of less than about 2 in., mixtures that fall to the right side of point B tend to show bulkiness due to entrapped air, as we brought out in the preceding section. However, bulkiness can readily be eliminated by proper use of vibrators made for the purpose.

Although concrete vibrators are widely used for compacting mixtures of the same type that are prepared for low-energy methods, they are most effective, and of greatest value, when they are applied to mixtures unsuitable for any other kind of compacting method. In terms of Fig. 3.4 such mixtures are represented by points in the nonplastic range to the left of the arrow. Lack of plasticity may be due to lack of water, or due to a relatively high proportion of coarse aggregate, as for example the mix 1:2.09:6.8 in Fig. 3.9. Mixtures that are both too dry and too "rocky" to be handled by low-energy methods may be used advantageously with vibrators.

There are two general types of vibrator: those that can be thrust into the mass of fresh concrete, the internal type; and those that vibrate the form, mold, or screed, the external type. In ordinary building construction the internal type is used almost exclusively; the other is used mostly in factories for precast building elements.

5.7.2 Voids-ratio diagrams for vibrated mixtures. Figure 3.19 presents three voids-ratio diagrams; two are for mixtures that before

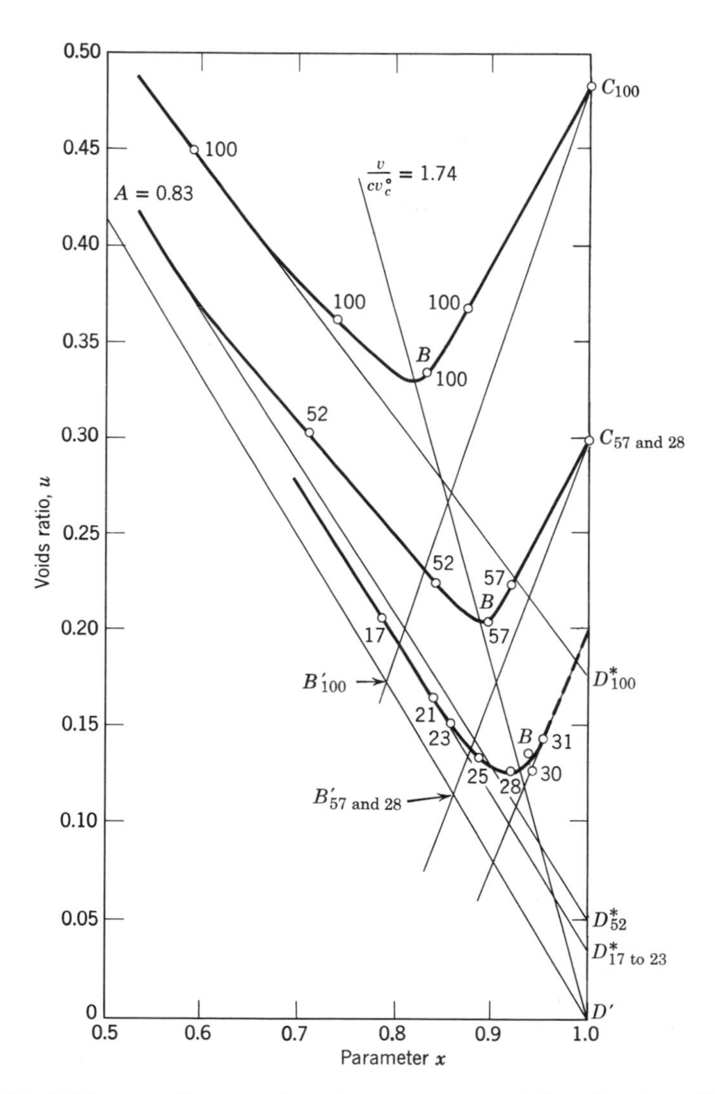

Fig. 3.19. Voids-ratio diagrams for mixtures compacted by vibration. Numerals appearing by the points in the diagrams are percentages of sand in the aggregate. (Data are from PCA Series 245.)

vibration were nonplastic for lack of water, and the third is for a series of mixtures that were nonplastic before vibration because of an excess of coarse material as well as lack of water. The most conspicuous effect of vibration on a dry, crumbly mixture is to transform it into a plastic mass.

The uppermost curve in Fig. 3.19 is an example of the results of vibration applied to a series of very dry mortars all made from the same lot of sand. The water content of each mixture was such as to give a "moist-earth" appearance and yet sufficient for obtaining compactness after about 60 sec of vibration in a $6 \times 8 \times 36$ in. rectangular mold, vibrated externally. The curve is like that for wetter mixtures compacted by rodding, except that B is somewhat closer to B'; the "normal" voids ratio at point B, calculated from the values of A and C_{100} by means of Eq. 3.40, is 0.37, whereas the observed value is 0.33, indicating something of the effectiveness of this relatively high-energy method of compacting.

The effectiveness of the vibrator is shown also by the value of v/cv_c^o at point B, 1.74; for similar material (but with more water) compacted by tamping the corresponding value is 1.44. This comparison shows that under vibration, point B occurs at a relatively lean mix. Using the coordinates of point B ($u_x = 0.33$; $x = 0.81$) and $C_{100} = 0.488$, we find from Eq. 3.49 that the computed value of t/d is 0.034; that is, the apparent clear distance between particles of aggregate is about 3.4 per cent of the mean diameter of the particles; a similar calculation applied to sand No. 24, in Fig. 3.17, $u_x = 0.37$, $x = 0.76$, and $C = 0.498$, gives t/d at point B to be about 6.3 per cent.

In contrast there appears to be a close similarity if not identity with respect to parameter D^*. The value of D_{100} is 0.18; the surface factor of the aggregate SM/ψ is 22, and from Fig. 3.13 (which was based on data not involving vibration) we find about the same value for that surface factor. This result indicates that when the water content is adjusted to permit compaction by a certain amount of vibration, each increment of aggregate added to the paste requires about the same increment of water as each does when the water requirement is adjusted at some higher level for a given slump, for mixtures to the left of point B.

In Section 5.3, we interpret parameter D^* as possibly involving structural distortions of the fine component (the same as the effects of increments of coarse aggregate in dry sand we discuss in Chapter 1) and involving also rheological effects, possibly predominantly. In the case above it seems likely that the effect is wholly rheological because vibration itself temporarily destroys paste structure, making the paste fluid. (See Chapter 10.)

Vibration as it is usually applied is not a simple jigging or mechanized tamping procedure. Applied to a mixture capable of becoming plastic but initially appearing loose and crumbly and without mobility under the force of gravity, it at first develops direct action between aggregate particles, but that stage is quickly replaced by a different mechanical effect. The vibrator sets up a complicated pattern of compression waves in the fluid, which gives the paste temporary fluidity and develops forces that tend to separate one particle from another and to keep them separated.

To demonstrate that vibration produces forces that separate the particles and keep them separated, the following experiment was carried out. A mixture was prepared of mortar and an aggregate of three sizes of steel balls with a maximum size of $1\frac{1}{2}$ in. The mortar appeared dry and crumbly, but contained enough water and cement to give plasticity under vibration. The mixture of mortar and steel aggregate was placed loosely in a 6×12 in. steel cylinder mold clamped to a table-type vibrator. When the vibrator was started, the clicking together of the steel balls could be heard distinctly, but in a few seconds, all such noise ceased, each steel ball becoming completely isolated by matrix material before consolidation was complete. After we crushed the hardened specimen, the interior appeared as shown in Fig. 3.20; the only material in direct contact with steel was the cement paste, and because each cement grain was also separated from its neighbors by water, we concluded that during vibration each steel ball was in direct contact with water only. Even at the bottom surface, the matrix enveloped each sphere, leaving no external evidence that the specimen contained steel balls.*

Thus we conclude that the primary function of vibration in concrete is to produce waves in the fluid and thus create relative motion of fluid with respect to the denser particles suspended in it. In a flocculent structure of concentrated cement particles such motion tends to open up each constriction in interparticle space and thus to increase the minimum distances between particles, making the minimum nearer to the average. Because interparticle attraction diminishes rapidly with the increase in interparticle distance (Chapter 9), the effect of vibration is to weaken interparticle attraction in the paste and thus to allow the whole mixture to become mobile under the force of gravity.

Thus when a mixture becomes consolidated during a given period of vibration, it shows its ability to flow and to expel air at the required rate. Therefore we should expect the aggregate particles to have the rheological effect observed; that is, they stiffen the mix relative to the consistency of the matrix by virtue of their rigidity, volume, and surface area and in very rich mixtures demand a certain addition of water indicated by D, if

* The experiment developed a comical aspect when the unwarned testing-machine operator, after testing several ordinary cylinders, attempted his customary one-handed lift of a cylinder having twice the usual weight.

Fig. 3.20. Concrete made from ordinary sand mortar with steel balls as the coarse aggregate. (Data from PCA Series 231.)

the mixture is to become consolidated at the same rate as the matrix alone. However, we could hardly have anticipated a priori that D or D^* would be about the same for vibration as for hand tamping.

The curve in Fig. 3.19, which terminates at C_{57}, represents a series of concrete mixtures made with the same sand mentioned above, and with gravel having a maximum size of $1\frac{1}{2}$ in. As indicated in the diagram, the sand constituted from 52 to 57 per cent of the total aggregate, the average surface factor of the combined aggregate being 12.6. Although each mixture appeared dry and crumbly before vibration, compaction was readily achieved. Equation 3.49 gives the voids ratio at point B to be 0.246, whereas that achieved by means of vibration was 0.205. Again, the degree of aggregate dispersion at point *B* is relatively small.

5.7.3 Stable mixtures and ultimate compaction. The lowest curve in Fig. 3.19 represents a series of concrete mixtures made with the same sand mentioned above and with gravel having a maximum size of $1\frac{1}{2}$ in., the same kind we used for the middle curve. Each point represents a limiting composition characteristic of the particular materials used and to some degree characteristic of the vibrating equipment. That is, each mixture contained the least amount of water and the greatest amount of coarse aggregate that could become fully compacted. If an increment of coarse aggregate is added to such a mixture, that increment will be rejected by the mixture during vibration, the coarse material remaining loose on top of the plastic mass. Such behavior seems to be a characteristic of stable mixtures as defined by Bergstrom [8].

We can see certain interesting aspects of compacting by vibration, not already discussed, by observing such mixtures during vibration under laboratory conditions. After a mold is loosely filled with a mixture of this kind and vibration of the mold begun, compaction begins in a certain part of the batch where a plastic mass quickly develops. This mass grows by incorporating loose material, the process continuing until all the material has become plastic. However, if the mixture contains an excessive proportion of coarse material, the excess will not become incorporated in the plastic part but will remain loose on top, as mentioned above. However, if downward force is applied to the excess loose material while vibration continues, by pressing it with a flat board, for example, it may become incorporated in the plastic, vibrating mass; yet when the pressure is removed the material will be ejected and again appear loose at the top surface. This observation shows that under given conditions, vibration maintains a certain average degree of separation of the solid particles, marking an equilibrium state between the interparticle pressures developed by the fluid wave motion discussed above and gravitational force that tends to cause the particles to settle. By augmenting gravity by the means just described, a new equilibrium can be established at a smaller average particle spacing, but when the extra force is removed, the system undergoes the change required to restore the original equilibrium state. The mixtures represented by the lowest curve of Fig. 3.19 are each very near the equilibrium composition just described; any greater proportion of coarse aggregate, that is, any smaller percentage of sand, would result in rejection of coarse material during vibration.

With mixtures of the kind dealt with above, plasticity or fluidity largely disappears the moment the vibrator is stopped. When the mass is quiescent a steel ball could rest at the top, but when the vibration starts, the ball sinks. The plasticizing or fluidizing effect of vibration can be experienced if you thrust your finger vertically into a vibrating mass of what would otherwise be described as a dry, rocky, impenetrable mix. As you insert your finger you feel the larger particles distinctly as they are crowded aside, but in a few seconds it feels like a soft bed of paste or mortar only. If you move your finger from side to side while holding it vertically, you find that it occupies a pocket larger than the finger itself, with rather definite boundaries. It is easy then to comprehend the situation of an individual particle, and how it is able to move under gravitational force and finally reach a position of mechanical equilibrium described above. Only under extreme conditions does this equilibrium involve point-to-point interparticle contact.

Effects described above are observed when the amplitude of vibration is properly adjusted. If the amplitude is too great, the material seethes and circulates in the mold, and consolidation cannot take place.

The lowest curve in Fig. 3.19 appears superficially to be like the others until we discover that point B occurs entirely outside the triangle $AB'_{28}C_{28}$. Note that because the percentage of sand at point B was 28, C should be the voids ratio of the aggregate having that sand content; however the actual voids ratio for 28 per cent sand is the same as that for 57 per cent, as shown. From the discussion of the voids-ratio diagram in Chapter 1 it would seem that the result obtained is a physical impossibility, but it seems necessary to conclude that under vibration a denser aggregate structure was produced than can be produced by rodding or vibrating dry aggregate. Apparently, while the aggregate is embedded in a matrix that is kept in a fluidized state by vibration, each aggregate particle is able to rotate in such a way that it interferes to the least possible degree with its neighbors. If such orientations could be produced without the aid of a fluidized matrix, the voids ratio of the aggregate would be some value such as that indicated by the termination of the broken line, about $u_{x=1} = 0.2$. This value of C corresponds to a voids content of 17 per cent, as compared with 23 per cent in the dry-rodded state. Russel [9] mentioned a similar observation: The voids content of aggregate in vibrated concrete was observed to be 0.18 as compared with 0.22 for the dry-rodded state.

Estimating the degree of dispersion at point B of the aggregate particles on the same basis as above, we have $u_x = 0.13$, $x = .92$ and $C_{28} = 0.20$, and obtain $t/d = 0.0077$, a dispersion of about 0.8 per cent of the average diameter, a small dispersion indeed. It is indicated also that in the two mixtures to the right of point B, dispersion was nil; in other words, the matrix material was entirely contained by the voids of the aggregate, with the aggregate particles virtually in contact. This result was made possible by mobility of the particles, which was imparted to them by vibration, and by the thinness of the paste; for the last plotted point, the water-cement ratio was about 1.0 by weight. As shown in the following section, the whole curve represents the greatest possible compaction that can be achieved by a particular type of vibration.

Even though the degrees of compaction are at the ultimate level, we find that the parameter D^* is normal for the specific surface area of the aggregate. Values of SM/ψ for this curve differ from point to point, but all are less than 10, and thus mostly the volume-displacement effect mentioned in Section 5.3.2 is apparent; accordingly $D^* = 0.04$.

There remains to be explained why the value of C for the dry-rodded state seemed to be the proper parameter for the upper two curves in Fig. 3.19, whereas a special state seemed to have been produced under the conditions represented in the lowest curve.

First to be considered is the method of evaluating C for a given aggregate, which was the standard method for unit weight (ASTM Designation C29) with the result expressed as a voids ratio; the method specifies compaction by rodding. The individual aggregate particles though rounded were irregular, and, considering the method of mixing, filling the measure, and the rodding, their principal axes can be expected to have random orientations, a factor that contributes to the voids content in the aggregate; if each particle could be reorientated in the most advantageous way, the voids content would diminish.

As may be seen in Fig. 3.19, effects of reorientation are apparent only when the percentage of sand is small, which is to say, only when the mixture is dominated by coarse material. This indicates that reorientation is caused during vibration by moments of force developed by points of pressure that in turn developed between particles of similar size. However, when the percentage of sand is large, the coarse aggregate particles are so dispersed that they have little or no effect on each other. The pressures they feel are developed mainly by fluid motion between them and cement particles. Such pressures would be too evenly distributed over the surface to develop moments of force able to rotate the particle, and thus random orientation persists.

5.8 Mixtures of Mortar and Coarse Aggregate

5.8.1 Mortar-voids-ratio diagram. When we deal with concrete as if it were a two-component system, the definitions of fine and coarse components are arbitrary. In the foregoing discussion we called cement the fine component; now we define the fine component as cement plus sand, which when mixed with water is called mortar.† Figure 3.21 is a mortar-voids-ratio diagram for a series of mixtures of mortar and gravel.

† Many aspects of concrete can be discussed without making a distinction between mortar and concrete because there is no fundamental difference; often, however, a distinction is convenient.

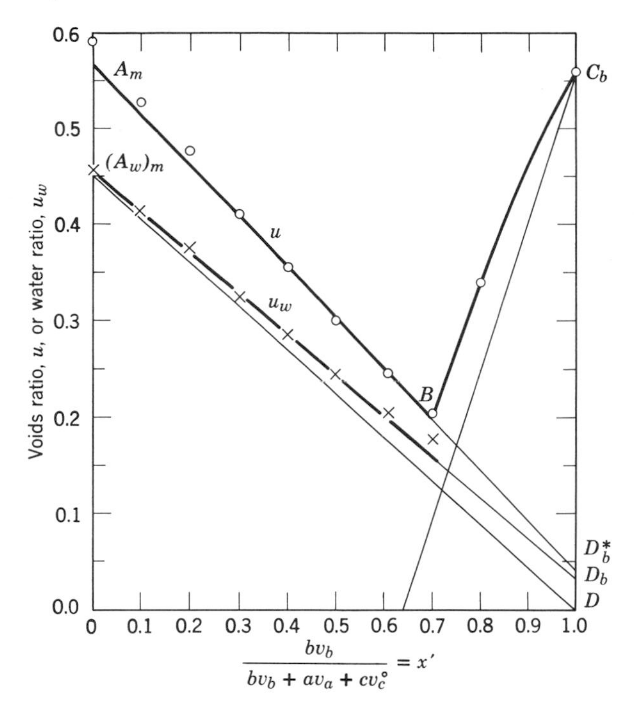

Fig. 3.21. A voids-ratio diagram for mortar and coarse aggregate. (Data are from Blanchette [3].)

In comparing Figs. 3.21 and 3.12, we see that all the features are qualitatively alike. Figure 3.21 is based on data such as those in Fig. 3.9. Each point represents a mixture that is consolidated by exactly the same mechanized procedure we describe in Section 4.2; it contains no more water than is necessary for consolidation by a definite amount of mechanical work.

When coarse aggregate is added to mortar of a given consistency, each increment requires an increase in the water content of the mortar, just as combined aggregate added to paste requires an increase in the water content of the paste.

The first three experimental points for the voids ratio lie above the straight line drawn from A_m to D_o^*, A_m being the voids ratio of the mortar at $x' = 0$, where x' is the ratio of coarse aggregate to total solids. As we discuss in Section 3.2, this particular mortar does not show normal compactness, but as increments of coarse aggregate are added normal compactness is gradually acquired. This phenomenon is shown here by the conformity of points to the straight line at values of x' from 0.3 to about 0.7.

Figure 3.21 contains unmistakable evidence of the volume-displacement effect on parameter D or D^* we discuss in Section 5.3.2. The surface factor of the aggregate is negligible, less than 1.0, but parameter $D_b^* = 0.04$, the value indicated in Fig. 3.13. Thus whether the matrix is paste or mortar, the value of parameter D is independent of particle size if the specified surface area is negligible.

5.8.2 Analytical relationships. The voids ratio of a mortar A_m is given by

$$A_m = \frac{v}{cv_c^o + av_a} = u_{x=0} \tag{3.45}$$

The voids ratio of the mortar in a mixture represented at x' is

$$(A_m)_{x'} = \frac{u}{1 - x'} \tag{3.46}$$

The void-cement ratio of a mixture represented at x' is given by

$$\frac{v}{cv_c^o} = \frac{u}{1 - x'}\left(1 + \frac{av_a}{cv_c^o}\right) \tag{3.47}$$

5.8.3 The Talbot and Richart parameter b/b_o. The Talbot and Richart [7] mortar-voids method for designing concrete mixtures is based on the two-component system, mortar and coarse aggregate. (See Chapter 5.) The method involves the assumption that for satisfactory workability, the coarse aggregate must be dispersed by mortar to a degree that can be prescribed in advance. Talbot and Richart expressed the dispersion in terms of the ratio b/b_o where b is the solid volume of coarse aggregate in a unit volume of concrete and b_o is the solid volume in a unit volume of coarse aggregate in the dry-rodded state. In our nomenclature

$$\frac{b}{b_o} = \frac{bv_b/V}{b_o v_b/V_o} \tag{3.48}$$

In terms of the mortar-voids-ratio diagram, a given value of b/b_o implies a corresponding relationship between C_x' and C_b. Using the analogue of Eq. 3.32 we have

$$b_o = \frac{x' = 1}{1 + u} = \frac{1}{1 + C_b} \tag{3.49}$$

Similarly,

$$b = \frac{x'}{1 + u} = \frac{1}{1 + (C_b)_{x'}} \tag{3.50}$$

Hence

$$\frac{b_o}{b} = \frac{1 + C_b}{1 + (C_b)_{x'}} \tag{3.51}$$

For example, in Fig. 3.21 at $x' = 0.7$, $u = 0.205$. Then from Eq. 3.50

$$1 + (E_b)_{x'} = \frac{1.205}{0.7} = 1.72$$

From the graph, at $x' = 1.0$,

$$1 + C_b = 1.56$$

Hence

$$\frac{b}{b_o} = \frac{1.56}{1.72} = 0.91$$

Such a value is what Talbot and Richart would have called the absolute limit of coarse aggregate concentration. (See also Chapter 5, Section 5.) Ordinarily, a lower value would be used, according to the nature of the work.

REFERENCES

[1] Walsh, H. M., *How to Make Good Concrete,* Concrete Publications, Ltd., 1947.

[2] Sears, F. W., *Mechanics, Heat and Sound,* 2nd ed., Addison-Wesley, Reading, Mass. 1958, pp. 319–330.

[3] Blanchette, W. A., "Some New Relations Bearing on Concrete Mixtures," *Public Roads,* **15,** 57–75, 1934.

[4] Plum, N. M., "The Predetermination of Water Requirement and Optimum Grading of Concrete under Various Conditions," Building Research Studies No. 3. The Danish National Institute of Building Research. (Table 11-b and 11-c were corrected by the author to eliminate indicated negative air contents in the original data.)

[5] Talbot, A. N., "A Proposed Method of Estimating the Density and Strength of Concrete and of Proportioning the Materials by the Experimental and Analytical Consideration of the Voids in the Mortar and Concrete," *Proc. ASTM,* **21,** 940, 1921.

[6] Richart, F. E., and E. E. Bauer, "Relations Between Voids and Plasticity of Cement Mortars at Different Relative Water Contents," *Proc. ACI,* **22,** Part II, 385–403, 1922.

[7] Talbot, A. N., and F. E. Richart, "The Strength of Concrete: Its Relation to the Cement, Aggregates, and Water," Bulletin 137, Engineering Experiment Station, University of Illinois, Urbana, Illinois, 1923.

[8] Bergström, S. G., "Stable Concrete Mixes," Bulletin No. 24, Swedish Cement and Concrete Institute, Royal Institute of Technology, Stockholm, Sweden, 1951.

[9] Russel, F. M., "Discussion of a Paper by C. T. Kennedy," *Proc. ACI,* **36,** 400-1 to 400-3, 1940.

4

Plastic Mixtures at Various Consistencies

1. INTRODUCTION

In Chapter 3 we have dealt mostly with mixtures containing just enough water to give a compacted specimen having a minimum voids content (air plus water), the compacting being done usually by a low-energy procedure. However, concrete is usually produced to have a specified consistency, and the consistency is generally softer than that given by the basic water content. Thus, in practical application, consistency is an important variable.

For the purposes of this discussion we may define the term *consistency* as the rheological characteristic of a plastic or quasiplastic mixture that changes with change of water content. (For a discussion of this subject, see Chapter 10, Rheology of Fresh Concrete.)

2. RELATION OF CONSISTENCY TO WATER CONTENT

As we show in Chapter 10, the only aspect of consistency that can be measured quantitatively is the resistance to deformation that a molded sample will have. When this resistance can be expressed in terms of shear strain, the resistance can be stated in terms of the shear stress required to produce a unit of shear strain, which is a number called the *modulus of stiffness*. Lacking a means of measuring stiffness in units such as psi or g/cm^2, we use devices that give values which are supposed to be proportional, or inversely proportional, to the stiffness. In the following discussion we use data mostly from the slump test and the remolding test.

2.1 Earlier Consistency Equations

Abrams stated in 1918 [1] that the relative consistency of an ordinary concrete mixture, as measured by the slump test, is a function of the relative water content of the mixture. Assigning a value of 1.0 to the water content giving a slump of $\frac{1}{2}$ to 1 in., he stated that a relative water content of 1.1 would give a slump of 5 to 6 in., and a factor of 1.25 would give 8 to 9 in.

Popovics [2] pointed out that if the relationship between change in consistency and change in water content is as was stated by Abrams, it may be stated as follows:

$$\frac{dS}{S} = n\left(\frac{dw}{w}\right) \tag{4.1}$$

where S is the slump and n is a constant.* That is to say, the fractional change in consistency is proportional to the fractional change in water content. The integral form of this relationship is

$$S = C\left(\frac{w}{V}\right)^{n} \tag{4.2}$$

where C is a constant of integration and V is the volume of the mixture. Literally, C would be the slump of a cone full of water and should therefore have only one value, virtually 12 in. Actually, Popovics found that C is a function of the composition of the mixture and that n is the instrument constant, it having a different value for each different apparatus.

Popovics showed that the relationships given above could be used for calculation of the amount by which the water content of a given mixture must be changed to obtain a desired change of consistency.

In 1944 Matern and Odemark [3] reported that consistency as measured by the Vebe apparatus is related to the water content in the way shown by the following equation:

$$VB^{o} = -A^{(w/V)-B} \tag{4.3}$$

or

$$\log VB^{o} = -[(w/V) - B]\log A \tag{4.4}$$

where VB^{o} stands for the consistency value in "Vebe degrees"; A and B are constants dependent on material characteristics but practically independent of the cement content of the mixture. It is shown in Chapter 10

* The terminology of Popovics was different from that used here.

that an exponential form of the consistency equation has been found successful for various materials, although the water ratio rather than water content seems best for plastic materials.

2.2 The Exponential Consistency Equation

Considering the effect of adding an increment of solid material to an already plastic mixture, we might reasonably make the assumption that the effect on consistency is proportional to the existing stiffness. If the stiffening effect were due entirely to the replacement of some of the plastic matrix by the rigid bodies added, we could expect the change in consistency to be proportional to the change in volume concentration of the added material, and the effect should be independent of the particle size. However, we have already seen in Chapter 3 that the stiffening effect of added material is a function of the average particle size, which suggests that the effect depends on the average distance between particles. Thus there is reason to assume that the effect might consist of two parts, one a function of the quantity and the other a function of the average distance between the particles of the material. However at the present stage of development of the subject, it proves to be feasible to assume that the effect is for given materials simply proportional to the ratio of the volume of the added material to the volume of the matrix material. For our present purpose, we consider water to be the matrix material and all the solids to be the added material (neglecting air for the present) and obtain the following differential equation:

$$\frac{dG}{d(V_s/wv_w)} = k_1 G \tag{4.5}$$

where G is the actual stiffness, expressed as stress (see Chapter 10), V_s is the total volume of the solid constituents, and k_1 is a constant. The integral of this relationship may be written as follows:

$$G = G_o e^{k_1(V_s/wv_w)} \tag{4.6}$$

or

$$\log \frac{G}{G_o} = k_2 \left(\frac{V_s}{wv_w}\right) \tag{4.7}$$

where G_o is a constant of integration and $k_2 = k_1 \log e = 0.4343k_1$.

The stiffness factor G becomes equal to G_o when the volume of solids per unit of water volume becomes zero. Thus, taken literally, G_o stands for the consistency of water. Whatever its physical meaning, it is independent of the characteristics of the solid material and of relative proportions of the ingredients of the mixture. As we shall see, experimental data are compatible with this assumption.

The value of k_1 or k_2 depends on the makeup of the solid material. For a given mix, it may be different for different materials, and for given materials it will depend on the ratio of aggregate to cement, the mix, M. Because

$$V_s = cv_c^o(1 + M) \tag{4.8}$$

we may write the consistency equation for any given mix as follows:

$$\log\left(\frac{G}{G_o}\right) = k_2(1 + M)\left(\frac{cv_c^o}{wv_w}\right) \tag{4.9}$$

Some data on the relationship between actual stiffness measured in units of strain and corresponding stress are given in Chapter 10, but the amount of data in terms of G now available is much too limited to be useful. It turns out that values given by some of the "workability" tests serve our purpose very well, making it possible to assume, for example, that $S_o/S \sim G/G_o$, where S stands for slump; or, we assume that $R/R_o \sim G/G_o$, where R stands for the number given by the remolding apparatus [4]. Similarly, data from the Vebe apparatus, and others, that give numbers that depend on the stiffness of the mixture might be used.

When using data from one of the arbitrary tests for consistency, it is necessary to use a proportionality constant to represent the relationship between the number from a given test, which we give the general designation Y, and the actual stiffness in fundamental units. Thus

$$k'_Y = k_3k_2 \tag{4.10}$$

where k'_Y is an empirical constant for a test designated Y, and k_3 is a proportionality constant connecting that empirical constant with the fundamental constant k_2 defined in Eq. 4.9. There is evidence, some of which is given in Chapter 10, that k_3 for a given apparatus is not independent of M; that is to say, a given slump, for example, does not indicate the same modulus stiffness for different mixes with the same materials.

In practice the subscript Y is replaced by one that identifies the particular test apparatus used; thus k_S indicates the slump test, or k_R indicates the remolding apparatus.

With the terminology presented above we may write the consistency equation for any suitable test apparatus as follows:

$$\log\left(\frac{Y}{Y_o}\right) = k_Y(1 + M)\left(\frac{cv_c^o}{wv_w}\right) \tag{4.11}$$

According to this equation, plotting experimental values of log Y against the corresponding ratios of cement volume to water volume should produce a straight line, which, by extrapolation, gives an intercept indicating the value of Y_o and a slope giving the value of $k_Y(1 + M)$. Examples are given in the following section.

3. CONSISTENCY CURVES

3.1 Remolding-test Data

Experimental data for seven different mixes, including neat cement, are plotted in Fig. 4.1 in terms of Eq. 4.11. Characteristics of the aggregate are given in Tables 3.1 and 3.2 of Chapter 3.

Some of the points in Fig. 4.1 do not conform exactly to the pattern of straight lines as drawn; yet the overall conformation seems to justify our considering Eq. 4.6 or Eq. 4.11 to be at least a good first approximation of a law for the effect on stiffness of changing the water content of a mixture. In Chapter 10 it is shown that an equation like Eq. 4.6 also represents the relationship between the plastic viscosity U of cement paste in its fluidized state and the ratio of cement to water in the paste. In that case the constant U_o comes out to have the correct value for the viscosity of water at the temperature of the experiment. At first thought it might seem that R_o is likewise proportional to the viscosity of water, but it is soon evident that this cannot be true. The remolding number R is the number of jolts of the apparatus required to complete a certain molding process, and for concrete the process is not one of viscous flow, as is shown in Chapter 10. If fluid water cannot support static shear stress, the physical significance of R_o is not obvious. However, we note that the value of R_o is 0.004,* literally, four-thousandths of one jolt; perhaps this corresponds to a very small interval during which fluid water is able to support static stress, and is thus the value toward which the curve for a concrete mixture trends as the water content of the mixture is increased.

3.2 Slump-test Data

As applied to a given mix, the slump test responds to changes of water content in conformity to the same exponential law discussed above, except that it shows some variations when applied to very rich mixtures not found when using the remolding test. A typical family of curves is shown in Fig. 4.2.

The data given in Fig. 4.2 conform to the exponential law for slump if the slump is not more than 6 or 7 in. However unlike those for the remold-

* This is the value for a half-size replica of the original apparatus; for the full-size apparatus $R_o = 0.02$.

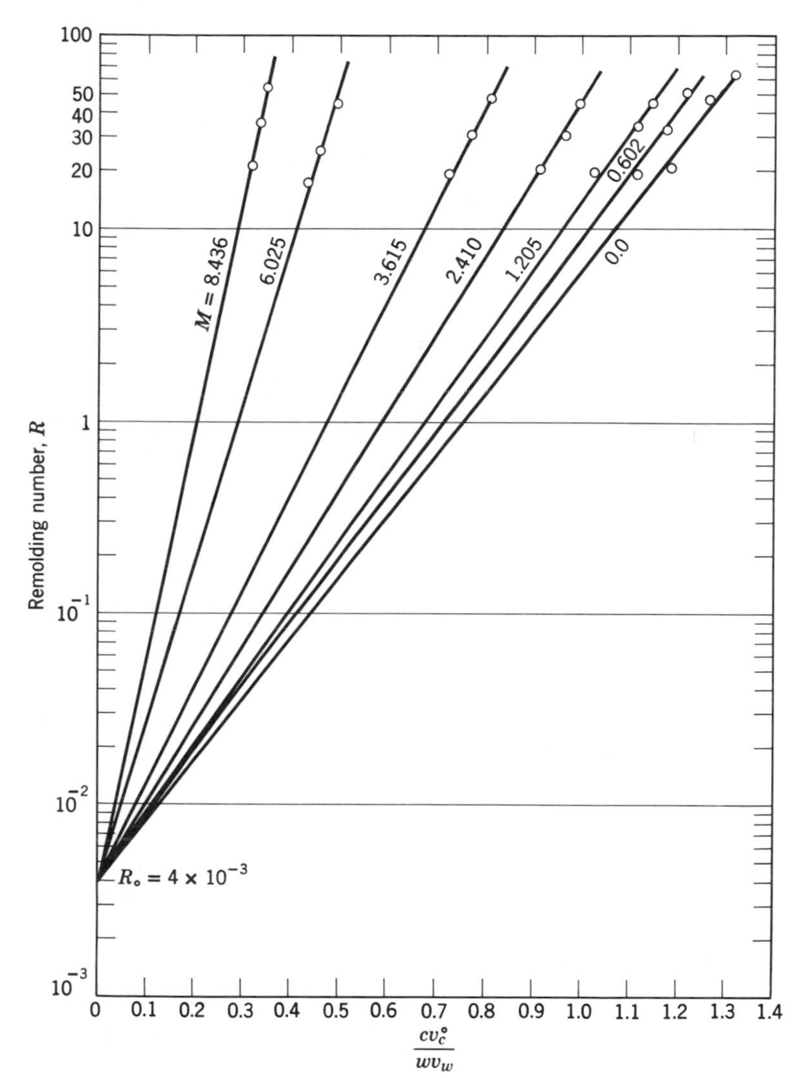

Fig. 4.1. Consistency curves on a remolding basis. (Data are from Table 4.2, Sand No. 265-11.) (PCA Series 265G.)

ing test, these data do not indicate the same values of S_o for all mixtures; for $M \geq 1.84$, $S_o = 10^6$, but for neat cement $S_o = 5 \times 10^3$, and S_o apparently increases progressively from 5×10^3 to 10^6 as M increases from 0.0 to 1.8. However in Fig. 4.2 all lines are drawn so they will converge at $S = S_o = 10^6$ in. Because for practical concrete M seldom if ever is as

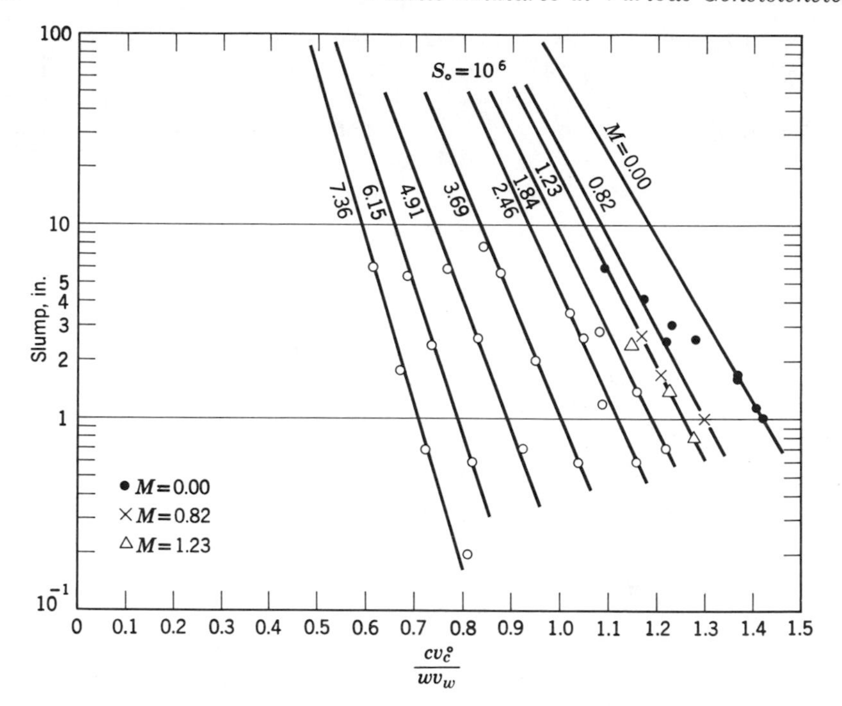

Fig. 4.2. Consistency curves on a slump basis. The sand and gravel aggregate used had a maximum size of $\frac{3}{4}$ in. and a sand content of 33 per cent. (Data from PCA Series 265.)

small as 1.8, the departures from constancy of S_o discussed above are not important, and we can use the exponential law for the slump test, provided that the law is not applied for slumps exceeding about 7 in.

If a literal interpretation of the constant R_o for the remolding test presented difficulty, what can one say of the value of S_o, which literally says what water has a slump of 1 million in., whereas a slump of 12 in. is the physical limit? We may note that vertical slump is connected with lateral spreading and that if the slump values were converted into terms of lateral increase, the conceptual difficulty would be somewhat reduced. Anyway, S_o is a factor which when multiplied by the exponential term gives the actual amount of slump in inches.

3.3 Uncertainties Concerning Y_o

From inspection of Figs. 4.1 and 4.2 and the like, it is apparent that the instrument factors R_o or S_o, or in general Y_o, have not been determined precisely. Indeed, it has already been pointed out that the basic assumption stated in Eq. 4.5 may be an oversimplification, and we may now note that if it is, the empirical values of the "constants" may not appear strictly constant. For the slump test, it was observed that the value of S_o for neat cement and exceedingly rich concrete is not the same as for ordinary mixtures, but even so it is not certain that fault lies in the assumed "law."

It is known that neat cement having a given water content can take on different consistencies, depending on the degree to which it has become homogenized by the mixer. A mixer designed for concrete is not efficient when used for cement and water only, and the efficiency is lower the higher the water content of the paste. Even a seemingly small difference in conditions can change consistency through change of mixing efficiency. For example, batches of neat cement were produced with a mixer of the open-tub type, using a 5-min mixing period; a certain curve of slump versus water content was established. Then the experiment was repeated using batches 50 per cent larger than before, and the pastes were found to be stiffer than before. The relative slumps were 93 and 88 per cent for stiff and medium consistency, respectively; the softer the consistency, the lower the efficiency. When the mix is composed predominantly of aggregate, such effects do not appear because even though the action of the mixer does not seem rapid, the paste in the interstitial space is subjected to high rates of shear, much higher than can be produced in the absence of aggregate.

Thus there is reason to suppose that nonconformity to an exponential consistency equation might be accounted for in terms of extraneous factors that introduce systematic variations that are functions of consistency itself, and not directly functions of water content, and that when the aggregate is dominant as it normally is, such factors are unimportant.

The best value of Y_o might be established by making a suitable number of tests within the working range of the consistency apparatus and within the mix range for uniformly efficient mixing, and then establishing the most probable value of Y_o by means of an electronic computer, using a program based on the exponential consistency equation. This has not been done. The values already given for S_o and R_o were established graphically on the more or less arbitrary assumption that the same constant is valid for all plastic mixtures of given materials. Although a better handling of this point is desirable, the procedure followed is justifiable from a practical point of view. On account of, for example, the very large difference between S_o and the working values of S, the value of S_o can be changed by a factor of 10 without making much difference in the practical result.

3.4 Uses of Consistency Equations

In the following sections we see that it is possible to identify the factors that determine the water requirements of different mixtures in a family of mixtures made from the same materials. Equation 4.11 reduces the problem of analysis to that of finding the determinants of k_Y.

Given the water ratio, or water-cement ratio, of a given mixture at a given consistency, it is a simple matter to determine the ratio corresponding to another consistency. By Eq. 4.11

$$\left(\frac{w}{c}\right)_2 = \frac{\log (Y/Y_o)_1}{\log (Y/Y_o)_2}\left(\frac{w}{c}\right)_1 = \frac{Q_1}{Q_2}\left(\frac{w}{c}\right)_1 \tag{4.12}$$

where Q is the consistency factor, that is, $Q = \log (Y/Y_o)$.

Relative water-cement ratios based on the slump test are given in Table 4.1. They are based on the water-cement ratio giving a slump of

Table 4.1 Relation between Slump and Relative-water Ratio or Water-cement Ratio[a]

Slump (in.)	log S −6	Relative w/c or Relative u_w
1.0	−6.000	0.962
1.7	−5.770	1.000
2.0	−5.699	1.012
3.0	−5.523	1.045
4.0	−5.398	1.069
5.0	−5.301	1.088
6.0	−5.222	1.105
7.0	−5.155	1.119

[a] Based on Eq. 4.12.

1.7 in. because, as discussed in Chapter 3, for rich mixtures that value corresponds usually to the basic water content.

Equation 4.12 provides an easy way to adjust a table of measured values of composition and consistency, in which the consistencies vary from a desired value, so that the composition corresponds to the desired consistency.

4. THE WATER REQUIREMENT FACTOR k_Y

Considering Figs. 4.1 or 4.2, and Eq. 4.11, we see that the slope of the straight line in the semilog plot is $k_Y(1 + M)$. Thus k_Y is a parameter comprising all the factors in a given mixture that determine the amount of water required to obtain a given consistency as measured by a given method; it may therefore be called the *water-requirement factor.*

4.1 Water-requirement Factor as a Function of M

Because in Figs. 4.1 or 4.2 the slopes of the lines appear to vary progressively with the value of M, the ratio of aggregate to cement, we may state that

$$k_Y(1 + M) = f(M) \tag{4.13}$$

or

$$k_Y = f\left(\frac{M}{1 + M}\right) = f(x) \tag{4.14}$$

The nature of $f(M)$ is shown when the slopes of the graphs are plotted against the corresponding values of M, the slopes being conveniently calculated from the average values of the products of Q and the water-cement ratio, as may be seen in Eq. 4.11. Figure 4.3 is such a plot for the data in Table 4.2, the same data that are plotted in Fig. 4.1. Each point is the average of three values.

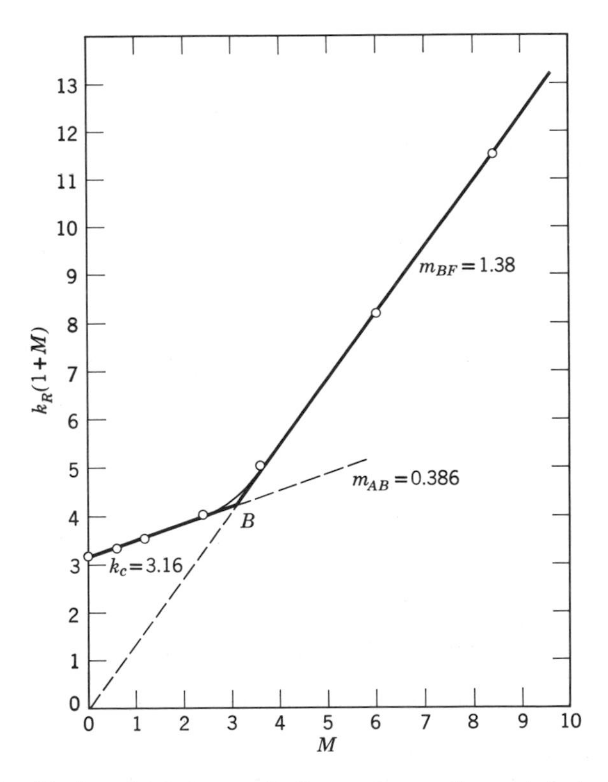

Fig. 4.3. Graphical method for evaluating water-requirement factors. (Data are from Table 4.2.)

Table 4.2 Water-requirement Factors for Various Mixes of Given Materials[a]

1[b]	2[c]	3[d]	4[e]	5[f]	6[g]	7[h]
Reference Number	$c{:}a$	$\dfrac{cv_c^o}{wv_w}$	R	M	$\log \dfrac{R}{4} + 3 = Q$	$k_R(1 + M)$
322	1:0	1.32	61	0	4.181	3.167
323	1:0	1.27	45	0	4.052	3.190
324	1:0	1.19	20	0	3.699	3.108
Average						3.16
505	1:05	1.22	50	0.602	4.097	3.358
506	1:05	1.18	32	0.602	3.903	3.308
507	1:05	1.12	19	0.602	3.677	3.283
Average						3.316
508	1:1	1.15	43	1.205	4.030	3.504
509	1:1	1.12	33	1.205	3.916	3.496
510	1:1	1.03	19	1.205	3.677	3.570
Average						3.523
511	1:2	1.00	44	2.410	4.041	4.041
512	1:2	0.970	30	2.410	3.875	3.994
513	1:2	0.918	20	2.410	3.699	4.029
Average						4.021
514	1:3	0.813	47	3.615	4.070	5.006
515	1:3	0.773	30	3.615	3.875	5.012
516	1:3	0.728	19	3.615	3.677	5.051
Average						5.023
517	1:5	0.496	44	6.025	4.041	8.147
518	1:5	0.463	25	6.025	3.796	8.198
519	1:5	0.439	17	6.025	3.628	8.264
Average						8.203
520	1:7	0.354	54	8.436	4.130	11.666
521	1:7	0.341	35	8.436	3.942	11.560
522	1:7	0.325	21	8.436	3.720	11.446
Average						11.56

[a] Data are for Figs. 4.1 and 4.3 and are from PCA Series 265G. The aggregate is sand No. 265-11, with a size range #200– #4. The voids ratio $C = 0.42$. See Chapter 3, Table 3.2.

[b] Values for each reference are the averages for three samples.

[c] The mix by weight.

[d] Volume ratio of cement to water.

[e] The remolding number by apparatus for which $R_o = 0.004$.

[f] The ratio of aggregate to cement by solid volumes.

[g] $\log (R/R^o) = \log (R/4) + 3 = Q$.

[h] (Col. 6)/(Col. 3) $= k_R(1 + M)$, the water requirement parameter for a given mix.

Deviations from the average among the values for a given mix probably would have been smaller if the value of R_o had been determined statistically rather than estimated graphically. However that may be, the deviations, which may be seen graphically in Fig. 4.1, are in only one or two cases greater than the expected variation of the test value, about 10 per cent, and are generally less than 10 per cent. Equation 4.11 does indeed closely estimate the effect of a change of water content on the change of consistency, at least within the range of consistencies dealt with here, $R = 20$ to 60. This range corresponds approximately to slumps from 1 to 6 in.

The remolding apparatus (see Chapter 10, Section 8.2) was a half-size replica of the original model, made for testing mortars; its instrument constant was $R_o = 0.004$ as compared with about 0.02 for the full-size apparatus.

The plotted points in Fig. 4.3 describe two linear relationships with perhaps a short transition section between them. The important feature is that for the mixtures represented to the left of point B, the relationship is characterized by two constants, whereas for those on the right side one constant is sufficient. The nature of these constants will be discussed further on. Point B is the same as the point so designated in the voids-ratio diagram. (See Chapter 3.)

Examples based on the slump test are shown in Fig. 4.4. Data for two series of mixtures are shown, one for mortar and the other for concrete. In the data for the mortar tests we see the same general characteristics as those found in Fig. 4.3. For the concrete mixtures, the linear relationship to the left of point B is shown clearly, but for mixtures to the right of point B the nature of the relationship is not clearly defined; however, other data, some of which will be examined, indicate that the relationship is linear, up to a certain point designated F. As we see further on, beyond that point a third kind of relationship appears.

The principal purpose of Figs. 4.3 and 4.4 is to demonstrate that the relationships to the left of point B is linear, and to indicate that the relationship on the right is linear also within the range shown. On examining data from various sources, the ranges of mixes in any one series is seldom wide enough to show the nature of the function unequivocally; for that reason it is necessary for us to examine several series.

4.1.1 Sources of systematic error. There are data from which point B cannot be designated clearly, and for which, if the points are to be approximated by a straight line, the line for the range of leaner mixes does not pass through the origin; in other words, such data indicate that at least two constants are required to approximate the data for the mixes on either side of point B. Why this kind of relationship appears for data

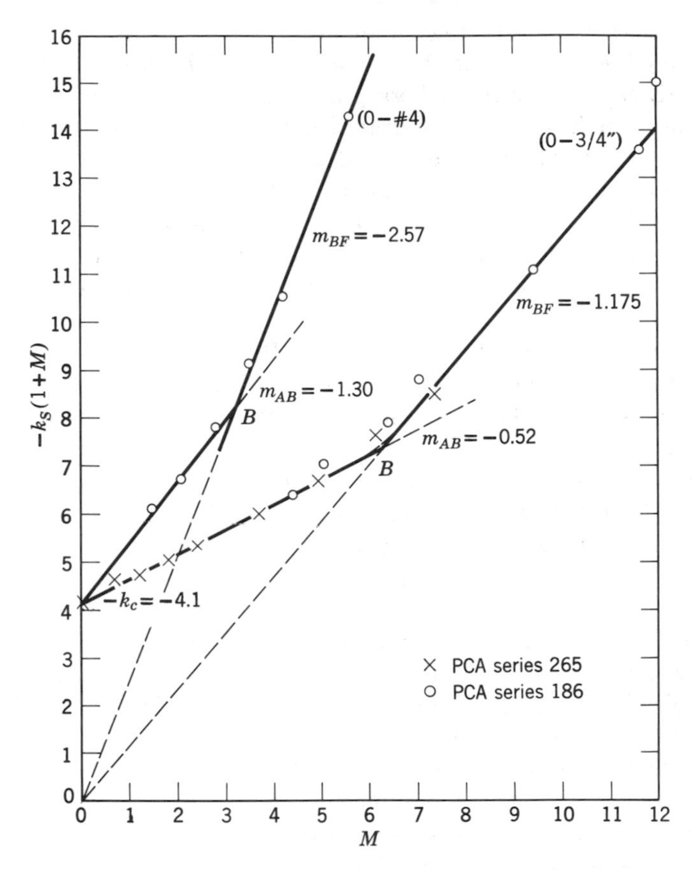

Fig. 4.4. Water-requirement factors of mortar and concrete on a slump basis. (Data are from Table 4.3.)

from some sources but not for data from others is not certain, but matters of laboratory technique are probably involved. For example, if batch temperature is not controlled, it will be higher the richer the mix, and the water requirements of the richer mixes will be relatively too high; the line the slope of which indicates the water requirement of the material will become tilted in the manner sometimes observed.

The magnitude of the effect described above differs among different cements. Especially cements having a relatively high specific surface area (ASTM Type III) liberate a relatively large amount of heat during the mixing period, and thus generally produce higher batch temperatures than do ordinary cements; also, at a given specific surface area, cements having a relatively high content of tricalcium aluminate liberate relatively large amounts of heat, rapidly.

It can be contended that such temperature effects as those described above are due to intrinsic properties of the mixture and therefore are properly included in the water-requirement factor for the mixture. However, if the purpose of experiments is to develop an understanding of the factors that influence the water requirement factor, it is obvious that temperature should remain a constant while other factors are under consideration, or at least the temperature of each batch should be observed and recorded if a basis for correcting for temperature effects has been established. When such a correction is not feasible, the correct laboratory technique is to precool the materials to such degree that after the period of mixing the temperature of each batch will be within one or two degrees of the temperature selected for the experiment. Under some circumstances we find it convenient to cool the mixing water only; when the cement content is high, and the cement gives a rapid initial reaction, it becomes necessary for us to use a mixture of ice and water, and sometimes even to cool the solids as well.

Another cause of distortion of relationships is the use of inaccurate figures on which to base the calculation of absorption of water by the aggregate. When in laboratory tests we use dry aggregates the consistency with a given total amount of water in the batch will be determined partly by the amount of water that remains after some has become absorbed by the rock particles. The quantity absorbed increases with time, and therefore the amount absorbed *up to the time when the consistency test is made* which might be 10 min after the water is added to the batch, must be ascertained as accurately as possible. If the amount absorbed is based on the ASTM Standard Methods (C127 and C128), it will be correct for the total amount that might be absorbed during the first 24 hr after immersion, but it will be incorrect for the amount that has become absorbed at the time the consistency test is made; which is to say, a correction so based does not give the net amount of water affecting the consistency at the time of test. Such an error does not change the trend of a curve in the manner we are concerned with now if the rates of absorption of water by fine and coarse aggregates are alike or if the same ratio of fine to coarse is used in every mix; but the rates are usually not alike, especially during the first 10 min, and usually the relative amount of coarse aggregate is greater the higher the cement content. Hence the error in net water-cement ratio changes progressively with the richness of the mix, and thus lines such as those shown in Fig. 4.3 could become tilted. (However, the laboratory procedure used to obtain the data shown in Fig. 4.3 was such as to preclude the error just described as completely as possible.)

Some of the data presented in this chapter were obtained from published reports that did not include complete descriptions of laboratory procedure and method of calculation, and some of it does not conform in

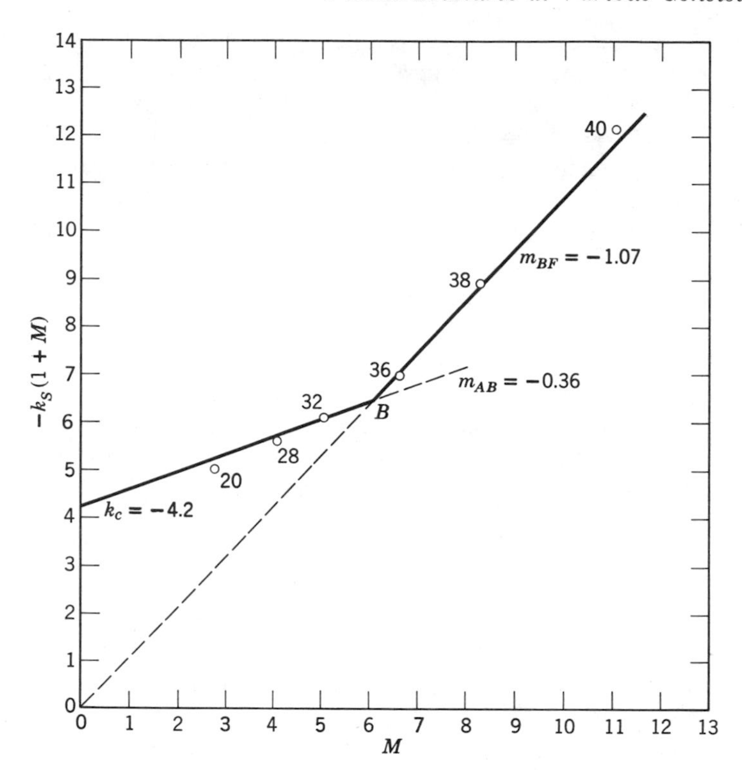

Fig. 4.5. The water-requirement diagram for a series of mixtures in which the percentage of sand in the aggregate was varied. The maximum size of the aggregate was 1½ in. Numerals appearing by the points are the percentages of sand. (Data are from PCA Series 231.)

all respects to the pattern that seems to be indicated by the data as a whole. It would have been desirable to present only the data known to be relatively free from systematic error, but unfortunately any one of the several test series of that kind was not of sufficient scope; it was necessary to use some questionable data, at least questionable because of insufficient information.

In Figs. 4.3 and 4.4 the aggregate grading was the same within each series of mixtures; in Fig. 4.5 the grading was caused to differ from point to point by changing the ratio of fine aggregate to coarse aggregate, this being indicated in the graph by numbers giving sand percentages. Table 4.3 gives the data for Fig. 4.5 and shows how mixtures having different consistencies can be used to construct a diagram.

For mixtures to the right of point B in Fig. 4.5, varying the percentage of sand as indicated does not at first seem to cause appreciable deviation

Table 4.3 Calculation of Mixture Composition and Calculation of $-k_S(1 + M)$

A: Composition of Mixtures

Reference Number	Per cent Sand in Aggregate	Volume Composition $\frac{cv_c^o}{V}$	$\frac{av_a}{V}$	$\frac{bv_b}{V}$	$\frac{wv_w}{V}$	$\frac{v}{V}$	M
36	40	0.068	0.304	0.455	0.151	0.173	11.1
37	38	0.090	0.284	0.462	0.146	0.164	8.3
38	36	0.111	0.264	0.470	0.142	0.155	6.66
39	32	0.137	0.221	0.470	0.154	0.172	5.06
40	28	0.161	0.184	0.474	0.167	0.181	4.10
41	20	0.213	0.116	0.465	0.204	0.206	2.73

B: Calculation of $-k_S(1 + M)$ by Eq. 4.11

Reference Number	$\frac{mv_w}{cv_c^o}$	Slump S (in.)	$\log S - 6$	$-k_S(1 + M) = Q$	M
36	2.202	3.0	−5.523	12.1	11.1
37	1.614	3.0	−5.523	8.9	8.3
38	1.275	3.1	−5.509	7.0	6.66
39	1.123	4.0	−5.398	6.1	5.06
40	1.036	3.7	−5.432	5.6	4.10
41	0.956	6.0	−5.222	5.0	2.73

Data are for Fig. 4.5 and are from PCA Series 231.

from the straight line; indeed, as compared with Fig. 4.4, the alignment seems somewhat improved. However there are effects of grading variations of this kind on water requirement, as we see later, but they are not large enough to be seen clearly in this general relationship. The percentages of sand given, in the range of mixtures to the right of point B at least, are those which by trial were found to require the least amount of water, and thus are called "optimum percentages." Actually, each point does fall on a different line, each line having a slope indicating the water requirement of the particular aggregate represented by the point; the line as drawn is a sort of average water requirement indicating the general level achievable with this material.

For mixtures to the left of point B in Fig. 4.5, the points fall higher on the graph the higher the specific surface area of the aggregate; hence, the line from k_c through the point for the aggregate having 32 per cent sand lies above the two points representing lesser percentages of sand.

Again we see that the water-requirement factor indicated under these circumstances does not apply accurately to some of the mixes but does serve to characterize the series.

4.1.2 Effect of cement fineness. Figure 4.6 represents two series of four mixes each, all mixtures falling to the right of point B. One series is made with cement having a specific surface area typical of ASTM Type I cements and the other with a finer cement, typical of ASTM Type III. The water-requirement factor for a given mix appears to be the same for the finer cement as it does for the coarser. This does not mean that the rheological characteristics of a given mix were the same with one cement as with the other; it means only that the consistency as measured by the remolding test is the same. Also, it does not mean that the same conclusion would be indicated for mixtures represented to the left of point B; on the contrary, for mixtures in that range the water requirement of the mixture is generally higher the higher the specific surface area of the cement.

4.1.3 Point *F*. In Figs. 4.3, 4.4, and 4.5 there is no clear indication of any departure of plotted points from the linear relationship to the right

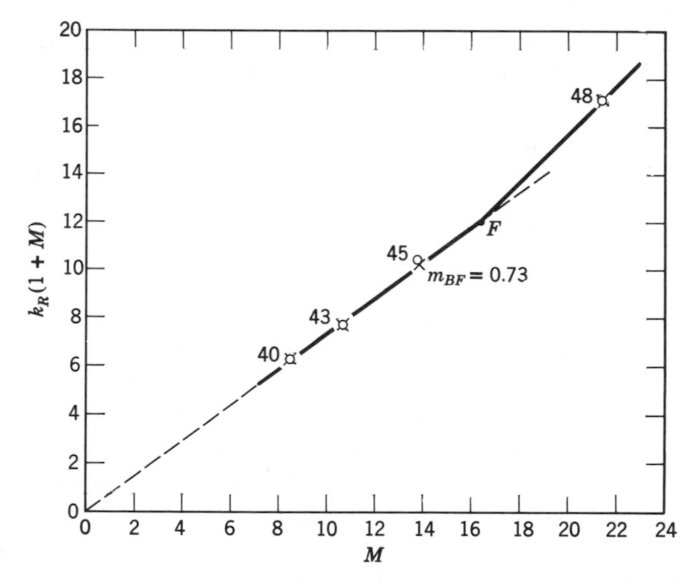

Fig. 4.6. The water requirement diagram for two series of mixtures made with cements having different specific surface areas. The size-range of the sand-and-gravel aggregate was (0–1½ in.). O stands for mixture made with a cement having a specific surface area of 1725 cm^2/g as measured with the Wagner turbidimeter, and × stands for a cement having a specific surface area of 2245. (The data are from Table 4.4.)

Table 4.4 Data for Figure 4.6[a]

1[b]	2[c]		3[d]		4[e]	5[f]
	wv_w/cv_c^o		$k_R(1 + M)$			
M	1725	2245	1125	2245	$100 \dfrac{a}{a+b}$	C
8.53	1.74	1.74	6.32	6.32	40	0.266
10.7	2.13	2.10	7.74	7.63	43	0.263
13.8	2.86	2.83	10.4	10.3	45	0.261
21.4	4.70	4.70	17.1	17.1	48	0.259

[a] Data are from PCA Series 267. The aggregate was sand and gravel. The fineness modulus of the sand was 2.92, the fineness modulus of the coarse aggregate was 7.25, the size range was (0–1½ in.). The consistency $R = 43 \pm 3$.
[b] The ratio of aggregate to cement by solid volumes.
[c] Water-cement ratios required for indicated consistency with cements having Wagner surface factors indicated.
[d] Water-requirement parameters for different cements having Wagner surface factors indicated.
[e] The percentage of sand in the aggregate.
[f] The voids ratio of the aggregate.

of point B, although it becomes clear on reflection that the range for conformity must be limited. In Fig. 4.6 there is an unmistakable departure from the straight line representing the average water-requirement factor of three of the mixtures, that being the point at $M = 21.7$. The point of departure is indicated on the diagram as point F; the method of determining it is described in Section 5.3.3.

Figure 4.7 gives other examples, one based on slump test data and the other on remolding test data for the same series of mixtures; the data are in Table 4.5. The two plots indicate similar but not identical values for points B and F. It is not certain that the values for points B are accurate, for in either case the one point that appears to be on the straight line to the left of point B may actually be on a transition section between the two straight lines. However it will serve our purpose further on to assume that the indicated values are correct, the purpose being mainly that of illustrating procedures.

4.2 The Three Categories of Concrete Mixtures

As was mentioned above, it is apparent that the water-requirement factor can be expressed as a function of M by two simple equations as

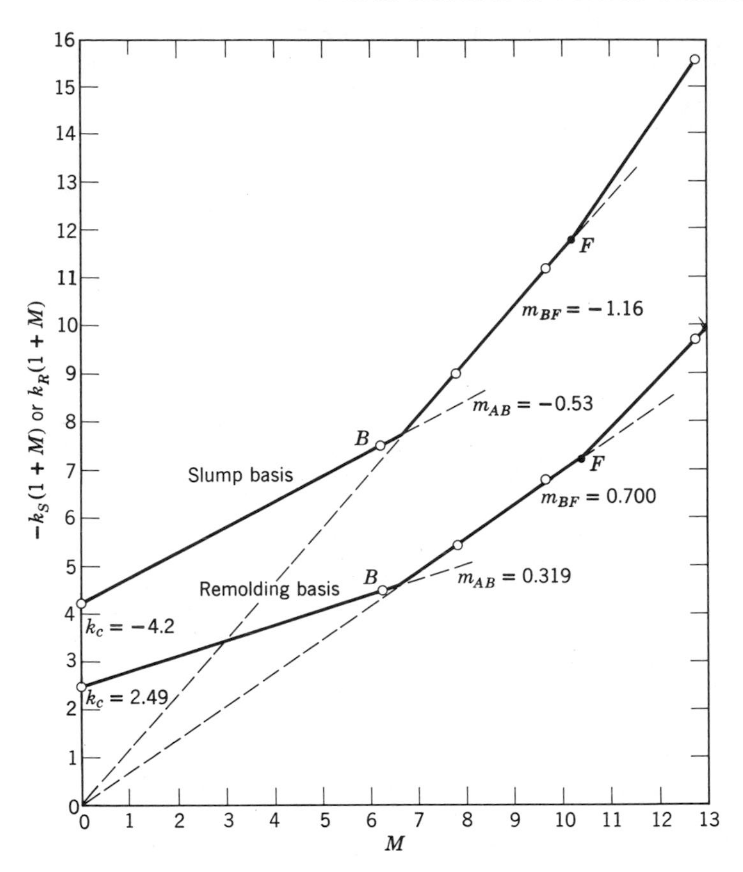

Fig. 4.7. Comparison of the water-requirement diagram on a slump basis with that on a remodling basis. The size-range of the sand-and-gravel aggregate was (0–$\frac{3}{4}$ in.). (Data are from Table 4.5, PAC Series 233-47.)

follows. For mixtures to the left of point B

$$k_Y(1 + M) = k_c + (m_{AB})M \tag{4.15}$$

or, considering the definition of x,

$$k_Y = k_c(1 - x) + (m_{AB})x \tag{4.16}$$

and for mixtures to the right of point B and up to point F,

$$k_Y(1 + M) = (m_{BF})M \tag{4.17}$$

or

$$k_Y = (m_{BF})x \tag{4.18}$$

Table 4.5 Data for Figure 4.7[a]

1	2[b]	3[c]	4[d]	5	6[e]	7[f]	8	9[g]	10[h]
Reference Number	M	$\frac{wv_w}{cv_c^o}$	R	$\log\left(\frac{R}{0.02}\right) = Q$	$k_R(1+M)$	S (in.)	$\log S$	$-k_S(1+M) = -Q$	$\frac{a}{a+b} \times 100$
239	12.75	2.651	62	3.491	9.25	0.5	−5.301	14.0	38
		2.853	42	3.322	9.48	4.0	−5.398	15.4	
		2.998	39	3.290	9.86	4.5	−5.347	16.0	
		3.171	38	3.279	10.40	5.2	−5.222	16.8	
Average					9.75			15.6	
244	9.65	1.917	46	3.362	6.44	1.3	−5.886	11.3	39
		2.027	42	3.323	6.74	5.0	−5.301	10.7	
		2.207	35	3.243	7.16	5.0	−5.301	11.7	
Average					6.78			11.2	
249	7.79	1.596	50	3.398	5.42	3.5	−5.456	8.71	41
		1.679	32	3.204	5.38	5.0	−5.301	8.90	
		1.770	27	3.130	5.54	6.0	−5.222	9.24	
Average					5.45			9.0	
253	6.23	1.313	49	3.389	4.45	2.0	−5.699	7.48	43
		1.402	32	3.204	4.49	5.0	−5.301	7.43	
		1.477	22	3.041	4.49	6.0	−5.222	7.71	
Average					4.48			7.5	
Estimate	0	0.733	50	3.398	2.49	1.7	−5.770	4.2	

[a] Data are from PCA Series 233-47. The aggregate used was sand and gravel, the fineness modulus of the sand was 2.92. The fineness modulus of the coarse aggregate was 6.40, the size range is (0–$\frac{3}{4}$ in.), and the voids ratio at the maximum density was $C = 0.302$.

[b] The ratio of aggregate to cement by solid volumes.

[c] The ratio of water to cement by volume.

[d] Remolding number $R_o = 0.02$.

[e] Product of columns 3 and 5; this is the water requirement parameter.

[f] Slump in inches.

[g] The data are the same as those in column 6 but on slump basis.

[h] The percentage of sand in the aggregate.

In these equations k_c is the left-hand terminus of the line and may be called the water-requirement factor of the cement: m_{AB} is the slope of the line between k_c and point B and is thus the water-requirement factor of the aggregate when it is used in mixtures within that range. Parameter m_{BF} is the proportionality constant between the water-requirement factor of the mixture and the fraction of aggregate in the total solids. As we see further on, its value depends on the characteristics of the cement paste as well as on those of the aggregate, but it will be convenient to speak of m_{BF} as the water-requirement factor of the aggregate, for with a given cement it varies with the nature of the aggregate.

Thus considering all the determinants of the water-requirement factors of all the mixtures that can be made of given materials up to point F, we find that the mixtures can be divided into two distinct categories. In one, the water requirement of the mixture is the weighted average of the water-requirement factors of the cement and the aggregate, and in the other it is simply proportional to the volume fraction of aggregate in the total solids, the proportionality constant being called the water-requirement factor of the aggregate as used in mixtures within the designated range of mixes. The water requirement factors for the mixtures beyond point F are also determined by the quantity of aggregate and one constant, and thus in that respect they may be included in the same category as those in the range BF; however, as is shown further on, the constant is different from m_{BF}, and thus we find it advantageous to consider such mixtures as a separate class.

For convenience, we speak of the mixtures to the left of point B as the AB category, and the other we call the BF category; the third class is called the FC class. The AB category comprises the very rich mixtures, the BF category comprises ordinary mixtures, and the FC class the very lean.*

4.3 Water-requirement Formulas

From Eq. 4.11, and Eqs. 4.14 and 4.16, we obtain the following expressions for the water requirements of concrete mixtures. For the AB category

$$\frac{w}{c} = \frac{v_c^o}{v_w}\left[\frac{k_c}{Q} + \frac{(m_{AB})}{Q} M\right] \tag{4.19}$$

For the BF category

$$\frac{w}{c} = \frac{v_c^o}{v_w}\left(\frac{m_{BF}}{Q} M\right) \tag{4.20}$$

* Previously, before the FC category was recognized, the author called the first two categories B minus and B plus.

We see further on, in the development of Eq. 4.30, that the water-requirement formula for the *FC* category is as follows:

$$\frac{w}{c} = \frac{v_c^o}{v_w} \frac{Q}{Q_b} (CM - 1)$$

where Q_b stands for basic consistency. *C* is the voids ratio of the aggregate in the dry-rodded state, as before.

4.4 Significance of Water-requirement Formulas

Although the practical meaning of the relationships shown by Eqs. 4.18, 4.19, and 4.20 becomes more apparent as the discussion proceeds, we should not pass these equations by without a preliminary appraisal. In the first place they reflect the existence of three distinct categories and point to simple relationships within each category that are at least close approximations of accurate experimental data. Such simple relationships cannot be discerned until the categorical differences are recognized. The line of demarcation between the different categories is not the same for different materials, and this is a matter of interest with respect to the relative merits of different materials and different methods of formulating mixes.

As we see below, point *B* represents an optimum combination of given materials, from some points of view; it is usually found near $w/c = 0.40$ (at an intermediate consistency), which means that relatively little such concrete is used in present-day practice. However the use of mixtures in the *AB* category is not negligible and may become more general in the future. Some specifications for paving concrete require mixtures that would be represented only a little to the right of point *B*; factory-made concrete products, including prestressed, reinforced-concrete building units, use mixtures in the *AB* category to some extent. Special mixtures for architectural concrete developed by Litvin and Pfeifer [5] turn out to be in that category, and relatively high-strength structural concrete made with lightweight aggregate requires such mixtures. (Most concrete produced in the United States for structural purposes is in the *BF* category, although some of it is in the *FC* category, or near the borderline.)

Concrete in the *BF* class is generally satisfactory, but that in the *FC* category has many undesirable properties: it lacks cohesiveness and is prone to segregate during handling; after hardening, the concrete tends to be full of fissures and other structural faults. (See Chapter 11.) We see further on that when designing mixes, it is highly desirable to ascertain what type of concrete a given mix produces and, when necessary, to take steps to obtain at least the characteristics of the *BF* category, if not the optimum characteristics.

In anticipation of Chapter 5 we may observe also that Eqs. 4.18, 4.19, and 4.20 make it clear why methods of mix formulation based on one theory or another have been only partly successful; all the methods developed from a theoretical approach have involved the assumption that one law should underly the formulation of all kinds of plastic mixtures. The present analysis, which is made without prejudice, that is, without any hypothesis concerning grading or packing of particles, shows that the water requirements follow distinctly different laws in each of the three categories. In the *FC* category, the voids content of the aggregate together with relative consistency determines the water requirement of the mixture; in the *BF* category, the voids content and the grading of the aggregate control the water requirement, there being an optimum combination of grading and void content. Only in the *AB* category does the water requirement of the cement need to be taken into account explicitly to arrive at a first approximation of the water requirement of a given set of materials.

It must be acknowledged that the experimental data we have considered up to this point could be represented by continuous curves with, on the whole, about the same degree of approximation as that achieved by using three linear functions. (Figure 4.3 would be an exception.) This is particularly true if each mixture in the series has a different, but optimum, aggregate grading. However as we proceed through this and other chapters, we find much evidence that the categorical differences are real and practically significant. Furthermore, the linear relationships lead to relatively simple methods of analytical study and are thus to be preferred as simple satisfactory approximations, at least.

5. CONSTRUCTION OF WATER-RATIO DIAGRAMS FROM WATER-REQUIREMENT FACTORS

The water-ratio diagram for any given set of materials can readily be derived from the consistency equations, which we now do for the first two categories, deferring the third category until after it has been discussed further. The first step is to convert the equations into terms of the coordinates of the water-ratio diagram.

5.1 Transformation of Equations

In Chapter 3 the following relationships are given:

$$u_w = \frac{wv_w}{cv_c^o} \cdot \frac{1}{1 + M} \tag{4.21}$$

$$\frac{wv_w}{cv_c^o} = u_w(1 + M) \tag{4.22}$$

$$x = \frac{M}{1 + M} \tag{4.23}$$

Substituting from Eq. 4.22 into Eqs. 4.19 and 4.20, dividing both sides by $(1 + M)$, and using Eq. 4.23, gives the following two equations:

$$(u_w)_{AB} = \frac{k_c}{Q}(1 - x) + \frac{m_{AB}}{Q}x \tag{4.24}$$

and

$$(u_w)_{BF} = \frac{m_{BF}}{Q}x = Ex \tag{4.25}$$

E is the water-requirement factor for a given consistency, and it is a terminal point on the water-ratio diagram.

Equation 3.33 of Chapter 3 is

$$u_w = A_w(1 - x) + Dx$$

Comparing this expression with Eq. 4.24 above, we see that

$$A_w = \frac{k_c}{Q} \tag{4.26}$$

$$D = \frac{m_{AB}}{Q} \tag{4.27}$$

Thus the relationships shown in Fig. 4.3 reveal the makeup of parameters A_w and D. The first consists of a factor dependent on the characteristics of the cement, mostly the specific surface area, and a function of consistency; similarly the second is made up of a factor which represents an intrinsic water requirement of rock particles in rich mixtures and the same function of consistency.

Equation 4.25 represents a straight line in the water-ratio diagram from $u_w = 0$ at $x = 0$ to the point $u_w = E$ at $x = 1.0$. We designate the second point as point E on the water-ratio diagram. The value of E, to reiterate, is the water-requirement coefficient of the mixtures within the BF category, and its value is larger the softer the consistency.

The factor m_{BF} depends mostly on the voids content of the aggregate and secondarily on the aggregate gradation. Because the voids content in aggregates is in general larger the smaller the range of sizes from minimum to maximum, the factor for a mortar is relatively high, as shown for example by the relatively large value of $-m_{BF}$ in Fig. 4.4.

Before leaving this subject it is interesting to consider the alternative of expressing the water ratio as a function of the ratio of *cement* to total solids, rather than the aggregate to total solids. Doing so leads to the following expressions:

$$(u_w)_{AB} = D + c_1 y$$

and

$$(u_w)_{BF} = E - Ey = E(1 - y)$$

where

$$y = \frac{cv_c^o}{av_a + bv_b}$$

and c_1 is a water-requirement factor for a given consistency; specifically,

$$c_1 = A_w - D$$

Notice that when considered this way, the water requirements of mixtures in the *BF* category seem to be a function of the cement content and that it increases as the cement content decreases, whereas for the *AB* category the water ratio increases as the cement content increases. This comparison underlines the point already made that there is a fundamental difference between the *AB* and *BF* categories of concrete mixtures.

5.2 Construction of a Water-ratio Diagram for Two Categories

Let us now construct a water-ratio diagram, using the water-requirement factors noted in the lower graph of Fig. 4.7, which is to say, data based on the remolding test for consistency. The values are

$$k_c = 2.49$$

$$m_{AB} = 0.319$$

$$m_{BF} = 0.700$$

The aggregate is sand and gravel, continuously graded, with a maximum size of $\frac{3}{4}$ in., its voids content in the dry rodded state is 0.232, the corresponding voids ratio being $C = 232/768 = 0.302$. These data apply to any consistency within the practical range of the remolding test, but a water-ratio diagram represents a specific consistency. We calculate the parameters of a diagram for $R = 50$:

$$\log \frac{50}{0.02} = 3.398 = Q_{50}$$

The instrument constant was $R_o = 0.02$. Then, using Eqs. 4.26, 4.27, and 4.25, we obtain the parameters of the diagram as follows:

$$(A_w)_{50} = \frac{2.49}{3.398} = 0.733$$

$$D_{50} = \frac{0.319}{3.398} = 0.094$$

$$E_{50} = \frac{0.700}{3.398} = 0.206$$

These three values give the terminal points for the water-ratio diagram.

We also make use of lines for different constant water-cement ratios, which are established as follows. We use Eq. 3.26 which is

$$\frac{w}{c} = \frac{v_c^o}{w_w} \cdot \frac{u_w}{1 - x} = 0.309 \frac{u_w}{1 - x} \qquad (4.28)$$

The numerical coefficient corresponds to $v_c^o = 0.31$, cm³/g, which according to Table 2.2 is average for Type I cement, and $v_w = 1.002$, cm³/g, which is the specific volume of water at room temperature. For construction of the diagram it is convenient to select a value of u_w, assign a value for w/c, and solve for x, thus, for $u_w = 0.3$,

$$x = 1 - \frac{0.094}{w/c}$$

For example, the line for $w/c = 0.35$ crosses the line $u_w = 0.3$ at

$$x = 1 - \frac{0.094}{0.35} = 0.731$$

and for $w/c = 0.8$, $x = 0.882$.

The two water-ratio diagrams defined by the numerical values given above, together with the water-cement ratio line for $w/c = 0.35$, are shown in Fig. 4.8. Comparing Fig. 4.8 with Fig. 4.7, we see graphically what has already been demonstrated analytically, that line OE in Fig. 4.8 corresponds to the straight line to the origin in Fig. 4.7, and the line from A_w to B (and on to D) corresponds to the line from k_c to B.

The water-cement ratio line in Fig. 4.8 is approximately the lower limit for practical concrete mixtures; that is to say, in nearly all concrete the paste has a water-cement ratio higher than 0.35. Therefore we are interested mostly in mixtures for which x is upwards of 0.80. However, the scale of Fig. 4.8 is too small for detailed consideration of that region. Therefore let us construct a diagram for values of x from 0.8 to 1.0. We

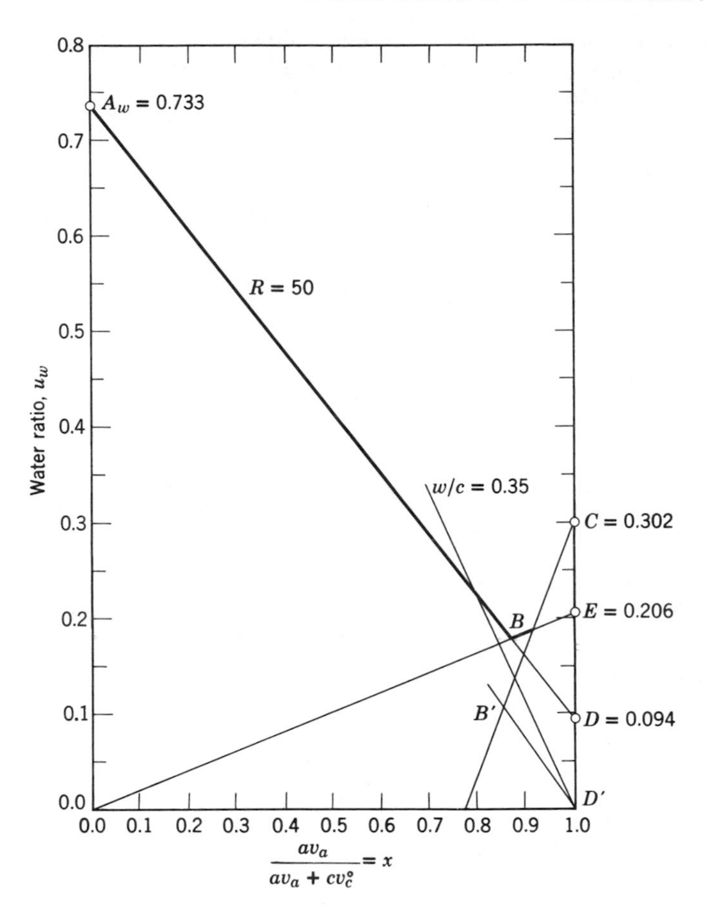

Fig. 4.8. A water-ratio diagram constructed from water-requirement factors.

can use the same values for E and D as before, but to establish slopes we need values of u_w at $x = 0.8$. By using Eq. 4.24 we obtain a point on the A_w to D line at $x = 0.8$:

$$(u_w)_{x=0.8} = 0.2A_w + 0.8D$$

For the other line,

$$(u_w)_{x=0.8} = 0.8E$$

For the data of Fig. 4.8, we have

for the point on line A_wD at $x = 0.8$, $u_w = 0.222$
for the point on line OE at $x = 0.8$, $u_w = 0.165$

Figure 4.9 was constructed with the data given above and is of course a magnification of part of Fig. 4.8; but it contains features not seen in Fig. 4.8, particularly the sector from F to C, and some plotted data points.

Once we have established the diagram for basic consistency ($R = 50$ for the apparatus used), it would be easy for us to establish the diagram for any other consistency or to find the value of the water ratio or water-cement ratio for a given mix at any other consistency. For example, if the desired consistency corresponds to $R = 20$, the consistency factor is $Q = \log(20/0.02) = 3.000$. For $R = 50$, $Q = 3.398$, and thus the desired water ratio or water-cement ratio for any specific value of x is $3.398/3.000 = 1.13$ times the value for $R = 50$, by Eq. 4.12. The result of such calculations is shown by the broken line in Fig. 4.9.

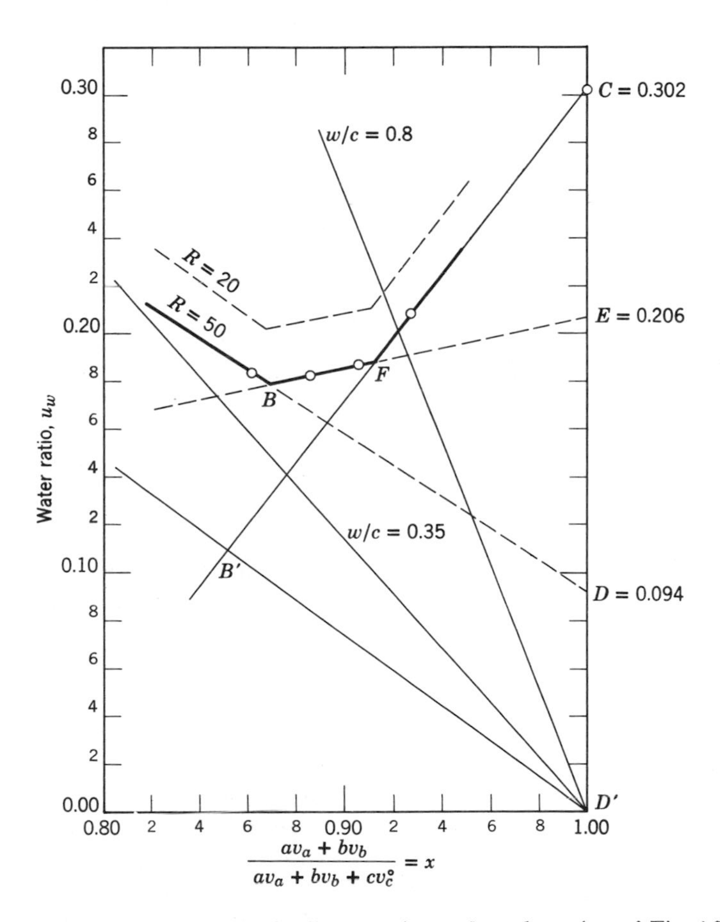

Fig. 4.9. A partial water-ratio diagram. (An enlarged portion of Fig. 4.8.)

5.2.1 Calculation of the mix at point *B*. The proportions for the mixture at point B are of special interest in that this mixture is the one that has the highest solids content and lowest water content for the given materials and given consistency; in some respects, it is the optimum mix. At point B, Eqs. 4.15 and 4.17 give the same value of u_w; hence

$$k_c + (m_{AB})M_B = m_{BF}M_B$$

and thus

$$M_B = \frac{k_c}{m_{BF} - m_{AB}} \tag{4.29}$$

where M_B is the mix in solid-volume units at point B. The corresponding value of x_B is readily computed from the relationship $x = M/(1 + M)$.

For the data in the lower graph of Fig. 4.4, for example,

$$M_B = \frac{-4.1}{-1.18 + 0.52} = 6.26$$

and

$$x_B = \frac{6.26}{7.26} = 0.862$$

which is to say the mix is 1:6.26 and the aggregate constitutes 86.2 per cent of the total solids.

It may be noticed that in the calculations above it is assumed that the coordinates of point B are those of the point of intersection of the line A_wD with the line OE, thus indicating that there is no transition section between the two curves. If there is a transition section, the calculated value of $(u_w)_B$ corresponding to the calculated value of x_B will be too small by some small amount.

Data show clearly that the *voids*-ratio diagram is rounded in the vicinity of point B, but there is some indication that the rounding is due to the increase in the air-water ratio as particle interference begins to influence the consistencies of the mixtures. If this is the case, the abrupt change in the *water*-ratio curve shown in a few diagrams, which at first seems unnatural, becomes more understandable and acceptable as a reliable observation; this is to say that there is some reason to believe that the point of intersection of the two branches of the water-ratio diagram may give a very close approximation of the actual point B on the water-ratio diagram. However if Z is the air-water ratio for the mixtures within the AB category, $(u_w)_B(1 + Z)$ does not give the correct value of the u_B on the voids-ratio curve; the calculated value will be somewhat too low.

Although this matter of uncertainty should not go unnoticed, we must keep in mind that our principal purpose is to identify the determinants of water requirement and voids content of fresh concrete by an analytical procedure. This being so, an approximation of the kind and degree under discussion is quite justifiable.

Anyone experienced with concrete knows that under field conditions now considered suitable for making concrete, the water requirement of a given mixture cannot be calculated exactly from data obtained in the laboratory, for it will vary with ambient conditions as well as with variations in the grading and perhaps other characteristics of the aggregate, from time to time. (Such variations can be controlled, but it usually considered not worthwhile to do so.) Therefore we may consider that the procedure under discussion is one that facilitates analysis and not be too much concerned with the refinements required for higher accuracy.

5.3 *Mixtures in the FC Category*

In Table 4.6 are figures derived from the data of Table 4.5. The third column gives the average value of three or four values of $k_R(1 + M)$. The fourth column gives the water-requirement factor from which the value of E can be calculated for any desired consistency. The values in column 7 were obtained by multiplying the values in column 5 by x, as per Eq. 4.25. It should be noted that when E is calculated for each mix indi-

Table 4.6 Data for Figures 4.8 and 4.9

1	2[a]	3[b]	4[c]	5[d]	6	7[e]	8[f]	9
						u_w		
Reference Number	M	$k_R(1 + M)$	$\frac{k_R(1 + M)}{M}$	E_{50}	x	$R = 50$	$R = 20$	Category
239	12.75	9.75	0.765	0.225	0.927	0.208	0.235	*AB*
244	9.65	6.78	0.702	0.206	0.906	0.187	0.211	*BF*
249	7.79	5.45	0.700	0.206	0.886	0.182	0.206	*BF*
253	6.23	4.48	0.719	0.212	0.862	0.183	0.207	*FC*

[a] The data are from column 2, Table 4.5.
[b] The data are from column 6, Table 4.5.
[c] For the *BF* category (see column 9) these ratios are m_{BF}; for the others, see text.
[d] The values are (Col. 4)/Q_1 = (Col. 4)/3.398 = E_{50} for R = 50, the basic consistency. For the *BF* category E_{50} = 0.206.
[e] The product (Col. 5) × (Col. 6), which is the water ratio.
[f] Calculated from column 7 by Eq. 4.12; the factor is 1.13.

vidually it is not necessary to consider whether the mix belongs in one category or another; the procedure amounts to establishing a consistency curve for each mix, as in Fig. 4.1, and then taking selected values from the curves, expressing the results in terms of x and u_w.

5.3.1 Relationships for basic consistency. We may note first that a set of four points for basic consistency ($R = 50$), as they appear in Fig. 4.9, correspond to the four points on the lower curve of Fig. 4.7. The line OE in Fig. 4.9 is based on the average water requirement of two mixtures in which the aggregate grading (per cent of sand) was not the same. The upward deviation of the point to the left of point B has already been explained: the mixture belongs in the AB category. The substantial upward deviation of the point farthest to the right, seen before in Fig. 4.7, will now be discussed.

In Fig. 4.9 we see at once that the point for $x = 0.927$ falls on the line from point B' to point C. That this is not a unique occurrence is shown by various other data. The data in Fig. 4.10 (from data by Swayze and Gruenwald [6]) particularly show the same kind of relationship. Figures 4.10*a*, 4.10*b*, and 4.10*c* contain three different series of sand and gravel mixtures, the aggregates having maximum sizes of $1\frac{1}{2}$, $\frac{3}{4}$, and $\frac{3}{8}$ in. Figure 4.10*d* represents mixes made with the same sand, but with crushed stone for coarse aggregate. All these data are based on the slump test.

Since a point on the line $B'C$ represents the voids ratio of the aggregate in its compacted state, it is evident that at point F, and at any other point between F and C, the sum of the volumes of cement and water are just equal to the void volume of the compacted aggregate. However as it actually exists in the concrete mixture, the aggregate cannot be in the compacted state; this is shown by the fact that the mixture was able to slump about 1.7 in., thereby indicating that aggregate particles have some freedom for movement and the mixture has some plasticity. Because the volume of the paste was only sufficient to fill the void volume of the aggregate in the compacted state, it follows that the aggregate dispersion and plasticity is due entirely to entrapped air, which is not shown in this diagram.

Weymouth [7] was perhaps the first to notice the role played by entrapped air in plastic mixtures; he spoke of "structural air." Now we see that for mixtures in the FC category at the basic consistency it plays the *essential* part as far as plasticity is concerned.

Another distinguishing characteristic of the FC category is the apparent interchangeability of cement and water, volume for volume, at constant consistency. At basic consistency a reduction in the solid volume of cement (increase of x) calls for an exactly equal increase in the volume of water. This means that the water-requirement equation for basic consistency is simply the equation of the line from B' to C, which is

$$u_w = (1 + C)x - 1 \tag{4.30}$$

Since

$$u_w = \frac{wv_w}{cv_c^o} \cdot \frac{1}{1 + M}$$

we may obtain an expression for the water-cement ratio as a function of M as follows:

$$\frac{(wv_w/cv_c^o)_b}{1 + M} = (1 + C)\frac{M}{1 + M} - 1$$

$$\left(\frac{wv_w}{cv_c^o}\right)_b = CM - 1 \tag{4.31}$$

where subscript b indicates basic consistency.

The conclusion is that the water requirement of a mixture in the FC category at basic consistency is determined by the voids content of the compacted aggregate. In the present case, where the compacting of the concrete was done by a low-energy procedure, the value of C is that obtained by the dry-rodding procedure, but when concrete is compacted by intensive vibration it is possible under certain circumstances to reduce C for the aggregate as it exists in concrete considerably below the figure for the dry-rodded state. (See Chapter 3, Section 5.7.3.)

In some series of mixtures the FC category does not appear even when the cement content is reduced below a reasonable practical limit; in such cases, the water-ratio diagram indicates that point E falls close to point C, making point F fall at a relatively large value of x or M. The factors producing this result are not entirely clear, but it seems that the grading of the aggregate is such that the aggregate voids are too small to accommodate more than a very small fraction of cement particles; such could be the case when there is a considerable percentage of material in the aggregate small enough to pass the No. 100 sieve, together with a normal fraction between the No. 100 and No. 50 sieves. There is some indication that coarsely graded fine aggregate causes point F to occur at relatively low values of x or M, which also means that it occurs at relatively low values of the water-cement ratio for basic consistency. If we consider the aggregate to be made up of two aggregates, fine and coarse, the voids content of the fine aggregate increases as the volume fraction of coarse aggregate in the mixture is increased as we have seen in Chapter 1. From this observation we may infer that with a given minimum particle size, the larger the maximum size of a graded aggregate, the larger the mean size of interstitial spaces and thus the larger the fraction of cement that

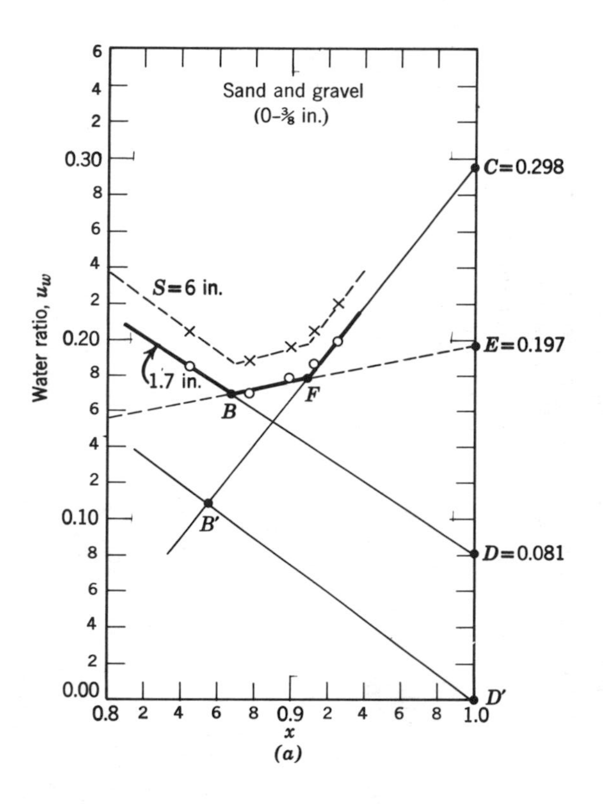
Sand and gravel
(0–⅜ in.)
S=6 in.
1.7 in.
C=0.298
E=0.197
D=0.081
D′
F
B
B′
Water ratio, u_w
0.30
0.20
0.10
0.00
x
(a)

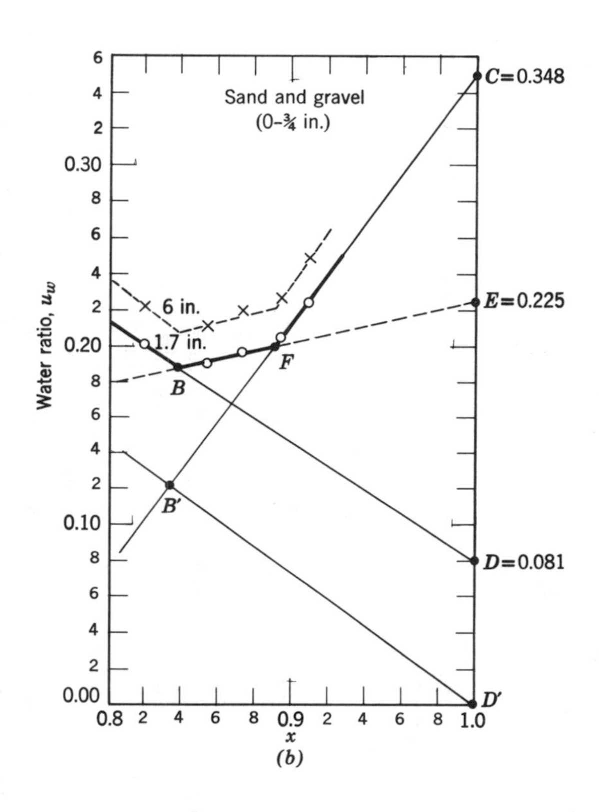
Sand and gravel
(0–¾ in.)
6 in.
1.7 in.
C=0.348
E=0.225
D=0.081
D′
F
B
B′
Water ratio, u_w
0.30
0.20
0.10
0.00
x
(b)

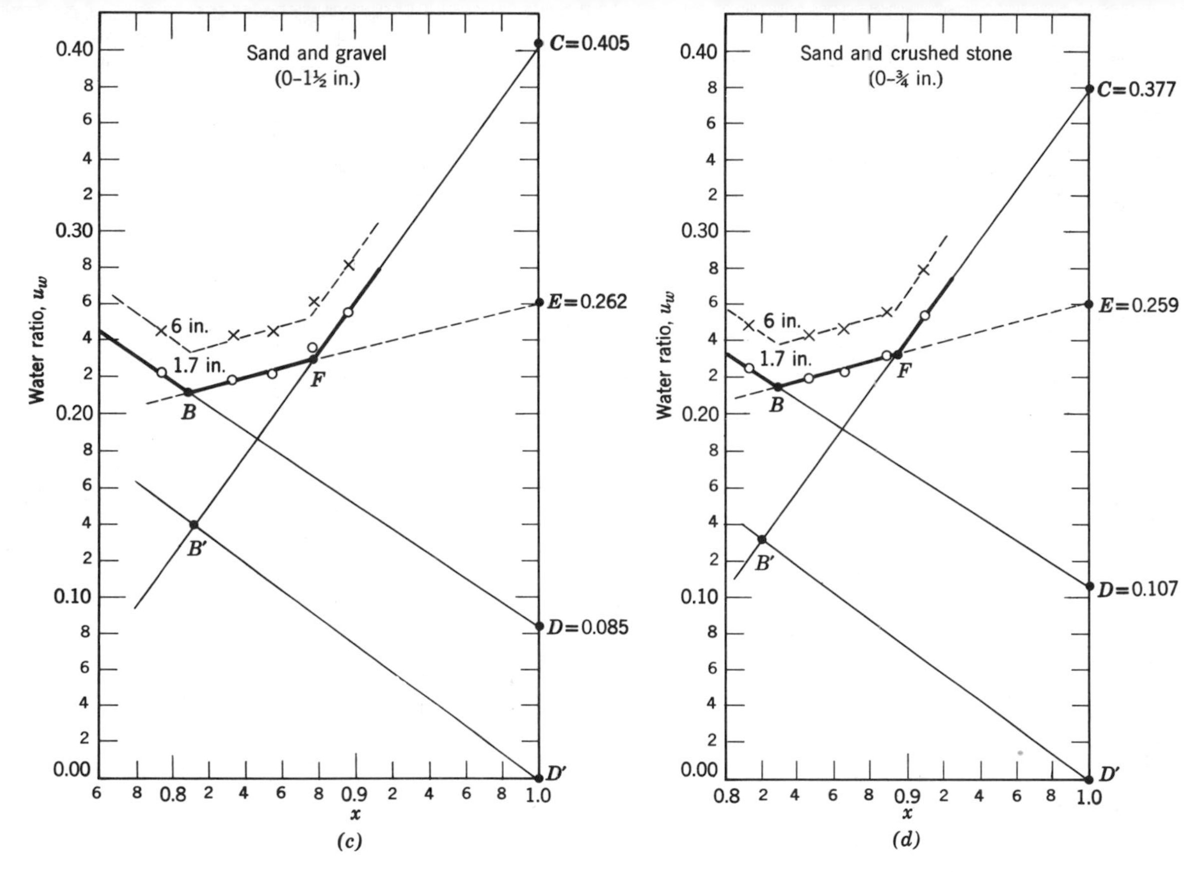

Fig. 4.10. Water-ratio diagrams for four series of concrete mixtures. (Data are from Swayze and Gruenwald [6].)

could be accommodated by the aggregate voids. (See also Chapter 2, Section 4.)

5.3.2 Relationships for softer consistencies. It would seem reasonable to expect point F to occur at the intersection of line OE with line $B'C$ regardless of the consistency, or, in Fig. 4.7, we might expect the point of departure F to be at a higher value of M the softer the consistency.

Experimental results show the contrary. Although the data are not entirely consistent, they indicate that point F occurs at the same value of M or x for all consistencies. Accordingly, in Fig. 4.9, point F for $R = 20$ is shown to be at the same value of x as for the basic consistency, $R = 50$.

The limit on the applicability of consistency tests is apparently somewhat beyond the range of the data shown in Figs. 4.9 and 4.10, and thus Eqs. 4.11 and 4.12 are applicable, making it possible to state the water requirement for any consistency of any mix within the practical range of category FC in terms of the water requirement at the basic consistency. Thus using the subscript b to indicate basic consistency, we find

$$\frac{w}{c} = \frac{Q_b}{Q}(CM - 1)\left(\frac{w}{c}\right)_b \tag{4.32}$$

In terms of the water ratio and x, the relationship is

$$u_w = \frac{Q_b}{Q}[(1 + C)x - 1](u_w)_b \tag{4.33}$$

It must be remembered that Eqs. 4.32 and 4.33 are for the FC category only.

5.3.3 Calculation of the mix at point *F*. The mixture at point F is of special interest not only because it marks the limit for certain empirical linear relationships already discussed, but also because it designates a boundary beyond which concrete mixtures have especially undesirable qualities. This mix is easy to determine if the parameters of the water-ratio diagram for the materials in question are known.

From Fig. 4.10d, for example, it is apparent that point F is the intersection of the line $B'C$ and the line OE_b. Thus, from Eqs. 4.22, 4.23, and 4.25 we may write for values of point F

$$\left(\frac{wv_w}{cv_c^o}\right) = E_bM_F$$

When $M = M_F$, Eq. 4.31 gives the same value; hence we find

$$CM_F - 1 = E_bM_F$$

and

$$M_F = \frac{1}{C - E_b} \tag{4.34}$$

or

$$x_F = \frac{1}{1 + (C - E_b)} \tag{4.35}$$

Using the data of Fig. 4.3 as an example, we have the following values: for basic consistency, $R = 50$ and, for the apparatus used, $R_o = 4 \times 10^{-3}$; $C = 0.42$; $m_{BF} = 1.381$. Then $Q = \log 50 - \log (4 \times 10^{-3}) = 4.097$, and by Eq. 4.25,

$$E_b = \frac{m_{BF}}{Q_b} = \frac{1.381}{4.097} = 0.337$$

and

$$M_F = \frac{1}{0.42 - 0.337} = 12.0$$

It is apparent that in Fig. 4.3 point F would have been found at a mix leaner than the leanest used in the series.

For another example, we may use Fig. 4.5, based on the slump test. For basic consistency, $S = 1.7$ in., $S_o = 10^6$, and $Q_b = 5.770$; with the percentage of sand giving minimum aggregate voids, $C = 0.26$; $m_{BF} = -1.07$. Then

$$E_b = \frac{-1.07}{-5.770} = 0.185$$

$$M_F = \frac{1}{0.26 - 0.185} = 13.3$$

In this case also point F is beyond the range of the data.

The water-cement ratio at point F, and for basic consistency, is found by using the following form of Eq. 4.31:

$$\left(\frac{w}{c}\right)_F = \frac{v_c^o}{v_w}(CM_F - 1) = 0.309(CM_F - 1) \tag{4.36}$$

Therefore, for Fig. 4.3, $M_F = 12.0$, $C = 0.42$, and $(w/c)_F = 1.25$. For Fig. 4.5, $M_F = 13.3$, $C = 0.26$, and $(w/c)_F = 0.76$. For Fig. 4.7, $M_F = 10.2$, $C = 0.302$, and $(w/c)_F = 0.64$ for the slump test basis, or for the remolding test basis $M_F = 10.4$, and $(w/c)_F = 0.66$. In each of these examples the water-cement ratio is so high that it indicates a thin, watery cement paste.

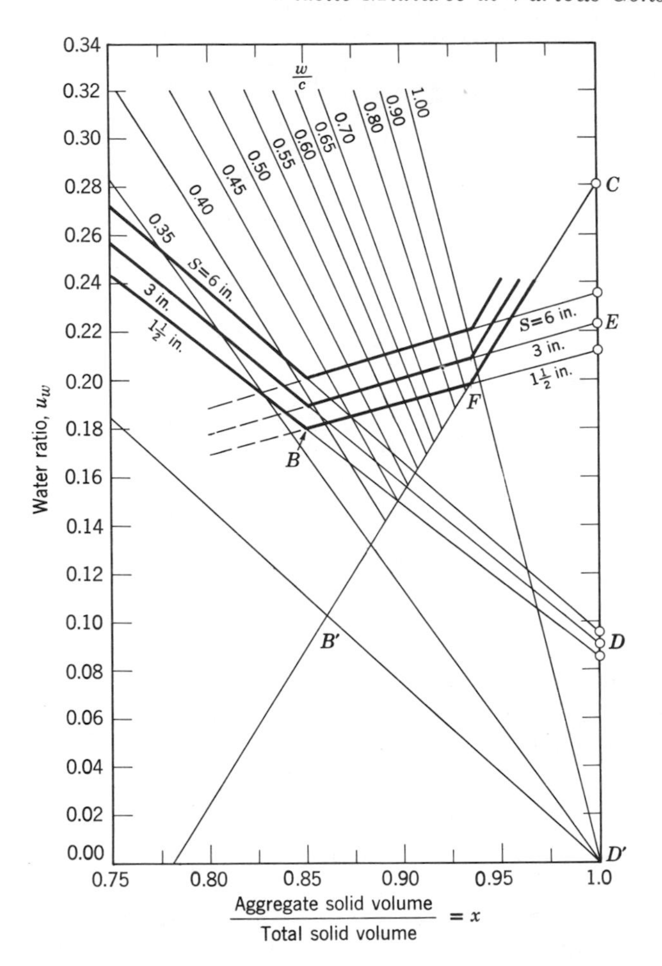

Fig. 4.11. A full-range water-ratio diagram.

5.4 A Full-range Water-ratio Diagram

By way of recapitulation, Fig. 4.11 is presented. It is a complete water-ratio diagram for three different consistencies. The basic data are as follows:

$$k_c = -4.2$$

$$m_{AB} = 0.49$$

$$m_{BF} = 1.23$$

$$C = 0.28$$

Derived parameters for *basic consistency* (slump = 1.7 in.) are

By Eq. 4.29

$$M_B = \frac{-4.2}{-1.23 + 0.49} = 5.675; \; x_B = \frac{5.675}{6.675} = 0.850$$

By Eq. 4.25

$$E_b = \frac{-1.23}{-5.770} = 0.213$$

and

$$(u_w)_B = E_b x_b = 0.213 \times 0.850 = 0.181$$

By Eq. 4.35

$$x_F = \frac{1}{1 + 0.28 - 0.213} = 0.937$$

By Eq. 4.25

$$(u_w)_F = E_b x_F = 0.213 \times 0.937 = 0.200$$

By Eq. 4.27

$$D = \frac{-0.49}{-5.770} = 0.085$$

Having obtained the parameters for basic consistency, we adjust those that are functions of consistency by means of the factors given in the last

Table 4.7 Parameters for Figure 4.11

	Values for Slump and Factor Indicated			
Parameter	1.5 in. 0.991	1.7 in. 1.00	3 in. 1.045	6 in. 1.105
$(u_w)_B$	0.179	0.181	0.189	0.200
E_b	0.211	0.213	0.222	0.235
$(u_w)_F$	0.198	0.200	0.209	0.221
D	0.084	0.085	0.089	0.094

column of Table 4.1. The resulting values are given in Table 4.7. With the values of D and the corresponding coordinates of points B given, the lines for the AB category are established. The coordinates of point F at basic consistency are given. The other points are assumed to fall at the same value of x but at lower or higher values of u_w, according to respective relative consistency factors. The lines of constant water-cement ratio were established by means of Eq. 4.28.

6. CONSTRUCTION OF A VOIDS-RATIO DIAGRAM

We learned in Chapter 3 that for mixtures in the AB category the ratio of the volume of air to that of water Z is a constant for a given consistency. Therefore, if the value of Z is known, the AB sector of a voids-ratio diagram can easily be constructed from the water-ratio diagram. The value of u at any convenient value of x to the left of B can be obtained from the value of u_w by multiplying it by $(1 + Z)$; the value of D^* is obtained from D in the same way. (See Eqs. 3.37 and 3.28, Chapter 3.)

However values for u in the BF category cannot be so estimated because Z increases with x; in fact, it begins to increase before point B is reached. A good approximation of the voids-ratio curve can nevertheless be constructed as follows. Assume that

$$u_B = (1 + Z)(u_w)_B \tag{4.37}$$

Having located u_B by means of Eq. 4.37 and the given value of $(u_w)_B$, we draw a straight line from that point to point C. That line is a good approximation of the locus of the voids-ratio curve in the BF region except at or very near point B, where the indicated value will be a little too low.

The value of Z we should use in Eq. 4.37 is the observed value for the AB range at the given consistency. For purposes of illustration, we may use the figures due to Weymouth given in the accompanying table.

Slump (in.)	Z
1	0.18
2	0.16
3	0.14
4	0.12
5	0.10
6	0.08

Figure 4.12 gives an example of a water-ratio diagram and a derived voids-ratio diagram based on the data in Fig. 4.11. Figure 4.13 is another

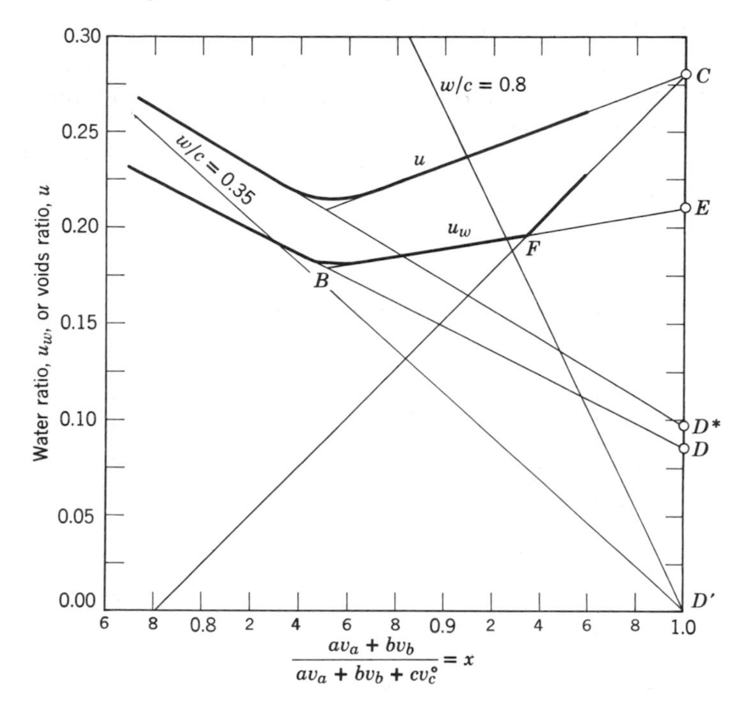

Fig. 4.12. A voids-ratio diagram (upper curve) derived from the water-ratio diagram (lower curve). The assumed slump is $1\frac{1}{2}$ in.; $Z = 0.167$; $A_w = 0.721$; $D = 0.84$; $E = 0.211$.

example based on the data of Fig. 4.3, and in this case the plotted points show the degree to which the method of construction just described represents experimental observations. It seems probable that the relationship between the water ratio and voids ratio assumed in making the constructions above is not equally valid for the *BF* and *FC* categories.

The air content at any value of x can be calculated from the difference between Eqs. 3.20 and 3.21 of Chapter 3, that is,

$$\frac{A}{V} = \frac{u - u_w}{1 + u} \tag{4.38}$$

For Fig. 4.12, based on a slump of 1.5 in., the calculated air content at point B is 2.5 per cent, and at $w/c = 0.8$, $x = 0.91$, it is 3.8 per cent; for a slump of 6 in., the corresponding values are 1.2 and 2.4 per cent. Although different air contents are to be expected with different aggregates (see Chapter 7), these estimates, based on assumed values of Z, are in line with common experience. As we shall see, the increase in air content accompanying a decrease in cement content (increase in x) indicates that the aggregate is not correctly graded for such mixtures. We shall see also that the aggregate was not correctly graded for very rich mixtures; the value of D in Fig. 4.12 is too large.

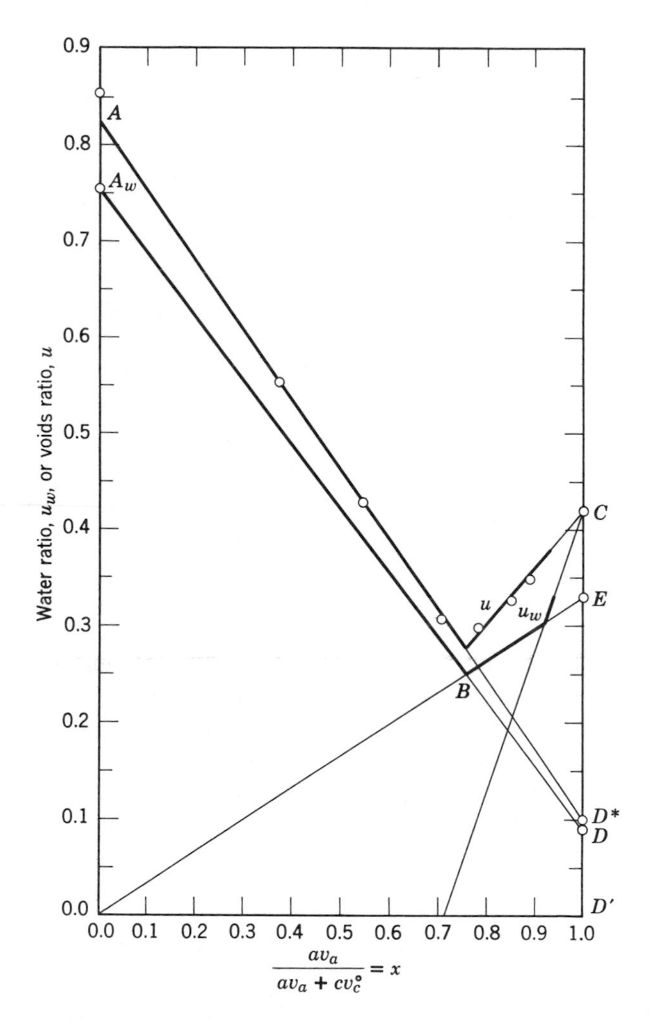

Fig. 4.13. A voids-ratio diagram for mortars derived from the water-ratio diagram and showing experimentally determined points. Aggregate No. 265-11 with a size range of (#200–#4) was used. The remolding number is 50; $C = 0.42$; $E = 0.331$; $D = 0.092$; $Z = 0.10$; and $A_w = 0.757$.

7. OPTIMUM RATIO TO FINE TO COARSE AGGREGATE

In some of the series of mixtures already discussed we note that the percentage of sand in the aggregate was not the same in all mixes, and we speak of the optimum percentage of sand, or optimum ratio of fine to coarse aggregate. We now consider that topic, defining the term and considering some of the methods used for finding the optimum value.

7.1 Definition

Changing the percentage of sand is one way of changing the grading of an aggregate. It has been found by experience that when the grading is changed by this means, the water requirement of a given mix is likely also to be changed. Abrams [1] expressed the grading change in terms of his fineness modulus, and by a large number of systematic tests he established the maximum permissible fineness modulus for various mixes, size ranges of aggregate, and various kinds of aggregate; in each case the maximum permissible fineness modulus corresponded to the percentage of sand that gave the lowest water requirement and highest potential strength. Powers [4] made systematic experiments in which the amount of cement paste of a given composition required for a given consistency, as determined by the remolding test, was studied as a function of the percentage of sand; the percentage requiring the least volume of cement paste was considered to be the optimum percentage. The different percentages of sand indicated in Figs. 4.5, 4.6, and 4.7 were established by the method just mentioned.

A variant of the method used by Powers was described by Wuerpel [8]. Keeping the mix and water-cement ratio fixed, Wuerpel varied the percentage of sand to find the combination requiring the least time of vibration; he used a prototype of the Vebe apparatus for measuring consistency. (See Chapter 10.) Swayze and Gruenwald [6] used another procedure. They made a series of hand-mixed trial batches for each of three classes of concrete, varying the percentage of sand, and selecting the optimum on the basis of specific weight, and on workability as judged by the worker; in general, the best workability, as indicated by the maximum unit weight (minimum water and air content), was found to be compatible with judgment. The mixtures represented in Fig. 4.10 were established by this procedure.

By any of the procedures described above the implied definition of optimum percentage of sand is the same: that percentage is optimum which gives a minimum water requirement for a given mix and consistency. The definition rests on the assumption that it is always desirable to make the water requirement as low as possible, at the same time recognizing that the consistency must be suited to the conditions under which the concrete is being worked.

Although the definition of optimum percentage of sand is applicable to all mixtures, the factors to be considered are somewhat different for the different categories; therefore we discuss the three categories in turn.

7.2 The AB Category

The optimum percentage of sand can be considered in terms of the water-requirement factor of the aggregate, or in terms of the voids ratio or water ratio of a given mix at a given consistency. It can be considered in terms of the paste requirement at a given water ratio and consistency or the water-cement ratio at a given cement content and consistency. For mixtures in the AB category, it is convenient to consider the water-requirement factor of the aggregate in terms of D, which together with the consistency factor Q leads to the water-requirement factor m_{AB}. In some cases when only the total voids ratio is given it is necessary to use D^* and to obtain D by means of an estimated value for Z.

In Chapter 3 we have seen that the parameter D^* for basic consistency is a function of the specific surface area of the aggregate, the function being such that the value of D^* increases with specific surface area when the specific surface area exceeds some threshold value; at low values of specific surface the value of D^* is apparently established by a function of the volume fraction of aggregate material per unit volume of mixture.

On the basis of what we know now it cannot be said whether the specific surface area correctly evaluates the combined effects of mean particle size and characteristic particle shape. The uncertainty cannot be cleared up until an unequivocal measure of surface area is at hand. There is also a question whether surface texture, or roughness, has an effect distinct from specific surface area, and this involves a related question about "molecular roughness" versus "hydrodynamic roughness."

The surface modulus and sphericity factor as applied to aggregates of particles having various shapes cannot give accurate values for surface area partly, because the mean size, that is, the mean volume diameter of a given size group, is a function of particle shape, and in extreme cases, the difference between nominal and actual mean size may be considerable. (See Section 3.5.5, Chapter 1.)

Until such questions as those raised above are cleared up, it seems necessary to deal with experimental values in terms of such measurable parameters as we are now able to produce, hoping that we shall not be led to false conclusions, but at the same time expecting not always to be able to decide whether an assumed "law" has proved false or whether the numbers used have not accurately represented what they were supposed to.

It should be understood that the questions raised above about the adequacy of specific surface area as a determinant of the water-requirement factor of an aggregate pertain to the AB category, within which the factors that influence the voids content of the aggregate have no bearing on the water requirement factor. However for the mix at point B and for those in the other categories, particle shape has an effect distinct from its effect on specific surface area, for it has a marked influence on the content of voids in the aggregate.

7.2.1 Maximum permissible water-requirement factor. As is shown in Chapter 3, Fig. 3.13, the specific surface area of the aggregate apparently has only a small effect on the water-requirement factor (as reflected in the value of D^* in this case) until the surface factor SM/ψ exceeds some small value, which might be called a "threshold value." This value should not be exceeded; otherwise the water-requirement factor of the aggregate will be higher than it needs to be. The threshold value of specific surface area cannot be told definitely on the basis of available data, mainly because of the scarcity of data on measured specific surface, and questions concerning the accuracy of such data as there are.

In terms of Loudon's data on specific surface area and the sphericity factor for individual size groups (Chapter 1, Section 5.1) it appears that the specific surface area should not exceed about 30 cm^2/cm^3, which is to say that the surface factor SM/ψ_L should not exceed about 5.38. (See Eq. 1.3.) With such a limit on the specific surface area of the aggregate, we should expect parameter D^* to be about 0.06 at basic consistency; the ratio of air volume to water volume would be between 1:6 and 1:7, which means that the value of Z would be about 0.15. Hence the value of D should be about $0.06/1.15 = 0.052$, say 0.05. Assuming the slump at the level of the basic water content to be 1.7 in., we find the consistency factor to be $\log 1.7 - 6$, or -5.770. (See Table 4-1.) Thus the maximum water-requirement factor for aggregate used in concrete of category AB should be

$$m_{AB} = 0.050 \times (-5.770) = -0.29$$

Such a water-requirement factor should be attainable by adjusting the percentage of sand, provided that not more than about 5 per cent of the sand passes the No. 100 sieve, and practically none passes the No. 200. It might be possible to attain even lower values for m_{AB} if practically all material passing a No. 50 sieve is eliminated.

The specific surface area of the combined aggregate can be made to meet the above requirement for specific surface area, or for surface factor

SM/ψ_L, provided that the maximum size is sufficiently large. Any single size group smaller than the fifth (#16–#8) has a specific surface area higher than the permissible limit for this category. Presumably, size group No. 6 could be used alone ($SM = 3.12$), but the leanest mix in the AB category would be a very rich one because of the large voids ratio any single size group, the value of C being upwards of 0.7. In other words the range of the AB category would be unduly limited. If the maximum size of the aggregate is increased to $\frac{3}{8}$ in. by using size group No. 7 as coarse aggregate (surface modulus = 0.78), the range of the AB category mixes may thereby be made sufficient for some purposes.

For example, if the fine aggregate has a surface modulus of 21, the permissible weight fraction of sand n_a would be

$$n_a = \frac{(SM) - (SM)_b}{(SM)_a - (SM)_b} = \frac{5.38 - 0.78}{21.0 - 0.78} = 0.227$$

or about 23 per cent sand. Such an aggregate would have a voids ratio of about $C = 0.4$, which is not very low. It is clear that to achieve a relatively low voids ratio for mixtures in the AB category, the value of C must be low, and so the size range must be as large as other conditions will permit.

7.2.2 Optimum percentage at point *B*. The leanest mix in the AB category is that which falls at point B. For this mix we may define the optimum percentage of sand for given materials as that giving the lowest value of the water ratio $(u_w)_B$ for the given maximum size. In Chapter 3 we give an empirical relationship in terms of the voids ratio (Eq. 3.40), which has the following equivalent in terms of water ratio:

$$(u_w)_B = \left(\frac{A_w C}{1 + A_w + C}\right)(1 + 0.01(\sigma_{ab})^{1/2})$$

A_w is a constant for a given cement. The variable factors are the voids ratio of the combined aggregate C and its specific surface σ_{ab}. Since the values of C and σ_{ab} are both functions of the percentage of sand, we see, as is discussed above, that finding the optimum percentage of sand for point B is a matter of finding the optimum combination of voids ratio and specific surface area; reducing the percentage reduces the specific surface area, but beyond a certain point it increases the voids ratio. (This subject is dealt with from another point of view in Chapter 6, Section 4.)

7.2.3 Use of vibration. We can achieve a lower value for the water-requirement factor than that corresponding to $D^* = 0.06$ (slump basis), which is discussed in Section 7.2.1, and a relatively wide range of mixtures in the AB category for a given aggregate by the use of intensive vibration for compacting concrete mixtures having a no-slump consistency, that is, mixtures that are not plastic except while in a state of vibration. As is shown in Fig. 3.19, the value of C is reduced, and a value for parameter D^* of 0.04 can be achieved by this procedure using a continuously graded sand-and-gravel aggregate comprising nine size groups. With Z about 0.10, the corresponding value of D is about 0.035. Figure 3.19 indicates that the AB category extended to about $x = 0.875$ or $M = 7.00$. The lowest voids-cement ratio was 1.20 cm^3 per cubic centimeter of cement, or assuming $Z = 0.1$, the water-cement ratio by weight was 0.336. The cement content was

$$\frac{cv_c^o}{V} = \frac{1 - x}{1 - u} = \frac{0.125}{1.15} = 0.109$$

or 0.109×5440 lb/yd^3 = 593 lb of cement per cubic yard of concrete. Under standard curing conditions, such concrete is able to produce a compressive strength of about 10,000 psi in 28 days.

The foregoing is an example of what can be achieved with mixtures of this type under laboratory or factory conditions where vibration can be made as intense as is required.

7.3 The BF Category

For mixtures within the *BF* category, the optimum percentage of sand is not that which gives negligible surface area. For a given proportion of cement, it is a definite value which gives a minimum water requirement: either to use more or less than the optimum percentage requires an increase of water content. In most cases the optimum is not a point, but a small range of percentages within which the water requirements are virtually the same.

Whether or not the optimum percentage of sand marks an optimum specific surface area has long been a matter of controversy. It was shown by Abrams [1] that specific surface area alone is not the sole determinant of the water-requirement factor; indeed Abrams seemed to feel that surface area was irrelevant, and he developed the fineness modulus, which has no general relationship to specific surface area. However there is a reasonable basis for assuming that specific surface area of an aggregate *together with its void content* constitute a determinant of the water-requirement factor; development of this idea by Charles T. Kennedy and others is taken up in Chapter 5.

7.3.1 Minimum paste content. When the composition of the cement paste in concrete is preselected on the basis of specifications or other considerations, the optimum percentage of sand may be selected on the basis of that which requires the least amount of paste for a given consistency. Examples of variation in paste content with change in percentage of sand are shown in Fig. 4.14 for three different water-cement ratios. The optimum percentage is generally found to be lower the lower the water-cement ratio of the paste, although the values for different pastes are often found to be not much different, as in the present case.

7.3.2 Minimum water ratio, for a given ratio of aggregate to cement. If the proportion of aggregate to cement is kept constant while the percentage of sand is varied, the optimum percentage will be that which gives the smallest water requirement for the chosen consistency. Figure 4.15 is an example that shows the effect of changing the proportion of sand from 100 per cent to none. For comparison, the voids-ratio curve for the aggregate alone is shown. The point on the water-ratio curve for no sand was estimated on the basis of the voids ratio of the coarse aggregate and the probable dispersion. This graph brings out the point clearly that the optimum percentage of sand in an aggregate for concrete in the *BF* category is smaller than the percentage giving a minimum voids content for the aggregate alone.

Although the ASTM specification (C33) requires that the fineness modulus shall not be less than 2.3, it has been shown repeatedly that still finer sands can be used without penalty, *provided that the optimum per-*

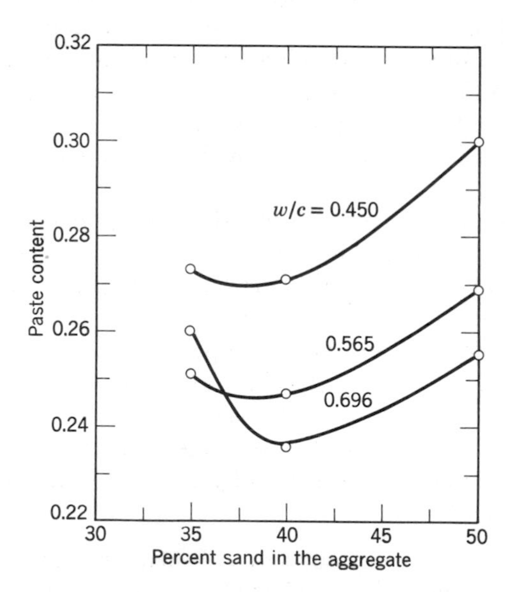

Fig. 4.14. An example of the influence of the proportion of sand in the aggregate on the paste content of concrete at different water-cement ratios. (The paste volume is the water volume plus the cement volume. The remolding number is 50, with the ring clearance set at 2.625 in.)

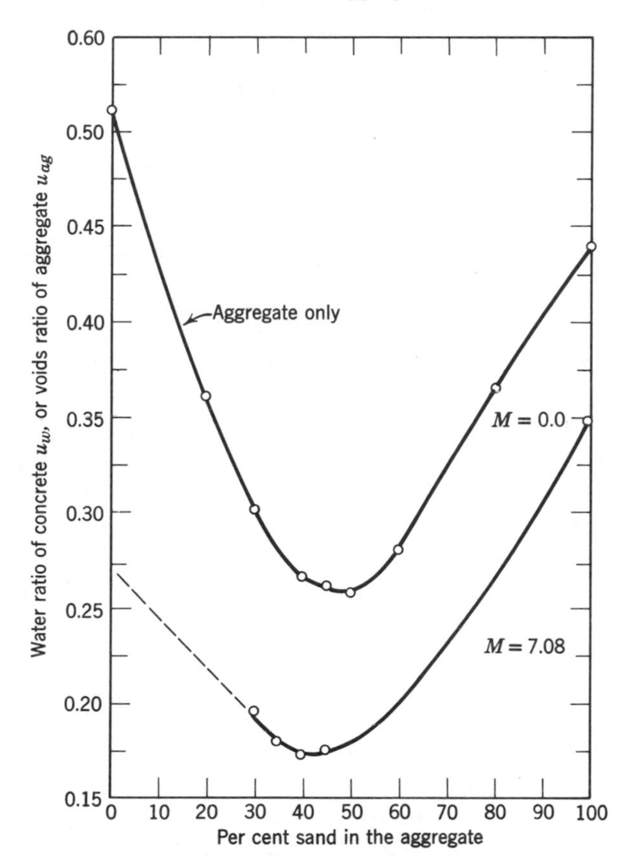

Fig. 4.15. The effect on the water ratio of varying the percentage of sand in a given mix, with the voids-ratio curve for the aggregates for comparison. The size-range of the sand-and-gravel aggregate was (0–1½ in.). (The data are from PCA Series 267.)

centage is used. With a given coarse aggregate and proportion of cement, the optimum percentage is smaller the finer the sand.

The ASTM specification also gives a maximum limit of 3.3 for the fineness modulus of sand. As applied to mixtures in the *AB* category, this restriction might be questioned, but for the other two categories it is well founded, for even at the optimum percentage the water requirement of the combined aggregate is likely to be higher than it is when a finer sand is used.

7.4 The FC Category

As is pointed out above, for mixtures in the *FC* category at basic water content, the volume of the cement-water paste is just equal to the volume of voids in the compacted aggregate; plasticity, such as it is in this category, is due to aggregate dispersion maintained by entrapped air. Data for systematic tests dealing expressly with mixtures of this nature are lacking. It seems reasonable to assume that, if the aggregate is clean, the optimum percentage of sand is that which gives a minimum voids content for the aggregate alone.

7.5 Recapitulation and Comparison of Requirements

In Section 7 we have seen that the same criterion for optimum percentage of sand is applicable to each category but that the optimum percentages are not the same for mixtures in different categories, except at the junction points or, perhaps more correctly speaking, the transition sectors. For the *AB* category, the specific surface area of the aggregate is the dominant factor; it must be kept at or below a certain threshold value which corresponds to a surface factor of about $SM/\psi_L = 5.38$. This figure might not be exactly correct for particles of different shapes and surface textures; presumably, a single figure would suffice if it were possible to measure the surface area exactly, and it would be correct for any mix within the category. Except that it defines the leanest mix that can be in the *AB* category, the voids content in the aggregate would be unimportant.

On the other hand, the quantity of sand needed in the *BF* category is generally greater than that needed to meet the surface area requirement for the *AB* category. Whether the surface area per se is a factor is a disputed point, but there is reason to believe that surface area and voids content together determine the optimum percentage of sand for any given ratio of aggregate to cement, with given materials. Generally, as the cement content is reduced, the optimum sand percentage increases toward the percentage that gives a minimum voids content in the aggregate alone, that percentage being indicated when point *F* is reached.

For the *FC* category, the considerations are similar to those for the *BF* category, except that the optimum percentage seems to be the same for all mixtures within the category. The optimum percentage is that which gives a minimum content in the aggregate alone, provided the aggregate is clean.

8. OPTIMUM PASTE CONSISTENCY

As shown above, any complete series of mixtures made with given materials and having a certain consistency will show one point at which the water ratio or the voids ratio is at a minimum, the point being designated *B* on the water-ratio diagram. The water content diminishes as the

The Materials Research Laboratory boasts excellent characterization facilities. Scanning electron microscopes (including special particle size analysis facilities) and x-ray diffraction apparatus are routinely available and combined ion scattering and Auger electron spectroscopy instrumentation. In addition to MRL facilities, which include UV, IR and Raman spectral instruments, a commonly used infrared spectrometer and other analytical facilities are housed in the Mineral Constitution Laboratory, where they are used on a service basis. It is expected to add an energy-dispersive x-ray fluorescence unit for chemical analysis of cements, rocks and reaction products during the immediate future months.

aggregate content increases up to the proportion at point B, but beyond that point the more aggregate we add, the higher the water-plus-air content becomes, and the lower the solids content; the effect becomes very pronounced when point F is passed.

The point of maximum-solids content is not the point of maximum strength for the hardened concrete, but it is a particularly advantageous point with respect to overall properties, and especially to the strength needed by hardened concrete members to withstand stresses due to the shrinkage accompanying the gradual drying out of concrete.

In this section we examine further some of the factors that cause the solids content to diminish as the aggregate proportion is increased beyond point B and consider how the solids content can be kept from falling when we are producing a mixture that has a water-cement ratio higher than that at point B.

8.1 *The Concept of Paste Consistency versus that of Total Grading*

In Chapter 2, Section 4, Fig. 2.7, it is shown that the voids-ratio diagram for mixtures of sand and dry cement is qualitatively indistinguishable from a water-ratio diagram for plastic mixtures. This comparison gives us reason to suppose that because the diagram for the mixtures made without water had the same characteristics as the diagram for plastic mixtures, the diagram for the plastic mixtures simply reflects the manner in which the fine and coarse components fit together in the various proportions. Such a point of view leads to the conclusion that for either dry or plastic mixtures we are concerned with the total particle-size gradation from the smallest to the largest particle. However, such a conclusion is a dubious one, for, as is emphasized in Chapter 2, very small particles are incapable of forming the same kind of aggregations that geometrically similar but coarser particles do. Thus when we discussed the diagram for dry mixtures we explained the appearance of points F and B in terms of the way the cohesive *cement powder* fitted into the interstices of the coarser component, or was able to disperse the coarse component; instead of gradation, we considered powder volume, which is determined by the forces of interparticle attraction as well as by the particle-size distribution of the material.

After there has been sufficient time for the air to bleed out, the resulting dry cement has one characteristic powder volume when handled in a certain way. (See Section 3.6, Chapter 2.) Apparently this volume is augmented slightly by structural disturbances caused by the presence of sand particles, as is indicated by the value of D in the diagram for the dry mixtures. However, in wet mixtures we consider paste volume instead of powder volume, and that volume for a given cement is controlled by the volume of water; thus the space occupied by cement in concrete is not determined by the grading of the cement but by its solid volume together with the volume of water. The fact that some of the water may be lost by bleeding is not relevant here, for we are considering the conditions that determine the volume and consistency of a mixture at the time the consistency test is made, which is before there has been time for an appreciable amount of bleeding, except in extreme cases.

8.2 Definition of Optimum Consistency

Thinking of freshly mixed concrete as aggregate particles in a matrix composed of cement paste and air-filled spaces, we conclude that the voids content of the whole mixture is equal to the volume fraction of matrix multiplied by the voids content of the matrix. Because the matrix is predominantly cement paste, the voids content of the mixture is determined mainly by the paste content and the water content of the paste. Also, except when air is purposely entrained by means of an air-entrained agent, in a given mix the amount of air in the matrix bears a regular relationship to the amount of water in the paste. Thus it is feasible to consider the voids content of the mixture as a whole as a function of the paste content and the water content of the paste.

However, with a given aggregate the water content of the paste and the paste content are not independent variables. As the water content of the paste increases, the volume fraction of paste in the mixture at a given consistency diminishes; that is to say, if the proportion of aggregate is increased, the solid content of the paste must be diminished, and it follows that at some intermediate proportion of aggregate the solids content of the mixture will be at a maximum: in other words, it may be said that the solids content of the mixture will be at a maximum when the water content of the paste has a particular value. This water content denotes a paste composition that may be considered to be an optimum composition for the existing conditions.

At a given temperature, the water content of the paste also is the principal determinant of the consistency of the paste. Therefore it seems that there is an optimum paste consistency, which may be defined as that consistency at which the solids content of the paste and the paste content of the mixture are such that they produce the maximum solids content possible with the given materials. Using popular terms for lubricants, we may say that pastes in mixtures to the left of point B are "heavier," and those to the right are "lighter," than the optimum lubricant; at point B the paste is neither too heavy nor too light.

8.3 Paste Composition at Optimum Consistency

The consistency of a paste is not determined by composition alone. One important factor affecting consistency is temperature. Most data are based on mixtures made at room temperature, probably about 70°F (21°C); at any lower temperature (above the freezing point) a lower water-cement ratio is required to obtain the same consistency as that found at room temperature. At higher temperatures the opposite is true. Consider, for example, Fig. 4.10*a* or 4.10*b*, and suppose that the mix at point B were produced at 40°F instead of 70°; the water required for basic consistency would normally be less than that indicated by the diagram for 70°, and thus point B would be found at a correspondingly lower level on the graph, although only slightly lower. Then, because the line OE also passes through point B, we see that temperature also has some effect on the value of E for a given consistency or on the more general factor m_{BF}.

At a given temperature the consistency of a given paste can be made thinner by reducing the cohesive strength of the paste. This can be done by reducing interparticle attraction by means of a suitable surfactant, currently called a "water-reducing agent." (However, reduction of interparticle attraction should be allowed only with full knowledge and control of undesirable side effects; surfactants should not be used in mixtures to the right of point B unless the cement is supplemented by another suitable paste-making material such as certain mineral powders or purposely entrained air bubbles.)

Even with a given cement, optimum paste consistency is not the same for different aggregates. For example, in Fig. 4.10 the water-cement ratios of the pastes at point B, for basic consistency, are $(w/c)_B = 0.403, 0.368,$ and 0.352 for maximum sizes $\frac{3}{8}$, $\frac{3}{4}$, *and* $1\frac{1}{2}$ *in.*, respectively.* The different water-cement ratios of course signify different paste consistencies.

In this case the water-cement ratio of the optimum paste was higher the higher the surface modulus of the aggregate; in other cases, where the aggregate gradings may be related to each other in a different manner, the values of $(w/c)_B$ may be the same for different maximum sizes, or the values may be different for the same maximum size if the gradings of the sand differ, particularly with respect to the proportion of material in size groups 1 and 2; a gap between the maximum size of the cement particles and the smallest size in the aggregate generally requires the optimum paste to be relative stiff.

* These values were calculated from the coordinates of the points by relationship

$$\frac{w}{c} = 0.309 \frac{u_w}{(1 - x)}$$

At a given paste consistency, paste composition is influenced also by the presence of fine material other than cement. Material in the aggregate fine enough to pass the No. 200 sieve usually can be made to form paste with water,* and even though it is not included in the calculation of paste volume, it has an influence on the consistency associated with a given water cement ratio and thus influences the experimentally determined value of the water-cement ratio at the optimum paste consistency. Hence it is not a simple matter to relate paste composition at optimum consistency with the sieve analysis of the aggregate.

Paste composition may be varied also by introducing mineral powders other than cement, such materials being usually regarded as admixtures.

9. ADMIXTURES OF MINERAL POWDER

9.1 General Considerations

Regulation of the percentage of sand in the aggregate is a way of establishing at least a good first approximation of the best grading for mixtures in the *AB* category, including point *B*, provided that the sand is free from material that would pass through a No. 200 sieve, especially clay and silt. In the foregoing section we have seen that when the mix at point *B* is established in this way the paste in that mixture is regarded as having optimum consistency. We have seen also that the optimum paste is likely to have a water-cement ratio not over 0.40 by weight. Under standard conditions such a mixture is capable of producing a 28-day compressive strength (6 × 12 in. cylinder) upwards of 6500 psi, which is more strength than is ordinarily called for; consequently, as has been said before, most concrete mixtures used in practice are in the *BF* category, and probably many are in the *FC* category. In all such mixtures the paste cannot have optimum consistency unless the cement is supplemented by another paste-making material. Although such supplementation is not generally practiced, its desirability is recognized by a few, and it might well become more general where economic considerations justify it.

For mixtures to the right of point *B*, increasing the sand content above the percentage best for point *B* is helpful, provided the sand is not too coarse, but it does not restore optimum conditions. When the water-cement ratio is upwards of 0.6 by weight, as it is in much concrete, the characteristics of such mixtures are far from ideal; the principal short-

* Many mineral powders are able to form pastes having rheological properties like those of fresh cement paste when they are mixed with lime water [a saturated solution of $Ca(OH)_2$] but not when mixed with pure water. When used with cement, lime for the solution is furnished by the cement, usually in sufficient quantity before the end of the mixing period.

comings are lack of cohesiveness, "harshness," development of internal fissures during bleeding, and, above all, the solids content is not as high as it might be. When mature, such concrete has less strength than it could have at the same water-cement ratio, and it is more permeable than it might be; it is likely to have a relatively unfavorable combination of strength and volume-change characteristics, leading to a tendency toward excessive cracking under some conditions.

Since the cost of cement under many circumstances is a small fraction of the cost of concrete in place, it might seem that no mixture should be leaner than that at point *B*, even when the strength produced by such mixtures is not fully utilized, and certainly this alternative should always be given careful consideration. However, there are circumstances where the use of relatively rich concrete is not indicated. For example, in massive sections the cement content must be kept as low as other considerations permit to control the temperature rise that occurs during the early stages of hardening. Also there may be circumstances where it is important to keep the cost of material as low as possible, and where the use of another paste-making material significantly cheaper than cement would be indicated.

Natural or artificially produced mineral powders can be used to supplement cement; so can entrained air, but that is usually done with a different purpose in mind, namely protection of cement paste from the action of frost. Here we confine our attention to mineral powders, which, incidentally, are usually not beneficial with respect to frost action. It should be kept in mind, however, that the very small, "tough," bubbles obtained with a suitable agent can produce most of the effects expected of mineral powders, and usually at less cost. Many of the considerations set forth below are applicable to entrained bubbles.

Properly selected and used in mixtures to the right of point *B*, and especially in mixtures of the *FC* category, mineral powders give a denser matrix, a lower overall voids content, lower bleeding rate and capacity, better rheological properties, and in mature concrete higher strength and less permeability to fluids than can be obtained from the same mix without a mineral powder supplement. Shrinkage properties too can be improved, provided that the mineral powder is practically inert chemically; a pozzolanic material is likely to influence volume-change characteristics unfavorably, although the undesirable effect is usually not large enough to become a decisive factor.

9.2 Mineral Powder Regarded as an Aggregate Component

Mineral powder other than portland cement is not by custom definitely classified with either of the main components of concrete; it is usually called an "admixture." We may therefore consider it either as a supplement to cement or to aggregate. The reasons given in Section 8.1 for not considering cement powder to be a means of extending the range of grading of the aggregate are applicable to any mineral powder, but it is instructive to observe the effect of mineral powder when it is arbitrarily considered to be a component of the aggregate. Some data are given in Table 4.8 pertaining to powdered limestone. Seven mixtures are represented, four without limestone powder and three with it. In column 2 the actual ratio of the solid volume of sand and gravel to the solid volume of cement M is given for each mix, and in column 3 values of M_q are given, the subscript indicating that the mineral powder is included with the aggregate; it is the sum of values in columns 2 and 5.

In Table 4.9 other aspects of the data of Table 4.8 are developed. Values in the first line, reference B, were estimated from the normal consistency of the cement ($NC = 0.235$ by weight, $A_w = 0.733$), assuming $D = 0.05$, and from the water requirement factor E for mixtures of basic consistency in the *BF* category, according to relationships already discussed.

Table 4.8 Effect of Limestone Powder on Composition of Fresh Concrete[a]

1	2[b]	3[c]	4	5[d]	6[e]	7[f]	8[g]
Reference Number	M	$M + \frac{qv_q}{cv_c^o} = M_q$	$\frac{wv_w}{cv_c^o}$	$\frac{qv_q}{cv_c^o}$	Z	$\frac{v}{cv_c^o}$	$\frac{p}{V}$
2	8.54	8.54	1.735	0	0.021	1.79	0.238
44	10.68	10.68	2.095	0	0.072	2.25	0.217
45	13.85	13.85	2.832	0	0.089	3.08	0.208
46	21.37	21.37	4.698	0	0.090	5.12	0.197
35	10.67	10.93	2.163	0.256	0.011	2.19	0.238
38	14.22	14.88	2.889	0.655	0.024	2.96	0.237
40	21.32	23.17	4.465	1.845	0.014	4.53	0.241

[a] Data are from PCA Series 267. The aggregate was sand and gravel; the size range was (0–1½ in.). The nominal consistency was $R = 43 \pm 3$; $R_o = 0.02$; $\log (R/R_o) = 3.332$.

[b] The ratio of sand-and-gravel aggregate to cement (solid volume).

[c] The ratio of aggregate and limestone powder to cement; q is the weight of limestone powder, v_q its specific volume.

[d] The ratio of limestone powder to cement (solid volumes).

[e] The ratio of air volume to water volume.

[f] The ratio of void volume (water and air) to cement volume.

[g] Paste content calculated as cement, limestone powder, and water.

Table 4.9 Effect of Limestone Powder on the Water Ratio at Basic Consistency (R = 50)[a]

1	2[b]	3[c]	4[d]	5[e]	6[f]
Reference Number	$\left(\frac{wv_w}{cv_c^o}\right)_{50}$	$\frac{qv_q}{cv_c^o}$	M_q	x_q	u_w
B[a]	1.035	0	4.89	0.830	0.166
2	1.707	0	8.54	0.895	0.179
44	2.061	0	10.68	0.914	0.177
45	2.786	0	13.85	0.933	0.187
46	4.621	0	21.37	0.955	0.208
35	2.128	0.256	10.93	0.916	0.179
38	2.841	0.655	14.88	0.937	0.179
40	4.392	1.845	23.17	0.959	0.180

[a] Data are from Table 4.8, but they are adjusted to R = 50 by means of Eq. 4.12.
[b] The adjusted water-cement ratio, R = 50.
[c] The ratio of limestone powder to cement (solid volumes).
[d] $M_q = M + (qv_q/cv_c^o)$.
[e] $x_q = M_q/(1 + M_q)$.
[f] $u_w = (wv_w/cv_c^o)(1 - x_q)$.
[g] Reference B gives estimated values; see text.

Figure 4.16 is the water-ratio diagram for the data in Table 4.9. For the mixtures without limestone powder, it is apparent that three are in the *BF* category and the fourth in the *FC* category. Each of the mixtures contains considerably less cement than the mix at point *B*.* Also, we can see clearly how using limestone powder keeps the water ratio at nearly the same level as that for the richest mixture of the series, even though the leanest mix is well within the *FC* category.

Because the powder discussed above has a high specific surface area, it increases substantially the specific surface area of the aggregate but

* There is reason to believe that the water-cement ratio presented in Reference 44 in Table 4.9 is inaccurate. The consistency was reported as R = 43 ± 3, based on preliminary trial batches. No consistency tests were reported for the actual batches represented by these data. Consequently it was necessary to calculate k_R for each mixture on the assumption that its consistency was exactly R = 43 in every case. Probably R was actually below average for the case in question.

Incidentally, Fig. 4.16 is a good illustration of the value of the water-ratio or voids-ratio diagram as a frame of reference. When the data plotted here were obtained, a water-ratio diagram was not made, and Reference 44 was consequently mistaken for the mix giving the minimum voids content for the series.

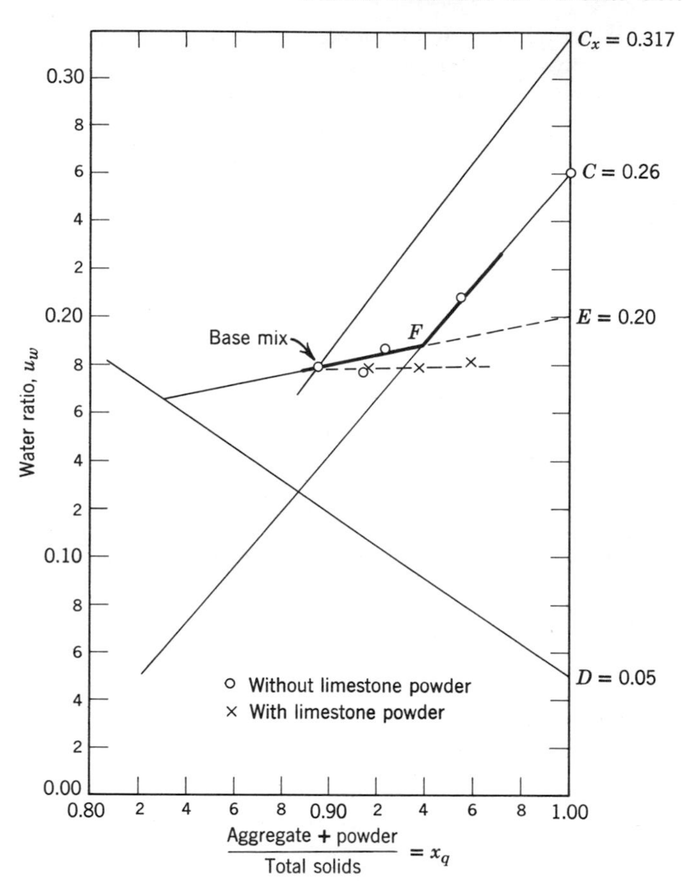

Fig. 4.16. The effect of powdered limestone on the water ratio for a series of concrete mixtures. The size-range of the sand-and-gravel aggregate was (0–$1\frac{1}{2}$ in.). The remolding number was 50. (Data are from Table 4.8.)

without increasing the water-requirement factor; indeed it reduces it. Thus the effect of surface area in mixtures in the *BF* category is apparently just the opposite of its effect in mixtures in the *AB* category.

The quantities of limestone powder in the leaner mixtures were so regulated that the degree of dispersion of the sand-and-gravel aggregate remained virtually unchanged as the cement content was reduced. The function of the added powder was actually not that of filling aggregate voids; on the contrary it was that of maintaining an arbitrarily selected degree of aggregate dispersion. Thus, even though the limestone powder is here considered to be a part of the aggregate, the degree of dispersion of the sand-and-gravel part of the aggregate is in each of the three mixes the same as that in the base mix; the degree is indicated by the value $C_x = 0.317$, as is shown on the graph. This would be true even if all the cement were completely replaced by the limestone powder so that $x_q = 1.0$; the water ratio would remain at virtually the same level, as is indicated in the graph.

The zero slope of the line through the points for mixtures containing limestone powder is due to the circumstance that the water requirement factor of the limestone powder was the same as that of the cement. Therefore, replacements on a solid-volume basis resulted in no change in the water-requirement factor for any given mixture. However, it is possible to select a powder that has a smaller water-requirement factor than that of the cement it replaces, in which case the slope in question would be downward, indicating a reduction in the water ratio as cement is replaced by such a powder, but of course not a reduction in the water-cement ratio. In contrast a mineral powder may have a water-requirement factor higher than that of the cement it replaces, in which case adding the powder is accompanied by an increase in the water ratio at a rate dependent on the difference between the two factors; in the present case, the difference was nil.

9.3 Mineral Powder Regarded as a Paste-making Material

In Table 4.10 the data of Tables 4.8 and 4.9 are reworked to include the limestone powder with the cement rather than with the aggregate. Actual water-cement ratios at basic consistency are given in column 6, and the corresponding values for the total paste-making material in column 7. Water-ratio values are given in the last column. One point to be noted especially is that the mixes given in column 4 for References 35 and 38 are virtually the same as that for Reference 2; that for Reference 40 is different, but the difference was probably unintentional, for apparently a mistake was made by the experimenter. The intention was for the last three mixtures to have the same paste content as the richest mixture of the series. Table 4.8, column 8, Reference 40, shows that too much limestone powder was used in the leanest mix, the paste content being 0.241 instead of 0.238; the value for Reference 38 is slightly low.

On the basis of the calculated values of x and u_w in Table 4.10 all the points in the water-ratio diagram for mixtures containing limestone powder would have almost coincided with the point for the richest mixture without limestone powder, had the original plan been executed accurately; two of the three actually do so, as may be seen by comparing the values of u_w given in column 8 of Table 4.10.

It has been found empirically (Williams [9], Powers [4], McMillan and Powers [10]) that if two pastes having the same consistency but made with

Table 4.10 Effect of Limestone Powder Considered as a Paste-making Material

1	2[a]	3[b]	4[c]	5[d]	6[e]	7[f]	8[g]
Reference Number	M	$\frac{qv_q}{cv_c^o}$	$\frac{M}{1+\frac{qv_q}{cv_c^o}}$	x	$\frac{wv_w}{cv_c^o}$	$\frac{wv_w}{cv_c^o + qv_q}$	u_w
2	8.54	0	8.54	0.895	1.707	1.707	0.179
44	10.68	0	10.68	0.914	2.061	2.061	0.177
45	13.85	0	13.85	0.933	2.786	2.786	0.187
46	21.37	0	21.37	0.955	4.621	4.621	0.208
35	10.67	0.256	8.50	0.895	2.126	1.693	0.179
38	14.22	0.655	8.59	0.896	2.841	1.717	0.179
40	21.32	1.845	7.49	0.882	4.392	1.544	0.180

[a] The ratio of sand-and-gravel aggregate to cement (solid volumes).
[b] The ratio of limestone powder to cement (solid volumes).
[c] The ratio of sand-and-gravel aggregate to cement and limestone powder.
[d] The values of x corresponding to column 4.
[e] The water-cement ratio at optimum consistency.
[f] The ratio of water to cement and limestone powder.
[g] The ratio of water to total solids.

different mineral powders are mixed together, the consistency remains approximately the same. This is true provided that when they are tested separately mineral powders other than portland cement are mixed with a saturated lime solution rather than with pure water. This is required by some of them to produce a normal flocculent structure, as is mentioned above.

To the extent that the indicated additivity of consistency is true, the cement paste in any given mix can be replaced, or partially replaced, with an equal volume of a different paste having the same consistency without appreciably changing the consistency of the mixture as a whole. Experiments showed that the assumption is sufficiently accurate with almost any powder. Therefore by measuring the water-requirement factor of a given mineral powder, it is possible to calculate how much paste it will produce per unit solid volume of powder and how much to use for replacing some of the cement without changing the paste content. If the solids content of the mineral powder paste is as high as that of the cement paste it replaces, the substitution can be made without changing the total solids content of the freshly mixed concrete. If such a powder is selected and if the cement paste is that in the mixture at point *B*, maximum solids content can be maintained regardless of how much the cement content is reduced below that of the mixture at point *B*. Powders that give pastes having relatively high solids content are of course to be preferred. The maximum particle size should be in the neighborhood of that which will pass through a No. 200 sieve, and the specific surface area should generally be of the same order of magnitude as that of portland cement.

9.3.1 Proportioning mineral powder. To assess the paste-making characteristics of different mineral powders, Powers [4] compared the relationships between solids content and consistency, using a crude viscometer. An example of results is shown in Fig. 4.17. McMillan and Powers [10] reported that among 12 different mineral powders used or considered for use in concrete, the solids contents of pastes ranged from 3 per cent to 45 per cent as compared with 41 per cent for a certain cement paste of the same consistency. Among the materials showing the lowest content of solids were colloidal clay and diatomaceous silica. Such materials not only have a specific surface area much too high, but also an unfavorable particle shape. The materials having outstandingly favorable characteristics are some lots of fly ashes recovered from the flue gases of furnaces fired with powdered coal. Although the specific surface area generally runs higher than that of portland cement, the majority of the fly-ash particles are spherical and therefore have relatively low water-requirement factors.

If we wish to produce a mixture having the same density and plasticity

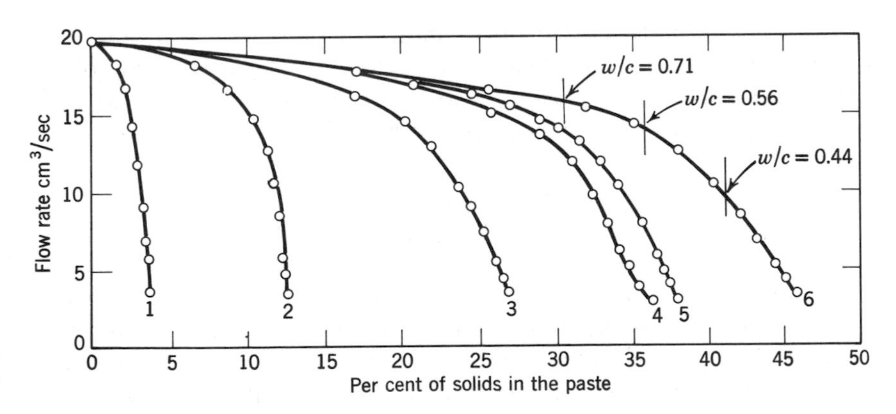

Fig. 4.17. Consistency (flow rate) curves for some paste-making materials. In the diagram 1 stands for colloidal clay; 2, diatomaceous silica; 3, hydrated lime; 4, a hydraulic lime; 5, pumicite; 6, portland cement.

as the mixture at point B, but with a lower cement content, we can calculate approximately the amount of mineral powder to add. We must first know the solid content of that mineral paste having the same consistency as the cement paste in the mixture at point B. As already mentioned, the procedure is simply to calculate the quantity of paste of optimum consistency that can be produced with the amount of cement to be used, and then to calculate the amount of additional paste needed to make up the optimum volume and at the same time maintain the same consistency. The calculation may be done as follows. Let

p_B = total paste volume (cement and water) in the mixture at point B
p_c = volume of cement paste in concrete containing mineral powder
p_q = volume of mineral-powder paste.

Then

$$p_B = p_c + p_q \tag{4.39}$$

and

$$\frac{p_c}{V} = \left(\frac{cv_c^o}{V}\right)\left[1 + \left(\frac{wv_w}{cv_c^o}\right)_B\right] \tag{4.40}$$

where cv_c^o/V is the desired cement content which is smaller than that of the mixture at point B; and $(wv_w/cv_c^o)_B$ is the water-cement ratio of the cement paste having optimum consistency. Let

$$C_q = \left(\frac{qv_q}{qv_q + wv_w}\right)_B \tag{4.41}$$

where C_q is the solids content of the mineral paste having optimum consistency, q is the weight of the admixture, and v_q its specific solid volume.

Having made the calculation by Eq. 4.40 and having evaluated C_q by experiment, we make the final step, given by Eq. 4.42:

$$\frac{q}{V} = \frac{C_q}{v_q}\left[\frac{p_B}{V} - \left(\frac{cv_c^o}{V}\right)\left(1 + \left(\frac{wv_w}{cv_c^o}\right)_B\right)\right] \tag{4.42}$$

The calculation indicated here gives the weight of mineral powder needed to produce the desired volume of paste having optimum consistency.

Comparisons of computed and experimentally determined amounts of mineral powder for four different materials are given in Table 4.11. None of these materials is ideal because, as is shown in the second column, none of them gives as high a solids content as does portland cement. Two of them, colloidal clay and diatomaceous silica, are quite unsuitable,

Table 4.11 Comparison of Computed and Experimentally Determined Amounts of Admixture (Mineral Powder) Required to Maintain Fixed-paste Content at Different Cement Contents[a]
$R = 35$

1	2[b]	3	4	5	6[c]	7	8
		Cement Content		Admixture Content			
				By Computation		By Experiment	
Material	C_q	$\frac{cv_c^o}{V}$	$\frac{\text{lbs}}{\text{yd}^3}$	Paste Content	$\frac{100\ q}{c}$	Paste Content	$\frac{100\ q}{c}$
Cement	0.41	0.1115	608	0.278	0	0.278	0
Colloidal clay	0.03	0.0896	489	0.278	1.3	0.278	1.2
		0.0724	395	0.278	2.9	0.278	2.8
		0.0621	338	0.278	4.3	0.278	4.3
Diatomaceous silica	0.12	0.0896	489	0.278	4.3	0.270	3.5
		0.0724	395	0.278	9.5	0.270	10.0
		0.0621	338	0.278	13.7	0.280	18.0
Hydrated lime	0.24	0.0896	489	0.278	17.0	0.278	18.0
		0.0724	395	0.278	38.1	0.277	38.0
		0.0621	338	0.278	55.0	0.279	62.0
Pumicite	0.34	0.0896	489	0.278	16.0	0.278	15.0
		0.0724	395	0.278	34.0	0.278	35.0
		0.0621	338	0.278	50.0	0.278	51.0

[a] See F. R. McMillan and T. C. Powers, "A Method of Evaluating Admixtures," *Proc. ACI*, **30**, 325–344, 1934.
[b] C_q is the solid content of paste having a given consistency, that is, that of the cement paste.
[c] The value of $100\ q/c$ is the amount of admixtures as a percentage of the weight of cement.

principally because as paste makers they furnish much more water than solid material. The accuracy of calculations based on the equations above may be judged by comparing the sixth and eighth columns. It seems good enough for practical purposes, especially as applied to materials having favorable characteristics.

It is not essential that the mixture at point B be chosen as the basis for proportioning mineral powder for leaner mixtures; for the experiments represented in Tables 4.8, 4.9, and 4.10, and in Fig. 4.16, the base mix was considerably leaner than that at point B.* However, it would not be reasonable to select a base mix beyond point F. All things considered, use of point B as the base has much to commend it if mixes leaner in cement are to be used. The grading of the sand would be selected to be suitable for such a rich mixture, and the optimum percentage of sand would be based on that mixture. It would be the same for all leaner mixtures, the missing cement being replaced with mineral powder by the procedure described above. Such a practice would assure the maximum possible solids content and the benefits to be derived therefrom, particularly with respect to volume stability and strength in relation to the tendency to develop cracks while drying.

9.3.2 Admixture or replacement? Sometimes it is said that powdered minerals might be approved for use as admixtures but not as substitutes for cement. Such a statement implies an oversimplified view of the question. To be considered carefully is the combination of properties the finished concrete is expected to have, alternative ways of obtaining it, and relative overall costs. The question of replacement of cement or addition to the mix may have very little meaning, as the following illustrates.

Consider in Fig. 4.16 the mixture without limestone powder at $x = 0.94$, point F. The water ratio is 0.188, the water-cement ratio is $0.188/0.06 = 3.133$ or $w/c = 3.133 \times 0.309 = 0.97$ by weight. From the corresponding point on the line for mixtures containing limestone powder we find the water ratio is 0.179, the water-cement ratio is 2.983, or $w/c = 0.92$. These figures indicate a significant improvement along with other improvements only implied, such as an improvement in rheological and bleeding properties. These improvements can be regarded as the effect of an admixture in the mix $M = x/(1 - x) = 0.94/0.06 \doteq 15.67$. Presumably this mixture is considered suitable for some purpose.

However, the same mixture can be thought of as the result of substituting limestone powder for cement in the mix at $x = 0.895$, or $M = 8.53$, the base mix for this series; that is to say, the mixture in question can be considered a 1:15.67 mix containing an admixture of limestone powder, or as a 1:8.52 mix with some of the cement replaced by limestone powder. The mixture in question has certain properties determined by its composition, and the question of admixture versus replacement is, after all, irrelevant, except perhaps when considering the relative costs of admixture and cement.

* Such was not the original intention; one of the mixtures gave a false indication of having reached the point of the minimum water ratio, leading to the selection of a base mix that was not what it seemed to be. This work was done before the categorical differences had become clearly established. (See also the footnote on page 191.)

9.3.3 Are solids content and paste consistency sufficient criteria? A certain inconsistency may have been noted in the foregoing discussions. Whereas two cements having considerably different specific surface areas seemed to give mixtures having the same water-requirement factors (Fig. 4.6) we have just seen that mineral powders can be used to reduce the water-requirement factor and that the amount of powder to produce a given effect is not the same for different powders. Some of this inconsistency is only apparent and readily explained; the rest is not so clear.

Although for mixtures in the *BF* category the water requirement of the mixture is simply proportional to the quantity of aggregate, we have seen that the cement is actually one of the determinants of the proportionality factor, that is, the water-requirement factor; the water requirement increases as the cement content decreases, or it increases as the aggregate content increases. Whichever way we look at it, it is a fact that if we add cement to a mixture in that category and think of that increment as an admixture, the water-requirement factor of the mixture will appear to have been reduced, just as it would have been by any other suitable mineral powder.

The question remaining is why, as shown in Fig. 4.6, adding a certain cement changed the water-requirement factor no differently from adding a considerably finer cement. The question is pertinent, for without doubt the finer cement produces the stiffer paste at a given water-cement ratio; in terms used in Section 9.3.1, and thinking of an increment of cement as an admixture to the existing mix, we can be sure that the value of C_q was not the same for the two cements, and we should therefore not expect them to give equal water ratios at the same value of x, as they actually did. We must conclude that there is at least one more factor than has been considered so far. The nature of that factor is at present a matter of speculation.

There is reason to believe that the effective *maximum size* of the particles in the powder may have a considerable influence on the way a mineral powder, cement or other, works with a given aggregate. The average clearance between the particles of the slightly dispersed aggregate must be at least equal to the largest cement particles that are present in significant amount. Because the effective maximum size of the particles is smaller for the finer cement, it follows that so far as particle interference is concerned, the aggregate need not be dispersed as much for the fine cement as for the coarser one. In other words, we suppose that the required degree of dispersion of aggregate for a given consistency depends not only on paste consistency, but also it may depend on the effective maximum size of particles in the paste. Thus, if two pastes of the same consistency differ with respect to maximum particle size, the one with the smaller particles may produce the concrete mixture having the softer consistency; conversely, if two pastes differ both in consistency and maximum particle size, they will not necessarily produce concretes of different consistencies. Although grinding a cement finer has the effect of stiffening the paste, it also reduces the maximum size and the resistance due to particle interference. In the particular case represented in Fig. 4.6 it appears that these opposite effects were practically in balance, leaving the water-requirement factor unaffected. In view of this explanation, the balance between opposite effects seems fortuitous; if the difference between the two cements had been greater, the water-requirement factor of the mixture would have reflected the difference.

We should expect also that if the fraction of fine particles in the sand were reduced below what it was in the particular sand used in the above-mentioned examples, thereby enlarging the interstitial spaces, the observed effect of cement fineness, or lack of effect, would not have been the same. These considerations have a bearing on the value of x at which point B is found. Because that point seems to mark the degree of aggregate dispersion at which the effect of interference between the aggregate screen and the largest cement particles becomes negligible, increasing the size of the aggregate voids should move point B to the right, although the countereffect of the increase in voids ratio C is to be considered also, as is indicated in Eq. 3.40.

Test results from a wide variety of mineral powders are generally compatible with the explanations given below. The most effective powders are those that have a specific surface area somewhat higher than that typical of a Type-I cement. The value of C_q in certain unsatisfactory mineral powders is relatively low, but that is due to the relatively large maximum size, and, according to the explanation above, the unsatisfactory characteristics are due to excessive interference between the largest particles in the powder and the aggregate particles. The best of all mineral powders, from the standpoint of water requirement, was a certain lot of carbon-free fly-ash. Although the specific surface of this material was relatively high, the solids content of its paste was higher than that of cement paste. The outstanding characteristic of this material is the sphericity of its particles.

The answer to the question raised in the heading of this section seems to be that considerations of consistency and the solids content of paste are sufficient for making a first approximation of the suitability and proper proportion of a given mineral powder, but a final evaluation would involve consideration of maximum particle size in relation to the void size of the aggregate, particle shape, and perhaps other secondary factors. Of course, when we are considering the ultimate effects on the properties of hardened concrete also, cementitious value, or pozzolanic activity with its desirable and undesirable effects, must be taken into consideration.

REFERENCES

[1] Abrams, D. A., "Design of Concrete Mixtures," Bulletin 1, rev. ed., Structural Materials Research Laboratory. Lewis Institute, Chicago, 1925.

[2] Popovics, S., "Relation Between the Change of Water Content and the Consistence of Fresh Concrete," *Magazine of Concrete Research,* **14** (41), 99–108, 1962.

[3] Matern, N. V., and N. Odemark, "Investigation as to the Consistency of Concrete Mixtures," Report 68, State Road Institute, Stockholm, 1944.

[4] Powers, T. C., "Studies of Workability of Concrete," *Proc. ACI,* **28,** 419–448, 1932.

[5] Litvin, A., and D. W. Pfeifer, "Gap-graded Mixes for Cast-in-place Exposed Aggregate Concrete," *Proc. ACI,* **62,** 521–536, 1965.

[6] Swayze, M. A., and E. Gruenwald, "Concrete Mix Design—a Modification of the Fineness Modulus Method," *Proc. ACI,* **43,** 829–843, 1947.

[7] Weymouth, C. A. G., "A Study of Fine Aggregate in Freshly Mixed Mortars and Concretes," *Proc. ASTM,* **38,** Part II, 354–372, 1938.

[8] Wuerpel, C. E., "Vibratory Remolding Test as a Measure of Concrete Workability." *J. ACI,* **15,** 70–75, 1943.

[9] Williams, G. M., "Admixtures and Workability of Concrete." *Proc. ACI,* **27,** 647–653, 1931.

[10] McMillan, F. R., and T. C. Powers, "A Method of Evaluating Admixtures," *Proc. ACI,* **30,** 325–344, 1934.

5

Mix Design by Empirical Methods

1. INTRODUCTION

As we reach this chapter it is apparent that more chapters on the properties of freshly mixed concrete lie ahead. It might seem that a chapter on the selection of the proportions of the ingredients for concrete might be presented at the end, after all aspects of the subject have been considered. It turns out, however, that prevailing practice involves explicitly only a few aspects of the total subject and that a description of it, which is the burden of this chapter, can be handled well enough before further complications are introduced.

In this chapter we trace some steps leading up to present methods and describe the most widely accepted procedures. Details of methods of computation are not given because they are readily available in convenient form from several different agencies.

A newcomer to concrete technology might soon gain the impression that the formulation of a mixture of cement aggregate and water that will prove to be plastic and to meet other requirements presents an intricate technical problem that has so far evaded satisfactory solution; such might be his impression if he happened to concentrate his attention on the papers discussed in Chapter 6. On the other hand, should he happen to study only the more popularized discussions of the subject, he might learn that almost anyone can design a satisfactory mix by following simple procedures after a brief period of training; with any reasonable amount of cement and a reasonably graded aggregate, plasticity seems to appear spontaneously as the solids are mixed with the water. All that remains for the "designer" to do is to adjust proportions so that workability becomes satisfactory and the water-cement ratio and air content assure proper strength and adequate durability.

Either impression is somewhat inaccurate. It is not true that mix design is an intricate unsolved problem, and neither is it true that optimum combinations of materials always result from methods now in vogue. To follow recommended practices of design does assure attaining a specified strength, acceptable workability, and reasonable promise of durability under severe conditions. However, uses of concrete are now so varied and the demands on concrete occasionally so severe that there is little to justify limiting our considerations to these usually easy-to-meet criteria. It is especially inadvisable to design concrete without giving full consideration to minimizing shrinkage and the cracking often resulting from it; in general, this consideration requires us to obtain the maximum possible total solid content compatible with other requirements. The indicated procedure is an analytical method combined with the trial method, with more emphasis on the former than is now generally practiced, and with more comprehensive criteria by which to evaluate the degree to which a given design approximates optimum conditions.

2. PRIMITIVE METHODS

2.1 *Some historical background*

There is evidence that 2000 years ago the Romans had a well developed although prescientific art of making concrete. They used a cementing material composed of certain volcanic slags (now called *pozzolanas*) mixed with burned lime [1]. The art apparently went down with the empire and did not begin to revive until the middle of the eighteenth century when natural cement (made by burning certain argillaceous limestones) was discovered. A little later a somewhat better cement known as hydraulic lime was developed. A rudimentary form of portland cement was known of in England early in the nineteenth century.

According to Bogue [2], use of cement in quantity in the United States began early in the nineteenth century. At first the principal cementing material was domestic natural cement, but during the last half of the century portland cement was imported; also a domestic portland cement industry began to develop. By 1910 portland cement had almost supplanted other cements, and soon the term "concrete" was usually understood to mean "portland cement concrete."

Along with the rediscovery and further development of cementing materials after the Middle Ages, there began a new development of the art of making concrete. It seems that during its early stages of development, portland cement concrete was not thought of as a plastic material that later becomes rigid; it was prepared in a nonplastic state, and was consolidated in layers by means of rammers. If, by ramming, the workman was just able to bring water flush with the surface, the water content was considered to be correct. For example a specification of 1890 read as follows [1]:

"The concrete shall be formed of sound broken stone or gravel . . . to be mixed in proper boxes, with mortar of the quality before described, (one of cement to two of sand, by measure, and a moderate dose of water) in the proportion of four parts of broken stone to one part of cement; to be laid immediately after mixing, and to be thoroughly compacted throughout the mass by ramming till the water flushes to the surface, the amount of water used for making concrete to be approved or directed by the Engineer."

Then, around the beginning of the twentieth century, the making of reinforced concrete increased; with that development the use of nonplastic mixtures began to be abandoned. Indeed, the pendulum of opinion about good practice swung to the opposite extreme, as exemplified by the following exerpt from a 1910 specification [1]:

"The concrete shall be composed of cement, sand, and crushed rock in the proportion of 1:3:6. The concrete shall be used as wet as possible without being so wet that the matrix separates from the rock. Any places that are porous due to matrix separating from the rock shall be removed."

It seems safe to surmise that during the periods mentioned above few engineers considered the possibility that the properties of concrete could generally be predicted and controlled; performance tests were uncommon, and not much was known about the properties of the material actually produced, except that it usually developed enough strength to meet the demands made of it; not much was expected as compared to present-day requirements.

As already indicated by the quotations from two specifications, the making of concrete was based on empirical formulae. The origin of these formulae is not entirely clear. One prominent engineer nearly a hundred years ago explained it as is told in the next section.

2.2 The Civil Engineer's Pocketbook

The Civil Engineer's Pocketbook was published by John Wiley & Sons in 1871; the author was John C. Trautwine. According to the publisher, the book was ". . . the foremost civil engineers pocketbook, not only in the United States, but in the English language." Although the book is obsolete, it seems worthwhile to quote at some length from some parts of it; it shows how unfirm was the foundation on which concrete practice was based, and it has a certain enjoyable quaintness. It was written about 19 years before the work of the French investigator Feret, who was apparently the first to publish the results of a scientific investigation of mixtures.

Trautwine in his preface said that his object was to ". . . elucidate in plain English a few important elementary principles. . . ." Turning to pp. 676–679, we find instructions for formulating mortar and concrete. The first consideration is the voids content of the sand, about which Mr. Trautwine wrote as follows:

"The voids in sand of pure quartz like that found on most of our sea-shores, when perfectly dry and loose, occupy from .303 of the mass in sand weighing 115 pounds per cubic foot, to .515 in that weighing 80 pounds. Usually, however, such dry sand weighs say from 105 pounds with voids of .364: to 95 pounds, with voids .424; the mean being 100 pounds, with voids .394.* But the wet sand in mortar occupies from 5 to 7 per cent less space than when dry; the shrinkage averaging say 6 per cent or $\frac{1}{17}$ part; thus making the voids .304 of the 105 pound sand when wet; and .364 of the 95 pound; the mean of which is .334. But to allow for imperfect mixing, etc., it is better to assume the voids at .4 of the dry sand. Moreover since the cements, as before stated, shrink more or less when mixed with water and worked up into mortar, it would be as well to assume that to make sufficient paste to thoroughly fill the voids, we should not use a less volume of dry common cement, slightly shaken, than half the bulk of the dry sand; or than .45 of the bulk if Portland. The bulk of the mixed mortar will then be about equal to or a trifle less than that of the dry sand alone . . . ."

* "If greater accuracy is desired, pour into a graduated cylindrical measuring-glass 100 measures of dry sand. Pour this out, and fill the glass up to 60 measures with water. Into this sprinkle slowly the same 100 measures of dry sand. These will now be found to fill the glass only to say 94 measures, having shrunk say 6 per cent; while the water will reach to 121 measures; of which 121 − 94 = 27 measures will be above the sand; leaving 60 − 27 = 33 measures filling the voids in 94 measures of wet sand; showing the voids in the wet sand to be $\frac{33}{94}$ = .351 of the wet mass. If the sand is poured into the water hastily, air is carried in with it, the voids will not be filled, and the result will be quite different."

As to the strength to be expected from concrete, the author says:

"The strength of concrete is affected by the quality of the broken stone, as well as by that of the cement, the degree of ramming, etc.

Cubes of either of the above with Portland, as well as one composed of one measure of good Portland to Five of sand only, well made, and rammed, should either in air or in water require to crush them at different ages, not less than about as follows:

Age in Months	1	3	6	9	12
Tons per Sq. Ft.	15	40	65	85	100

Under favorable conditions of materials, workmanship and weather the strength may be from 50 to 100 per cent greater. . . . If not rammed the strength will average about one-third part less.*"†

* "*Weight of good concrete* 130 to 160 lbs. per cubic foot dry."

Regarding water content, we find this statement in connection with mortar:

"The highest strength was obtained by the use of *just enough* water to thoroughly dampen the cement. An excess of water retards setting."

Nothing was said explicitly about water in connection with the formulas for concrete.

"The best sand is that with grains of very uneven sizes, and shape. The more uneven the size the smaller are the voids and the heavier is the sand. It should be washed if it contains clay or mud, for these are very injurious to mortar or concrete."

Having considered the sand, and the amount of cement required to make mortar, Mr. Trautwine applied the same principle to formulation of concrete as follows:

"*Cement concrete or Beton,* is the foregoing cement mortar mixed with gravel or broken stone, brick, oyster shell, etc., or with all together. In concrete as in mortar, it is advisable on the score of strength that all the voids be filled or more than filled. Those of broken stone of tolerably uniform size and shape are about .5 of the mass; with more irregularity of size and shape they may decrease to .4. Those of gravel vary like those of sand, and had likely be taken at .5 when estimating the dry cement. We shall then have as follows.

"*For 1 Cubic Yard of Concrete of Stone, Gravel and Sand, Without Voids*
1 cu. yd. broken stone with .5 of its bulk voids, requiring .5 cu. yd. gravel
0.5 cu. yds. gravel with .5 of its bulk voids, requiring .25 cu. yd. sand

† Fifteen tons per square foot being about 200 psi, it is apparent that not much was expected of concrete at early ages; even at the age of one year, the indicated strength is only about 1400 psi. (TCP.)

0.25 cu. yds. sand with .5 of its bulk voids, requiring .125 (or $\frac{1}{8}$) cu. yd. dry cement."

"For 1 Cubic Yard of Concrete of Broken Stone and Without Voids
1 cu. yd. broken stone with .5 of its bulk voids, requiring .5 cu. yd. sand
.5 cu. yd. sand, with .5 of its bulk voids, requiring .25 cu. yd. dry cement."

The expression "without voids" as used here probably refers to spaces resulting from lack of complete consolidation.

The instructions for making concrete as given above result in bulk volume proportions as follows, taking 1 ft^3 of cement powder as unity. For the first example,

1 cement:2 sand:4 gravel:8 broken stone

For the second example,

1 cement:2 sand:4 broken stone

From the rather unclear remarks preceeding the data on strength, above, it appears that with a given cement Trautwine would expect the same strength from the 1:2:4:8 mix as from the 1:2:4.

It is apparent that the approach to mix design described above was based on deductive reasoning, the reasoning being based on the premise that intermediate aggregate fills the voids in the coarse aggregate, the fine aggregate fills the voids in the intermediate; and the cement, that is, its powder volume in a "shaken down" state, fills the voids in the fine aggregate. There is good reason to believe, on the basis of what we have seen in the foregoing chapters, that the actual relationship of the separate aggregate volumes to the voids in the coarser material was far different from that implied by that premise. Nevertheless the 1:2:4 mix was usable. Presumably the 1:2:4:8 mix would be usable too, provided that the coarsest aggregate was sufficiently large relative to the mean diameter of the next smaller aggregate, a necessity not mentioned by Mr. Trautwine.

Obviously the state of knowledge reflected in Trautwine's book left much to be desired and constituted a challenge to the profession, a challenge that gave rise to various theoretical and laboratory investigations.

In a book published in 1845, Wright [1] stressed the importance of proper grading of concrete aggregates. During the last decade of the nineteenth century the publications of Feret, already mentioned, uncovered the basic principles that underly modern design. Unfortunately, many did not understand its full implications, and it had little immediate influence on American practice.

A step beyond the restricted concept of 1:2:4 was taken by those who realized that using either more or less cement than the amount apparently needed to fill the voids in the sand was feasible and often desirable. Hence, it became common to specify such mixes as 1:1:2; 1:1½:3; 1:2:4; or 1:3:6, the tendency to use one volume of fine aggregate to two of coarse still persisting, at least on paper.

Another forward step began when more attention was paid to units of measure. The need for this was stated by Sabin [3], writing in 1907 as follows:

"The usual method of stating proportions in concrete is to give the number of parts of sand and aggregate to one of cement. These parts usually refer to volumes of sand and stone, measured loose, to one volume of packed cement. However, there is no established practice in regard to this, and a "1:2:5 concrete" may mean five volumes of loose stone, to two volumes of loose sand, to one volume of loose cement, or any one of several combinations."

It became the practice in some places to specify proportions in terms of aggregate volumes in the "dry-rodded" state, and 94 pounds of cement was considered to be one cubic foot. This practice made it necessary to establish factors for converting dry-rodded volumes into volumes as measured in the field, and in time the bulking effect of moisture in sand became recognized and concrete engineers began taking it into account. Not until comparatively recently did gravimetric proportioning in the field become common.

A great impetus forward followed the publication by Abrams in 1918 of a method of proportioning based on what became known as the *water-cement ratio law* for concrete strength [4]. Although it was based on original research, it amounted to a popularization, or perhaps a practical implementation, of the principles announced by Feret.

These and other developments are reviewed and discussed in the following sections.

3. USE OF TABULATED DATA

A step beyond the use of arbitrary proportions is made if we make some attempt is to adjust the proportions according to the characteristics of different aggregates and to the characteristics wanted for the final mixture. One procedure is to provide the designer with tabulated data from which appropriate values may be selected. An outstanding example of this kind is a publication (1947) of the Road Research Laboratory, Department of Scientific and Industrial Research, England [5].

Using a variety of aggregates, and a wide range of mixtures, the Road Research Laboratory carried out a sufficient number of laboratory tests to establish the proportion of aggregate to cement appropriate for almost any combination of materials and strength requirements that might be encountered in ordinary English practice. To provide concrete of a certain strength, it is considered sufficient to select an appropriate ratio of water to cement according to findings previously published by Abrams and others [4]. The next step is to select an appropriate degree of workability: Very Low, Low, Medium, or High. The maximum size of the aggregate is selected, presumably on the basis of the conditions of the work, and the characteristic particle shape of the available aggregate is noted. The designer then consults one of several tables of data, the table being selected according to aggregate size and particle shape. He finds in the table a section corresponding to the selected degree of workability and within that section from one to four figures corresponding to the selected water-cement ratio, each figure being a proportion of aggregate to cement, that is, the mix.

Where more than one mix is shown for a given water-cement ratio, each corresponds to a different type of grading for the given maximum size, ranging in four steps from finest to coarsest. If the aggregate must be used as received, presumably its grading is compared with the four types, and the mix selected according to the type it resembles. Or, if the aggregate is supplied as separate fine and coarse aggregates, it is necessary to select a proportion of fine to coarse that will give a combined grading as close as possible to a selected one of the four types. (A variation of this procedure is described by Neville [6] whereby three aggregate sizes are proportioned so as to conform as closely as possible to a selected type curve. Neville did not advocate this as an essential refinement, however.)

In any case, a decision must be made as to which of the four types of grading is wanted. The tabulated data indicate that for a given water-cement ratio one type might be preferred on the basis that it requires less cement than the others; indeed in some cases, only the finest grading is recommended. However, in many cases two or more gradings are considered usable with little or no difference in the cement requirement.

When cement requirements are alike, a preference for the coarser grading is indicated. For example, for an aggregate having a maximum size of $\frac{3}{4}$ in., and for medium workability when using a water-cement ratio of 0.55 by weight, the mix would be 1:5.4 with either of two of the four type gradings, corresponding to 35 and 42 per cent of sand in the aggregate. According to these instructions, the lower percentage would be preferred.

When selecting proportions by the method described above, it is expected and recommended that the mix would be subject to final adjustment according to the judgment of the designer after he has observed the way the mixture performs under job conditions; indeed, the same is true of any of the methods reviewed in the following pages of this chapter.

4. ABRAMS' FINENESS MODULUS METHOD

In 1918 a systematic method of formulating concrete mixtures was published by Abrams [4]. It was not the first work of its kind, but it is discussed here before the others because its most important features are so well established that it has become a criterion in some respects, making it necessary to refer to it repeatedly while discussing other methods.

An outstanding characteristic of the Abrams method is that it is almost completely empirical and is based on a large number of systematic tests.

4.1 The Water-cement Ratio "Law" for Strength

Abrams presupposed that a concrete mixture should be formulated so that it would be workable under the given conditions and so that it should be able to develop a specified compressive strength. He enunciated the "law of strength so far as the proportions of materials are concerned," as follows: ". . . for given materials, the strength depends only on one factor—the ratio of water to cement."*

The observed relationship between comprehensive strength after a certain period of standard curing could be expressed by a formula having the following form:

$$f_c = \frac{A}{B^X} = \frac{A}{B^{1.5(w/c)}} \tag{5.1}$$

where f_c is the compressive strength of concrete at a stated age, X is the water-cement ratio in water volume per unit of standard powder

* Some 20 years earlier the French investigator Feret had reported that strength depended on the ratio of the quantity of cement to the total volume of voids, voids being defined as air-filled space as well as water-filled space. Also, in 1908, Zielinski, a Hungarian, published extensive work from which the dominant role of the water-cement ratio was correctly deduced.

volume of cement, w/c is the water-cement ratio by weight; A is an empirical constant, usually 14,000 psi, and B is a constant that depends on the characteristics of the materials, especially the cement, and on the age at test. For 28-day strength, Abrams found $B = 7$. (Using cements of the 1940's Blanks, et al. [1], found the value to be about 4.)

With a formula such as Eq. 5.1 established, it was possible to select a water-cement ratio for a desired strength.

4.2 Abrams' Water Requirement Formula

Having established the empirical relationship between strength and the water cement ratio described above, it was necessary only to find the combination of cement and aggregate that would give the desired water-cement ratio, adequate workability, and contain no more cement than necessary. To this end Abrams developed a water-requirement formula involving his grading factor called the *fineness modulus* (which is described in Chapter 1, Section 4). The formula is

$$\frac{W}{C} = R\left[\frac{3}{2}p + \frac{0.3n}{1.26^{FM}}\right] \tag{5.2}$$

where W is the volume of water in cubic feet; C is the powder volume of cement in cubic feet, 94 pounds being taken as one cubic foot; p is the weight ratio of water to cement at normal consistency; n is the ratio of aggregate to cement in cubic feet of combined aggregate in the dry-rodded state, to the powder volume of cement; R is the "relative consistency," that is, the ratio of the amount of water actually used to that amount giving a slump of about 1 in.; and FM is the fineness modulus, as before.

With p known (p being the normal consistency value from a standard test on the cement to be used), with the water-cement ratio selected on the basis of the desired compressive strength, and with the proper value for fineness modulus determined, the required proportion of aggregate could be calculated. According to the formula, for a given water-cement ratio the mix n is leaner the larger the value of the fineness modulus.

So far as the formula is concerned there is apparently no limit to the degree to which the cement content can be reduced by increasing the fineness modulus. However the formula is empirical, and there are certain limiting factors not explicitly in it. Specifically, for any given maximum size of aggregate and given value of n, there is an upper limit for FM beyond which the mixture becomes unworkable; the fineness modulus can be increased without limit only if the maximum size can be increased without limit. Therefore to use the water-requirement formula, we must know the maximum permissible fineness modulus.

4.3 Maximum Permissible Fineness Modulus

An elaborate series of laboratory experiments was carried out entailing numerous systematic tests which, among other things, established maximum permissible fineness moduli for a wide variety of conditions. These experiments did not involve the water-requirement formula, although presumably they produced the data to which the formula pertained.

The experimental method was about as follows. With a fixed proportion of aggregate to cement, and with a certain maximum size, a series of mixtures would be made in which the fineness modulus of the aggregate was progressively increased until the limit for adequate workability had been exceeded considerably. Test cylinders were cast from each mixture and tested for compressive strength after a term of curing under standardized conditions. The results were plotted, as shown, for example, in Fig. 5.1, for four mixes and one maximum size. For each mix its maximum permissible fineness modulus was selected. The limit was not based on the requirement for maximum strength, but on the least proportion of sand giving acceptable workability. Line *B* in the graph indicates the limits that seemed suitable where a minimum of plasticity is needed; line *A* gives limits that seemed suitable for concrete that is to be placed in narrow forms and actually corresponds to the figures given in Table 5.1.

Limits were established for maximum sizes ranging from sieve No. 30 upwards, and for mixes ranging from 1:1 to 1:9 in terms of the powder volume of the cement and the dry-rodded volume of the combined aggregate. The results of these experiments, converted to terms now commonly used by Walker and Bartel, are given in Table 5.1. In the notes accompanying the table we see that when we use a given natural sand the value of the maximum permissible fineness modulus must be adjusted according to the voids content of the coarse aggregate. For a given maximum size the voids content is controlled principally by the particle shape. The more angular the particles are the higher is the voids content, and the lower the maximum permissible fineness modulus for given conditions.

The principal advantage claimed for this method of proportioning was that it provided a way to use a wide range of different gradings with equally good results. "Any sieve analysis," said Abrams, "which will give the same . . . fineness modulus . . . will require the same quantity of

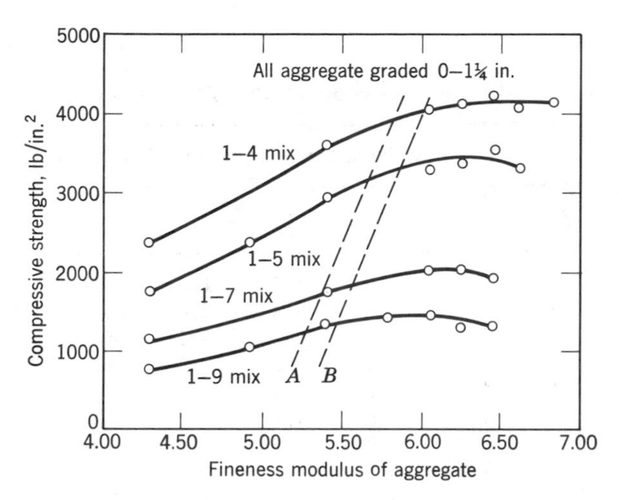

Fig. 5.1. Example of Abrams' method to determine the maximum permissible fineness modulus. Line *A* describes the limiting values from Table 5.1. Line *B* describes the maximum permissible fineness moduli. Strength tests were made on 6 × 12 in. cylinders 28 days old.

The sieve analyses of the aggregate are given below:

Range in Size	Fineness Modulus	Per Cent Coarser than Each Sieve								
		100	48	28	14	8	4	$\frac{3}{8}$	$\frac{3}{4}$	$1\frac{1}{2}$
$0–1\frac{1}{4}$	4.30	89	82	72	62	51	38	25	11	0
$0–1\frac{1}{4}$	4.93	95	89	82	73	61	47	32	14	0
$0–1\frac{1}{4}$	5.40	98	94	88	80	69	55	38	18	0
$0–1\frac{1}{4}$	6.04	99	98	95	90	81	68	49	24	0
$0–1\frac{1}{4}$	6.25	100	99	97	92	85	72	53	27	0
$0–1\frac{1}{4}$	6.45	100	99	98	95	88	77	58	30	0
$0–1\frac{1}{4}$	6.60	100	100	99	96	91	80	62	32	0
$0–1\frac{1}{4}$	6.82	100	100	99	98	94	86	68	37	0

water to produce a mix of the same plasticity . . . and concrete of the same strength, so long as it is not too coarse for the quantity of cement used."

Abrams' work was done before the introduction of vibrators for placing concrete. Systematic experiments have demonstrated that when vibrators are effectively used the best results are obtained when the percentage of sand in the aggregate is reduced below that which is best when placing

Table 5.1 Abrams' Maximum Permissible Values of Fineness Modulus[a]

Maximum Size of Aggregate	Sacks of Cement per yd^3 of Concrete							
	3	4	5	6	7	8	9	10
No. 30	1.4	1.5	1.6	1.7	1.8	1.9	1.9	2.0
No. 16	1.9	2.0	2.2	2.3	2.4	2.5	2.6	2.7
No. 8	2.5	2.6	2.8	2.9	3.0	3.2	3.3	3.4
No. 4	3.1	3.3	3.4	3.6	3.8	3.9	4.1	4.2
$\frac{3}{8}$ in.	3.9	4.1	4.2	4.4	4.6	4.7	4.9	5.0
$\frac{1}{2}$ in.	4.1	4.4	4.6	4.7	4.9	5.0	5.2	5.3
$\frac{3}{4}$ in.	4.6	4.8	5.0	5.2	5.4	5.5	5.7	5.8
1 in.	4.9	5.2	5.4	5.5	5.7	5.8	6.0	6.1
$1\frac{1}{2}$ in.	5.4	5.6	5.8	6.0	6.1	6.3	6.5	6.6
2 in.	5.7	5.9	6.1	6.3	6.5	6.6	6.8	7.0
3 in.	6.2	6.4	6.6	6.8	7.0	7.1	7.3	7.4

[a] Edited by S. Walker, and F. F. Bartel, "Discussion," *Proc. ACI*, **43**, 844-1–844-10, 1947. Values were derived from D. A. Abrams, "Design of Concrete Mixtures," Bulletin 1, Structural Materials Research Laboratory. Values for aggregate having a maximum size of No. 4 and smaller are based on natural sand. Values for aggregate having a maximum size of $\frac{3}{8}$ in. and greater are based on mixtures of natural sand with rounded gravel having voids of about 35 per cent in the dry rodded condition. Values should be reduced about 0.1 for each increase of 5 in the percentage of "dry-rodded" voids in the coarse aggregate. The maximum size is represented by that sieve next larger than the sieve on which 15 per cent of the total aggregate is retained. Mixes represented by these fineness moduli of combined aggregates are of about the harshness permissible for concrete pavement construction. They should be reduced about 0.2 to 0.3 for reinforced concrete construction.

"*Important Note:* The maximum permissible fineness moduli should be used only as a guide. They can be attained only with well graded aggregates. All portions of the combined aggregate from 'zero' size to the various intermediate sizes should have fineness moduli not greater than permitted for those intermediate sizes, if workability is to be assured. For example, consider the 6.6 permissible fineness modulus for the 3 in. maximum size and the '5-sack' mix; that portion of the combined aggregate finer than $1\frac{1}{2}$ in. should not have a fineness modulus greater than 5.8; that portion finer than $\frac{3}{4}$ in. not greater than 5.0; and so on all the way down; *including the fine aggregate and its fractions.* Fine aggregates should be examined from this viewpoint with particular care. Except in rare cases of ideal grading, a sand with a fineness modulus of, say, 3.4 is too coarse for ordinary mixes because *its finer portions are too coarse.*" (From Walker and Bartel.)

concrete without vibration [8]. This means that the maximum permissible fineness modulus of concrete that is to be placed with vibrators may be considerably higher than is indicated in Table 5.1; however, no table of values suitable for vibrated concrete has been produced.

4.4 *The Proportion of Fine Aggregate*

With the maximum size of the aggregate established by specification or by the conditions of the work, with the maximum permissible fineness modulus selected as described above, and with the fine and coarse aggregates supplied separately, the next step is to determine the proper proportion of fine to coarse aggregate that is required to produce the proper fineness modulus for the combined aggregate. This is done on the basis of the fineness modulus of each of the aggregates. The required percentage of sand is computed by the law of mixtures as follows:

$$\text{per cent sand} = 100\,\frac{FM_o - (FM)_b}{(FM)_a - (FM)_b}$$

where FM_o stands for the maximum permissible value, and subscripts a and b designate fine and coarse aggregates.

Certain limits were given for the fineness modulus of the sand. First, a value below 1.5 is undesirable (and seldom found in ordinary sources of aggregates). Second, the value should not be greater than the maximum permissible fineness modulus for the given mix and the maximum size of the sand.

The principle of proportioning described above has been applied when more than two sizes of aggregate are supplied, for example when making concrete for dams and other massive structures. It has been found feasible to vary the proportions of intermediate sizes of coarse aggregate, in order to accommodate the source of supply, without changing the combined fineness modulus and without appreciably changing the characteristics of the mixture.

4.5 *Discussion of Abrams' Method*

The water-cement ratio "law" as stated by Abrams has proved to be somewhat of an oversimplification, although it is still useful. The water-cement ratio itself is certainly a valid way to express the physical composition of cement paste made with a given cement. Much more questionable is the water-requirement formula, as will be seen in the ensuing discussion.

4.5.1 The water-requirement formula. When the water requirement formula is converted to the nomenclature of this book, we have

$$\frac{wv_w}{cv_c^o} = R\left[A_w + \left(\frac{0.398}{1.26^{FM}}\right)M\right] \tag{5.3}$$

This may be compared with Eq. 4.19, rewritten as follows:

$$\frac{wv_w}{cv_c^o} = \frac{1}{Q}(k_c + m_{AB}M) \tag{5.4}$$

It will be recalled that k_c is the water-requirement factor for the cement and that $k_c/Q_b = A_w$ is the water-cement ratio of neat cement at the basic consistency. As the expression appears here, consistency is indicated by the slump; m_{AB} is the water requirement of the aggregate when it is used in mixtures within the AB category; for a given consistency m_{AB}/Q appears as D in the water-ratio diagram. That is to say, Eq. 5.4 may be written

$$\frac{wv_w}{cv_c^o} = A_w + DM \tag{5.5}$$

Equation 5.5 may be compared directly with Equation 5.3.

It is apparent that the Abrams formula has the form that is appropriate only for mixtures in the AB category. However, Abrams stated that it would apply to all workable mixtures; it appears, therefore, that in this aspect of his analysis of test results, empiricism did not prevail. Perhaps the form of the equation was adopted on the basis of its reasonableness.

As an example of the application of his formula, Abrams gave in his 1918 paper a table of calculated water requirements for a wide range of mixes and for different values of the fineness modulus. The results, in converted units, for one value of FM are shown in Fig. 5.2, along with a graph of the kind developed in Chapter 4 for materials comparable to those used in most of Abrams' experiments. It seems clear that a series of mixtures containing the amounts of water corresponding to Eq. 5.2 or Eq. 5.3 could not give mixtures having the same slump.

So far as this author knows, the Abrams water-requirement formula is seldom if ever used; perhaps the intrinsic flaw pointed out above is the reason for it. Other procedures for finding the workable mix for a given water cement ratio have been developed, and some of these make use of Abrams' maximum permissible fineness modulus.

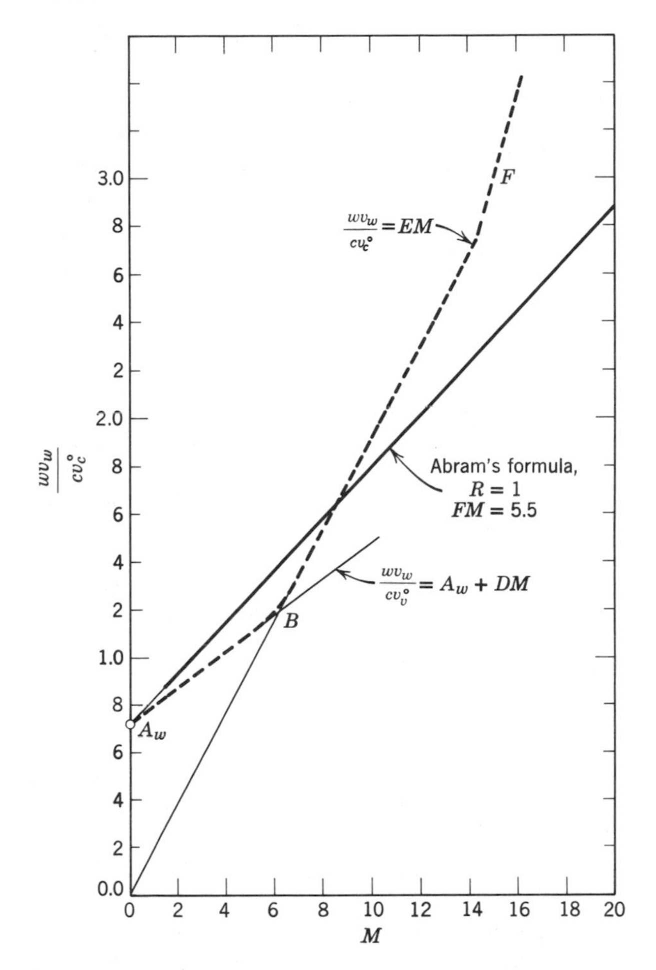

Fig. 5.2. Comparison of Abrams' water-requirement formula with empirical relationships given in Chapter 4. $A_w = 0.72$; $D = 0.078$; $E = 0.191$; $M_f = 1/(C - E) = 14.5$; $C = 0.26$; $Q = 5.770$. The maximum size of the aggregate used was $1\frac{1}{2}$ in. The slump is 1.7 in.

4.5.2 The fineness modulus. Although Abrams presented data showing that difference in grading per se was not important if the maximum size and the fineness modulus were kept constant, the method has been subjected to the criticism that the fineness modulus is inadequate because it does not define a definite grading. Such criticism seems to be based on the assumption that grading per se is crucial, whereas it was Abrams' contention that a wide variety of gradings are equally suitable for a given situation, provided that they all have the same, suitable fineness modulus.

A comparatively recent study by Gaynor [9] supports Abrams' contention and at the same time gives an indication of the magnitude of the effect of grading variation at a constant fineness modulus. Gayor separated one lot of sand into size groups, and recombined the groups in the 17 different ways, as shown in Table 5.2. The first block comprises gradings related to ASTM specifications. The second block comprises

Table 5.2 Sand Gradations[a]

Grading Number	Description		Per Cent Passing Sieve Number						Fineness Modulus
			4	8	16	30	50	100	
1[b]	Coarsest ASTM		100	80	50	25	10	2	3.33
2	Median ASTM		100	90	68	42	20	6	2.74
3	Reduced	−100	100	90	67	41	18	4	2.80
4	Reduced	−100	100	90	67	40	17	2	2.84
5[c]	Reduced	−100	100	89	66	38	15	0	2.92
6[d]	Finest ASTM		100	100	85	60	30	10	2.15
7	Maximum	−100	100	90	69	45	23	10	2.63
8	Belly	100-50	100	91	72	49	30	2	2.56
9[e]	Belly	50-30	100	93	78	60	10	3	2.56
10[e]	Belly	30-16	100	95	85	25	12	4	2.79
11[e]	Belly	16- 8	100	100	50	31	15	4	3.00
12	Maximum	8- 4	100	80	60	37	18	5	3.00
13	Gap	100-50	100	89	64	35	10	10	2.92
14	Gap	50-30	100	87	59	25	25	8	2.96
15	Gap	30-16	100	84	50	50	24	7	2.85
16	Gap	16- 8	100	85	85	52	25	8	2.45
17	Gap	8- 4	100	100	76	47	22	7	2.48

[a] The sand used was a nearly pure quartz bank sand of subangular particle shape. The Sand was separated into individual sizes and recombined to indicated gradings

Except as noted, all gradings comply with limits of "Tentative Specifications for Concrete Aggregates." (ASTM Designation: C 33-55T.)

Data are from R. D. Gaynor, "Effect of Fine Aggregate on Concrete Mixing Water Requirement," unpublished paper of The National Sand and Gravel Association, 1963.

[b] The fineness modulus exceeds the ASTM maximum of 3.1.

[c] The ASTM minimum of 2 per cent passing a No. 100 sieve is not met.

[d] The fineness modulus is below the ASTM minimum of 2.3.

[e] Belly (excess) exceeds the ASTM maximum of 45 per cent permitted between two successive sieves.

Table 5.3 Characteristics of Concretes Made with Sands Listed in Table 5.2[a]

Grading Number	Fineness Modulus	Per Cent Sand	Cement Sacks per yd³	Gallons of Water per yd³	Gallons of Water per sack	Slump (in.)	Air (per cent)	Specific Weight (lb/ft³)	28-Day Compressive Strength (psi)
1	3.33	43.4	5.48	36.2	6.60	3.1	0.9	147.9	4670
2	2.74	37.3	5.53	35.2	6.37	3.4	0.4	149.0	5120
3	2.80	37.8	5.53	35.3	6.39	3.6	0.4	149.0	5210
4	2.84	38.2	5.53	35.1	6.34	3.4	0.4	149.0	5180
5	2.92	39.0	5.52	35.2	6.37	3.3	0.6	148.8	5060
6	2.15	32.7	5.54	35.6	6.43	3.5	0.1	149.4	5230
7	2.63	36.3	5.53	35.1	6.35	3.6	0.6	149.3	5230
8	2.56	35.7	5.54	35.3	6.37	3.5	0.3	149.3	5240
9	2.56	35.7	5.52	35.1	6.37	3.3	0.7	148.7	5250
10	2.79	37.7	5.47	35.7	6.52	3.2	1.0	147.7	4830
11	3.00	39.7	5.50	35.8	6.51	3.4	0.6	148.4	4980
12	3.00	39.7	5.54	34.6	6.26	3.1	0.5	149.2	5280
13	2.92	39.0	5.53	35.5	6.42	3.6	0.6	148.6	5120
14	2.96	39.4	5.54	35.2	6.35	3.1	0.3	149.1	5330
15	2.85	38.3	5.54	34.7	6.26	3.4	0.5	149.2	5370
16	2.45	34.8	5.52	36.0	6.51	3.9	0.2	149.1	5160
17	2.48	35.1	5.54	35.5	6.41	3.6	0.3	149.2	5180

[a] All the concretes were designed to contain 5.5 sacks of cement per cubic yard with a slump of 3–4 in. and a *fineness modulus of combined aggregate of 5.25*.

The *coarse aggregate* was quartz gravel, graded to consist of 25 per cent each of 1–$\frac{3}{4}$ in., $\frac{3}{4}$–$\frac{1}{2}$ in., $\frac{1}{2}$–$\frac{3}{8}$ in., and $\frac{3}{8}$ in.–No. 4.

such gradings altered by introducing excesses of particular size groups. In the third block certain size groups are omitted.

Gaynor combined each of the 17 sands with a coarse aggregate of controlled and constant grading to give a combined fineness modulus of 5.25 in every case. Each combined aggregate was made into concrete containing about 5.5 sacks of cement per cubic yard, and enough water to produce a slump of about 3.5 in. The results as reported by Gaynor are shown in Table 5.3. It is apparent that there were no large differences in slump and water-cement ratio among the various gradings. Adjusting the percentage of sand to obtain the same combined fineness modulus seems to have nullified the possible effects of differences in the sand gradings.

However there is some variation in water content and consistency, leaving open the question as to whether the variation would appear smaller or larger if the figures intended to be alike had been exactly alike. This writer therefore converted the data in Table 5.3 into terms of the water-requirement factor m_{BF}, which for a given category is independent of consistency and mix. This was done by using the following arrangement of Eq. 4.20.

$$m_{BF} = \left(\frac{wv_w}{cv_c^o}\right)\frac{Q}{M}$$

where $Q = (\log S) - 6$. To evaluate M the values of aggregate contents were calculated by difference from the given volumes of cement, water, and air per unit volume of concrete. The consistency factors Q were calculated from the recorded average slumps.

The results of the conversion described above are given in Table 5.4. Figures in column 6 can be taken as values of merit of the different gradings, each one having a fineness modulus of 5.25. Looking at column 7 we see that the largest deviations occurred with gradings 1, 10, and 15; most of the rest showed deviations from the average of the order of 1 per cent, whereas these three deviated 4 per cent, 2 per cent and $2\frac{1}{2}$ per cent, respectively. Two of the three factors were higher than the average, and the third was lower. It appears that grading No. 1 was too coarse to work well with the coarse aggregate used, at least not with the proportion calculated from its fineness modulus; grading No. 10 had an excess of the (#30–#16) group, and as proportioned it required more than the average amount of water; grading No. 15 had a gap at the same group, and as proportioned it required less than the average amount of water.

It thus seems clear that under the conditions of these experiments above, proportioning on the constant-fineness-modulus basis held the water-requirement factor constant within a small tolerance, even for the more extreme gradings; but there is a discernible effect of grading independent of fineness modulus when certain bounds, not clearly defined, are exceeded.

Other experiments have been made using variable coarse-aggregate grading with similar results, the principal controlling feature being the effective maximum size. (There is more on this in Section 5.3, in connection with Table 5.5.)

Although the data discussed above indicate some limitations of the fineness modulus as a means of controlling the water requirement of a workable mixture, there are other and possibly more important limitations that these data could not show. As already indicated, the mixture chosen for these comparisons was in the *BF* category. If the mixture had been in the *AB* category, the water requirement would have been equal to the water requirement of the cement plus that of the aggregate, and the latter would have been proportional to its specific surface area, or surface modulus. The fineness modulus probably would not have appeared as satisfactory as it does in Table 5.4.

To illustrate the point, let us consider gradings 5 and 6 showing values of $-m_{BF}$ of 1.318 and 1.321, practically the same. The first is made with a coarse sand, with a surface modulus of 18.5, and the second with a fine sand, with a surface modulus of 31.6. None of one and 10 per cent of the other pass through a No. 100 sieve. Thus, even though gradings 5 and 6 had the same fineness modulus, they had different

Table 5.4 Water-requirement Factors m_{BF} Calculated from Gaynor's Data

1	2	3[b]	4[c]	5[d]	6[e]	7
Grading Number[a]	Slump (in.)	$\frac{wv_w}{cv_c^o}$	M	$-Q$	$-m_{BF}$	Deviation from Average
1	3.1	1.896	7.590	5.509	1.376	+0.054
2	3.4	1.827	7.623	5.468	1.311	−0.011
3	3.6	1.833	7.618	5.444	1.310	−0.012
4	3.4	1.823	7.618	5.468	1.308	−0.014
5	3.3	1.830	7.611	5.482	1.318	−0.004
6	3.5	1.845	7.616	5.456	1.321	−0.001
7	3.6	1.823	7.608	5.468	1.310	−0.012
8	3.5	1.829	7.610	5.456	1.311	−0.011
9	3.3	1.824	7.606	5.482	1.315	−0.007
10	3.2	1.874	7.625	5.495	1.351	+0.029
11	3.4	1.869	7.616	5.468	1.342	+0.020
12	3.1	1.794	7.625	5.509	1.296	−0.026
13	3.6	1.844	7.586	5.468	1.329	+0.007
14	3.1	1.824	7.568	5.509	1.328	+0.006
15	3.4	1.799	7.620	5.468	1.291	−0.031
16	3.9	1.872	7.611	5.409	1.331	+0.009
17	3.6	1.840	7.600	5.468	1.324	+0.002
Average					1.322	
Standard deviation					0.02 or 1.5%	

[a] See Table 5.2.
[b] The ratio of volume of water to solid volume of cement.
[c] The ratio of aggregate to cement in solid volumes.
[d] Consistency factor is $Q = \log(S/S_o)$.
[e] The water-requirement factor for the *BF* category.

surface moduli, actually about 7.6 and 10.6, respectively. We can estimate the water requirements of these aggregates when used in mixtures of the *AB* category as follows.

From Fig. 3.13 we can estimate the parameter D^* and from that the parameter D, which is the water requirement of the aggregate when it is used in mixtures in the *AB* category at basic consistency. For the surface factor, SM/ψ_L, we estimate $1/\psi_L$ to be 1.2, the aggregate having been described as "subangular." This gives surface factors of 9.1 and 12.7, respectively. The corresponding values of D^* are

for No. 5,

$$D^* = 0.05$$

for No. 6,

$$D^* = 0.08$$

At basic consistency the air-water ratio Z may be assumed to be 0.17, and on that basis

for No. 5,

$$D = \frac{0.05}{1.17} = 0.043$$

for No. 6,

$$D = \frac{0.08}{1.17} = 0.068$$

The water requirement of the cement at basic consistency may be assumed to be $A_w = 0.75$, and thus the water requirement for the whole mixture at basic consistency can be given on the basis of Eq. 5.5. Thus

for grading No. 5,

$$\frac{wv_w}{cv_c^o} = 0.75 + 0.043M$$

and for grading No. 6,

$$\frac{wv_w}{cv_c^o} = 0.75 + 0.068M$$

For a maximum size of 1 in. we may choose $M = 6$ as being within the *AB* category. On that basis we find

for grading No. 5,

$$\frac{wv_w}{cv_c^o} = 1.01$$

$$\frac{w}{c} = 0.323$$

for grading No. 6,

$$\frac{wv_w}{cv_c^o} = 1.16$$

$$\frac{w}{c} = 0.37$$

For a slump of 3.5 in., these amounts would be raised by the factor Q_b/Q, which is $-5.770/-5.456 = 1.058$; thus the water-cement ratios by weight at a slump of 3.5 in. would be 0.342 and 0.391, respectively, the difference being 15 per cent of the smaller value.

Thus we find that the coarser of the two sands would have been preferable for rich mixtures if the proportion was established on the basis of a combined fineness modulus of 5.25. However the results would probably have been practically equal if the two aggregates had been combined to produce the same surface factors. It turns out that to obtain the same surface modulus as that obtained with the coarse sand, the percentage of fine sand would have been 22 per cent rather than the 32.7 per cent arrived at by the fineness modulus method.

4.5.3 Maximum permissible fineness modulus. During the years since Abrams first published his table of maximum permissible values for the fineness modulus, several investigators have made similar experiments and have obtained similar values. On the basis of an extended study, Popovics [10] concluded that a properly graded aggregate must necessarily have the optimum (maximum permissible) fineness modulus, but some gradings meeting that requirement are not necessarily suitable.

As we have seen in Section 4.5.2, a maximum permissible fineness modulus established by trial for a mixture in the *AB* category should not be expected to have the same degree of reliability as one established for a mixture in the *BF* category; indeed the optimum *FM* for one grading cannot be the optimum for another grading unless the physical properties of the individual particles and the specific surface area are alike. In the *FC* category the water requirement seems to depend primarily on the voids content of the aggregate, and because the voids content is not necessarily the same for different gradings at the same fineness modulus, it would seem that the optimum value of *FM* for one grading may not be optimum for another grading. However because mixtures in the *FC* category are not likely to be tolerated where quality is a matter of concern, this limitation of the fineness modulus method is not of much practical importance.

Such matters considered, it seems that the maximum permissible fineness modulus is a reliable factor for aggregates of a given shape factor when the aggregates are used in mixtures in the *BF* category. Also using it for mixtures in the *FC* category would not introduce significant inaccuracies. It might be used for the *AB* category, provided that the specific surface area of the sand is not significantly different from that of the sand which was used when the experimenters were establishing the maximum permissible values given in the original table.

It should be noticed that Abrams' method aims to use any given pair of aggregates to obtain the best results possible from those aggregates *as found by the mix designer,* with only two restrictions: (1) the sand must not be too fine, and (2) it must not be too coarse, the first being given in terms of a minimum permissible fineness modulus, and the second as the maximum permissible fineness modulus for the given mix and the maximum size of the sand. The method does not lead to prescribing additional fine material when the amount of cement is insufficient to give the minimum possible voids content; in other words, the method is not concerned with point *B* of the voids-ratio diagram. Modifications of the original method have appeared which have in one way or another been concerned not only with the material retained on the No. 200 sieve in the aggregate as normally produced, but also with the amount of material fine enough to pass that sieve.

4.6 Thaulow's Modification

Thaulow [11] used the Abrams fineness modulus, but he used it with a modification of the Abrams design method. The modification had several aspects, the one of interest here being that the maximum permissible fineness modulus (optimum fineness modulus) was based on the water-cement ratio of the paste rather than on the cement content of the mixture. Thaulow found that the optimum value was nearly independent of the shape and surface roughness of the aggregate particles, which is not so when it is based on the cement content. He recognized that use of the optimum fineness modulus did not assure the best results obtainable under all conditions; particularly, he found that a mixture having a water-cement ratio higher than about 0.45 by weight would be "unstable," which is to say, it would bleed excessively and show other undesirable characteristics unless the sand contained a certain amount of material that would pass through a No. 100 sieve; specifically, he stated that the quantity of "fines" in the sand should be such that the following relationship would be satisfied:

$$\frac{(\text{``fines''})}{c} = \frac{(w/c - 0.45)0.85}{0.45} \tag{5.6}$$

where 0.85 is an average ratio of the specific volume of cement to that of the fine material.

For example, if the water-cement ratio of a given mixture is 0.55, the weight ratio of "fines" to cement should be $(0.1/0.45)0.85 = 0.19$. This procedure is to be compared with the development in Chapter 4, Section 9.

4.7 Modification by Swayze and Gruenwald

In 1947 Swayze and Gruenwald [12] studied several series of concrete mixtures for which the aggregate gradings were expressed in terms of fineness moduli. They confirmed, in a general way, Abrams' findings on the maximum permissible value for the fineness modulus of the aggregate. Going farther, they discovered that by basing the fineness modulus on the *total* solid material rather than on aggregate alone, the optimum fineness modulus became virtually independent of the cement content of the mixture, within certain limits of cement content. For example, for sand-and-gravel aggregate with a maximum size of $1\frac{1}{2}$ in., they found that the optimum fineness moduli for different cement contents were as follows:

Sks./yd^3	3.5	4.25	5.0	6.0	7.5
$(FM)_o$	4.80	4.83	4.85	4.84	4.75

where $(FM)_o$ is the optimum or maximum permissible fineness modulus. A similar degree of invariance was found for smaller maximum sizes, and when crushed stone was used as coarse aggregate.

In calculating the fineness modulus of total solids Swayze and Gruenwald expressed the amounts in solid volumes rather than weights, and the fineness modulus of the cement was taken as 0.00, in line with Abrams' original definition. For example, if the fineness modulus of the aggregate is 5.65, and if the cement amounts to 15 per cent of the total solid material, the fineness modulus of the total solid material is $5.65 \times 0.85 = 4.80$.

If the proper fineness modulus for the total solids has been established for one mixture, it is a simple procedure to calculate the proper fineness modulus for the aggregate when a different cement content is used. For example, if the cement is to be 10 per cent of the total solids instead of 15 per cent as in the example above, the fineness modulus of the aggregate should be $4.80 \times 1/0.90 = 5.33$, as compared with the 5.65 for the richer mix. It may be noted that in the nomenclature of this book

$$FM_o = \frac{(FM)_t}{x}$$

where $(FM)_t$ is the optimum fineness modulus of the total solids.

The reasoning leading to the development described above was expressed by Swayze and Gruenwald as follows:

"Abrams said regarding fineness modulus, 'The fineness modulus may be considered an abstract number; it is in fact a summation of volumes of material.' With this concept it is apparent that when two or more aggregates of different specific gravity are proportioned to give a certain combined fineness modulus, the proportions should be in relation to the volumes they occupy in the mix, rather than to the weights they contribute to a unit volume of concrete. Further, since cement is an essential ingredient in concrete, the volume which it occupies in the mix must have an important bearing on the relative proportions of the fine and coarse aggregates."

In another place they say,

". . . we suggest that the concept of cement and water in a concrete batch as forming a fluid medium of varying density and viscosity as the water cement ratio of the batch is changed. In this fluid medium, which has no fineness modulus of its own (FM = 0.00), the particles of aggregate are suspended for a sufficient length of time to allow transportation, placing and finishing of the surface before settlement of the aggregate and interlock of particles destroys the essential workability of the batch."

4.8 *Modification by Walker*

An earlier modification similar to that made by Swayze and Gruenwald was made by Stanton Walker, which was brought out in an extended discussion of the Swayze and Gruenwald paper by Walker and Bartel [7]. The fineness modulus of the total solids was computed by assigning a negative fineness modulus to the cement. In the examples given, the fineness modulus of the cement was taken to be —2.5. A recalculation of Abrams' table of maximum permissible fineness moduli gave similar values for all mixes from about 1:4 to 1:9 by bulk volumes.

Figure 5.3 is a plot of the maximum permissible fineness moduli for various mixes with (0–1½ in.) aggregate, recalculated from the Abrams table in terms of both modifications. The upper curve gives the values as calculated by Walker, using —2.5 as the fineness modulus of the cement. The curve below is that calculated from the same data assuming the fineness modulus of the cement to be 0.00. It is apparent that neither basis will indicate a constant fineness modulus for the whole range of mixtures and that each gives substantially one value for a limited range; this range includes the most commonly used mixes. For the range 1:3

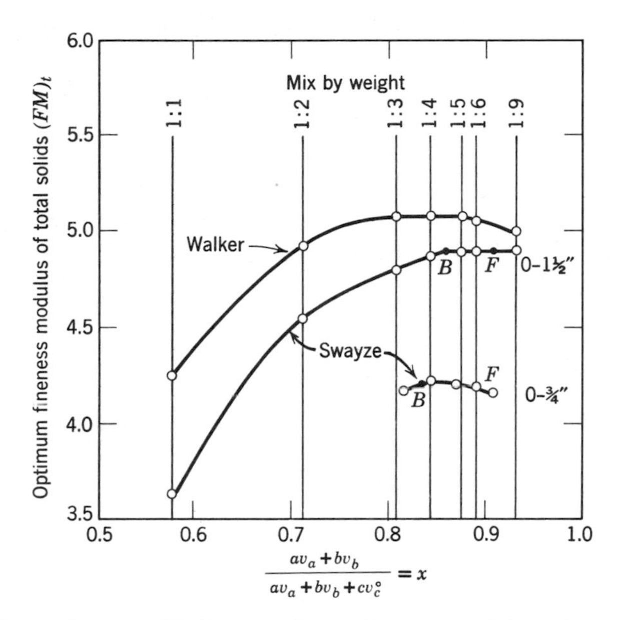

Fig. 5.3. Comparisons of Walker's optimum fineness modulus of total solids with that of Swayze.

to 1:6, the Walker modification seems to be slightly better than the other; and vice-versa for the range 1:4 to 1:9.

The short curve in Fig. 5.3 for (0–$\frac{3}{4}$ in.) sand and gravel aggregate represents some of the data of Swayze and Gruenwald; the curve shows a tendency to drop on either side of a substantially flat middle section. (These data are the same as those in Fig. 4.10b.) It turns out that the nearly flat portion represents mixtures in the *BF* category. The same is true for the (0–1$\frac{1}{2}$ in.) curve (which may be compared with the Fig. 4.10a, except that there is no indication of a drop at the first point beyond point *F*.

We must keep in mind that an average value based on the last four or five points of either of the upper curves in Fig. 5.3 will serve the purpose of calculations based on the maximum permissible fineness modulus, which purpose is to arrive at a reasonably good first-approximation of a workable mix.

4.8.1 Empirical formulas. Popovics [10] found that the maximum permissible fineness modulus of the total solids as computed by Walker is given by the following formula:

$$(FM)_t = 2.45 \log D + 1.05 \tag{5.7}$$

where D is the average nominal size of the largest size group, in millimeters, and the constants pertain to data derived from sand and gravel aggregate; actually they are the data used by Walker and Bartel, which were due to Abrams. Of course, the formula is valid only for the substantially flat part of curves such as the upper one in Fig. 5.3, a sufficient range for most practical purposes.

Popovics also gave a formula for finding the maximum permissible fineness modulus of the aggregate after the value for total solids has been established:

$$(FM)_o = \frac{(FM)_t - (FM)_c}{(v_{ag}/v_c)N} + (FM)_t \qquad (5.8)$$

where N is the mix by weight. With the usual values of specific volumes, the ratio $v_c/v_{ag} = 0.85$, and with $(FM)_c = -2.5$, the formula takes the following form:

$$(FM)_o = \left(1 + \frac{0.85}{N}\right)(2.45 \log D_{\max} + 1.05) + \frac{2.1}{N} \qquad (5.9)$$

5. THE TALBOT AND RICHART MORTAR-VOIDS METHOD

Three years after Abrams' fineness modulus method of mix design appeared, Talbot [13] published a paper on what became known as the mortar-voids method of formulating mixtures, and in 1923 Talbot and Richart [14] published the comprehensive study on which the first paper was based. Like Abrams, Talbot and Richart based their method on compressive strength.

5.1 Strength and the Cement-space Ratio

Citing an 1897 paper by Feret, Talbot and Richart presented strength data in terms of the ratio of the solid volume of cement to the volume of space in concrete outside the aggregate, which they called the cement-space ratio. Their experimental results showed that for mixtures at basic consistency, the observed strengths after 28 days of curing agreed well with the following formula:

$$f_c = 32{,}000 \left[\frac{cv_c}{(cv_c + v)}\right]^{2.5} \qquad (5.10)$$

or

$$f_c = \frac{32{,}000}{(1 + v/cv_c)^{2.5}} \qquad (5.11)$$

The second expression shows strength as a function of the voids-cement ratio and is thus comparable with Abrams' strength equation (Eq. 5.1), except that the volume of air as well as the volume of water is represented. It should be noted that the specific volume of the cement is its true value v_c and not the water-displacement factor v_c^o, used in preceding chapters. At the time, the inaccuracy of v_c for computing the actual amount of space occupied by cement in freshly mixed concrete had not been recognized.

For mixtures containing more than the basic water content, Talbot and Richart found strengths at a given cement-space ratio generally to be lower than those indicated by Eq. 5.11. It seemed, therefore, that for a given voids-cement ratio, the strength was lower the smaller the fraction of void space occupied by air. Talbot and Richart chose to base their mix design on mixtures having the basic water content, and the corresponding curve for the strength versus cement-space ratio. If the mixture actually was intended to contain more than the basic water content, the strength to be expected at a given cement-space ratio was scaled down by means of an empirically established reduction curve. For example, if the relative water content was 1.2, the curve indicated the relative compressive strength would be 0.85, or 85 per cent of that obtained at basic consistency and with the same cement-space ratio. Practice developed later, particularly that developed by the Division of Highways, Department of Public Works, State of Illinois, was based on strength curves established by tests of specimens having the desired relative water content, and made of the specific materials to be used in the field.

Thus the first step of the mortar-voids method was essentially like that of Abrams, but the cement-space ratio or the voids-cement ratio corresponding to the desired strength and consistency was chosen instead of the water-cement ratio, which, for a given strength, Abrams had assumed would be practically independent of consistency.

5.2 Mortar-voids Curves

The next step in the mortar-voids method is based on the concept that concrete is essentially a coarse aggregate in a matrix of mortar and that the coarse aggregate functions only as an "inert filler." For each sand to be considered, a series of mixtures is made. Each mixture is tamped into a small cylindrical mold, using a prescribed technique to establish a curve such as one of those shown in Fig. 3.4. (In routine practice the lowest water content of the series is only a little less than that required for minimum voids content, the object being to find the basic water content and to establish the trend of the curve as the water content is increased from that point.)

Having established the curve for voids versus water for each of several mixes of a given sand and cement, the next step is to plot the relationship between the voids content at basic water content, or at some higher relative water content, against the ratio of sand to cement in the mortar. In our nomenclature this is a plot of $(wv_w + \text{A})/V$ versus av_a/cv_c, or, v/V versus M. From the resulting curve, the sand-cement ratio corresponding to the selected cement-space ratio is noted, which completes the second step.

5.3 Amount of Coarse Aggregate

The final step of the mortar-voids method is to establish the proportion of coarse aggregate. As originally proposed, it is assumed that "below a reasonable limit for the amount of the coarse aggregate, the strength of the concrete is identical with the strength of the mortar comprising it." Reasonable limits were established by trial, and it was found that the solid volume of coarse aggregate per unit volume of concrete was usually between 65 and 75 per cent of the solid volume in a compacted unit volume of the coarse aggregate alone. Talbot and Richart let b stand for the solid-volume fraction of coarse aggregate in concrete and b_o the solid-volume fraction of a unit volume of coarse aggregate alone. Hence it was said that the proportion of coarse aggregate should be at the practical limit for the materials and conditions of the work and that the limit would probably correspond to a value of b/b_o between 0.65 and 0.75. In practice the water content of the mortar must be higher than its basic water content if concrete of basic consistency is required. (See Chapter 3, Section 5.8.) However, the original theory gave no guidance on this point. It appears from the practice of the Illinois Highway Division that the relative water content of the mortar must be 1.4 or higher to obtain concrete having a consistency suitable for paving.

Although the use of mortar-voids method of mix design has not become widely adopted, the proportioning of coarse aggregate on the basis of b/b_o has become a part of other methods. In 1940 Dunagan [15] proposed establishing the optimum value of b/b_o on the basis of Weymouth's theory of particle interference. In 1942 (revised in 1949) Goldbeck and Gray [16] published a method in which the selection of the coarse aggregate content of the concrete was made the first step, this selection being based on b/b_o. On the basis of experimental work, they compiled a table showing that the optimum values of b/b_o vary systematically with the range of grading of the coarse aggregate and with the fineness modulus of the sand. For example, with sand having a fineness modulus of 2.80, b/b_o ranged from 0.67 for (#4–$\frac{3}{4}$ in.) aggregate, to 0.74 for (#4–$2\frac{1}{2}$ in.) aggregate. For (#4–$1\frac{1}{2}$ in.) coarse aggregate, b/b_o ranged from 0.74 for $FM_a = 2.40$, to 0.67 for $FM_a = 3.10$.

The finding that the optimum value of b/b_o is different for different sands is in line with the observation already pointed out that we must finally deal with the properties of the combined aggregate, regardless of how the aggregate may have been separated for purposes of processing and proportioning. But it is not clear why for the Goldbeck and Gray method the coarse aggregate is characterized only by its size range and b_0, or its voids content, whereas the fine aggregate is characterized by its fineness modulus. It can be contended with reason that if the fineness modulus is suitable for the fine aggregate, it is suitable also for the coarse aggregate and that two aggregates can be combined to best advantage on the basis of the optimum combined fineness modulus; Bloem and Walker [17] found the latter to be the more accurate.

Data in Table 5.5, compiled from experimental data obtained by the author in 1932, are derived from tests on concrete mixtures made with sand and gravel aggregates, the same sand being used in each mixture, whereas the coarse-aggregate grading was varied systematically. The coarse-aggregate gradings shown in the table were each used with the optimum percentage of sand, the optimum being taken as that which required the least volume of cement paste at a given water-cement ratio and a given consistency as indicated by the remolding test; in general the figures given correspond to points from well-established smooth curves. The figures given in the first block of data pertain to coarse aggregates graded from No. 4 to $1\frac{1}{2}$ in., whereas those in the second block pertain to single size groups.

For the data in the first block in Table 5.5, the figures in column 8 show that the deviation from the average value of b/b_o is negligible. The values for optimum fineness modulus shown in column 6 are also nearly alike, except that there is some indication that FM_o begins to diminish when the largest size becomes less than about 45 per cent of the total coarse aggregate, which suggests that the effective maximum size begins to diminish before the largest size is completely eliminated, as might be expected. (Effective or practical maximum size was defined by Abrams about as follows: If the largest size group contains 15 per cent or more of the total aggregate, the top sieve size of that group shall be considered the maximum size; otherwise, maximum size is the top size of the next smaller group.) Thus the data in the upper block seem to support the advocated use of either the optimum fineness modulus or the optimum value of b/b_o as a means of estimating the best proportion of coarse aggregate in a mixture. (The mixtures were in the *BF* category.)

The lower block of data in Table 5.5 illustrate the point that either basis described above for estimating the proper quantity of coarse aggregate to be used can be expected to be reasonably accurate for only

Table 5.5 Optimum Values of Fineness Modulus and b/b_o for Mixtures Made with Various Gradings of Coarse Aggregate

1	2	3	4	5	6	7	8	9
Grading of Coarse Aggregate (per cent)				Sand[b] (per cent by weight)	*FM* combined aggregate			Cement Content[c] (sacks/yd³)
#4–$\frac{3}{8}''$	$\frac{3}{8}''$–$\frac{3}{4}''$	$\frac{3}{4}''$–$1\frac{1}{2}''$	*FM*[a]			b_o	$\frac{b}{b_o}$	
				Maximum Size $1\frac{1}{2}$ in.				
35.0	00.0	65.0	7.30	37.5	5.65	0.687	0.691	4.95
30.0	17.5	52.5	7.22	38.0	5.50	0.667	0.705	4.95
25.0	30.0	45.0	7.20	38.5	5.55	0.663	0.701	5.00
20.0	48.0	32.0	7.12	39.0	5.43	0.660	0.697	5.05
				Single Size Groups				
100	0	0	6.00	46.5	4.56	0.642	0.653	5.75
0	100	0	7.00	44.5	5.17	0.626	0.679	5.10
0	0	100	8.00	49.5	5.42	0.615	0.611	5.30

[a] The fineness modulus of the coarse aggregate; the aggregate was composed of rounded pebbles.

[b] The percentage of total aggregate. The fineness modulus of the sand was 2.90.

[c] The cement content was computed on the assumption that air content was nil.

Data are from Powers, T. C., "Studies of Workability of Concrete," *Proc. ACI* **28**, 419–448, 1932.

limited conditions, although the limits may be wide enough for most practical purposes. The last item shows that the single size group with a $1\frac{1}{2}$ in. maximum size showed a distinctly lower optimum fineness modulus for the combined aggregate and a lower optimum value of b/b_o than the values for the corresponding graded coarse aggregates. However, in this connection the nature of the remolding test should be kept in mind. (See Chapter 10, Section 8.2.) In the course of the test the mixture is required to flow under a barrier (the ring) which clears the bottom by only about 2.75 in. The hindrance to flow through a clearance of this dimension should be expected to be greater the more numerous the largest particles, assuming the actual stiffness of the mixtures (as indicated by resistance to shear strain) is the same. (See Chapter 10, Section 5.2.1.) Therefore, although the limits on *FM* and b/b_o seem relatively low for the condition of this test, under a condition of less hindered flow, as for example the flow required when a batch is flattened to make a pavement slab, the deviations from the optimum values for the more ordinary gradings might not appear.

Nonetheless, it is evident that the limiting values are not independent of the maximum size. According to the fineness modulus method they are not supposed to be, but by the original Talbot and Richart proposal it would be supposed that b/b_o would be the same for different average sizes. It has already been pointed out that subsequent experimental investigations, particularly those by Goldbeck and Gray, showed that the optimum value of b/b_o is greater the larger the maximum size of a coarse aggregate.

5.4 *Discussion of the Mortar-voids Method*

The mortar-voids method as proposed by Talbot and Richart is open to criticism on at least two counts: It seems more laborious and complicated than other methods that seem to give results as good, and it is based on a questionable concept, discussed below. Perhaps these are the reasons the method has not been widely adopted.

The assumption that coarse aggregate can be added to mortar without affecting the properties of the mortar is not true to fact. As we have seen, when dry aggregates are mixed together, each alters the properties of the other, and in concrete the relatively small particle dispersion effected by the paste does not eliminate this effect. After the concrete is made, the water requirement of the mixture cannot properly be said to depend on the cement and fine aggregate only; it depends on the aggregate as a whole, and as we show in Chapters 3 and 4, the water-cement ratio, or the voids-cement ratio, seems to be practically independent of the characteristics of the cement in two of the three categories of concrete mixtures. Nevertheless, the mortar-voids method can produce well designed mixtures; it is essentially empirical.

Perhaps the contribution of Talbot and Richart that has proved of greatest value is the concept underlying b/b_o, which is to say, the idea that an expression of the degree of dispersion (or concentration) of an aggregate in concrete can be based on the bulk density of that aggregate. This idea influenced subsequent developments in mix design.

6. LYSE'S RULE OF CONSTANT WATER CONTENT

Along with some of the ideas discussed above, current practice in mix design makes use of the *Lyse rule of constant water content* [18], sometimes referred to as the *Slater-Lyse rule* [19]. Lyse observed that when we make a series of mixtures with given materials, the consistency of one mixture as measured by the slump test is nearly the same as that of another if the total water contents of the mixes are the same. He pointed out that if the water content and water-cement ratio for a given mix has been determined, the required cement content for a mixture of the same consistency but different water-cement ratio can be estimated closely by assuming that the total water content will remain unchanged when the cement content is changed.

On the basis of the water-ratio diagrams studied in previous chapters, we can see that the rule of constant water content is not exact and that it can be applied safely only to a limited range of mixtures, points well understood by Lyse. Mixtures within the *AB* and *FC* categories are likely to have considerably higher water contents than the average for the *BF* category, and within the *BF* category there is an upward trend as the cement content is reduced, except perhaps for a short transition sector near point *B*. However, within the *BF* category the air content usually increases faster than the water content, thus giving an increase of water content smaller than the increase in water ratio. An example of the near constancy of the total water content within the *BF* category is given in Table 5.6. Although these numbers are nearly alike, they are in fact derived from a smooth curve that passes through a minimum point within range of the values tabulated, and for that reason we must conclude that exact equality of water content for different mixtures is

Table 5.6 Water Content at a Given Consistency[a]

1	2	3[c]	4[d]	5[e]	6	7	8	9
							Water Content	
Reference Number	Category[b]	u	$1+Z$	u_w	Slump	$\frac{w}{c}$	$\frac{wv_w}{V}$	gal/yd³
101	*AB*	0.208	1.082	0.192	3.1	0.361	0.159	32.1
106	*BF*	0.188	1.098	0.171	2.8	0.435	0.144	29.1
111	*BF*	0.188	1.121	0.168	2.7	0.507	0.142	28.7
116	*BF*	0.199	1.137	0.175	2.3	0.666	0.146	29.5

NOTES:
[a] The Data are from PCA Series 303.
[b] Classification of the mixture in the voids-ratio diagram.
[c] The voids ratio.
[d] Z is ratio of air volume to water volume.
[e] The water ratio.

fortuitous. Nevertheless, within the *BF* range the amount of the departure from an average water content is so small as to make the practical application of Lyse's rule quite feasible.

7. ACI STANDARD RECOMMENDED PRACTICE

The developments described in Sections 3, 4, and 5 contain the roots of the most widely accepted American mix design practice, which is the American Concrete Institute Standard Recommended Practice for Selecting Proportions for Concrete. The National Ready Mixed Concrete Association, the Portland Cement Association and other organizations publish and promote the use of similar methods. The ACI method [20] amounts to a consensus. Because this publication is readily available, we consider only the principal features, omitting ways and means.

7.1 *Consistency*

The Committee recommends using the stiffest consistency that will permit satisfactory and "efficient" placement. (It does not specify the

Table 5.7 Recommended Slumps for Various Types of Construction[a]

Types of Construction	Slump in. Max	Slump in. Min
Reinforced foundation walls and footings	5	2
Plain footings, caissons, and substructure walls	4	1
Slabs, beams, and reinforced walls	6	3
Building columns	6	3
Pavements	3	2
Heavy mass construction	3	1

[a] The slumps were recommended by ACI Committee 613. When high-frequency vibrators are used, the values given should be reduced about one-third.

kinds of placing tools considered acceptable and efficient.) Recommended consistencies in terms of slump are given in Table 5.7.

7.2 *Maximum Size*

Recognizing the relationship between the maximum size of an aggregate and its water-requirement factor, the Committee recommends using

Table 5.8 Maximum Sizes of Aggregate Recommended for Various Types of Construction[a]

Minimum Dimension of Section (in.)	Maximum Size of Aggregate[b] (in.)			
	Reinforced Walls, Beams and Columns	Unreinforced Walls	Heavily Reinforced Slabs	Lightly Reinforced or Unreinforced Slabs
$2\frac{1}{2}$–5	$\frac{1}{2}$–$\frac{3}{4}$	$\frac{3}{4}$	$\frac{3}{4}$–1	$\frac{3}{4}$–$1\frac{1}{2}$
6–11	$\frac{3}{4}$–$1\frac{1}{2}$	$1\frac{1}{2}$	$1\frac{1}{2}$	$1\frac{1}{2}$–3
12–29	1 –3	3	$1\frac{1}{2}$–3	3
30 or more	$1\frac{1}{2}$–3	6	$1\frac{1}{2}$–3	3 –6

[a] Recommended by ACI Committee 613.
[b] Based on square openings.

the largest size the type of construction will permit. The specific recommendations are given in Table 5.8.

7.3 Estimation of Water Content

After selecting consistency and aggregate size, the first step toward selecting the proportions of cement, aggregate, air, and water is to estimate the water content. This is done by following Table 5.9 (from [21]), which is based on the Lyse rule. That is to say, the Committee assumes that the quantity of water in a unit volume of concrete will be the same at a given consistency and maximum size for all cement contents.

7.4 Estimation of Cement Content

The second step is to select the appropriate cement content. This may have been fixed by specification, in which case no alternatives are involved. More commonly, the cement content is based on a specified upper limit for the water-cement ratio.

7.4.1 The water-cement ratio. The maximum permissible water-cement ratio may be based on the requirement for strength, either compressive or flexural, the latter usually being specified for highway concrete. But questions of durability bring into consideration the permeability of the hardened cement paste, which controls the rate of ingress of water and injurious solutes. Because the coefficient of permeability is higher the higher the water-cement ratio of the paste and because

Table 5.9 Approximate Mixing Water Requirements for Different Consistencies and Maximum Sizes of Aggregates[a]

Consistency				Relative Water	Water for Indicated Maximum Sizes of Coarse Aggregate (gal/yd³)							
Description	Slump (in.)	Vebe (sec)	Compacting Factor	Content (per cent)	3/8 in.	1/2 in.	3/4 in.	1 in.	1 1/2 in.	2 in.	3 in.	6 in.
Nonair-entrained Concrete												
Extremely dry	—	32-18	—	78	36	34	32	30	28	27	25	22
Very stiff	—	18-10	0.70	83	38	37	34	32	30	28	27	23
Stiff	0–1	10- 5	0.75	88	40	39	36	34	32	30	28	24
Stiff plastic	1–2	5- 3	0.85	92	42	40	37	36	33	31	29	25
Plastic	3–4	3- 0	0.91	100	46	44	41	39	36	34	32	28
Flowing	6–7	—	0.95	106	49	46	43	41	38	36	34	30
Approximate Amount of Entrapped Air in Nonairentrained Concrete					3	2.5	2	1.5	1	0.5	0.3	0.2
Air-entrained Concrete												
Extremely dry	—	32-18	—	78	32	30	28	27	25	23	22	19
Very stiff	—	18-10	0.70	83	34	32	30	28	27	25	23	20
Stiff	0–1	10- 5	0.75	88	36	34	32	30	28	26	24	21
Stiff plastic	1–2	5- 3	0.85	92	37	36	33	31	29	27	25	22
Plastic	3–4	3- 0	0.91	100	41	39	36	34	32	30	28	24
Flowing	6–7	—	0.95	106	43	41	38	36	34	32	30	26
Recommended percentage of the average total air content[b]					8	7	6	5	4.5	4	3.5	3

[a] This table is an augmented form of the table originally given. It was prepared by a special subcommittee on designing mixtures for use with high-energy placing methods, particularly mechanical vibration. (Subcommittee No. 2, ACI Committee 211, "Recommended Practice for Selecting Proportions for No-slump Concrete," *Proc. ACI*, **62,** 1–22, 1965.)

The quantities of mixing water are for use in computing cement factors for trial batches. They are for reasonably well-shaped angular, coarse aggregates graded within limits of accepted specifications.

If *more* water is required than the amount shown, the cement factor, estimated from these quantities, should be increased to maintain the desired water-cement ratio, except as otherwise indicated by laboratory tests for strength.

If *less* water is required than the amount shown, the cement factor, estimated from these quantities, *should not* be decreased except as indicated by laboratory tests for strength.

[b] For consistencies below 1 in. slump, the volumes of air entrained by an air-entraining cement or the usual amount of air-entraining admixture used for more plastic mixtures may be significantly lower than those shown. A suitable adjustment of the mix proportions can be made after the actual air content of the first trial mix is determined.

Table 5.10 Maximum Permissible Water-cement Ratios (Gal per Bag) for Different Types of Structures and Degrees of Exposure[a]

Type of Structure	Exposure Conditions[b]					
	Severe Wide Range in Temperature, or Frequent Alternations of Freezing and Thawing (Air-entrained Concrete Only)			Mild Temperature Rarely below Freezing, or Rainy, or Arid		
	In Air	At the Water Line or within the Range of Fluctuating Water Level or Spray		In Air	At the Water Line or within the Range of Fluctuating Water Level or Spray	
		In Fresh Water	In Sea Water or in Contact with Sulfates[c]		In Fresh Water	In Sea Water or in Contact with Sulfates[c]
Thin sections, such as railings, curbs, sills, ledges, ornamental or architectural concrete, reinforced piles, pipe, and all sections with less than 1 in. concrete cover over reinforcing	5.5 (0.49)	5.0 (0.44)	4.5[d] (0.40)	6.0 (0.53)	5.5 (0.49)	4.5[d] (0.40)[f]
Moderate sections, such as retaining walls, abutments, piers, girders, beams	6.0	5.5	5.0[d]	[e]	6.0	5.0[d]
Exterior portions of heavy (mass) sections	6.5	5.5	5.0[d]	[e]	6.0	5.0[d]
Concrete deposited by termie under water	—	5.0	5.0	—	5.0	5.0
Concrete slabs laid on the ground	6.0	—	—	[e]	—	—
Concrete protected from the weather, interiors of buildings, concrete below ground	—	—	—	[e]	—	—
Concrete which will later be protected by enclosure or backfill but which may be exposed to freezing and thawing for several years before such protection is offered	6.0	—	—	[e]	—	—

[a] Recommended by ACI Committee 613.
[b] Air-entrained concrete should be used under all conditions involving severe exposure and may be used under mild exposure conditions to improve workability of the mixture.
[c] Soil or ground water containing sulfate concentrations of more than 0.2 per cent.
[d] When sulfate resisting cement is used, the maximum water-cement ratio may be increased by 0.5 gal per bag.
[e] The water-cement ratio should be selected on basis of strength and workability requirements.
[f] Figures in parentheses are w/c by weight.

under some conditions a very low coefficient of permeability is required, the control of permeability rather than the requirement for strength sometimes determines the maximum permissible water-cement ratio.

Table 5.10 gives the Committee's recommendations for situations where the concrete is to be subjected to moisture and frost action, or to sulfate ions in ground water or, if the concrete is reinforced concrete, to sea water with its attendant hazard of steel corrosion. For situations

Table 5.11 Compressive Strength of Concrete for Various Water-cement Ratios[a]

Water-cement Ratio		Probable Compressive Strength at 28 Days (psi)	
gal/bag	$\frac{w}{c}$	Nonair-entrained Concrete	Air-entrained Concrete
$3\frac{1}{2}$	0.31	6800	5500
4	0.36	6000	4800
5	0.44	5000	4000
6	0.53	4000	3200
7	0.62	3200	2600
8	0.71	2500	2000
9	0.80	2000	1600

[a] These average strengths are for concretes containing not more than the percentages of entrained and/or entrapped air shown in Table 5.9. For a constant water-cement ratio, the strength of the concrete is reduced as the air content is increased. For air contents higher than those listed in Table 5.9, the strengths will be proportionally less than those listed in this table. Strengths are based on 6 × 12-in. cylinders moist-cured under standard conditions for 28 days. See "Method of Making and Curing Concrete Compression and Flexure Test Specimens in the Field" (ASTM Designation C 31).

where the concrete is to be subject to frost action, the Committee recommends not only certain limits on the water-cement ratio of the mixture, but also a certain controlled amount of entrained air. (See Chapter 7.) The recommended limits on the water-cement ratios are given in Table 5.10. Normal quantities of entrapped air, which are supposed to be smaller the larger the maximum size of the aggregate, are given in Table 5.9.

When strength is the governing factor, the Committee refers to the relationships given in Table 5.11, but it recommends obtaining data for the specific materials to be used, rather than using this table.

7.4.2 Cement content. Having selected the water-cement ratio, the designer next calculates the cement content from the estimated water content. The cement content is obtained by multiplying the water content by the water-cement ratio.

7.5 *Quantity of Coarse Aggregate*

The next step is to calculate the quantity of coarse aggregate per unit volume of concrete. This is done according to the Golbeck and Gray adaptation of the Talbot and Richart method, the quantity of coarse aggregate being expressed in terms of b/b_o. The recommended quantities are given in Table 5.12.

Table 5.12 Recommended Values of b/b_o[a]

Maximum Size of Aggregate (in.)	Values of b/b_0 for Different *FM*'s of Sand[b]				
	2.40	2.60	2.80	2.90	3.00
$\frac{3}{8}$	0.46	0.44	0.42	0.41	0.40
$\frac{1}{2}$	0.55	0.53	0.51	0.50	0.49
$\frac{3}{4}$	0.65	0.63	0.61	0.60	0.59
1	0.70	0.68	0.66	0.65	0.64
$1\frac{1}{2}$	0.76	0.74	0.72	0.71	0.70
2	0.79	0.77	0.75	0.74	0.73
3	0.84	0.82	0.80	0.79	0.78
6	0.90	0.88	0.86	0.85	0.84

[a] Recommended by ACI Committee 613.
[b] These values are selected from empirical relationships to produce concrete with a degree of workability suitable for usual reinforced construction. For less workable concrete, such as that required for concrete pavement construction they may be increased about 10 per cent.

Table 5.13 is an extension of Table 5.12, giving the data in relative terms and extending the range of the data to include mixtures suitable cnly for placement by high-energy methods, particularly mechanical vibration, as recommended by a special subcommittee of ACI Committee 211 [21].* This table is based on the quantities of coarse aggregate that are appropriate when using sand having a fineness modulus of 2.90; presumably the same *relative* values are appropriate for different sands.

7.6 *Quantity of Fine Aggregate*

To compute the quantity of fine aggregate, the designer first converts The quantities of cement and coarse aggregate to solid-volume units,

* Sometime after 1959, ACI Committee 613 was renumbered Committee 211.

Table 5.13 Relative Values of b/b_o for Concretes of Various Consistencies[a]

Consistency				Relative Values of b/b_o for Maximum Sizes Shown (per cent)				
Description	Slump (in.)	Vebe (in.)	Compacting Factor	$\frac{3}{8}$ in.	$\frac{1}{2}$ in.	$\frac{3}{4}$ in.	1 in.	$1\frac{1}{2}$ in.
Extremely dry	—	32-18	—	190	170	145	140	135
Very stiff	—	18-10	0.70	160	145	130	125	125
Stiff	0–1	10- 5	0.75	135	130	115	115	120
Stiff plastic	1–2	5- 3	0.85	108	106	104	106	109
Plastic	3–4	3- 0	0.91	100	100	100	100	100
Flowing	6–7	—	0.95	97	98	100	100	100

[a] Based on tests of nonair-entrained concretes made with a natural sand having a fineness modulus of 2.90 and a rounded gravel, containing some crushed over-size. Maximum sizes used were $\frac{3}{8}$ in., $\frac{3}{4}$ in., and $1\frac{1}{2}$ in. Values for $\frac{1}{2}$ in. and 1 in. are interpolated. Absolute values for different maximum sizes are given in Table 5.12.

Three slump ranges were used in these tests: 0–1 in., 1–2 in., and 4–5 in. Other values were obtained by interpolation and extrapolation.

These values are intended as a guide in establishing the first trial mix. Further adjustments will be necessary.

and expresses the quantities of air and water in volume units. Then the sum of these volumes is subtracted from the total volume, the difference being the volume of sand required. Since the quantity of coarse aggregate was selected partly on the basis of the fineness modulus of the sand, as shown in Table 5.12, it turns out that for a given maximum size the finer the sand the smaller the percentage in the combined aggregate, which is at least qualitatively in line with the finding of Abrams and others.

It may be observed that this method of determining the proportion of sand treats entrained air as if it were equivalent to sand with respect to its effect on consistency and workability; that is to say, for each increment of air above its normal percentage in concrete, the solid volume of sand is reduced by an equal amount. For the basis of this procedure see Chapter 7.

7.7 *Laboratory Data*

To carry out the procedure outlined above, a certain minimum of test data from the materials are required. These are the following:

Specific gravity of cement.
Specific gravity and water absorptivity of the aggregates.
Dry-rodded unit weight of the coarse aggregate.
Sieve analysis and fineness modulus of the fine aggregate.
Maximum size of the coarse aggregate, which may require a sieve analysis to determine the effective maximum size.

7.8 Final Adjustment of the Mix

It is assumed by the Committee that the procedure described above will result in a mixture that will prove to be approximately but not exactly in accord with specifications; therefore, the final or at least semifinal step is to make adjustments according to the actual amount of water required for the desired consistency. This amounts practically to retracing the steps outlined above on the basis of the newly established water requirement.

Presumably, the final adjustment is made while the concrete is being produced on the site, taking into account the difference between the characteristics of large and small batches of the same mixture and the specific condition under which the concrete is being produced.

7.9 Structural Lightweight Concrete

The mix-design procedure outlined above is given for use when the aggregates are ordinary sand and gravel or crushed stone. When lightweight aggregates are used, special difficulties are encountered mostly because of the difficulty of obtaining reliable laboratory data of the kind mentioned in Section 7.7 for the aggregates; it is especially difficult to determine how much of the water in a mixture is within the boundaries of the particles and how much is in the matrix. A subcommittee worked out a special practice for this kind of material [22].

Besides the difficulty just mentioned of determining the water-absorption factor of lightweight aggregates, a subcommittee of ACI Committee 211 pointed out that sieve-analysis data as usually reported are questionable; they should not be reported by weight, but by volume, because the bulk specific gravities are not the same for different sizes of grains, the small particles being generally more dense than the large ones. For example, a certain lightweight sand showed a fineness modulus of 3.08 on the weight basis, but on the volume basis the value was 3.23. The recommended method is a trial procedure, aided by a few points of guidance.

The cement content is selected on the basis of the strength required. Since, in general, the water-cement ratio will not be known, and since also the strength might be influenced in an unpredictable manner by the structural weakness of the light-weight aggregate particles, it is usually necessary to establish the required cement content by means of trial mixes and strength tests. The subcommittee recommends using at least three trial mixes having different cement contents. The first mix is made entirely on the basis of preliminary estimates; subsequent trials can be based partly on information gained from the first.

For the first trial mix, the quantity of aggregate is estimated on the assumption that the sum of the separate, *loose* volumes of the dry fine and coarse aggregates contained in a cubic yard of concrete will be between 31 and 33 ft^3, 32 being a good first-approximation. Because the optimum percentage of sand in the combined aggregate usually is found between 40 per cent and 60 per cent by loose volumes, equal volumes are used for the first mix.

The cement content is chosen more or less arbitrarily, 6 sacks per cubic yard for structural concrete being a reasonable first choice. Enough water is used to produce the desired consistency, which must be fairly stiff to avoid segregation of the light particles from the matrix. If the mixture proves to have too little or too much fine aggregate, as judged by the designer, the work must be repeated with suitable adjustments. Finally, after measuring the volume of concrete produced, and testing the concrete for air content, the designer calculates the quantities of ingredients by weight in a unit volume of concrete.

7.9.1 The specific-gravity factor. The making of mixtures required to establish the strength curve is facilitated by using a specific gravity factor [23] derived from the data of the first trial; it is derived as follows. The volumes of air and total water are added to the solid volume of cement, and the sum is subtracted from the total volume of the trial batch. The remainder is the volume of space *apparently* occupied by the known weights of dry fine and coarse aggregates. Dividing the dry weight by the weight of the volume of water (at 4°C) equal to the apparent volume of the aggregate gives an apparent specific gravity called the *specific gravity factor*. Actually the space is prorated to the fine and coarse aggregates according to their weight percentages, and a separate factor is derived from each.

With the factors mentioned above at hand, the ordinary procedure for determining the amount of aggregates to be used is followed; that is, the volume of total water and air and the volume of coarse aggregate in a unit volume of concrete are assumed to remain unchanged when the cement content is changed. Thus when the cement content is increased the volume of fine aggregate is reduced, in this case by an equal volume as computed from its specific-gravity factor.

Experience with the method showed that it is satisfactory if the moisture content of the aggregate at the time it is used is the same as it was when the specific-gravity factor was established, but when the initial moisture content is different, the amount of water for a given consistency was not the same as that predicted. In other words, the Lyse rule did not hold. Landgren investigated the problem and developed a method of determining the conventional values for bulk specific gravity of lightweight aggregate, thus making it possible to use the same design procedure as that outlined above for ordinary aggregate [24]. However that method has not yet replaced the original recommendations of the subcommittee based on the specific gravity factor. More recently, Landgren, Hanson, and Pfeifer [25] described a modification of the method of establishing the specific-gravity factor that seems to preclude the difficulty encountered with the original method.

7.9.2 Modified specific-gravity factor. When a lightweight aggregate contains absorbed moisture before the mixing water is added, the amount of water added to a batch of concrete for a given consistency is smaller than the amount added when the aggregate is initially dry. Landgren, Hanson, and Pfeifer [25] found that if the specific-gravity factor of an initially moist material was based on the moist weight rather than the dry weight, the difficulty found with the original method was practically eliminated. Moreover, they found that the factor could be determined by direct tests on the aggregate by means of a picnometer, or by measurements of buoyancy, to determine the water displacement. The specific gravity factor so determined is a linear function of the initial water content of the particles; it may increase or decrease with water content according to the change in the relative amount of air and water within the immersed particles.

REFERENCES

[1] Blanks, R. F., E. N. Vidal, W. H. Price, and F. M. Russel, "The Properties of Concrete Mixes," *Proc. ACI,* **36,** 433–473, 1940.

[2] Bogue, R., *The Chemistry of Portland Cement,* 2nd ed., Rheinhold, New York, 1955.

[3] Sabin, L. C., *Cement and Concrete,* McGraw-Hill, New York, 1907.

[4] Abrams, D. A., "Design of Concrete Mixtures," Bulletin 1, Structural Materials Research Laboratory, Lewis Institute, Chicago, 1st ed., 1918; rev. ed., 1925.

[5] Anon. "Design of Concrete Mixes," Road Note No. 4, DSIR, Road Research Laboratory, England, 1947.

[6] Neville, A., *Properties of Concrete,* Wiley, New York, 1963, Chap. 10.

[7] Walker, S., and F. F. Bartel, "Discussion," *Proc. ACI,* **43,** 844-1–844-10, 1947.

[8] Powers, T. C., "Vibrated Concrete," *Proc. ACI,* **29,** 373–381, 1933.

[9] Gaynor, R. D., "Effect of Fine Aggregate on Concrete Mixing Water Requirement," unpublished paper of the National Sand and Gravel Association, 1963.

[10] Popovics, S., "Comparison of Several Methods of Evaluating Aggregate Gradings," RILEM Bulletin No. 17, 13–21, December 1962.

[11] Thaulow, S., *Concrete Proportioning,* Norwegian Cement Association, Oslo, 1955.

[12] Swayze, M. A., and E. Gruenwald, "Concrete Mix Design—A Modification of the Fineness Modulus Method," *Proc. ACI,* **43,** 829–843, 1947.

[13] Talbot, A. N., "A Proposed Method of Estimating the Density and Strength of Concrete and of Proportioning the Materials by the Experimental and Analytical Consideration of the Voids in the Mortar and Concrete," *Proc. ASTM,* **21,** 940–969, 1921.

[14] Talbot, A. N., and F. E. Richart, "The Strength of Concrete, Its Relation to the Cement Aggregates and Water." Bulletin No. 137, Engineering Experiment Station, University of Illinois, Urbana, 1923.

[15] Dunagan, W. M., "The Application of Some of the Newer Concepts to the Design of Concrete Mixes," *Proc. ACI,* **36,** 649–684, 1940.

[16] Goldbeck, A. T., and J. E. Gray, "A Method of Proportioning Concrete for Strength, Workability, and Durability," Bulletin No. 11, National Crushed Stone Association, Washington, D.C., 1949.

[17] Bloem, D. L., and S. Walker, "Proportioning Ready Mixed Concrete," Circular No. 91, National Sand and Gravel Association, Silver Spring, Md., 1963.

[18] Lyse, I., "Tests on Consistency and Strength of Concrete Having Constant Water Content," *Proc. ASTM,* **32,** Part II, 629–636, 1932.

[19] Slater, W. A., and I. Lyse, "Compressive Strength of Concrete in Flexure as Determined from Tests of Reinforced Beams," *Proc. ACI,* **26,** 831–874, 1930.

[20] ACI Committee 613, Walter H. Price, Chairman; William Cordon, Secretary. ACI Standard Recommended Practice for Selecting Proportions for Concrete" (ACI 613-54), 1954.

[21] Subcommittee No. 2, ACI Committee 211, Paul Kleiger, Chairman. "Recommended Practice for Selecting Proportions for No-slump Concrete," *Proc. ACI,* **62,** 1–22, 1965.

[22] ACI Committee 613, J. J. Shideler, Chairman, Subcommittee on Proportioning Lightweight Aggregate Concrete, ACI Standard 613A-59, 1959.

[23] Nelson, G. H., and O. C. Frei, "Proportioning and Control of Lightweight Structural Concrete," *Proc. ACI,* **54,** 605–622, 1958.

[24] Landgren, R., "Determining the Water Absorption of Coarse Lightweight Aggregates in Concrete," *Proc. ASTM,* **64,** 846–865, 1964.

[25] Landgren, R., J. A. Hanson, and D. W. Pfeifer, "An Improved Procedure of Proportioning Mixes of Structural Lightweight Concrete," *J. PCA Research and Development Laboratories,* **7**(2), 47–65, 1965.

6

Theoretical Aspects of Mix Design

1. INTRODUCTION

In Chapter 5 current practice in mix design is described. In tracing its development we found it unnecessary to deal with the subject theoretically. However, much thought has been given the theoretical aspects of formulating mixtures, and papers of lasting value were produced even earlier than most of those mentioned in the last chapter. In this chapter we review some of this theoretical work, for it leads to improved understanding of the subject.

2. THEORIES OF IDEAL AGGREGATE GRADING

Some if not most early attempts to put the making of concrete on a scientific basis were based on considerations of the packing characteristics of particulate material. Perhaps the most influential work of this kind was that of Fuller and Thompson [1], published in 1907.

*2.1 The Fuller and Thompson Grading Principle**

The work of Fuller and Thompson was such that it could have been considered along with other empirical methods in the preceding chapter, but it is given here because it proved to be the starting point of subse-

* I am here following the summary of the Fuller and Thompson work given by N. M. Plum [2].

quent theoretical developments. On the basis of their experiments, Fuller and Thompson concluded that there are certain ideal grading curves that consist of ellipses at the lower end, merging into straight lines tangent to the elliptical part. An ideal curve pertained to all the solid material, and it was stipulated that at least 7 per cent of the aggregate plus cement should be finer than the No. 200 sieve. Their equation for the elliptical part of the curve was the following:

$$\frac{(y - 7)^2}{b^2} + \frac{(x - a)^2}{a^2} = 1 \tag{6.1}$$

where y is the percentage of material passing the sieve having opening x; a and b are the axes of the ellipse and have values in accord with the maximum size of the aggregate and the shape of the particles. The ellipse starts at $y = 7$, $x = 0.074$ mm (Sieve No. 200) and runs to a value of x equal to about 10 per cent of the maximum size, at which point the curve continues as a straight line to $y = 100$ per cent, $x = X$, the maximum size. The constants a and b were prescribed in such way that the more angular the particles the larger the percentage of material smaller than one-tenth the maximum size.

This work began to take on theoretical overtones when Fuller and Thompson discovered that their ideal grading curve, usually called the Fuller curve, could be closely approximated by a parabola, which, in our nomenclature, could be expressed as follows:

$$p_t = \left(\frac{d}{D}\right)^{1/2} \tag{6.2}$$

where p_t is the fraction of the *total* solids finer than size d, and D is the maximum particle size.

Although Fuller and Thompson called their curve a parabola, it turns out that they must not have meant exactly that, for they also specify a minimum quantity of fine material that will not fit a parabolic grading. For example, an aggregate having a maximum size of 38 mm (1½ in.) and having a parabolic grading would call for 4.4 per cent of the total solid material to be finer than No. 200 (0.074 mm). If the fine material is composed of cement only, the mix $M = 1/0.044 = 22.7$, a very lean mixture. The total material passing No. 4 (4.76 mm) and retained on the No. 200, the sand would constitute about 31 per cent of the material retained on No. 200, and it can be shown that the fineness modulus of that sand would be about 3.0. These figures describe an unworkable combination and show one reason why Fuller and Thompson stipulated that at least 7 per cent of the total material must be finer than No.

200. For the example above, this requirement would correspond to $M = 14.3$, still a rather lean mixture and still undersanded by modern criteria.

2.2 Parabolic Gradings

The parabolic grading, which requires that the fraction finer than a given size be equal to the square root of the relative size d/D, is only one of a family of similar relationships comprising parabolas of various degrees, which (after Popovics [4]) may be written

$$p_t = \sqrt[j]{d/D} = \left(\frac{d}{D}\right)^q \tag{6.3}$$

where $j = 1/q$. Such gradings, for which q may have values other than $\frac{1}{2}$, are called *parabolic.*

2.2.1 Density. A theoretical aspect of parabolic gradings was made clear by Andreasen and Anderson [3] who showed that when all sizes of aggregate material down to zero are present the voids content depends only on the value of q, sometimes called the *grading ratio,* and is independent of the maximum size D. The smaller q is the smaller the content of voids. As q approaches zero, so does th' voids content. At the other extreme, as q approaches infinity, the voids content approaches that of a single-size aggregate. It might seem therefore that the making of a dense mass of granular material is simply a matter of adopting a parabolic grading with q assigned a suitably small value.

However, this is far from true, as Andreasen and Anderson themselves pointed out. The principal obstacle is the inability of the smallest particles to form dense aggregates; the smaller the average size, the less dense the packing, as is discussed in Chapters 1 and 2. The densest aggregates were found at an intermediate value of q; for their materials, Andreasen and Anderson found the lowest practical value to be about $\frac{1}{2}$.

2.2.2 Pararolic grading and fineness modulus. Popovics [4] showed that the fineness modulus of any parabolic grading is given by the following equation:

$$FM_F = 3.32[\log 10D - 0.4343j(1 - (10D)^{-1/j})] \tag{6.4}$$

or, approximately,

$$FM_F = 3 \log (10D) - 0.95j \tag{6.5}$$

where subscript F designates a parabolic (Fuller) grading; D is the maximum size in millimeters, and j is the degree of the parabola.

In Chapter 5 it is shown that for any given mix and maximum

size of aggregate there is a maximum permissible, or optimum, value of the fineness modulus that cannot be exceeded without loss of workability and that if the cement as well as the aggregate is included when we compute the fineness modulus, the optimum value for a given maximum size is the same for different mixes, within a limited but practical range.

Popovics found that for parabolic gradings, the optimum fineness modulus is related to the mix proportion and maximum size as follows (Eq. 5.9):

$$(FM)_o = \left(1 + \frac{0.85}{N}\right)(2.45 \log D + 1.05) + \frac{2.1}{N}$$

where N is the ratio of aggregate to cement by weight. He compared this empirical relationship with the relationship between optimum fineness modulus and maximum size for different values of j (or q) as given in Eq. 6.3. He found that the curves were not parallel. The empirical curve for $N = 6$, for example, crossed the curve for $j = 2$, a parabola, at about $D = 75$ mm or 3 in., but at no other points would that value of j give the optimum fineness modulus. At smaller values of D, the mixtures tend to be oversanded, although in this case the discrepancy would not become noticeable unless D was made smaller than $1\frac{1}{2}$ in. (40 mm). This means that to obtain the optimum fineness with a parabolic grading, the value of j must be selected to fit each particular combination of mix and maximum size. Plum [2] reported similar results. He found experimentally that even when the proportion of cement (or cement plus filler) was determined by the requirements of a percentage grading curve, the optimum grading percentage varied systematically with the maximum size, diminishing as the size increased.

Thus we see that whatever merit a parabolic grading may have, it offers no way for us to make an a priori prediction of what the optimum grading is for a given combination of cement content and maximum size; the optimum value of q must be established empirically. Popovics showed that if the optimum value of q is taken to be that which gives the optimum value of fineness modulus $(FM)_o$ it can be evaluated as follows:

$$\left(\frac{1}{q_o}\right) \cong 3.15 \log (10D) - 1.05(FM)_o \tag{6.6}$$

where D is in millimeters.

2.2.3 Percentage gradings. If separate size groups, as they are defined in Chapter 1, are combined in such proportions that the amount of material in each successive group is a fixed percentage of the amount in the next larger group, the aggregate is said to have a percentage grading. Thus if G is the constant percentage,

$$\frac{G}{100} = \frac{1}{r} = \frac{p_{i-1}}{p_i} \tag{6.7}$$

The reciprocal of $G/100$ is called the grading ratio r. For an aggregate so composed, the sum of all the quantities smaller than the $(i-1)$th group is related to the sum of all quantities smaller than the ith group in the same way. Hence

$$p_t = \left(\frac{G}{100}\right)^{I-i} = \left(\frac{1}{r}\right)^{I-i} \tag{6.8}$$

where I is the group number of the largest size group of the series.

The following expression is the equivalent of Eq. 6.8, provided that the size quotient of the size groups is 2 [4]:

$$p_t = \left(\frac{d}{D}\right)^{\log r/\log 2} \tag{6.9}$$

Comparing Eqs. 6.9 and 6.3, we see that the percentage grading is a parabolic grading, the logarithm of the grading ratio r being proportional to q of the parabolic grading formula.

Popovics gives the following expression for obtaining the optimum value of the grading ratio for a given maximum size and corresponding optimum fineness modulus:

$$\log r_o = \frac{0.3}{3 \log (10D) - (FM)_o} \tag{6.10}$$

2.3 The Bolomey-Fuller Grading Formula

Recognizing that the Fuller-Thompson grading gives relatively harsh mixtures, Bolomey [5] developed the following modification of the Fuller parabola:

$$p_t = f + (1 - f)\left(\frac{d}{D}\right)^{1/2} \tag{6.11}$$

where f is an empirical constant and the other symbols have the same meanings as before. The value of f is chosen so that it is larger the higher the degree of workability, thereby providing relatively larger proportions of the smaller size groups; for a given degree of workability, its value is larger for angular than for rounded material.

Like the original Fuller curve, Eq. 6.11 is applicable to all the solid material in a mixture, but because the grading curve of cement is not variable in practice, Bolomey applied the equation only to sizes larger than about 0.1 mm, which is about the average size of size group No. 1. For a certain weight fraction of cement, C, (fraction of the total solid material) the grading of the aggregate alone is given as follows:

$$p = (p_t - C)\frac{1}{1 + C} \tag{6.12}$$

or

$$p = f - C + (1 - f)\left(\frac{d}{D}\right)^{1/2} \tag{6.13}$$

where p is the weight fraction of the *aggregate* smaller than d.

Using volume fractions instead of weight fractions, and our nomenclature, we find

$$C = \frac{cv_c^o}{av_a + bv_b + cv_c^o} = \frac{1}{1 + M} = 1 - x \tag{6.14}$$

and

$$p = \left[f - (1 - x) + (1 - f)\left(\frac{d}{D}\right)^{1/2}\right]\frac{1}{x} \tag{6.15}$$

where f is a solid-volume fraction.

In Eq. 6.14 it may be noticed that our nomenclature is not quite appropriate in this case, for it gives the quantity of the aggregate as the sum of the fine and the coarse aggregates, whereas the requirement of the theory is that each particle size, or at least each size group,

Table 6.1 The Constants f for Bolomey's Formula

	Values of f for Consistency Indicated		
	Damp Earth; Vibrated Concrete	Soft; Reinforced Concrete	Fluid; Poured Concrete
Rounded particles	0.06 – 0.08	0.10	0.12
Angular particles	0.08 – 0.10	0.12 – 0.14	0.14 – 0.16

is to be proportioned in conformity with the curve. However, for practical reasons, those who advocate the theoretical curve regard it as an ideal, and usually attempt to reproduce it as closely as possible by blending the conventional two aggregates, supplemented in some cases by a relatively fine sand.

According to Bolomey, values of f given in Table 6.1 are appropriate.

Values of C of Eq. 6.13 of course depend on the richness of the mix. The weight fractions of cement of the total solid material are related to the cement content of the finished concrete about as follows:

$$250 \text{ lbs/yd}^3;\ C = 0.068$$
$$340 \text{ lbs/yd}^3;\ C = 0.091$$
$$420 \text{ lbs/yd}^3;\ C = 0.113$$
$$505 \text{ lbs/yd}^3;\ C = 0.136$$

Comparing these figures with those in Table 6.1, we see that for the richer mixes, $f - C$ or $1 - x$ is negative, meaning that each value given by the second term inside the brackets in Eq. 6.15 is to be reduced by the amount of the difference between f and C. On the other hand, for leaner mixtures, and particularly when the aggregate is composed of angular particles, the difference between f and C is positive, calling for increases in the amounts given by the second term inside the brackets. Thus Bolomey's formula calls for different gradings for different circumstances. When the cement content is relatively low, it calls for a relatively finer aggregate with more material finer than the No. 100 sieve; but it does not explicitly call for the addition of paste-making material, as is discussed in Chapter 4, Section 9.

For all proportions of cement smaller than $C = f$, Bolomey's formula calls for the same grading, and if the fine material other than cement $(f - C)$ is classed with the cement, it provides for only one mix. In this respect the method is in harmony with the considerations of optimum mixtures set forth in Chapter 4. However it would seem that if the criterion is that mixtures should have the minimum voids content, the value of f should be such that when $C = f$ the optimum mix should be at point B on the water ratio or voids-ratio diagram. But, this is not the case. Table 6.1 shows a typical value of f to be 0.10, which, when $f = C$ corresponds to $x = 0.90$ on the voids, or water-ratio diagram. Reference to Fig. 4.9 of Chapter 4 shows that, for an aggregate with a $\frac{3}{4}$ in. maximum size (19 mm), the point is closer to point F than to point B.* It seems that the criterion applied by Bolomey was such that it precluded mixtures in or near the FC category rather than to produce mixtures having minimum voids content.

Since the supplemental fine material represented by $(f - C)$ is apparently that normally found in sands and is therefore material that would mostly be retained by the No. 200 sieve, it cannot fill the role of paste maker, as discussed in Chapter 4, Section 9.3. Bolomey's method

* Since we are here referring to a water-ratio diagram, it is necessary to point out that the C of Bolomey's formula is not the C of the water-ratio diagram.

probably gives the best result obtainable with such materials or nearly so; but the density at point B would not be attainable when the cement content is low.

The values of f given by Bolomey would preclude FC mixtures for only a limited range of maximum size. In general, the smaller the maximum size the larger f should be, but this is not indicated in the Bolomey formula. Additional discussion of the Bolomey method will be found in Section 5.4.

2.3.1 Optimum Bolomey grading. Popovics [4] found that the Bolomey equation is capable of producing the optimum fineness modulus for the total solids $(FM)_t$ if the optimum value of f and the optimum degree of the parabola j are used. On the basis of empirical Eq. 5.8 in Chapter 5, Section 4.8.1, which pertains to rounded aggregate, he found the optimum values to be $f_o = 0.15$ and $j_o = 2$ (or $q_o = \frac{1}{2}$). Thus the optimum grading curve is

$$p_t = 0.15 + 0.85\left(\frac{d}{B}\right)^{1/2} \tag{6.16}$$

This equation is valid for any mix within the range for which Eq. 5.8 is valid, which is about $4 \leq M < 10$. It may be noted that $q_o = \frac{1}{2}$, is the value most frequently used, but it should be noted also that this is an optimum value only if 15 per cent of the solid material is finer than 0.1 mm.

Since f_o is the optimum fraction of material smaller than 0.1 mm, it represents cement and other fine material, the proportion of cement in that material depending on the ratio of aggregate to cement. Therefore, for a certain range of mixes, the optimum aggregate grading is different for each mix; also, the value of j must be correspondingly different.

It is interesting that the optimum value of f as evaluated by Popovics is considerably higher than the comparable value given by Bolomey in Table 6.1, 0.15 as compared with 0.10. In fact, comparison with Fig. 4.9 indicates that for $(1 - x) = C = f_o$, and for $\frac{3}{4}$-in. gravel aggregate, it designates a mix near to point B, actually slightly to the *left*. This is about as it should be because the empirical values for optimum fineness modulus represented in the formula used by Popovics were based on mixtures having approximately the maximum density for a given mix and maximum size.

However, $f_o = 0.15$ is indicated to be independent of maximum size, and in this respect the formula is in conflict with the empirical data given in Chapter 4. This shows that in general the volume of paste, and hence the value of f_o, should be larger the smaller the maximum size, at least if maximum density is taken as the criterion.

Equation 6.16 embodies the assumption that properly graded sand can be substituted for cement, solid volume for solid volume, without altering the density of the mixture. The assumption is approximately but not exactly true to fact; but in its application, no attention is paid to factors other than density, particularly cohesiveness and rheological properties, which certainly do not remain unaffected as sand is substituted for cement.

2.4 Plum's Grading Equation

Plum [2] considered the fact that in real aggregates the sizes do not approach zero very closely and decided that a theoretical grading curve should be based on a finite minimum size. He found that to fulfill the requirement that the grading ratio be the same for all successive size groups, when the minimum size is finite, the grading curve should conform to the following formula:

$$p_t = \frac{r^i - 1}{r^I - 1} \tag{6.17}$$

The logical extension of the theoretical relationship was simple. Plum estimated that the size range of portland cement comprises nine size groups, the largest being (#200–#100), with an upper limit of 149 μ, and the smallest (0.291–0.582 μ). The smallest group was designated $i = 1$ and the largest $i = 9$. If all the material smaller than group 9 is cement, the proportion of cement required is

$$p_9 = \frac{r^9 - 1}{r^I - 1} \tag{6.18}$$

Plum reasoned that if the full amount of cement called for by Eq. 6.18 is not supplied, the difference should be made up of a suitable filler, ideally one having a proper percentage grading. In any case the composition would be expressed in our nomenclature as follows:

$$p_9 = \frac{cv_c^o + qv_q}{(cv_c^o + qv_q) + av_a + bv_b} = \frac{1}{1 + M} = 1 - x \tag{6.19}$$

where q and v_q are the weight and specific volume of the filler.

For example, suppose that $I = 17$ corresponding to a maximum size of 38 mm or $1\frac{1}{2}$ in. (and corresponding to $I = 9$ in our nomenclature) and assume $r = 1.25$, or $G = 80$ per cent. Then,

$$p_9 = \frac{1.25^9 - 1}{1.25^{17} - 1} = \frac{6.46}{43.4} = 0.149 = 1 - x$$

That is to say, of the total solid material, the amount smaller than the (#200–#100) group should be 14.9 per cent. Theoretically that material should have an 80 per cent grading down to its minimum size; practically, this requirement is ignored.

The value of x or p_9 arrived at in this way is very sensitive to the chosen value of r. For example, if r had been 1.3 instead of 1.25 (a 77 per cent grading instead of an 80 per cent), the required amount of fine material would have been 11.2 per cent instead of 14.8 per cent, the mix 1:8.92 instead of 1:6.76.

Whatever the chosen value of r, Plum's theory leads to the formulation of one mix only, the filler, if any, being classed with cement. If for any reason less cement than that required by theory is to be used, the logical procedure is to replace some of the cement with a filler material constituted to leave the grading unchanged; that is to say, a filler having the same grading curve as the cement would be substituted for some of the cement. Such a substitution in a base mix is to be compared with the substitution of a filler paste for some of the cement paste described in Chapter 4, Section 9.

Comparing values of x arrived at in the examples above, 0.852 and 0.888, with the value of x at point B in Fig. 4.10a, we find that the smaller value is nearly the same as that of x_B. It thus might appear that Plum's theoretical equation is a means of calculating proportions for the mix at point B of the voids-ratio diagram. However there is no theoretical basis for choosing the value of r or q. In his investigation Plum established optimum values empirically.

It should be said that Plum was aware of discrepancies between theoretical assumptions entailed in the above development and the real behavior of mixtures, and he did not offer Eq. 6.18 as a practical solution of the problem of formulating concrete mixtures. (His method for using semitheoretical grading and an empirical procedure is described in Section 5.5.)

2.5 Comments on Parabolic or Percentage Gradings

Ideal grading equations have been put forward on the grounds that they are expressions of natural law. Plum [4] attributed the following "proof" to Steffensen:

"A single piece of rock is enclosed in a steel cylinder and subjected to increasing pressure until it is crushed and the fragments occupy a volume virtually the same as the solid volume of the original particles, i.e., the void content of the crushed material is virtually zero. The size-grading of thc crushed material is then determined. Regardless of the kind of rock used, it always turns out that the grading is parabolic, with only a small variation in q."

It must be pointed out, however, that to demonstrate that crushing produces a parabolic grading and the lowest possible voids content, is not the same as demonstrating that a parabolic grading is ideal for concrete mixtures. It is obvious that an aggregate packed together as was that produced in the steel cylinder is totally without plasticity. Since the usefulness of concrete depends on the plasticity of the fresh mixture, the solid particles must be dispersed in water to some degree. Even though a parabolic grading might be the best for a nonplastic, solidly packed body, when it is required that the particles are not to be in contact with each other an additional degree of freedom is thereby introduced which allows departure from the ideal without penalty.

In Section 3 it is shown that the Weymouth theory of particle interference results in a modified percentage grading. The data and discussion leading to the conclusion that the Weymouth theory is untenable will be found pertinent to the broader question of ideal gradings in general.

The hypothesis that there is an ideal size gradation for concrete aggregate, or for all the solid material in concrete, has now become almost if not entirely abandoned. It is now held that primary considerations pertain to the amount of water required to convert the dry material to a plastic mass during mixing, perhaps with the final aid of vibration. Such considerations involve not only size gradation, but also physicochemical factors that influence rheological properties, a subject discussed in Chapter 10.

3. THE WEYMOUTH THEORY OF PARTICLE INTERFERENCE

In 1933 C. A. G. Weymouth [6] published a theory about concrete mixtures which has proved to have a considerable influence on subsequent thinking about the subject. His theory, along with subsequent modifications, as it pertains to particle interference as a determinant of the voids content of aggregates, is discussed in Chapter 1, Section 8.6. Now we shall consider it further in connection with its original application, the water requirement, workability, and strength of plastic concrete mixtures. Weymouth was mindful also of the connection between particle interference and air content.

3.1 Weymouth's Model

Although Weymouth was originally interested in concrete mixtures, and was concerned with the effect of particle interference on water re-

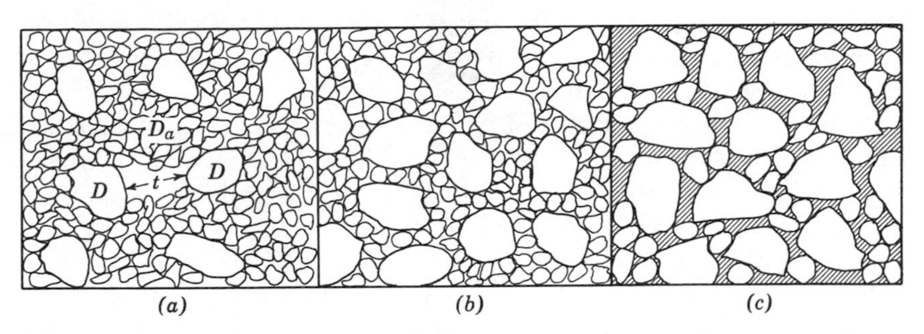

Fig. 6.1. Weymouth's model for particle interference. (*a*) $t > D_a$; (*b*) $t = D_a$; (*c*) $t < D_a$. (D indicates large particles and D_a small particles.)

quirement and workability, including the tendency of different-sized particles to segregate during handling, he illustrated his concept in terms of dry mixtures of aggregates, using the model shown in Fig. 6.1 [7]. Figure 6.1*a* represents, in two dimensions, a mixture of two sizes of particles. The larger particles are few and are widely separated by the smaller particles; the average clear distance between them t is considerably greater than the diameter of the smaller particles D_a. In diagram (*b*) the relative number of the larger particles is greater, and the average distance between them is supposed to be just equal to the diameter of the smaller particles. According to Weymouth, for the composition represented by either A or B, the mixture can be stirred without changing the uniformity of the "void pockets" defined by the smaller particles.

In Fig. 6.1*c* the concentration of the larger particles is such that the average clearance between them is less than the diameter of the smaller particles, making it impossible for the interstitial spaces of the larger particles to be filled uniformly with the smaller. Weymouth said that when such a one-layer mixture on a tray is stirred, there is a tendency for the two sizes to run into separate groups, each of its own kind; in other words, stirring such a mixture tends to produce segregation of the two sizes. To apply this observation to a deep mass, he visualized a given size group as forming a sort of grid structure through which the smaller particles move both horizontally and vertically during manipulation of the mixture; so long as they can move freely, the mass remains homogeneous, but if the larger particles interfere with the movement of the smaller, segregation occurs, and large void pockets are developed "with a great loss of strength and workability."

As direct proof of the validity of his criterion for particle interference, Weymouth cited data such as those shown in Fig. 1.10, which shows that when two size groups of dry aggregate are mixed in various proportions, the effect of particle interference can be observed to appear when the average clearance between particles of the larger group becomes less than the average diameter of the smaller. As was pointed out in Chapter 1, data on mixtures of two size groups do confirm Weymouth's theory, although for other types of mixture it is not confirmed. It seemed that a better application of the original concept is to follow the suggestion of Butcher and Hopkins which is to consider the mixture to be composed of two components only, fine and coarse, the division point being arbitrarily selected. But, it was found that the average clearance between the particles of the coarse component must often be much larger than the average particle size of the smaller component to eliminate the appearance of particle interference. Now, however, we shall follow Weymouth's original method of considering each size group in relation to the rest.

3.2 Concrete without Particle Interference by Weymouth's Criterion

Weymouth concluded that each and every size group making up the aggregate should be dispersed to such a degree that the clearance between its particles is at least equal to the average diameter of the next smaller group. Accordingly, he stipulated that the composition should be such as to satisfy the following equation for all consecutive size groups:

$$D_{i-1} = t = \left[\left(\frac{d_o}{d_i}\right)^{1/3} - 1\right] D_i$$

or (6.20)

$$\left(\frac{d_o}{d_i}\right)^{1/3} = 1 + \frac{D_{i-1}}{D_i}$$

where D_i is the nominal average diameter of the ith size group, and D_{i-1} is that of the next smaller group; d_i is the density of the ith group as it exists in the concrete mixture, and d_o is its density in the dry, compacted state. (Compare Eq. 6.20 with Eq. 1.17.)

The definition of a size group being such that $(D_{i-1}/D_i) = \frac{1}{2}$, it follows that

$$\left(\frac{d_o}{d_i}\right)^{1/3} = \frac{3}{2}$$

and

$$\left(\frac{d_i}{d_o}\right)_{cr} = \left(\frac{2}{3}\right)^3 = 0.296 \qquad (6.21)$$

That is to say, the density of any size group as it exists in the mixture must not be more than 29.6 per cent of the density of the same group by itself in the compacted state. In calculating d_i, the quantity of that size is divided by the volume of space available to it.

The amount of space available in a mixture for a given size group is considered to be that fraction of the unit volume that is not occupied by larger groups. Thus the largest group is dispersed throughout the volume of the mixture. If, for example, the solid-volume fraction of this group is 0.10, the value of d_i for this group is $d_i = 0.10/1.0$. The next smaller size group is considered to be dispersed throughout the matrix of the largest group. Hence the volume of space available to this group is $1.00 - 0.10 = 0.90$; thus if the volume fraction of the next-to-largest group is 0.20, the value of d_i for that group is $0.20/0.90 = 0.22$. The amount of space available to the next group is $1.00 - 0.20 - 0.10 = 0.70$, and d_i is computed on that basis and so on for the entire series of sizes. In principle this theory applies to all sizes, but Weymouth applied it only within the range of aggregate sizes.

Plum [2] pointed out that for a continuously graded aggregate in which d_o is the same for all groups, Weymouth's criterion requires that the solid-volume fractions be related to each other as follows:

$$\frac{p_{i-1}}{p_i} = 1 - \frac{d_i}{d_o} d_o = 1 - 0.296d_o \tag{6.22}$$

This means, of course, that for an aggregate in which each size group has substantially the same value of d_o, the grading ratio is constant; thus we see that the grading called for by Weymouth's theory is essentially a parabolic or percentage grading. If d_o varies from group to group, it is a modified percentage or parabolic grading. The value of d_o is usually found to be between 0.5 and 0.65, depending on the angularity factor of the particles. Thus the grading ratio, according to Weymouth's theory, should usually be between 1.18 and 1.24, the corresponding grading percentages being 81 and 85.

Weymouth applied his theory not only to continuous grading but also to gap-graded aggregates. When one size group is missing t is required to equal D_{i-2}, and if two groups are skipped, D_{i-3}. For such gaps Weymouth's criterion becomes $d_i = 0.513d_o$ and $d_i = 0.703d_o$ instead of $0.296d_o$ where the grading is continuous.

Such criteria would be applied by Weymouth only when the consistency is so soft as to be called "wet." For medium and dry consistencies, he considered the diameter of each particle to be augmented by a layer of cement paste, for which allowance must be made. That is to say, the clearance must be such as to exceed $(D_{i-1} + C)$ where C is a constant representing extra clearance required by the layer of paste. On this basis, Eq. 6.21 becomes replaced by

$$\left(\frac{d_i}{d_o}\right)_{cr} = \frac{1}{(3/2 + C/D_i)^3} \tag{6.23}$$

For a gap of one group the numerical factor becomes $\frac{5}{4}$, and for a gap of two, $\frac{9}{8}$, etc.

The grading factor thus becomes a function of the size of the group as follows:

$$\frac{p_{i-1}}{p_i} = 1 - \frac{d_o}{(3/2 + C/D_i)^3} \tag{6.24}$$

where the numerical constant $\frac{3}{2}$ is for continuous grading; for gap gradings, the numerical coefficient is different, as is shown above.

Although we have already questioned the assumption that a given size group interferes only with the next smaller group, which was the reasoning back of Weymouth's theoretical mixture composition, we may nevertheless consider the theoretical compositions and observe the consequences of introducing variations from theoretical composition.

3.3 Experimental Test of the Theory

Before the theoretical deficiencies pointed out in Chapter 1, Section 8.6, were recognized, this writer and his colleagues carried out many systematic experiments to test the Weymouth theory. Since these data, obtained in 1936 and 1937, have not heretofore been published, and because they are more complete and definitive than most, they shall be presented here in some detail.

The plan of the experiments just mentioned was to establish a mixture meeting Weymouth's requirements, and then to make systematic departures from the "ideal" mixture. The ideal composition for the aggregate used in our experiments (#200–#4) was calculated as shown in Table 6.2 and the accompanying notes. The sixth column gives the ideal amount of each size group in the mixture, and column 7 gives the corresponding size-distribution of the aggregate. It will be seen that Weymouth's method is like that of Bolomey (Section 2.3) or that of Plum (Section 2.4) in that it leaves a certain part of the unit volume to be filled with matrix material, that is, paste and air bubbles. To meet this requirement and at the same time to produce a mixture having the desired consistency, the proportion of cement to water must have a certain value, and since the amount of aggregate is theoretically established, this means also that the ratio of aggregate to cement must have a certain value.

Table 6.2 Calculation of Grading No. 1 and Mixture Composition Corresponding to $d_i = \dfrac{d_o}{(3/2 + 110/D_i)^3}$

1[a]	2[b]	3[c]	4[d]	5	6[e]	7
Group Number	D_i μ	d_o	Volume Available to the Group	$\left(1.5 + \dfrac{110}{D_i}\right)^3$	d_i	Size Distribution (per cent)
6	3540	0.590	1.000	3.59	0.164	31.5
5	1760	0.585	0.836	3.81	0.128	24.6
4	878	0.580	0.708	4.29	0.096	18.5
3	442	0.620	0.612	5.36	0.071	13.6
2	221	0.615	0.541	8.00	0.042	8.1
1	110	0.605	0.499	15.63	0.019	3.7
Total aggregate					0.520	100.0
Matrix					0.480	

[a] Group 1 is #200–#100, etc.
[b] The arithmetic mean of sieve openings, based on now obsolete sieve data: after Weymouth.
[c] The density of the group in the dry compacted state.
[d] Each group is supposed to occupy all the space not occupied by larger groups: see column 6.
[e] The solid volume of the group in a unit volume of mixture. The sum of these volumes is the maximum permissible aggregate content.

We had to determine the proportion of cement to aggregate by trial. It came out to be 1:2 by weight, or $M = 2.45$; the water-cement ratio was 0.354 by weight for a remolding number of 50; the slump was about 2 in.

In the experiments with variations in aggregate grading, we used the 1:2 mix but we replaced 30 per cent of the cement with an equal solid volume of pulverized silica to avoid producing what we considered to be excessively high strengths. The mixture of portland cement and pulverized silica was regarded as the cementing medium.

3.3.1 Effect of increasing the proportion of a size group. Let us first consider the results of experiments in which the proportion of group 3 was progressively increased while the total aggregate content was kept constant. The gradings are shown in Fig. 6.2, and the test results in Table 6.3. In column 3 of Table 6.3 we see that the proportion

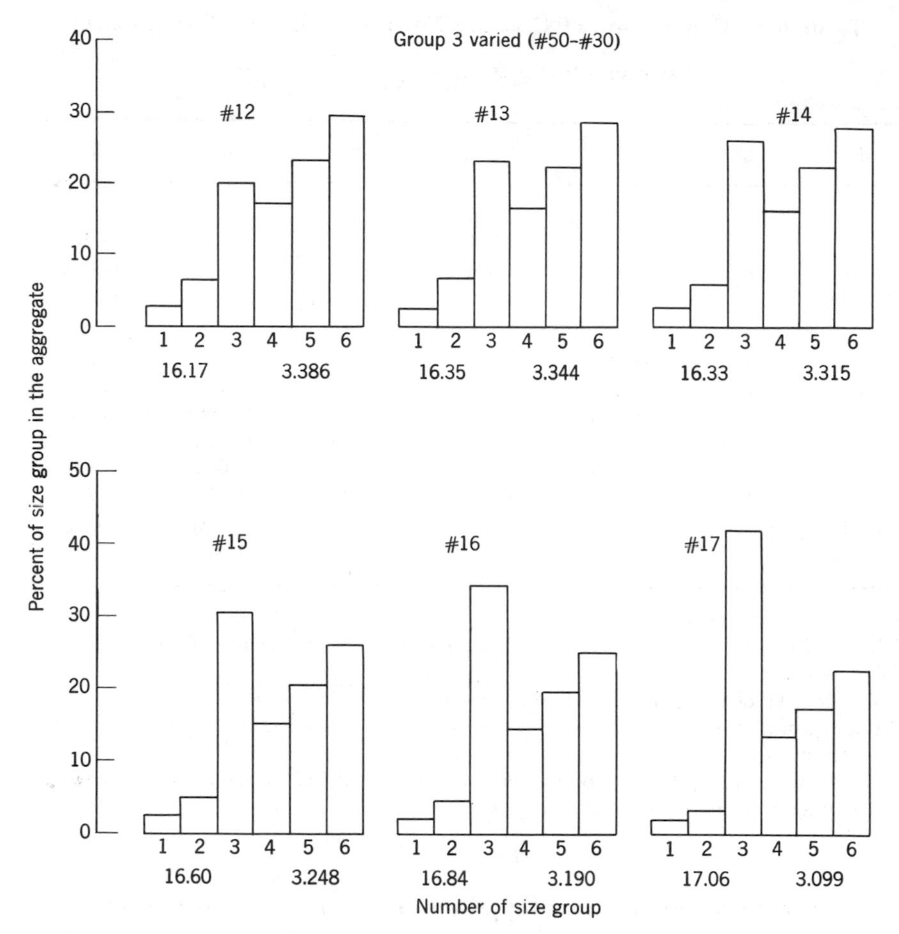

Fig. 6.2. Size distributions of experimental gradings. (The numbers below each histogram indicate the surface modulus and the fineness modulus.)

of group 3 was increased to as much as 2.5 times the proper amount indicated by Weymouth; in other words, according to the Weymouth criterion, using grading No. 12 introduced some particle interference between group 3 and group 2, and for the higher grading numbers (column 2) the degree of interference increased. As is shown in column 11, all mixtures had nearly the same consistency. Column 8 shows that the total aggregate content was slightly higher than the theoretical limit of 0.520 (see above), but it was virtually the same in all mixtures, as intended.

Table 6.3 Effect of Varying the Proportion of Group 3 (#50–#30)[a]

1[b]	2	3[c]	4	5	6	7	8	9	10[e]	11[f]	12[g]
			Composition of Test Specimens[d]								
Reference Number	Grading Number	Excess of Group 3	$\frac{cv_c}{V}$	$\frac{\text{Silica}}{V}$	$\frac{wv_w}{V}$	$\frac{A}{V}$	$\frac{av_a}{V}$	$\frac{v}{V}$	$\frac{cv_c}{v}$	R	Twenty-eight-day Strength of 2-in. Cubes (psi)
169	1	1.00	0.1568	0.0672	0.2248	0.0206	0.5306	0.2454	0.639	44	8400
151	12	1.48	0.1570	0.0673	0.2209	0.0234	0.5314	0.2443	0.643	48	8460
154	13	1.70	0.1571	0.0673	0.2187	0.0248	0.5318	0.2437	0.645	49	8530
157	14	1.88	0.1573	0.0674	0.2196	0.0234	0.5323	0.2430	0.647	52	8800
202	15	2.25	0.1562	0.0669	0.2247	0.0234	0.5287	0.2481	0.630	46	8100
205	16	2.54	0.1563	0.0670	0.2227	0.0250	0.5290	0.2477	0.631	46	8200

The following notes pertain to Tables 6.3, 6.4, and 6.5.

[a] The cement was No. 13280; its specific surface area was 1725 cm^2/g by the Wagner turbidimeter.

[b] Data from PCA Series 265G, 1936–1937.

[c] The excess of the designated group according to the Weymouth theory, expressed as the ratio of amount present to theoretical limit.

[d] Each value is the average of three determinations. Values for cement volume and air volume are somewhat inaccurate, owing to the use of v_c instead of v_c° in the computations, air volume being obtained by difference.

[e] Ratio of cement volume to volume of total voids.

[f] Consistency as measured by the remolding test. Each value is the average of at least three determinations on different batches.

[g] Each cube-strength value is the average of six tests.

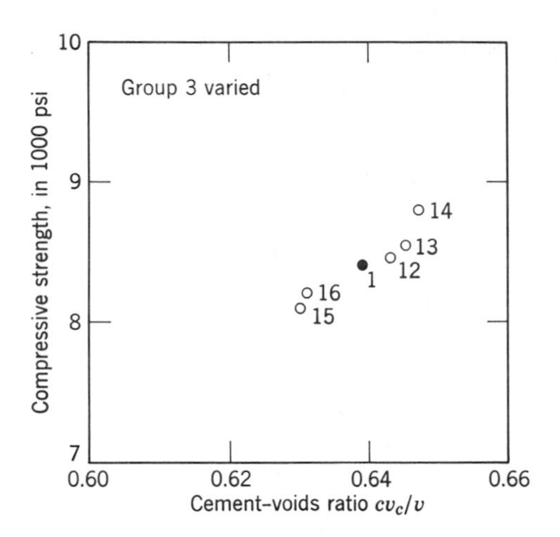

Fig. 6.3. Effects of certain gradings on the cement-voids ratio and compressive strength (2-in. cubes). Numerals are grading numbers. (Data are from Table 6.3.)

These data show that the variations in aggregate grading did have some effect on composition and strength of test specimens, but it is not thereby proved that the effects are due to particle interference as Weymouth conceived it. The effect of change in grading shows up in two ways: (1) the variation in voids content (column 9), particularly in the variation of the ratio of cement volume to void volume (column 10);* and (2) in the variation in compressive strength of the hardened specimens.

The "interference" in gradings 12, 13, and 14 had the effect of reducing the water content. This *increased* the cement-voids ratio, and thereby the strength. This is shown in Fig. 6.3: numerals by the points are the grading numbers, grading No. 1 being the reference point. There also we see that for gradings 15 and 16, the cement-voids ratios were lower, indicating an increased voids content; the strengths were correspondingly lower.

The data mentioned above show that for the particular mode of variation of grading used for this series of tests, there is an optimum proportion of group 3 corresponding approximately to that in grading 14. (See Fig. 6.2.) However, grading 14 gives a volume concentration of group 3 that is 1.88 times the maximum allowed by the Weymouth theory. Indeed, the theory gives no basis for anticipating an optimum degree of "interference."

Table 6.4 gives data (like those in Table 6.3) for variations of groups

* Although the pulverized silica was counted as a part of the matrix, cv_c/v was based on the actual portland cement content.

Table 6.4 Effect of Varying the Proportions of Groups 2, 4, and 6[a]

1[b]	2	3[c]	4	5	6	7	8	9	10[e]	11[f]	12[g]
			Composition of Test Specimens[d]								
Reference Number	Grading Number	Excess of Varied Size	$\frac{cv_c}{V}$	$\frac{\text{Silica}}{V}$	$\frac{wv_w}{V}$	$\frac{A}{V}$	$\frac{av_a}{V}$	$\frac{v}{V}$	$\frac{cv_c}{v}$	R	Twenty-eight-day Strength 2-in. Cubes (psi)
					Group 6 Varied (#8–#4)						
133	2	1.05	0.1571	0.0673	0.2216	0.0223	0.5317	0.2438	0.644	48	8480
136	3	1.22	0.1578	0.0676	0.2191	0.0212	0.5342	0.2403	0.657	49	8760
139	4	1.33	0.1582	0.0678	0.2184	0.0200	0.5355	0.2384	0.664	55	8980
190	5	1.56	0.1575	0.0675	0.2251	0.0167	0.5331	0.2418	0.651	48	8580
193	6	1.78	0.1576	0.0675	0.2266	0.0148	0.5334	0.2414	0.653	51	8550
					Group 4 Varied (#30–#16)						
142	7	1.22	0.1568	0.0672	0.2214	0.0239	0.5307	0.2451	0.640	54	8200
145	8	1.40	0.1568	0.0672	0.2218	0.0239	0.5306	0.2454	0.639	51	8390
148	9	1.54	0.1567	0.0672	0.2233	0.0223	0.5305	0.2456	0.638	50	8270
196	10	1.83	0.1559	0.0668	0.2276	0.0221	0.5276	0.2497	0.624	49	8340
199	11	2.93	0.1550	0.0665	0.2297	0.0242	0.5246	0.2540	0.610	46	7990
					Group 2 Varied (#100–#50)						
160	19	2.11	0.1566	0.0671	0.2222	0.0241	0.5300	0.2462	0.636	49	8270
163	20	2.41	0.1569	0.0672	0.2226	0.0223	0.5310	0.2449	0.641	49	8410
166	21	2.70	0.1566	0.0671	0.2224	0.0242	0.5299	0.2465	0.635	55	8560
208	22	3.20	0.1557	0.0667	0.2266	0.0240	0.5270	0.2506	0.621	49	7870
211	23	3.68	0.1553	0.0665	0.2277	0.0254	0.5252	0.2530	0.614	48	7880
					Grading No. 1						
169	1	1.00	0.1568	0.0672	0.2248	0.0206	0.5306	0.2454	0.639	44	8400

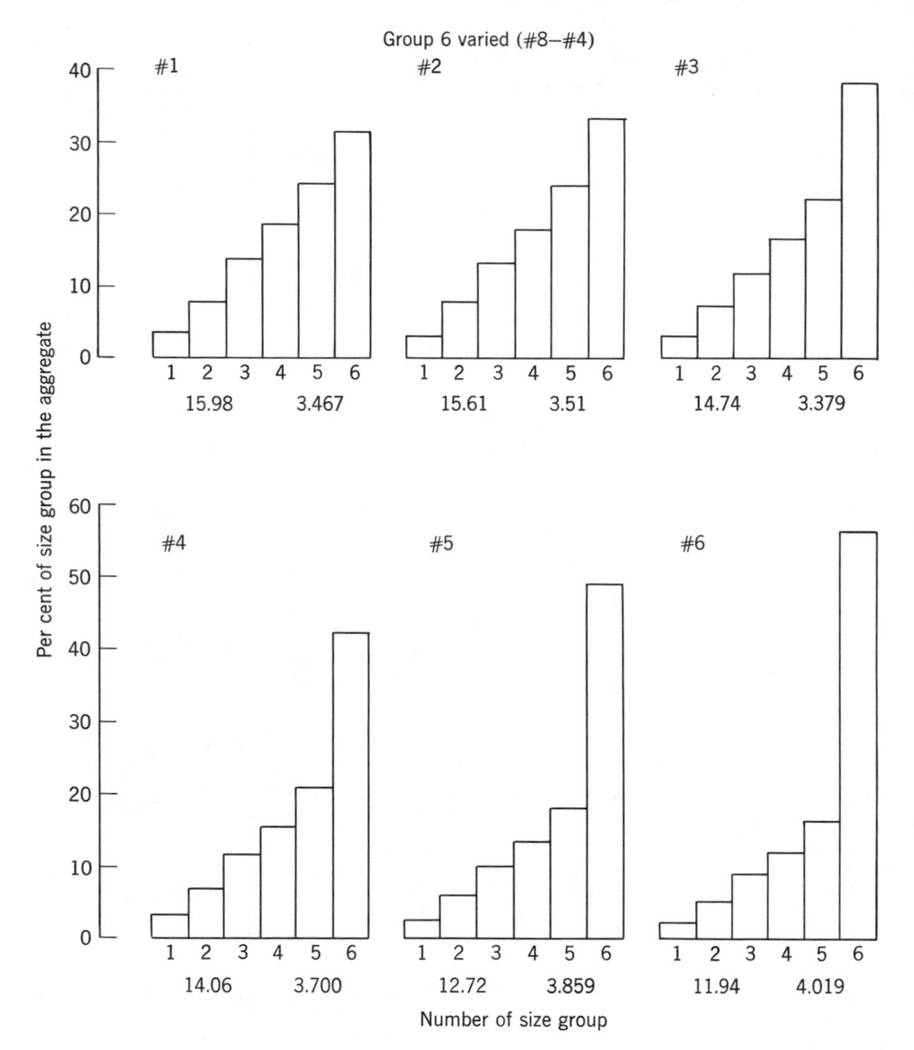

Fig. 6.4. Size distributions of experimental gradings. (The numbers below each histogram indicate the surface modulus and the fineness modulus.)

2, 4, and 6, the variations being shown in Figs. 6.4, 6.5, and 6.6. Figure 6.7 shows the relationships between the cement-voids ratio and strength.

We see that introducing "interference" by means of group 6 resulted in *reductions* in the total voids content and thus an *increase* in the cement-voids ratio and corresponding increases in strength with each of the gradings. Grading 4 gave the maximum strength, and gradings 5 and 6 gave lesser increases.

For variations of group 4, gradings 7, 8, and 9 gave virtually the same results as grading 1, which means that "excesses" up to about 1.5 times the theoretical limit for no interference had no appreciable effect. However, for gradings 10 and 11, the cement-voids ratio was reduced slightly, and the strength likewise.

For variations of group 2, increasing the proportion generally decreased the cement-voids ratio by increasing the voids content. This reduced the strength slightly, but the effects were hardly in excess of experimental error except where the concentration of the group was more than three times the theoretical limit.

Other tests were made, using different patterns of grading variation,

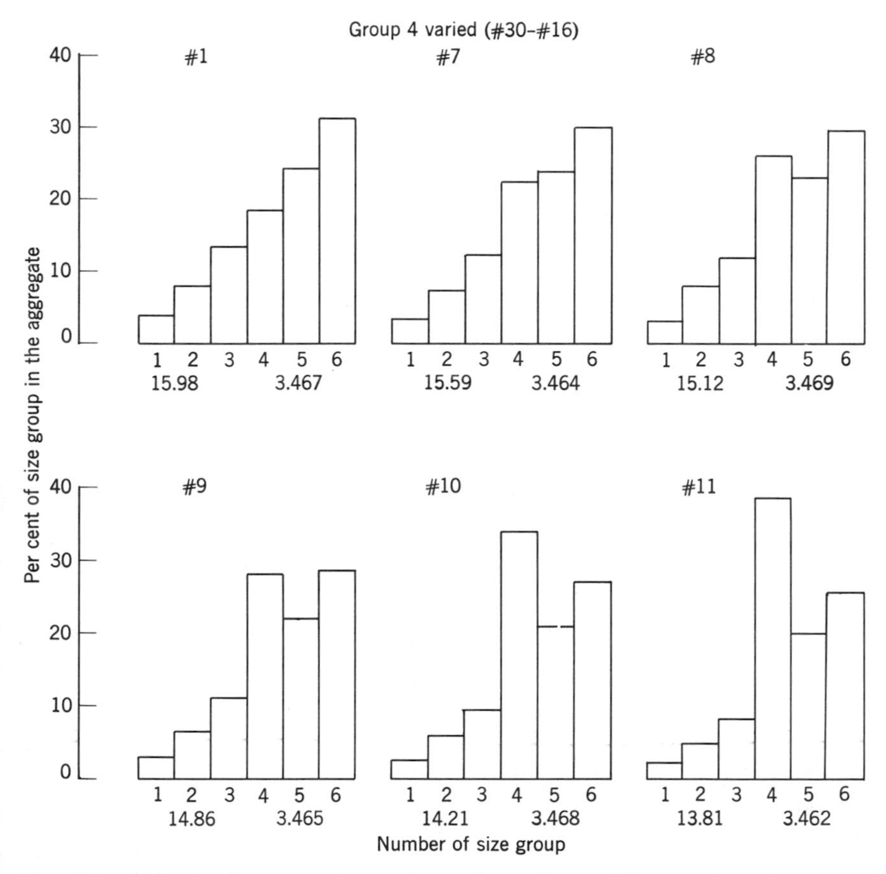

Fig. 6.5. Size distributions of experimental gradings. (The numbers below each histogram indicate the surface modulus and the fineness modulus.)

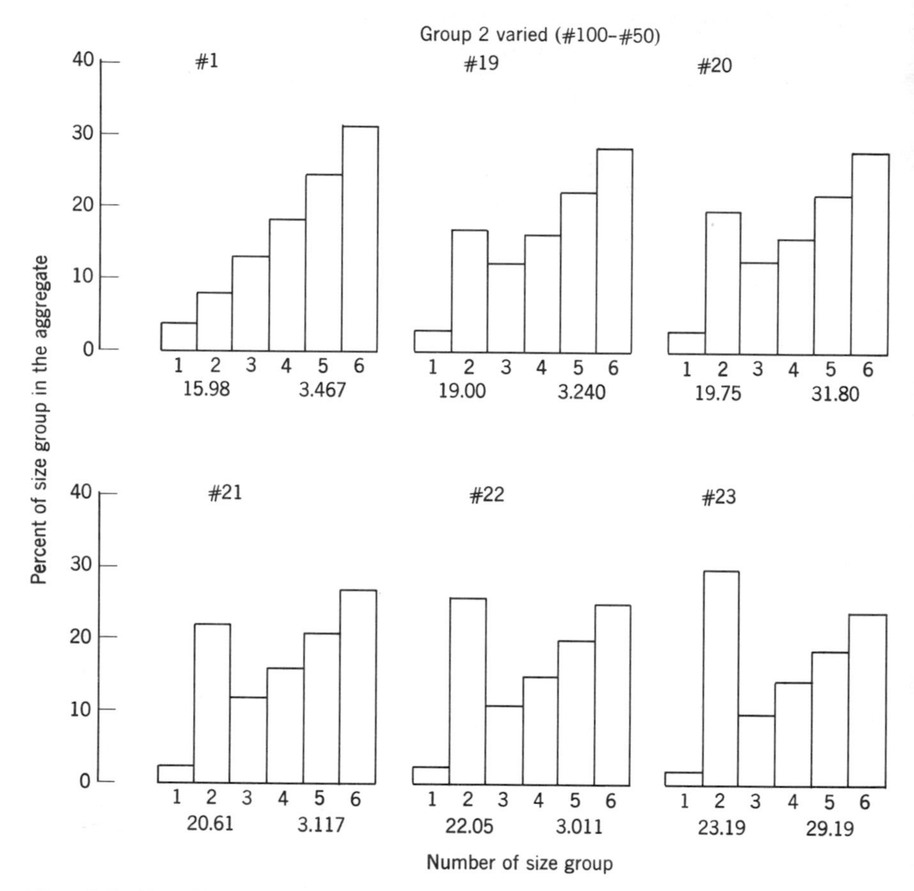

Fig. 6.6. Size distributions of experimental gradings. (The numbers below each histogram indicate the surface modulus and the fineness modulus.)

with the same general results: small variations in strength which had no relation to the degree of particle interference as computed by the Weymouth method. In terms of Eq. 6.21 or 6.24, there is no indication that any single value of d_i/d_o can be specified as critical. Also, it cannot be said that a grading showing no particle interference according to Weymouth's criterion is the equal of any other; in some cases variations from the ideal produced improvements in density and strength.

3.4 Discussion of Weymouth's Concept

Weymouth's concept of the aggregate in plastic concrete as constituting "a sort of grid structure through which the smaller particles move both horizontally and vertically during manipulation of the mixture" apparently accounts for his stipulation that each size should be so dispersed that it provides clearance for the next smaller size. If the aggregate is composed of definite sizes and if the grading and dispersion is such as to conform to Weymouth's stipulation, the mixture is capable of unlimited shearing without particle interference, or in other words, without dilatation. However, since only a small amount of shear strain is normally produced during the placing of concrete, it is not necessary to provide a capacity for unlimited shear strain to obtain sufficient workability. In the discussion below, when a mixture is said to be free of particle interference it is implied that the mixture is able to undergo the shear strain required of it without evidence of interference.

3.4.1 Detection of particle interference. In most writings on the subject of particle interference the presence or absence of interference in a mixture has been indicated only by the calculated clearances according to Weymouth's theory. However, Weymouth himself showed how the

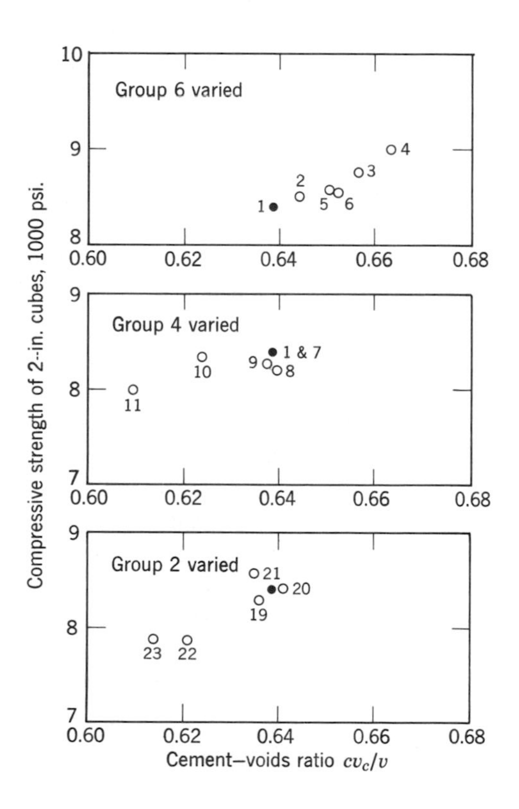

Fig. 6.7. Effects of certain gradings on the cement-voids ratio and compressive strength (2-in. cubes). Numerals are grading numbers. (Data are from Table 6.4.)

effect of particle interference could be observed experimentally. In Fig. 1.10 his method of finding the point of incipient particle interference, point I, in a series of mixtures of fine and coarse aggregate is illustrated; the point can be located rather accurately. In Figs. 1.6 and 1.11 we may see, for a series of mortar or concrete mixtures, how point I can be located on the voids ratio diagram, it being just to the left of point B.

Although Weymouth used only single size groups in his illustrations, he contended that the size range of the size group need not be restricted; that is to say, without violating the principles of Weymouth's method we can divide the total range of sizes at any arbitrarily selected point into one group of fine particles and another of coarse particles, and make mixtures from these two components. We can therefore name cement as the fine component and the aggregate as the other, which is virtually the same as taking the No. 200 sieve as the dividing point between the two components. Thus we may use for analysis with respect to particle interference any of the series of mixtures we have dealt with in Chapters 2, 3, and 4. There we found that point B was usually clearly indicated. The coordinates of point I can be estimated from either the voids-ratio diagram or the diagram for $k_Y(1 + M)$ versus M for mixtures of cement and aggregate, each different aggregate grading producing a characteristic diagram. Since cement was used as the fine component in these mixtures, graphical indication of particle interference signifies that the average clearance between the aggregate particles is too small to accomodate the cement particles. It follows, therefore, that mixtures in the AB category are those free from particle interference, or, since the category is defined by point B, we should say that mixtures in this category are free or practically free from particle interference. It follows also that in mixtures in the other categories, those to the right of point B, particle interference is a dominant determinant of the voids content of the mixture and of the voids-cement ratio and strength.

Thus we see that Weymouth was actually searching for a single criterion for aggregate grading and aggregate content that would always assure mixtures belonging to the AB category. He did not see clearly, however, that to accomplish this he would have to use mixtures relatively rich in cement, or rich in cement plus other paste-making material. In the early stages of his thinking he considered only the grading of the aggregate, trying to produce a grading that gave the smaller particles freedom to move among the coarser ones. Later he realized that some allowance must be made for the cement particles, but he regarded this as a matter of providing clearance for a paste-coating on the coarse particles. The formula used for grading 1 of the experimental study discussed above represented the most advanced form of Weymouth's theory.

We should be able to evaluate Weymouth's ideal mixture, whether or not it actually was without particle interference, by making a series of mixtures with grading 1 and analyzing the results in terms of the voids-ratio diagram, or the diagram of $(w/c)Q$ versus M or a/c. (See Eqs. 4.19 and 4.20.) Unfortunately, at the time the tests discussed above were made this method of analysis had not been developed, and no such series with grading 1 was made. However, three mixes with grading

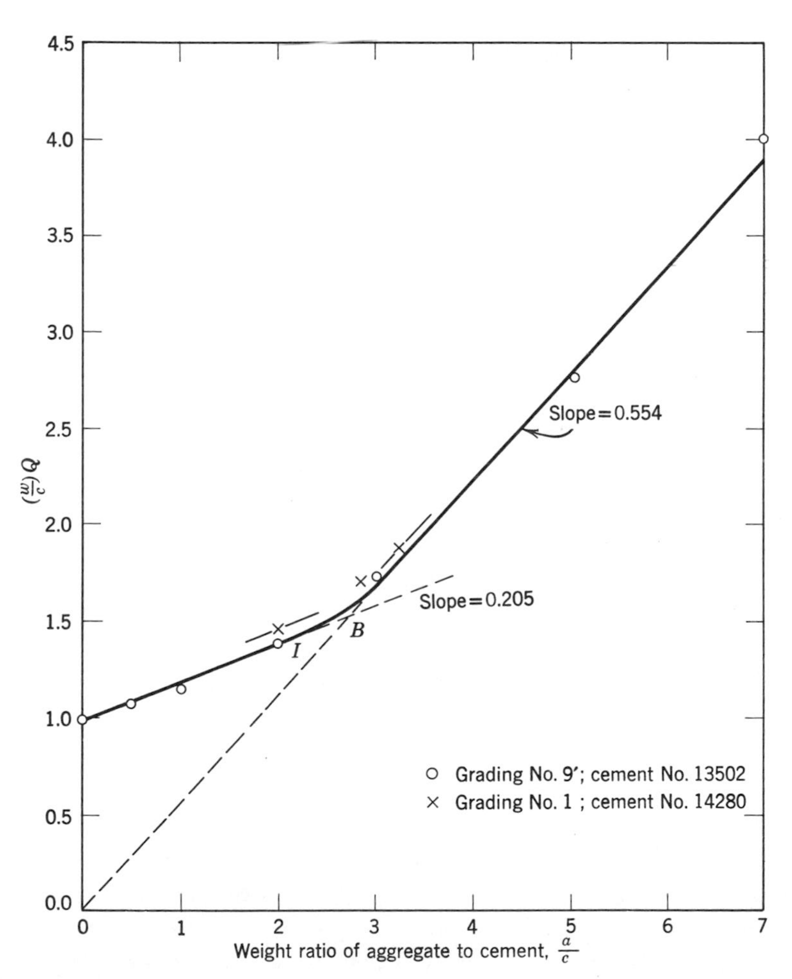

Fig. 6.8. Estimation of point B for mixtures made with gradings No. 9′ and No. 1.

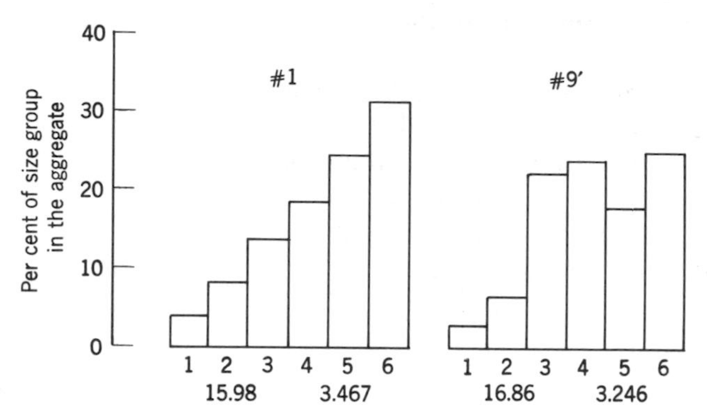

Fig. 6.9. Size distribution for grading 9′ and 1. (The numbers below each histogram indicate the surface modulus and the fineness modulus.)

1, and eight each of three other not too different gradings were made, and from some of these data we can arrive at a reasonable conclusion about the "ideal" mixture.

Data are shown in Fig. 6.8 for a series of mixtures made with grading 9′ and grading 1. Their size distributions as well as their surface moduli and fineness moduli may be compared in Fig. 6.9. The parameter $(w/c)Q$ is plotted against the mix by weight a/c. For grading 9′, point B is indicated at $a/c = 2.7$, and point I is estimated to be at $a/c = 2.4$. The points for grading 1 are at $a/c = 2.0$, 2.85, and 3.25. These mixtures were made with a cement having a slightly higher water requirement than that of the cement used with the other grading; hence the point for the 1:2 mix is considered to be on a line above and approximately parallel to the line for grading 9′.

With the considerations described above in mind, inspection of Fig. 6.8 gives a basis for assuming that the mixes at points B and I for grading 1 are about the same as for the other grading. Thus it appears that Weymouth's ideal mix (1:2, grading 1) actually was free from particle interference; it was in fact a little richer than necessary. In other words, our analysis shows that his method of calculation resulted in a higher degree of aggregate dispersion than was actually needed to prevent particle interference between aggregate and cement particles.

3.4.2 Necessary clearance for no interference. The data of Fig. 6.8 raise again (see Chapter 1, Section 8.6) the question of the relation between the average size of the fine component and the minimum clearance required between aggregate particles to avoid particle interference. The relationship between the size of the cement particles and the clearance between aggregate particles at point I can be calculated from the data given in Fig. 6.8, using reasonable estimates where specific data are missing. We shall make the computation for the consistency $R = 50$. The data necessary for the calculations are as follows:

Voids ratio of the aggregate: $C = 0.40$*

Specific volume of the cement: $v_c = 0.3179$

Apparent sp. vol. of cement: $v_c^o = 0.3123$*

Specific volume of water: $v_w = 1.002$

Specific volume of aggregate: $v_a = 0.383$

Ratios: $\dfrac{v_w}{v_c^o} = 3.21$

$$\frac{v_a}{v_c^o} = 1.226$$

Point I: $\left(\dfrac{a}{c}\right)_I = 2.4;\ M_I = 1.226 \times 2.4 = 2.94$

$$x_I = \frac{2.94}{3.94} = 0.746$$

Consistency factor: $Q = \log \dfrac{50}{(4 \times 10^{-3})} = 4.0969$

Water-cement ratio at point I: $\left(\dfrac{w}{c}\right)_I Q = 1.46$ (from Fig. 6.8)

$$\left(\frac{wv_w}{cv_c^o}\right)_I = \frac{1.46 \times 3.21}{4.0969} = 1.144$$

$$\left(\frac{w}{c}\right)_I = 1.144 \times 0.313 = 0.45$$

Air water ratio: $Z = 0.143$*

Voids-cement ratio at point I: $\left(\dfrac{v}{cv_c^o}\right)_I = 1.144 \times 1.143 = 1.308$

Voids ratio at point I: $u_I = 1.308 \times 0.254 = 0.332$

Average particle size of aggregate, grading 9′: $d = 285\ \mu$

* Estimated.

The clearance t between aggregate particles at point I is given by

$$t = \left[\left(\frac{1 + u}{x(1 + C)}\right)^{1/3} - 1\right] d$$

Using values given above gives

$$t = 24\ \mu$$

Making different reasonable estimates of the values that were estimated, including the coordinates of point I, indicates that the clearance between the particles in the mix at point I is between 24 and 30 μ. The average size (volume diameter) of an ordinary cement is about 30 μ. Hence particle interference between aggregate No. 9′ and cement appeared when the average clearance became less than the average size of the cement particles, a criterion proposed by Butcher and Hopkins. (See Chapter 1, Section 8.6.4.) However we may also see in Chapter 2, Section 4.2, that the critical clearance for mixtures of dry cement and standard sand is about 60 μ, and for plastic mixtures of the same materials at basic consistency, data given in Fig. 3.12 indicate the critical clearance to be about 90 μ.

The remarks above remind us that there is no single criterion by which to predict critical clearance between the particles of the fine and coarse components of a mixture. When the two granular materials are geometrically similar, the critical clearance may be expected, on the basis of limited evidence, to be equal to the average diameter of the particles of the fine component, but this is not so if the two components are geometrically dissimilar.

The mixtures represented in Fig. 3.12 were of cement and standard sand, the cement comprising eight or nine size groups, and the aggregate comprising a smaller range of sizes than one size group. On the other hand, grading No. 9′ comprised six size groups, and therefore, so far as size range is concerned, it was much less different from cement than was standard sand. However, since cement and grading No. 9′ are not actually similar, we can only suppose that other factors such as differences in grading, particle shape, and perhaps effects of interparticle forces, must be combined in such a way as to result in a critical clearance about the same as would be expected from a mixture of geometrically similar materials. It is important to keep in mind that in general the critical clearance for a given fine component is a function of the size and grading of the coarse component.

3.4.3 Actual and virtual interference. From the data in Fig. 3.16 we can calculate that the critical clearance in a mixture of cement and standard sand without water is about 58 μ, and that when the mixture contains the amount of water giving basic consistency, the calculated critical clearance is 107 μ. Although there is considerable uncertainty as to the exact coordinates of the critical point for the plastic mixture, it is certain that the critical clearance is definitely greater than it is in the waterless mixture.

In a dry mixture interference is an actual phenomenon; the dry particles are in contact with their neighbors, and the mass cannot undergo shear deformation without dilatation even though the coarse component is somewhat dispersed by the fine. But, in a plastic mixture the existence of plasticity signifies that such particle interference has been prevented or mitigated by a slight extra clearance between *all* particles, the extra clearance being effected by water and air bubbles. Since the required *total* clearance between aggregate and cement for plasticity is determined primarily by the degree to which the fine component is able to disperse the coarse in waterless mixtures, the voids-ratio curve for a series of plastic mixtures is shaped by the curve for the same series of mixtures without water. However, the two curves are not parallel, and the points I do not have the same coordinates.

The amount of water in a plastic mixture is that which is necessary to prevent actual interference during moderate shear deformations produced in handling, and the increase in water requirement to the right of point I, and especially to the right of point B, is due to *virtual* rather than actual particle interference. We can regard point I of a series of plastic mixtures as the point beyond which particle interference becomes a determinant of the water requirement and total-voids content of the mixtures. Strictly, when we speak of plastic mixtures we should use the term "virtual particle interference," but we shall not insist on the distinction because there is little likelihood of confusion if less discriminating language is used.

3.4.4 Grading requirements. Having found that Weymouth's criterion for particle interference is inadequate, it is pertinent to consider such information as we have to see if we can establish what actually determines the coordinates of point I, or more practically, point B. Presumably, the designer is seeking to produce a mixture having the least possible voids content, and to do this he must have an aggregate that will provide adequate clearance for the cement particles at a given consistency and at the same time provide the highest possible aggregate content.

The major factor is, of course, the range of sizes in the aggregate, or practically speaking, the maximum size; the larger the maximum size the higher the maximum possible aggregate content of the mixture at point B. The main question is, What is the best grading for a given maximum size? Weymouth and others have explored the possibility that there is one superior grading, or at least one attribute that might be had by various gradings, but we have already seen much evidence that one attribute, or one criterion is not enough. In Chapter 3 we found that the value of the voids ratio at point B is a function of the specific surface area of the aggregate, of its grading insofar as grading influences the voids ratio of the aggregate, and on the water requirement of the cement. See Eq. 3.40, following.

$$u_B = \frac{AC}{1 + A + C}\left[1 + 0.238\left(\frac{SM}{\psi}\right)^{1/2}\right]$$

Although that equation is entirely empirical, and thus subject to factors limiting its accuracy, it serves to show that with a given cement we should expect the relative merit of a given grading for the mixture at point I or point B to depend on the degree to which it approaches an optimum combination of specific surface area and voids ratio.

In the data on the gradings shown in Figs. 6.2, 6.4, 6.5, 6.6, and 6.9, we included the values of the surface moduli. If these values for a given group of grading variations are considered in connection with results obtained, we find some evidence of the interplay of surface area and the voids ratio of the aggregate discussed above, A being a constant for a given cement and consistency. Unfortunately, the values of C are available for only a few combinations, and these were obtained by recreating the gradings many years after the original work. In general, it is necessary to use conjecture where data are lacking, the conjectures being guided somewhat by the few data mentioned.

In Fig. 6.4 and the uppermost diagram of Fig. 6.7 we see that as the surface modulus of the aggregate was reduced the voids ratio of the mixture became smaller. This is evidenced by the increase in the cement-voids ratio, the cement content being a constant (see Table 6.3). However, a limit was found with grading 4 beyond which the voids ratio began to rise above the minimum point. Looking at Fig. 6.4, we see that if we regard group 6 as the coarse aggregate and groups 1 to 5 (#200–#8) as the fine aggregate, the experiment consisted of progressively increasing the ratio of coarse to fine aggregate and observing the effect on voids content. Grading 4 turned out to have the optimum percentage of sand for the particular fine and coarse gradings represented here. As the percentage of coarse aggregate increased, the surface modulus diminished; this change tended to reduce the value of D^* in the voids-ratio diagram, but at the same time the voids ratio of the aggregate, C, was affected. Presumably, beyond the optimum percentage of sand (grading 4), C was caused to increase sufficiently to reverse the trend.

The other groups are less easily analyzed in the terms used above, but yet the results seem explicable. In Fig. 6.5 we see that the variation may be thought of as changing the grading of the coarse aggregate while at the same time changing its proportion in the aggregate. In this case the fine aggregate comprises the first three groups. The changes of grading were such as to reduce the surface modulus, but the corresponding diagram in Fig. 6.7 indicates that there were concomitant increases in C such that for the first three gradings they offset the benefit of reduced surface area, and for gradings 10 and 11 they more than offset the effect of reduced surface area.

For the group of gradings shown in Fig. 6.2, we may think of size groups 4, 5, and 6 as the coarse aggregate, and the first three groups as the fine aggregate, the grading of the coarse aggregate remaining constant. The grading changes reduced the surface modulus of the "sand" while they increased the proportion of sand in the aggregate. The result was that gradings 12, 13, 14, and 15 had smaller surface moduli than that of grading 1, whereas gradings 16, 17, and 18 had surface moduli equal to or greater than that of grading 1. There were also effects on C, for which there were no direct data. Figure 6.3 shows that gradings 12, 13, and 14 were superior to grading 1. Since the changes in surface modulus were small, it seems that these gradings were more dense than grading 1. That is, the values of C were smaller. However, gradings 15 to 18 were inferior, presumably mainly because their values of C were relatively high.

Finally, in Fig. 6.6, we see a coarse aggregate composed of size groups 3 to 6 and a fine aggregate that had a relatively high surface modulus. In general, the surface moduli of the total aggregates were relatively high, but evidently the values of C were relatively low, as we see in Fig. 6.7. Only in the extreme cases were the results significantly and adversely affected by the changes of grading.

3.4.5 Mixes in the *BF* and *FC* categories. By definition, all mixtures in the *BF* and *FC* categories have particle interference between cement particles and the aggregate. Therefore, particle interference cannot be eliminated by a change of aggregate grading ("aggregate" being here defined as material retained on the No. 200 sieve), unless the mixture is very close to point *B* before the change is made. That is to

Table 6.5 Effect of Varying the Proportion of Group 3 in Mixtures of the *BF* Category[a]

1[b]	2	3[c]	4	5	6	7	8	9	10[e]	11[f]	12[g]
			Composition of Test Specimens[d]								
Reference Number	Grading Number	Excess of Varied Group	$\frac{cv_c}{V}$	$\frac{\text{Silica}}{V}$	$\frac{wv_w}{V}$	$\frac{A}{V}$	$\frac{av_a}{V}$	$\frac{v}{V}$	$\frac{cv_c}{v}$	R	Twenty-eight-day Strength of 2-in. Cubes (psi)
					$a/c = 2.00$						
19	1	1.00	0.2206	0.0000	0.2327	0.0240	0.5226	0.2567	0.859	47	10890
55	1	1.00	0.2199	0.0000	0.2360	0.0218	0.5237	0.2578	0.853	47	10720
					$a/c = 2.85$						
172A	1	1.00	0.1742	0.0000	0.2085	0.0277	0.5896	0.2362	0.738	46	9340
					$a/c = 3.25$						
172	1	1.00	0.1578	0.0000	0.2056	0.0294	0.6072	0.2350	0.671	46	8640
175	12	1.48	0.1572	0.0000	0.2067	0.0312	0.6050	0.2379	0.661	45	8120
178	13	1.70	0.1574	0.0000	0.2047	0.0317	0.6061	0.2364	0.666	46	7790
181	14	1.88	0.1567	0.0000	0.2069	0.0331	0.6032	0.2400	0.653	48	7810
184	15	2.25	0.1563	0.0000	0.2043	0.0383	0.6015	0.2422	0.645	46	7570
187	16	2.54	0.1553	0.0000	0.2055	0.0412	0.5957	0.2467	0.630	48	7240

say, a certain change of grading might give favorable changes of surface modulus and C, as in some of the cases discussed above, and thus eliminate a preexisting small degree of interference, and change the classification from BF to AB. However, many if not most commonly used concrete mixtures are so far to the right of point B that particle interference can be eliminated only by increasing the dispersion of the aggregate, as by introducing a sufficient quantity of a suitable material smaller than No. 200. All such mixtures not so corrected require an extra volume of water and air; and because of that, after hardening they are not as strong and impermeable as they might be. Such mixtures may be relatively weak not only because of the extra volume of voids, but also because at a given voids-cement ratio (or cement-voids ratio) they may show somewhat inferior strength.

Take for example the mix $a/c = 3.25$ for grading 1 shown in Fig. 6.8, which is in the BF category. A series of mixtures was made using the same mix, and with the grading variations shown in Fig. 6.2, that is, the proportion of group 3 was progressively increased. The resulting compositions and strengths are given in Table 6.5.

In columns 4 and 6 of Table 6.5 we see that the cement and water contents of the hardened specimens were practically the same in all specimens, but the air content and total voids content increased as the proportion of group 3 increased. Compressive strength decreased as the total voids content increased, the lowest strength being about 84 per cent of that of the mixture made with grading 1. More about the nature of the decrease in strength can be ascertained from Fig. 6.10.

Figure 6.10 shows relationships between the cement-voids ratio and

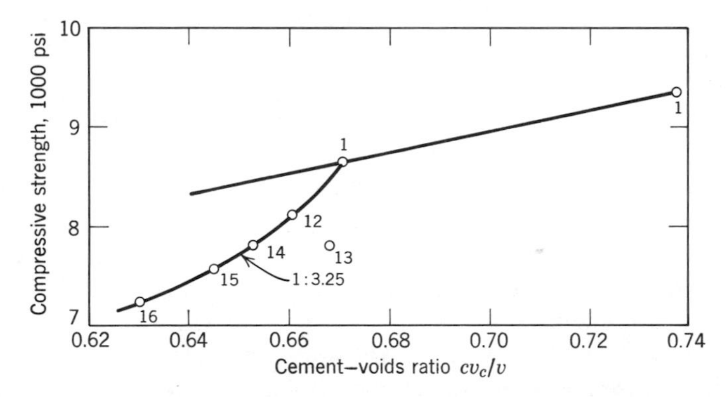

Fig. 6.10. An example of the influence of aggregate grading on the relationship between strength and the cement-voids ratio. Numerals are grading numbers. (Data are from Table 6.5.)

the strength for two different mixes made with grading 1 (two of the three in the *BF* category shown in Fig. 6.8). It is apparent that this curve is different from the one described by the points for the 1:3.25 mix made with different gradings. A decrease in cement-voids ratio caused by a change of grading is accompanied by a larger decrease in strength than that accompanying the same change in cement-voids ratio when the change is due to a change of mix with the grading constant. It may be noted in Table 6.3 that when the decrease in cement voids ratio was due to a change in grading, the water-cement ratio remained constant. On the other hand, when the change was due to a change of mix with grading constant, the principal change was in the water-cement ratio. It may be that the difference in the trends of the two curves reflects only an intrinsic difference between the effect on strength of void space due to entrapped air and void space due to water.

The example shown in Fig. 6.10 should not be taken as typical; it represents only the result of changing grading 1 in a certain way and in a certain mix. We have already seen that the same changes of grading in a mix belonging to the *AB* category gave different results. Up to a certain point, the grading changes permitted a reduction of water content, and there was a corresponding increase in strength, and of course the effect of change in grading will in any case depend on the nature of the grading that is being changed.

When we try to explain why the change of grading was detrimental in the case discussed above, we still cannot ascribe it to particle interference according to Weymouth's method of computing it, because there is no assurance that the computed interference actually occurs. Thus, although Weymouth's computation would indicate that the average distance between the particles of group 3 would be less than the average diameter of the particles of group 2, there is no reason to suppose that interference involves only the group 2 particles.

Indeed, even if there were no objection to considering only the group 2 particles, we could not expect the computed clearance alone to indicate the severity of particle interference. It also is necessary to consider the relative amounts of the two groups. This point is well illustrated in Fig. 6.2. For grading 12, calculation by Weymouth's method indicates that the clearance between the particles of group 3 was 73 per cent of the average diameter of the supposedly paste-coated particles of group 2, whereas for grading 16 it was only 42 per cent. Although the difference is of the proper kind to explain the observed effect on voids content and strength, it must be considered insufficient in view of the considerable difference between quantities involved in the two cases. From the data in Fig. 6.2 it can be calculated that in grading 12 group 2 constituted

26 per cent of the total of the two groups, whereas in grading 16 group 2 constituted only 10 per cent.

If we assume that there are eight times as many group 2 particles per gram as there are group 3 particles, it follows that for grading 12 there are 1.54 group 2 particles for each group 3 particle, whereas for grading 16 there is about 0.9 group 2 particle per group 1 particle. Hence on this basis as well as others it does not seem proven that clearance of group 3 particles with respect to group 2 particles could alone be the explanation of the observed effect.

Data are lacking, but it seems probable that the principal effect of the change of grading in this particular category (BF) was to increase the average size of the particles and, at least for gradings 15 and 16, to increase the voids content of the aggregate, that is, the values of C. Calculation by Eq. 3.44 on that basis shows that there would have to be an increase of the voids ratio u to provide the same clearance for the cement as that found with grading 1. Presumably, when other factors are as near alike as they are among the mixtures of this group, obtaining the same consistency entails providing about the same clearance between the aggregate particles and the cement particles.

3.4.6 General indications. In general, we see that particle interference in concrete mixtures is usually due to an insufficient clearance between aggregate particles, the aggregate considered as a whole. For a mixture to be practically free from particle interference it must appear at point B of the voids-ratio diagram, or to the left of it.

The clearance required by a given cement is not necessarily the same for different sizes and gradings of aggregate, and at present it is not predictable on a theoretical basis. However, it can readily be determined empirically for any given aggregate by analytical procedures that have already been described. The best aggregate grading for a mixture at point B seems to be that which for a given maximum size has a certain optimum combination of specific surface area and voids ratio.

If mixtures in the BF category are used, tolerance of some particle interference between cement and aggregate is implied, and if mixtures in the FC category are used, tolerance of extreme interference is necessary. The degree of interference between cement and aggregate in such mixtures can be influenced by changes of aggregate grading, and perhaps there can be interference between fine and coarse parts of the aggregate. But, the idea of particle interference between successive size groups of the aggregate does not seem to be useful in this connection.

It may be noticed that in approaching the subject from Weymouth's point of view we finally conclude that point B marks the limiting aggregate concentration for practically no particle interference, whereas in Chapter 4, Section 8, that point is regarded as the mixture containing a paste having optimum consistency. On the one hand we think of rheological characteristics, and on the other, packing characteristics.

The concept of paste consistency is defended in Chapter 4, and it does not seem to be weakened by the considerations of the present chapter. When dealing with dry mixtures, application of the packing concept is straightforward, but when plastic mixtures are considered, it becomes necessary to introduce the idea of virtual rather than actual interference as one of the determinants of the voids ratio of a mixture.

4. THE EXCESS PASTE THEORY

Several authors (Marquardsen [8]; Smith [9]; Katoh [10]; Kennedy [11]) have advocated methods of mix formulation based on the assumption that the consistency of concrete depends on two factors: the volume of cement paste in excess of the amount required to fill the voids of the compacted aggregate, and the consistency of the paste itself.

4.1 Kennedy's Consistency Factor

As expressed by Kennedy, but restated in our nomenclature, the relationship between the amount of paste and the amount of aggregate is as follows:

$$\frac{p}{V} \equiv \left(\frac{av_a + bv_b}{V}\right)(C + K\sigma_{ab}) \equiv \frac{cv_c^o + wv_w}{V} \tag{6.25}$$

where p is the volume of paste, that is, $cv_c^o + wv_w$; C is the voids ratio for the combined aggregate, cm^2/cm^3 solid volume; σ_{ab} is the specific surface area of the combined aggregate cm^2/cm^3; and V is the volume of the concrete. (Kennedy and the others, except Katoh, ignored the air content.) Kennedy called K the consistency factor; it is the volume of excess paste per unit surface area of the aggregate. (See Chapter 3, Section 5.6.1.)

Dividing both sides of Eq. 6.25 by the cement content we obtain

$$\frac{p}{cv_c^o} = 1 + \frac{wv_w}{cv_c^o} = M(C + K\sigma_{ab}) \tag{6.26}$$

At first glance, Eq. 6.26 might appear to be a water-requirement equation, but it is only an expression of identity. K is not independent of the water-cement ratio of the paste.

The relationship between the consistency factor K and consistency is different for each different water-cement ratio. Kennedy determined the relationship between K and slump from published data,* one curve for each selected water-cement ratio. Apparently he made no attempt to find a general expression for the group of curves so obtained.

4.2 Optimum Proportion of Coarse Aggregate

Referring to Eq. 6.25, we see that the volume fraction of paste in a mixture at a given value of K is higher the higher the specific surface area of the aggregate, provided also that the voids ratio of the aggregate C remains unchanged as the specific surface area changes. However, the voids ratio is *not* independent of specific surface area, and it is not true, as the expression seems to say, that any increase in specific surface area results in an increase in the paste requirement at a given water-cement ratio. For a given pair of fine and coarse aggregates, both factors are functions of the relative proportions of fine and coarse, typical relationships being such as those shown in Fig. 6.11.

It is apparent that the specific surface area of the combined aggregate is a linear function of the volume fraction of the coarse material such that

$$\sigma_{ab} = \sigma_a(1 - \mathbf{n}) + \sigma_b\mathbf{n} \tag{6.27}$$

where $\mathbf{n}$ is the solid-volume fraction of the coarse material in the combined aggregate, σ_{ab} is the specific surface area of the combined aggregate, and σ_a and σ_b are those of the fine and coarse aggregates, units of area per unit solid volume of material.

The voids ratio is a nonlinear function of $\mathbf{n}$; for our present purpose, it can be indicated as follows:

$$C = f(\mathbf{n}) \tag{6.28}$$

where $f(\mathbf{n})$ is such as to describe the voids-ratio curve. Because the voids-ratio curve has a minimum at an intermediate value of $\mathbf{n}$, we assume that for given aggregates, a given water-cement ratio, and a given consistency, there will be an optimum value of $\mathbf{n}$, optimum in the sense that the volume of paste in the mixture is at a minimum and the mixture has the lowest possible cement content for the conditions stipulated.

The optimum value of $\mathbf{n}$ can be ascertained as follows. Substituting Eqs. 6.27 and 6.28, into 6.26, we obtain

$$\frac{p}{cv_c^o} = Mf(\mathbf{n}) + MK[\sigma_a(1 - \mathbf{n}) + \sigma_b\mathbf{n}] \tag{6.29}$$

* PCA Series 186.

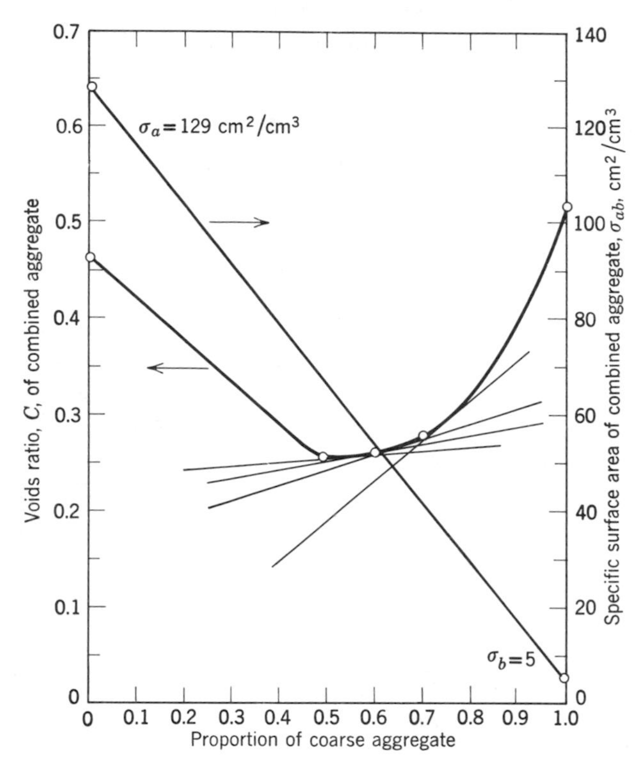

Fig. 6.11. Specific surface area and voids ratio of aggregate as functions of the proportion of coarse aggregate in the mixture.

Differentiating with respect to $\mathbf{n}$, and equating the result to zero, gives the optimum value of $\mathbf{n}$, as follows:

$$\frac{d(p/cv_c^o)}{d\mathbf{n}} = Mf'(\mathbf{n}) + MK(\sigma_b - \sigma_a) = 0$$

or (6.30)

$$f'(\mathbf{n}) + K(\sigma_b - \sigma_a) = 0$$

This means that the optimum proportion of coarse aggregate is that value of $\mathbf{n}$ at which the slope of the voids-ratio curve $f'(\mathbf{n})$ is opposite to that of the surface area curve $(\sigma_b - \sigma_a)$ and has a value K times the slope of the surface area curve. Since the slope of the surface area curve is negative, σ_a being larger than σ_b, the slope of the voids-ratio curve at the optimum point must be positive. Thus we see that the optimum proportion of coarse aggregate is somewhat greater than that giving minimum voids content for the aggregate alone.

The conclusion reached by the theory above, that the optimum proportion of coarse aggregate is larger, or the optimum percentage of sand is smaller, than that giving minimum voids content for the aggregate alone, is quite in line with experimental observations. For example, Abrams [12] reached that conclusion in connection with the experiments made while he was developing the fineness modulus method of formulating mixtures. Also, Powers [13] obtained that result by making systematic experiments in which the composition of the paste was kept constant while the ratio of coarse to fine aggregate was varied, consistency being regulated on the basis of the remolding test.

According to Eq. 6.30, the slope of the tangent to the voids-ratio curve at the optimum value of $\mathbf{n}$, will for a given value of K be greater the higher the specific surface area of the fine aggregate, the specific surface area of the coarse aggregate being small in any case. If K is independent of aggregate characteristics, as is assumed, it follows that the finer the sand the higher the optimum proportion of coarse aggregate, which also is in agreement with experience.

Since for a given consistency K is larger the lower the water-cement ratio of the paste, the optimum proportion of coarse aggregate is higher the lower the water-cement ratio of the paste. When the water-cement ratio is made higher by reducing the cement content in successive mixtures of nominally the same consistency, K approaches zero, and the optimum value of $\mathbf{n}$ approaches the value that gives the minimum voids content for the aggregate alone. In very lean mixtures it is possible for K to become zero or even negative as the cement content is reduced, the volume of paste being equal to or smaller than the voids content of the compacted aggregate. This result arises from the fact that the matrix of the aggregate is composed of paste and air voids, not paste alone as Kennedy assumed. In Chapter 4, Section 5.3, it is shown that for mixtures in the *FC* category at basic consistency, the water and cement barely fill the voids in the aggregate, the plasticity being due to the dispersion produced by the entrapped air. If the excess paste theory were to be developed further, it would be advisable to redefine K, or in some other way take the air content into account.

4.3 *An Application*

4.3.1 Evaluating *K*. The method of proportioning based on the excess paste theory cannot be applied until K has been evaluated for the materials and consistency to be used. For this purpose we may use data from Table 3.3 and additional data from the same source, by means of the following relationship, which is Eq. 6.26 solved for K:

$$K = \frac{(1 + wv_w/cv_c^o)1/M - C}{\sigma_{ab}} \tag{6.31}$$

All values on the right-hand side are measurable quantities; hence K can be evaluated for any given mixture and correlated with the consistency of that mixture. This has been done in Table 6.6 for four different mixtures.

With the K's evaluated, it is possible to calculate $f'(\mathbf{n})$, the slope of the tangent to the voids-content curve for the aggregate at the optimum value of **n** for each of the mixtures, as is shown in column 4 of Table 6.7. In column 5 the estimated tangent points are given, these being the theoretically optimum proportions of coarse aggregate in the combined aggregate, and in the final column the optimum values as determined empirically by the method described by Powers [13] are given for comparison. We can see that the theoretical procedure leads to higher proportions of coarse aggregate in the richer mixes than were arrived at by the empirical method, and that for the leanest mixture, less coarse material is called for than was found best by trial. By either method there is considerable uncertainty owing to the difficulty of determining

Table 6.6 Evaluation of K by Eq. 6.31

1[a]	2[b]	3[c]	4	5	6[d]	7[e]	8[f]
Reference Number	**n** Observed	M	$\frac{wv_w}{cv_c^o}$	$\frac{1 + \frac{wv_w}{cv_c^o}}{M}$	C	σ_{ab} (cm²/cm³)	K (μ)
101	0.64	4.99	1.154	0.432	0.266	49.6	34
106	0.62	7.14	1.389	0.335	0.265	52.1	13
111	0.60	8.69	1.622	0.302	0.263	54.6	7
116	0.57	11.3	2.129	0.277	0.260	58.3	3

[a] Data from PCA Series 303; slump 3–4 in. Sand and gravel aggregate (0–1½ in.).
[b] Fraction of coarse aggregate in the combined aggregate; optimum proportion as determined by trial, using the remolding test.
[c] The ratio of aggregate to cement, by solid volumes.
[d] The voids ratio of combined aggregate.
[e] Specific surface area of combined aggregate as computed from the surface modulus, with $1/\psi = 1.1$.
[f] Kennedy's consistency factor by Eq. 6.31.

Table 6.7 Calculation of Optimum Proportion of Coarse Aggregate in the Combined Aggregate by Eq. 6.30

1	2[a]	3[b]	4[c]	5[d]	6[e]
				Optimum **n**	
Reference Number	K (cm)	$\sigma_a - \sigma_b$	$f'(n)$	Theoretical	Empirical
101	0.0034	124	0.42	0.72	0.64
106	0.0013	124	0.16	0.64	0.62
111	0.0007	124	0.09	0.60	0.60
116	0.0003	124	0.04	0.50	0.57

[a] Data are from Table 6.6.
[b] $\sigma_a = 129$ cm²/cm³, and $\sigma_b = 5$ cm²/cm³, approximately, as computed from surface moduli.
[c] The slope of the tangent to the voids-ratio curve of the aggregates by Eq. 6.30.
[d] Optimum proportion of the coarse aggregate in the combined aggregate, by Eq. 6.30, and corresponding tangent points in Fig. 6.11.
[e] Optimum proportion established by trial.

the exact shapes of the curves, the voids-ratio curve for the aggregate in one case, and the water-ratio curve of the concrete mixtures in the other.

4.4 *Questionable Aspects of the Theory*

From the practical point of view, the procedure described above seems to be somewhat more complicated than can be justified by the degree of accuracy attained. To apply this method, it is necessary to have established an empirical relationship between the consistency factor and the consistency for each of the water-cement ratios that are to be considered and for each of the aggregates. It is necessary also to establish by experiment a diagram such as Figure 6.11 for each kind of the fine and coarse aggregate combinations to be considered. Also, as for any method, we must select the cement content, or the water-cement ratio of the paste according to specifications or other pertinent considerations. Thus, experiments must be made with the aggregates alone to obtain a basis for calculating, or graphically estimating, the optimum ratio of coarse to fine aggregate, and other experiments must be made with concrete mixtures to establish the consistency factor, or factors. It is not clear that the result is in any way superior to that obtained by systematic experiments made directly on concrete mixtures.

Another limitation is that this theory, at least as applied by Kennedy, does not point to the deficiency of fine material common to mixtures in the *BF* and *FC* categories. Kennedy pointed out that, "a well graded aggregate is one that combines the primary requirement of low percentage of voids with the secondary requirement of low surface area." This is true for mixtures at point *B* and to the left of it, but in leaner mixtures material of relatively high specific surface area can be incorporated advantageously.

It has been pointed out that many aggregates contain a considerable amount of material small enough to pass the No. 100 sieve, yet in computing the surface area of the aggregate from its sieve analysis as that was done by Kennedy, such material is ignored. Kennedy defended this omission on the ground that the only surface area to be considered is that of particles large enough to bear a coating of paste; in concrete a cement particle, for example, can hardly be thought of as being coated with cement paste. If we accept this point of view, it seems that we should expect the volume of the material in the aggregate that passes the No. 100 sieve (if that is the limit for coatable particles) to be added to the paste when we are computing the consistency factor, but in fact the fine material was considered to contribute to the solid-volume but not to the surface area of the aggregate. These considerations are of course related to the subject of optimum paste volume and optimum paste consistency discussed in Chapter 4, Section 8.

Since mixtures are not considered workable unless the paste volume exceeds the volume of voids in the compacted aggregate, it was inferred that workability is due only to the excess paste; Kennedy speaks of the lubricating film around each particle. This is a questionable concept. If the mixture is plastic, it is so because the aggregate is somewhat dispersed by the interstitial material, but there is no reason to assume that the deformations in the matrix material incident to the deformation of the concrete mass as a whole is confined to a boundary layer just equal to half the average distance between points of near contact among the aggregate particles. On the contrary, it seems reasonable to suppose that particle-size grading is ordinarily such that the difference between minimum and maximum clearances between particles is kept within certain bounds such that it is not too far fetched to think of the paste as having fairly uniform thickness, the layers being "dented" at the points of near contact.

5. WATER-REQUIREMENT FORMULAS

A major shortcoming of the early studies pertaining to ideal gradings was that they did not deal expressly with plasticity, or with the amounts of water and air required to transform an aggregation of dry particles into a plastic mass under the action of a concrete mixer. However, various studies of this question have been carried out with results reported in terms of water-requirement formulas. Some of these formulas were predicated on the use of ideal gradings; others were not. We now examine some of the results of these studies.

5.1 The Edwards-Young Formula

It seems logical that the surface area of materials should be considered in connection with the formulation of concrete mixtures, especially with respect to the quantity of water required to make the mixture plastic. In a 1918 paper Edwards [14] announced that "a natural solution" to the problem of concrete proportioning had presented itself; it appeared to him this way:

"The strength of mortar is primarily dependent upon the character of the bond existing between the individual particles of sand aggregate. Upon the total surface area of these particles depends the quantity of cementing material.

"The amount of water required to produce a 'normal' uniform consistency of mortar is a function of the cement and of the surface area of the particles of sand aggregate to be wetted."

Edwards carried out laboratory tests which seemed to him, but not to all others, to confirm his premises.

On the basis of the second of the statements given above, Young [15] stated that ". . . the quantity of water sufficient to give a workable consistency depends on the quantity required to reduce the cement to a paste, plus the quantity necessary to wet the surface of the particles of aggregate." The following is a formula expressing this idea:

$$wv_w = \left(\frac{wv_w}{cv_c^o}\right)_{NC} cv_c^o + k\sigma_{ab}(av_a + bv_b) \tag{6.32}$$

where k is defined as the quantity of water required to wet a unit area of aggregate surface. Young's original expression was in terms of surface modulus. The subscript NC signifies that the water-cement ratio is that of neat cement at normal consistency.

If we divide both sides by the total volume of solids, we obtain, in terms of symbols defined in preceding chapters,

$$u_w = A_w(1 - x) + k\sigma_{ab}x \tag{6.33}$$

Since $u_w/(1 - x) = wv_w/cv_c^o$ and $x/(1 - x) = M$, we can also write

$$\frac{wv_w}{cv_c^o} = A_w + k\sigma_{ab}M \tag{6.34}$$

These equations are identical with Eqs. 3.33 and 3.35 if $k\sigma_{ab} = D$. Thus it is clear that the premise adopted by Edwards and Young leads to the water-requirement equation for basic consistency of mixtures in the *AB* category, as is discussed in Chapter 4. It is equally clear that it is not applicable to mixtures in the other categories.

5.1.1 Experiments of Shacklock and Walker. The premise of Edwards and Young has been revived in recent years, giving rise to several papers. That of Shacklock and Walker [16] is especially noteworthy, for it is based on measurements of the specific surface area of aggregates using the permeability method, as is discussed in Chapter 1, Section 5. They tested the premise by careful experiments and found that the amount of water required by a given mix for a given consistency as measured by the compacting factor was directly proportional to the specific surface area of the aggregate for some mixes but not so for others.

Some of their data were replotted by the present author in the form of water-ratio diagrams; four mixes for each aggregate, and hence for each specific surface area, were represented. It was found that with each of the aggregates the richest mix was in the *AB* category; the second richest was either in that category, or very close to point *B* in the *BF* category (one exception); the other two were clearly in the *BF* category. Thus about half the mixes were in the *AB* category, or virtually so, and these conformed to the premise that the increased water requirement due to the aggregate was proportional to the surface area only. However, the results from the rest of the mixes showed that specific surface area was *not* the controlling factor. Thus the experiments of Shacklock and Walker are in agreement with the analysis presented in Chapters 3 and 4.

5.1.2 Comments on the particle-wetting theory. The premise stated by Young [15] is a recurrent one and thus deserves consideration. It seems logical, but actually it cannot be advanced as a deduction from experimental observation. As we have seen in Chapter 4, such a hypothesis, applied as a general rule, is untenable. This fact was recognized as long ago as 1919. Abrams [17] contended that the amount of water required to wet the aggregate is negligible relative to the amount by which the water content is increased when the aggregate content is increased. He cited work by Pettijohn [18] who measured the quantities of water and other liquids which would cause particles to cling to a glass vessel. For standard Ottawa sand, for example, the amount was reported as 3.74×10^{-4} cm^3/g or about 0.001 g/cm^3 solid volume. In Chapter 3, Fig. 3.12, we find that for basic consistency of mixtures made with standard Ottawa sand, the value of D is 0.11, which means that the water requirement of standard Ottawa sand in mixtures in the *AB* category is 0.11 g/cm^3 as compared with the 0.001 found by Pettijohn. Furthermore, in mixtures of the *BF* category, the apparent water-requirement factor of the same aggregate is different, and still larger.

Actually, the concept of wetting a surface can be a very vague one. The quantity measured by Pettijohn was approximately the volume of what the surface physicist calls a "duplex film." A duplex film may be thought of as a sheet of water held between the solid surface and the liquid surface where liquid-surface tension is manifested. (See Chapter 3, Section 2.2.) As such, the quantity will depend not only on the apparent surface area of the solid particle, but also on the texture of the surface, the rougher the surface the greater the amount of water held in the "pockets" due to the roughness and the tendency of the liquid-surface film to bridge over such pockets. By a more strict definition of the term, one could say the amount of water required to wet a solid surface is that required to form a film of adsorbed water at the solid surface when the solid is immersed in water. Such a film is perhaps 10 molecules thick; it would weigh about 2.6×10^{-7} g/cm^2 of surface. Since the surface area of a standard-sand particle (diameter = 0.07 cm) is about 8.6 cm^2/cm^3 the amount required to wet it is about $2.6 \times 10^{-7} \times 8.6 = 2.3 \times 10^{-6}$ g/cm^3 solid volume, or less than one-millionth of a gram per gram of sand. Some such amount is the quantity virtually immobilized by the aggregate surfaces. In plastic concrete duplex films exist only around air bubbles. Moreover, as brought out above, even if there were duplex films around the aggregate particles in concrete, they would not account for the actual water requirement factors observed.

We do observe that for mixtures in the *AB* category, the water-requirement factor is larger the higher the specific surface area of the aggregate, but we find reason to attribute the water increase either to the effect of the added particles on the structure of the matrix, or to rheological effects, or to both.

5.2 Abrams' Formula

Abrams' water-requirement formula is presented and discussed in Chapter 5, Section 4.2, in connection with the fineness-modulus method of formulating mixtures. It is based on the premise that the water requirement of a concrete mixture can be expressed as the sum of the water requirement of the cement and that of the aggregate, the latter being proportional to the amount of aggregate, and the proportionality constant being an inverse function of the fineness modulus. It was shown that this formula does not conform to the experimental data in any category, but it shows the least discrepancy for mixtures in the *BF* category.

5.3 Singh's Formula

On the basis of experimental work done at the Building Research Station (England) Singh [19,20] produced the formula below:

$$\frac{w}{c} = \frac{K_1\left(\frac{w}{c}\right)_{NC} + K_2(a + b/c)S_o}{\frac{1}{CF} - K_3(a + b/c)} \tag{6.35}$$

where the K's are empirical constants; $(w/c)_{NC}$ is the water-cement ratio of neat cement paste at normal consistency as determined by the British method; S_o is the specific surface area of the aggregate, area per unit weight; CF stands for the compacting factor by the Glanville-Collins test; all other symbols are as previously defined.

The rationale of Eq. 6.35 was not fully explained. The formula is based on the assumption that the water requirement of a mixture is determined by the water requirement of the cement and by the quantity and specific surface area of the aggregate, but the reason for the particular form of the formula is not clear to the present author.

5.4 Bolomey's Formula

In 1925 Bolomey published a water-requirement formula based on the premise that the water requirement of a concrete mixture in excess of that of the cement is the sum of the specific water requirements of the individual size fractions of the aggregate. The specific water requirement of each size fraction was taken to be an inverse function of the geometric mean size of the group. Thus,

$$u_w = \sum \frac{k_1 x_i}{(d_i d_{i+1})^{1/3}} \tag{6.36}$$

where x_i is the volume fraction of the ith size group of the total volume of solid material; k_1 is an empirical constant the value of which depends on the desired consistency and on the shape of the particles.

It was stipulated that Eq. 6.36 should not be used for calculating the water requirements of particles smaller than 0.2 mm, which is about the opening of the No. 70 sieve. However for our present purpose it will be convenient to let the lower limit be the No. 100 sieve and assume that cement constitutes all the finer material.

For the water requirement of cement, Bolomey allowed the water requirement for normal consistency, and thus in our terminology, it would be $(A_w)_b$. Hence we may write,

$$u_w = (A_w)_b(1 - x) + \sum_{i=2}^{i} \frac{k_1 x_i}{(d_i d_{i+1})^{1/3}} \tag{6.37}$$

where the subscript b indicates basic consistency.

Equation 6.37 looks like a water-requirement equation for mixtures in the *AB* category, with the second term on the right standing for Dx, but there are certain differences. The water requirement of the cement is indicated to be constant at the value for normal consistency, whereas in the corresponding equation of Chapter 4, A_w varies with consistency. The second term varies with consistency (k_1 varies), as does D.

In Chapter 3 it was indicated that the water requirement of the aggregate when used in mixtures of the *AB* category is an inverse function of the mean specific surface diameter; indeed it was indicated that the water requirement varies inversely with the square root of the mean specific-surface diameter, whereas in Eq. 6.37 it is indicated to vary inversely with the cube root of the square of the geometric-mean diameter.

As we saw in Section 2.3, Bolomey's method calls for using a certain minimum amount of cement, or a combination of a filler material with a lesser amount of cement. We saw also that the minimum amount was such as to give a mix in the *BF* category. It appears, therefore, that the Bolomey water-requirement formula is related to the water-requirement equations developed in Chapter 4 in about the same way as the Abrams formula discussed in Chapter 4, Section 4.2. That is to say, for a given aggregate, the plot of Eq. 6.37 (u_w versus x) would be one straight line from $(A_w)_b$ but not passing through point B, whereas the relationship for mixtures having basic consistency is approximated more closely by one straight line from $(A_w)_b$ to point B and another straight line from the origin through point B to point F, as is shown in Fig. 5.2.

5.5 Plum's Formula

Plum adopted as a working hypothesis the following expression:

$$u_w = k \sum w_i p_i + f(r, d_{\max}) \tag{6.38}$$

where w_i is the water-requirement factor of the ith size-group; p_i is the volume fraction of that group, of the total solid material, and k is a constant; as before, r is the grading ratio, and d_{max} is the maximum size in millimeters. The first term is a statement basically like that of Bolomey; that is, the water requirement of the mixture is the sum of the water requirements of the individual size fractions. The second term was inserted as a correction, on the basis of discrepancies between the first assumption and experimental results.

To test the hypothesis, Plum carried out systematic experiments using various values of d_{max} and aggregate gradings based on Eq. 6.18 for different grading ratios. Analysis of the results led to the conclusion that the first term of Eq. 6.38 accounts for that part of the total water requirement that is due to the surface area of the solid material. Since practically all the surface area is that of the cement, this meant that that term represented the water requirement of the cement or of cement plus filler material when such material is present. The rest of the water requirement was found, by curve fitting, to be a function of maximum size and cement-plus-filler content.

Plum's water-requirement equation for basic water content, in our terminology as far as possible, is as follows:

$$u_w = A_w(1 - x) + \frac{k}{1 + (10 + 0.9d_{max})(1 - x)} \tag{6.39}$$

In this equation $(1 - x)$ is the solid-volume fraction of material that will pass through a No. 100 sieve, which in Plum's terminology is p_9; k is an empirical constant, and for the materials used, its value was 0.40. Implicit in this equation is the condition that the grading ratio r has the optimum value for the given value of d_{max}.

In terms of the water-cement ratio and mix, in our terminology, Plum's equation can be written as follows:

$$\frac{wv_w}{cv_c} = A_w + \frac{k(1 + M)^2}{(1 + M) + K} \tag{6.40}$$

where k and K are constants, the latter being $(10 + 0.9d_{max})$.

Equation 6.40 describes a nonlinear relationship for a given aggregate beginning at point A_w. It was derived from data pertaining to mixtures in which the aggregate gradings conformed to Plum's theoretical grading, the grading ratio being the optimum value for a given maximum size. To test its applicability to ordinary mixtures in which the aggregates do not comply closely with the theoretical grading, the author used the data of Table 6.6. Each value in column 4 was divided by 1.045, according to Table 4.1, so as to find the values corresponding to a 1.7-in. slump, basic consistency. Then, using the adjusted value of the water-cement ratio and the corresponding value of M for reference No. 111, the constant k of Eq. 6.40 was evaluated and found to be 0.471. The maximum size being 38 mm, and with A_w at basic consistency, assumed to be 0.73, the formula turned out to be as follows:

$$\frac{wv_w}{cv_c^o} = 0.73 + \frac{0.471(1 + M)^2}{(1 + M) + 44.2}$$

The curve for this formula is shown in Fig. 6.12 along with the four adjusted data points from Table 6.6. Also shown are straight lines corresponding to the formulas developed in Chapter 4.

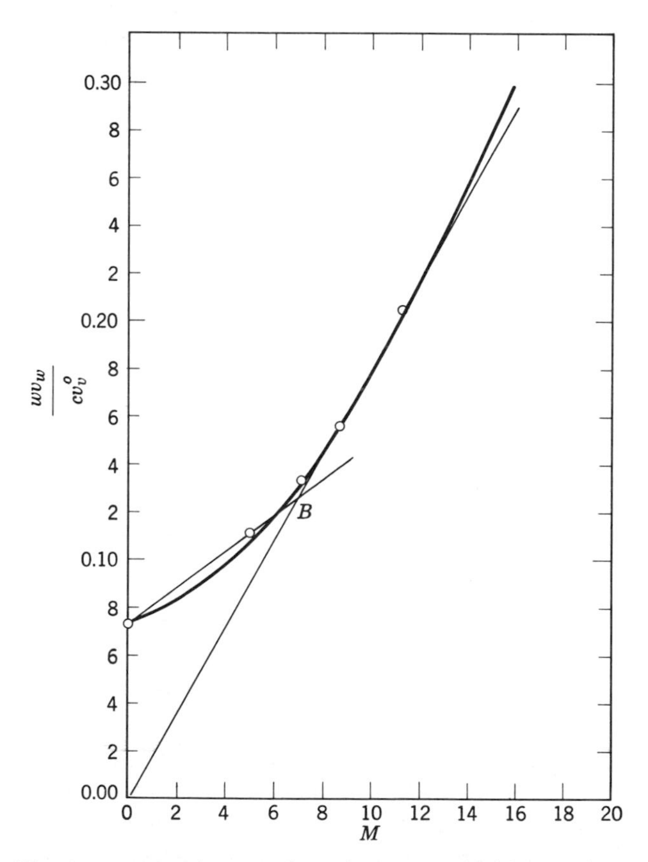

Fig. 6.12. Plum's water-requirement formula compared with experimental data. The maximum size of the aggregate is 1½ in. (Data are from Table 6.6, adjusted for a slump of 1.7 in.)

We see that by choosing the value of k that gives agreement with the data for one mix, a curve was obtained which approximated the other three points closely, the largest discrepancy appearing with the one mix in the AB category. This result is not particularly surprising since Plum's formula as well as those in Chapter 4 are essentially empirical. It is interesting, however, that mixes in which the aggregate grading was varied only by adjusting the percentage of sand to the optimum value, as determined by a trial method already described, conform closely to a formula derived from data pertaining to semitheoretically graded aggregate in which the grading ratios for the whole aggregate were adjusted to optimum values by trial.

In Eq. 6.39 we see that when $x = 1.0$, $u_w = k$, which is to say that the value of k is apparently the water ratio for the aggregate alone. It should therefore correspond to point C of the water-ratio diagram as shown for example in Fig. 4.10a. But, the value of C given in Table 6.6 is about 0.26, whereas from the same table we found k to be 0.471. It seems, therefore, that Plum's equation may be accurate for mixtures in the BF category, but is inaccurate in the FC range.

5.6 *Murdock's Formula*

In 1960 Murdock [22] proposed the following water-requirement formula, based on his grading index (Chapter 1, Section 4.2):

$$\frac{w}{c} = \phi(CF)f_s f_a(M - 2) + 0.25$$

and (6.41)

$$\phi(CF) = 0.135CF - 0.067$$

where f_s is Murdock's grading index, CF is the Glanville-Collins compacting factor as a measure of consistency, and f_a is Murdock's angularity index, a modification of Shergold's. (See Chapter 1, Section 7.2.) The term $(M - 2)$ is based on the conclusion that ". . . when the aggregate/cement ratio is reduced to 2, the effects of grading and angularity become negligible." According to this formula, when $M = 2$, $w/c = 0.25$ regardless of the characteristics of the aggregate or the consistency; obviously the formula was not intended for use with very rich mixtures.

Actually, as with various others, this formula is intrinsically restricted to a narrow range of mixtures. It provides only one factor for a given aggregate, whereas as shown in Chapters 3 and 4, three different factors are required for the whole range, or, as is shown above, the formula must be nonlinear.

5.7 Popovics' Formula

In 1966 Popovics [23] published the following nonlinear water-requirement formula for basic consistency, in terms of fineness modulus, cement content, and constants to be established empirically.

$$\left(\frac{w}{c}\right)_b = \frac{W_o}{c - c_o}(FM_o - FM)^2 + a + W_1 \tag{6.42}$$

where W_o, c_o, $(FM)_o$, a, and W_1 are constants to be established for a given set of materials; w and c are weights of cement and water. The subscript b indicates basic consistency. To find the water requirement for any other consistency, the following relationship was offered:

$$\left(\frac{w}{c}\right) = \left(\frac{Y}{Y_b}\right)^{1/n}\left(\frac{w}{c}\right)_b \tag{6.43}$$

where Y is the consistency value for a given method, and Y_b is the value at basic consistency; n is a constant characteristic of the measuring device; for the slump test for example $n = 10$.

Popovics stipulated that Eq. 6.42 is valid for cement contents ranging from three and a half to nine U.S. sacks of cement per cubic yard with continuously graded aggregate having a maximum size of about 1 in., and for a limited range of consistency. For consistency as measured by the slump test, the formula for basic consistency is as follows:

$$\left(\frac{W}{C}\right)_b = \frac{1.8}{C - 0.9}(5.9 - FM)^2 + 14.6 + 1.2 \tag{6.44}$$

where $(W/C)_b$ is the water-cement ratio in gallons per sack (U.S.) at basic consistency, and C is the cement content in sacks per cubic yard. To use this formula, the value of FM for the desired cement content might be taken from Table 5.1 of Chapter 5, using the line for the 1-in. maximum size.

Popovics' formulas are not based on the assumption that the water content is constant for a given aggregate grading and consistency; on the contrary, they agree with the observation that the water content is lowest in intermediate mixtures.

5.8 Equations of Chapter 4

Most of the foregoing water-requirement equations express some theoretical premise about the water requirements of the separate ingredients, individual particles or size groups sometimes being considered to be separate ingredients. On the other hand, the formulas developed in Chapter

4 are wholly empirical, being the outgrowth of a systematic graphical analysis of the relationships between various mixtures at various consistencies. None of the coefficients appear as functions of grading, specific surface area, voids content, or maximum size. The equations are reproduced here for convenient comparison.

For mixtures in the *AB* category

$$\frac{w}{c} = \frac{v_c^o}{v_w}\left[\frac{k_c}{Q} + \left(\frac{m_{AB}}{Q}\right) \cdot M\right] \tag{6.45}$$

For mixtures in the *BF* category

$$\frac{w}{c} = \frac{v_c^o}{v_w} \cdot \left[\left(\frac{m_{BF}}{Q}\right) M\right] \tag{6.46}$$

For mixtures in the *FC* category

$$\frac{w}{c} = \frac{v_c^o}{v_w} \cdot \frac{Q}{Q_b}\,(CM - 1) \tag{6.47}$$

where Q is the consistency factor, that is,

$$Q = \log\left(\frac{Y}{Y_o}\right)$$

and Q/Q_b is the relative consistency, Q_b being the consistency factor at basic consistency.

It will be recalled that k_c is the water-requirement factor of the cement and k_c/Q is the water-cement ratio of the paste at a given consistency. Likewise, m_{AB} is the water-requirement factor of the aggregate when it is used in mixtures of the indicated category, and m_{AB}/Q is the increase in water cement per unit increase in M at a given consistency; m_{BF} appears as a coefficient of M only, and it is to be regarded as the water-requirement coefficient of the whole mixture; in the *FC* category the water requirement is a function of the voids ratio of the aggregate and the relative consistency, that is, C and Q/Q_b.

6. THE STATE OF THE ART

In this chapter we have reviewed some of the contributions to concrete technology based on the assumption that some particular particle-size gradation, or type of gradation, of the granular ingredients of concrete is superior to any other. The hypothesis sometimes appeared as a stated conviction, perhaps strengthened by mankind's predilection to seek order in nature. It is conceivable that some hypothesis of this kind would have been successful if the packing of fine particles depended on geometrical factors only, but, as we have seen repeatedly, this is far from the case, and theories of ideal total grading break down at this crucial point. The best that can be claimed for "ideal" gradings is that they may equal the best that has been produced by nontheoretical procedures, but in this chapter we have seen evidence that such a claim is questionable. Also, some of those who espouse gap gradings believe them to be superior to any continuous grading.

The case for gap gradings is by no means clearly established. Too often the data given to support a conclusion favoring (or sometimes not favoring) gap grading are based on tests in which one coarse aggregate grading is simply substituted for another, with no demonstration that the ratio of fine to coarse aggregate was at the optimum point for either of the combinations compared. In some cases improvement due to omitting a certain size-group has been reported, but without an indication as to whether or not the particles eliminated were more angular than the rest.

In any case, the indications of experiments carried out to test a hypothesis may depend to a considerable degree on the design of the experiment. Sometimes, to test a particular point, a single mixture will be chosen. For example, to demonstrate that the fineness modulus is a reliable basis for establishing the ratio of fine to coarse when various sands are used with a given coarse aggregate, one investigator used a mixture in which the water-cement ratio of the paste was about 0.57 by weight. He obtained only small variations from the average water requirement, but since his mixture was definitely in the *BF* category, it is safe to say that if a mixture in the *AB* category had been chosen, the results might have been less favorable.

It will be remembered that for mixtures in the *AB* category, the specific surface area of the aggregate is one of the two main determinants of water requirement and that the fineness modulus is not closely related to specific surface area. Therefore successful use of the fineness modulus for mixtures in the *AB* category would seem to be contingent on the existence of a close correlation between fineness modulus and surface area for the particular combinations of fine and coarse aggregates used in the experiments. Gradings established on a theoretical basis would be likely to show such a correlation, but there is no generally reliable correlation. Thus, for a "spot check" of the kind just discussed, the choice of mixture may have an important bearing on the outcome, and if different experimenters make different choices, the results may not agree.

More commonly, test programs comprise two or even three categories of mixtures, but because the categorical differences brought out in Chapter 4 have not been recognized, attempts to generalize the results have produced some sort of average figures weighted according to the distribution of mixtures among the categories. For example, attempts to express water requirement in terms of the specific surface area of the aggregate have been strongly influenced that way. If the mixtures are all in the *AB* category, or predominently so, the coefficient of correlation will be satisfactorily high, but if a considerable fraction of the data are obtained from mixtures in the *BF* category, correlation will appear less satisfactory, even quite unsatisfactory. In any case, the numerical value of the coefficient will depend on the makeup of the data as regards the different categories.

It is not surprising, therefore, that investigators working in different places only occasionally come up with complementary results; we have come to expect similarity at best. Results found in European investigations are likely to differ from those in the United States because in Europe it seems to be habitual to use relatively more concrete in the *FC* category, at least in laboratory investigations.

It seems that the various theoretical studies have contributed significantly to our understanding of concrete mixtures, but they have not served to define practical limits for useful gradings of fine and coarse aggregates. The limits that now appear in standard specifications are broad, and according to the experience of some, they might be even broader without necessarily sacrificing economy or quality. If there is any unnecessary restriction, it is possibly a vestigial influence carried along from the days when it was common to use arbitrary proportions such as 1:2:4 or 1:3:5. For such practice, the present tolerances are too broad. The looseness of present aggregate-grading specifications is justifiable to the extent that the various possible gradings are used in proportions that are optimum for the particular materials brought together and the desired characteristics of the concrete.

It now appears that by systematic trials, suitable proportions for any given set of materials can be established more satisfactorily than by any other method yet devised. Some of the methods we have reviewed in Chapter 5 are at least desirable means of systematizing mix design, and some offer a useful means of using the first mix established by trial as a base for estimating equally satisfactory proportions for other mixtures of the same materials; this is notably so for the fineness modulus method. However, a question remains as to the criteria to use for assessing the result of applying any procedure of mix design. Present criteria are not very demanding, considering the possibilities at hand for producing mixtures having optimum properties for given situations.

To meet conventional requirements a mixture must have a suitable consistency and at least a minimum of plasticity. Furthermore, the water-cement ratio must be compatible with a specified strength requirement, or must comply with the limit fixed by consideration of impermeability and durability. Such requirements can be met with a wide variety of materials and mixtures, and although the specified characteristics may be obtained within reasonable tolerances, other important characteristics may differ widely among the various combinations.

More rigorous standards would require that we not only meet conventional requirements but also design concrete mixtures to obtain the highest possible solids content regardless of the composition of the paste and to have the same plasticity as well as same consistency at any cement content. Moreover, our selection of materials may have to be considerably more restrictive than is necessary when we are meeting only conventional requirements; it may be necessary to select the aggregate on the basis of the average mechanical compressibility of its particles and the specific strength of adhesion of hardened paste to particle surface as a means of reducing drying shrinkage and of lessening the probability of crack development due to shrinkage. In fact, the usually specified strength requirements are so easily met that a policy of design based *primarily* on control of shrinkage and cracking could be advocated with considerable justification. In general, proper design on that basis calls for a cement paste having a low water-cement ratio for high strength (particularly tensile strength) and for low paste shrinkage, and for a high aggregate content to restrain paste shrinkage. When such a requirement is met as well as possible, the chances are good that strength requirements will have been exceeded and some of the requirements for durability also.

However concrete is adaptable to many situations, and various adaptations require compromises among incompatible requirements. For example, to obtain lightweight concrete, we accept the higher compressibility of porous aggregate particles. Nevertheless, for many applications selection of materials and proportions could well be made to obtain an optimum combination of tensile strength, volume stability, and paste-impermeability, rather than on the basis of strength and durability alone; when compromises must be made for specific purposes, the designer can be aware of the price paid for a specific property in terms of degradation of another property.

Recognition of the three categories of concrete mixtures and the simple water-requirement equations for each class opens the way to a more critical evaluation of any given concrete mix design whatever the procedure of design. The water-ratio diagram, or the voids-ratio diagram, and the analytical methods presented in preceding chapters should make it easy to determine whether or not the materials at hand have physical characteristics as favorable to optimum results as another material might have.

The possibility of expressing the interrelationship of mix, consistency, and water-cement ratio in simple mathematical expressions might encourage the practice of designing concrete for an optimum combination of properties. We see the future development of electronic computer programs for the simultaneous weighting of various properties with respect to specific uses, a development becoming practicable as the fundamental factors controlling the properties and behavior of concrete in the fresh and hardened states become more fully understood and reduced to terms that can be expressed mathematically.

If we move toward designing on the basis of optimum paste consistency, we may expect further development in cement manufacture such that the rheological properties of cement paste of standard solids content are kept within specified tolerances. Such a development could involve the controlled use of surfactants, and it might also lead to the production of cheaper "diluted" cements that produce a paste having standard solids content and rheological characteristics. Diluted pastes would be used where the high strength of point-*B* concrete is not needed or where the somewhat high cement content is undesirable because of an excessive temperature rise during the early stages of hardening. In any case, the mixture would meet the compositional requirements associated with point *B*: optimum paste consistency for the given aggregate, which also means little or no particle interference between the particles in the matrix and the particles of the aggregate.

In this connection the possibility of using air bubbles as a cement diluent should not be overlooked. Although "pneumatic particles" reduce strength more than do mineral particles when they are used as a diluent, they do not affect volume stability adversely; they are here considered for situations where strength would be unnecessarily high without cement dilution. Properly produced and controlled (see Chapters 7 and 8) air bubbles enable us to use a point-*B* composition for the cement-paste part of the aggregate matrix, thereby obtaining a matrix that undergoes relatively low shrinkage during drying. Bubbles have the very great advantage of being virtually costless.

Along with any developments in the production and control of matrix material such as those described above would be the development of aggregate specifications especially for point-*B* mixtures. This might involve defining all material smaller than No. 100 as *matrix material,* and fixing the tolerable limit from zero per cent upward (of the aggregate), depending on the extent to which the cement is to be diluted; it would involve also the specification of acceptable kinds of such material, and the variation in quantity that could be tolerated, the latter requirement leading toward separation of the production and control of matrix material from the production and control of aggregate.

In present-day practice, by and large, we tacitly accept mixtures in which there is particle interference between matrix particles and the aggregate. In other words we accept paste consistencies thinner than the optimum. We make an important exception for concrete for highway paving. When such concrete meets state and federal specifications its composition is likely to be close to point *B*. It seems significant that such a composition evolved from experience.

REFERENCES

[1] Fuller, W. B., and S. E. Thompson, "The Laws of Proportioning Concrete," *Trans. ASCE,* **59,** 67–143, 1907.

[2] Plum, N. M., "The Predetermination of Water Requirement and Optimum Grading of Concrete," Building Research Studies No. 3. The Danish National Institute of Building Research, Copenhagen, 1950.

[3] Andreasen, A. H. M., and J. Anderson, "The Relation of Grading to Interstitial Voids in Loosely Granular Products (With Some Experiments)," *Kolloid-Z.,* **49,** 217–228, 1929.

[4] Popovics, S., "Comparison of Several Methods of Evaluating Aggregate Gradings," RILEM Bulletin No. 17, New Series 13-21, 1962.

[5] Bolomey J., "The Grading of Aggregate and Its Influence on the Characteristics of Concrete," *Revue Mater. Construct. Trav. Publ., Edition C,* 147-149, 1947.

[6] Weymouth, C. A. G., "Effect of Particle Interference in Mortars and Concrete," *Rock Prod.,* **36**(2), 26–30, 1933.

[7] Weymouth, C. A. G., "A Study of Fine Aggregate in Freshly Mixed Mortars and Concretes," *Proc. ASTM,* **38,** Part II, 354–372, 1938.

[8] Marquardsen, "Design of Concrete Mixtures," *Concrete,* 31, August 1929.

[9] Smith, A. A., "Proportioning—Its Fundamental Principles," *Concrete,* **44,** 7, 1936.

[10] Katoh, J., *The Constituents of Concrete,* Review of 23rd Annual Meeting of the Association of Japanese Portland Cement Engineers, 62, November 1936.

[11] Kennedy, C. T., "The Design of Concrete Mixtures," *Proc. ACI,* **36,** 373–400, 1940.

[12] Abrams, D. A., "Design of Concrete Mixtures," Bulletin 1, Structural Materials Research Laboratory, Lewis Institute, Chicago 1918.

[13] Powers, T. C., "Studies of Workability of Concrete," *Proc. ACI,* **28,** 419–448, 1932.

[14] Edwards, L. N., "Proportioning the Materials of Mortar and Concretes by Surface Areas of Aggregates," *Proc. ASTM,* **18,** Part II, 235–283, 1918.

[15] Young, R. B., "Some Theoretical Studies of Proportioning by the Method of the Surface Area of Aggregates," *Proc. ASTM,* **19,** Part II, 444–457, 1919.

[16] Shacklock, B. W., and W. R. Walker, "The Specific Surface of Concrete Aggregates and Its Relation to the Workability of Concrete," Research Report No. 4, Cement and Concrete Association, 1958.

[17] Abrams, D. A., Discussion of a paper by R. B. Young, *Proc. ASTM,* **19,** Part II, 480, 1919.

[18] Pettijohn, E., "Measurement of the Thickness of Film Formed on Glass and Sand," *J. Am. Chem. Soc.,* **41,** 477, 1919.

[19] Singh, B. C., "Effect of the Specific Surface of Aggregate on the Consistency of Concrete," *Proc. ACI,* **53,** 989–997, 1957.

[20] Singh, B. G., "Specific Surface of Aggregates as Applied to Mix Proportioning," *Proc. ACI,* **55,** 893–901, 1959.

[21] Bolomey, J., "Determination of the Compressive Strength of Mortars and Concretes," *Bull. Tech. Suisse Romande* (11, 14, 15, and 17), 1925.

[22] Murdock, L. J., "The Workability of Concrete," *Magazine of Concrete Research,* **12**(36), 135–144, 1960.

[23] Popovics, S., "Concrete Consistency and Its Prediction," RILEM Bulletin No. 31, 235–252, 1966.

7

Mixtures Containing Intentionally Entrained Air

1. INTRODUCTION*

1.1 Purpose

Much concrete is made in which the total air content of the plastic mixture is kept as close as possible to a preselected amount, such as $5\frac{1}{2}$ per cent, of the overall volume. The practice of deliberately entraining air bubbles in concrete was introduced in the mid-1930's, and it has since become one of the most important developments in concrete technology. Entrainment of air is accomplished by use of a suitable agent which stabilizes bubbles formed from some of the air incorporated during the mixing process.

The main purpose of introducing air is usually to protect hardened concrete from frost action. Sometimes it is done because of the beneficial effect on the workability of plastic concrete having a relatively low cement content. The latter purpose is paramount in making concretes suitable for gravity dams.

The practice of using entrained air has in general been highly successful, but there are exceptions. In these exceptional cases there is often evidence of a lack of understanding of the nature of the air-entrainment

* Some readers may find it helpful to study Chapter 9 before this one.

process and the identity of the essential factors to be controlled. We therefore examine this subject in detail.

1.2 Definitions

Since the discussion pertains to plastic mixtures, it deals with air in the form of bubbles. (See Chapter 3, Section 4.1.) In general, the term "bubble" applies when the water content of the mixture is such that the water is not in a state of tension. This condition can be recognized by appearance; if fresh concrete emerging from a mixer has a relatively dark hue and shows evidence of forming cohesive lumps or segments that do not readily coalesce with others, cohesiveness may be attributed in part to hydrostatic tension. On the other hand, if the hue is relatively light, if the mass pours out so as to form one body, and if level surfaces appear bright and shiny, signifying a flat liquid surface, there is no hydrostatic tension. Sometimes a mixture shows hydrostatic tension at first, but the evidence disappears when the mixture is puddled or vibrated. In the absence of hydrostatic tension the force due to surface tension and hydrostatic pressure which tends to collapse the bubble is balanced by compressed air in the bubble. The bubbles found in plastic concrete are not necessarily spherical, but most are nearly spherical.

The air content of concrete containing intentionally entrained air is considered by some observers to comprise two categories: entrained air and entrapped air. Entrapped air is usually taken to be that which is found in a given mixture when an air-entraining agent is not used, and accordingly, entrained air is said to be that in excess of entrapped air, present because of the air-entraining agent. Although such a distinction seems useful, it is probably not accurate. All air that is an ingredient of the mixture (the distinction excludes air in voids produced by segregation of coarse material or by incomplete filling of forms by a stiff mixture) is held by the same mechanisms, and when an air-entraining agent is present we may safely assume that all the bubbles carry films due to the agent, whether the bubbles appear spherical or not. Therefore, in this discussion the term "intentionally entrained air" only indicates that the air content has been increased to a preselected value by means of a suitable agent. As we shall see below, some of the "entrained" air seems to be held by entrapment.

In this chapter we are principally concerned with the properties of *freshly* mixed concrete containing less than 10 per cent air. The manufacture of some lightweight or insulating concrete involves entrainment of large quantities of air or other gas, but such practice will not be dealt with explicitly.

2. METHODS FOR MEASURING AIR CONTENT

2.1 Gravimetric Method

The air content of fresh concrete can be calculated from its measured unit weight and from the weights and densities of its ingredients. A standard method is given in ASTM Designation C 138. The accuracy of the standard method is limited by the recommendation that the solid volume of cement be calculated from its specific gravity and that of the liquid from the density of pure water. Such a practice gives values for air content lower than the actual amount, the inaccuracy being greater the greater the cement content. This source of inaccuracy can practically be eliminated by using the displacement factor of cement in water, also called the apparent density in water. (For further discussion, see Chapter 2.) Even with this improvement, it is not easy to obtain accurate results, mainly because of the difficulty of obtaining sufficiently accurate data for the solid volumes of rock particles, particularly when the aggregate is not mineralogically homogeneous, and especially when the particles are vesicular, as they are in some lightweight materials.

2.2 Direct Method

Another method for determining the air content of freshly mixed concrete is to remove the air from a measured volume of concrete and measure the volume of air directly. The method, originated by Pearson [1] and developed by Menzel [2], is described in ASTM Designation C 173; it involves mixing a volume of concrete with a similar volume of water in a closed container designed to serve as a picnometer. The separate volumes of concrete and water at first fill the container, but after mixing the extra water and concrete by shaking and rolling the container, air in the concrete becomes released and collected in the top, calibrated, part of the container. That part of the total air content that was held in porous rock particles is mostly if not entirely retained by the particles. This is a desirable feature, making the apparatus especially useful for concrete made with lightweight aggregate.

2.3 Pressure Method

Introduced by Klein and Walker [3] in 1946, this method is based on Boyle's law, which for the present purpose may be paraphrased as follows: At a given temperature the volume of a given mass of air varies inversely with the pressure applied to the air, provided that the pressures are not much above 1 atm. Since air is the only appreciably compressible ingredient of concrete, any reduction in the volume of a sample of fresh concrete due to an increase of external pressure may be ascribed to the air in the specimen. By increasing pressure on a sample in a closed container, and measuring the resulting decrease of volume, the quantity of air in the sample can be calculated.

Any air held in the pores within permeable rock particles is included in the calculated amount, and since air so situated is not considered the subject of control procedures, the gross amount is subject to a correction factor. To establish the correction factor, it is necessary to treat a sample of inundated aggregate in the same way as the concrete, thus ascertaining the air content of the particles composing the aggregate. Obviously if the test is to be applied under field conditions, the initial state of the sample of aggregate used for establishing the correction factor should be the same as it is when the concrete is being produced under field conditions. Use of an oven-dried sample, for example, even after prolonged soaking, might contain a different amount of air than does the same material as received from pit or quarry. (Details of the apparatus and test procedure are given in ASTM Designation C231.)

2.4 Linear Traverse Method (Rosiwal Traverse)

Although we are here principally concerned with the air content of fresh concrete, some information about it can be obtained only after the concrete has hardened. However, the air content of fresh concrete may not be the same as that after hardening; certainly this is true if manipulation after the first measurement is made is such as to expel some of the bubbles. When such losses are prevented, however, very little change is to be expected. Data giving contrary indications are questionable in view of the uncertainty about the accuracy of methods, particularly as applied in the early years of the development of air-entrained concrete. (Possible changes in void-size distribution are discussed in Chapter 8.)

One method of measuring the air content of hardened concrete involves cutting the specimen, polishing the cut surface, and measuring the fraction of the total area occupied by sections of air bubbles; it was first applied to concrete by Verbeck [4], but it has been mostly if not entirely replaced by a procedure described by Rosiwal in 1898 [5]. This method is based on the principle that, provided that the bubbles are randomly arranged, a straight line passed through a sufficiently large specimen, or a sufficient number of lines through any given specimen, will encounter a representative sample of the air bubbles, and, of the total length of the lines, that fraction which falls within voids is the same as the volume fraction of air in the sample. (The same is true of other components; the fraction of line traversing paste and the fraction traversing rock

Fig. 7.1. Apparatus for measuring the air content and void characteristics by the linear traverse method. (Portland Cement Association Laboratories.)

are the same as the volume fractions of paste and aggregate.) Application of this principle to the measurement of entrained air was first proposed by Rexford in 1947 [6], and was developed by Brown and Pierson in 1951 [7].

In practice, the sample is sawed into two or more pieces; the surfaces are then made plane, preferably with an automatic lapping machine; then the lines are "drawn," in effect, by a specially constructed traveling stage moving at a controllable rate under a fixed binocular microscope equipped with cross-hairs to indicate the line being traversed. (See Fig. 7.1.) The lengths of the chords that this line makes with the void sections are recorded on an accumulator; the total length of the line and number of voids intercepted are recorded also. (Essential features of the apparatus are described in ASTM Designation C457-60T.)

From the data mentioned above, the air content of a sample can be calculated by means of Eq. 7.1:

$$\frac{\text{A}}{V} = \frac{\Sigma\, \Delta l}{l} \tag{7.1}$$

where A is the volume of air; V is the volume of the sample; Δl is the length of a chord across an individual bubble in the line of traverse; l is the total length of traverse. (Determination of the void-size distribution is discussed in Chapter 8.)

2.5 *Point Count Method*

Another method based on statistical considerations and requiring a finely ground plane cross-section of the specimen, consists, in effect, of placing a rectangular grid on the plane surface and counting each grid intersection that falls within a void section. The air content is equal to the number of such coincidences with voids divided by the total number of grid intersections.

In practice the grid is created optically, point by point, by means of a mechanical stage capable of bilateral stepwise movements, mounted under a fixed binocular microscope. In typical use the stage is moved laterally by equal steps of about 0.05 in. for a distance of at least 5 in., counting the total number of steps, and the number of times the index point in the reticle of one of the eyepieces is seen to be within the boundary of a void section. Such traverses are repeated on parallel lines about 0.2 in. apart until the grid is complete. (Details of this method and the linear traverse method are given in ASTM Designation C457.)

2.6 High-pressure Method

Another method for determining the air content that is applicable to hardened concrete was introduced by Lindsay [8]. Like the pressure method for fresh concrete, it involves compressing the air in an oven-dried and presoaked (for 48 hr) specimen by means of hydraulic pressure, but instead of applying about 10 psi as for fresh concrete, a pressure of 5000 psi is used. After applying correction factors, a value for air content is obtained which, according to Erlin [9], is in good agreement with that given by the pressure method applied to fresh concrete.

On specimens taken from the same cylinders used for the high pressure method, Erlin found the linear traverse values to agree with those given by the high-pressure method for air contents up to about 5 per cent, but at higher air contents the linear traverse method gave relatively low values, the difference increasing with the air content. There is reason to believe that in such cases the discrepancy is due mostly to the difficulty of observing all the bubble sections, although it is possible that some of the fault may be in the high-pressure method.

3. AIR-ENTRAINING AGENTS

The air content of fresh plastic concrete can be caused to increase during mixing by adding an air-entraining agent or by using cement already containing one. Such a material may be a surface-active agent but not necessarily so, as is shown further on. A surface-active material (*surfactant*) is usually composed of long-chain molecules that are hydrophobic at one end and hydrophillic at the other. Such molecules tend to become concentrated and form a film at the boundaries presented to the water in which the agent is dissolved, the process of concentration at an interface being called *positive adsorption,* or just *adsorption.** Adsorption at an interface always is accompanied by reduction of the free-energy content of the interfacial region. When the interface is that between air and water, the loss of interfacial energy due to adsorption shows up experimentally as a reduction of the surface tension of the water. Backstrom [10] reported that a reduction of as much as 18 dyn from about 72 dyn/cm might be expected at concentrations of surface-active agents used in practice.

Substances that can be adsorbed promote the formation of bubbles that are more stable than those formed without such an agent. Reducing surface tension lessens the work required to subdivide a large bubble into smaller ones, and the adsorbed substance forms a film having some strength and elasticity; it may also give the bubble an electrostatic charge. The combined effect is to prevent small bubbles from coalescing when they collide.

However, some of the most commonly used air-entraining agents do not appreciably reduce the surface tension of water in concrete unless the quantity used is considerably more than that generally found necessary. From that fact we may conclude that such materials are not adsorbed and, for this particular use, are not to be classed as surface-active agents. Nevertheless, they are able to promote the formation of stabilized air bubbles. Powers [11] as well as Mielenz [12] observed that such materials produce an insoluble precipitate when mixed with a calcium hydroxide solution, or with the solution found in freshly mixed cement paste, which contains calcium hydroxide. The insoluble product is evidently an amorphous, probably colloidal, water-repellent material. The term water-repellent signifies that the surface of the substance has such small attraction for water molecules that when the surface is placed in contact with a drop of water, the water is unable to spread over the surface. Instead the water surface forms a definite angle of contact with the solid surface, leaving some of the solid exposed to air. Presumably, when such a material is caught in the interface between water and air, a part of each particle of it tends to remain dry, that is, exposed to air, while at the same time the force due to surface tension of water holds the particle in the interface.

The insoluble product discussed above thus forms a film in the interfacial region described by Mielenz [12] as follows: ". . . a solid or gelatinous film . . . characterized by finite thickness, strength, elastic-

* In some cases, not pertinent here, the solute is repelled from the surface or interface and accordingly is said to be negatively adsorbed.

ity, permeability, and other properties which, together, control the subsequent response of the air bubble to mechanical or physical-chemical changes in the system." He succeeded in photographing such solid-stabilized bubbles.

Agents that remain soluble and become adsorbed at the air-liquid interface, as shown by the lowering of surface tension of water, may at the same time be adsorbed at the solid-liquid interface, that is, on the cement particles and possibly on aggregate particles too. Such adsorption reduces interparticle attraction, and there is an effect on the consistency of a concrete mixture arising from a decrease in stiffness, or increase in quasifluidity, of the cement paste.* In contrast, agents that produce an insoluble product, and become virtually removed from solution by precipitation, are not adsorbed by cement; they have no effect on consistency due to reduction of interparticle attraction.

Air-entraining agents that produce insoluble products with calcium hydroxide and are thus not likely to function as surface-active agents in concrete at normal dosage are given by Mielenz as follows:

Sodium soap of wood resin
Sodium abietate
Sodium soaps of lignin derivatives, rosin, or fatty acid
Triethanolamine salts of sulfonic acid

Agents based on sulfonates do not form insoluble precipitates and thus are able to function as surfactants in concrete.

At present there is no clear evidence that one class of air-entraining agent is superior to another. However, among various materials tested, some agents produce relatively large bubbles and are inferior for protection from frost action. On comparing surfactants, both Lauer [13] and Bruere [14] reported the anionic to be superior to the cationic type, and the latter to be superior to the nonionic.

Bruere [18] drew conclusions somewhat different from those made above about the nature of agents which produce insoluble products when they are in contact with the cement solution, and later Greening [25] found data that seemed to support Bruere's view. Bruere concluded that with at least one exception the actual promotion of air entrainment is due to the small residue of soluble (or colloidally disperse) agent remaining after the insoluble material has been produced. This alternative conclusion is considered again further on, but we may note here

* Such agents also tend to interfere with the chemical processes, and thus extend the dormant period; they have come to be called "water-reducing, set-retarding admixtures." Some agents of this kind promote air entrainment; some do not.

that there is some basis for doubting the validity of this interpretation. For example, Bruere's data show that sodium abietate, which is like the commonly used NVR (neutralized, that is, saponified, Vinsol resin) was able to entrain air in cement paste, but the filtrate of that paste was able to produce virtually no foam when tested by itself. This agrees with earlier experiments with NVR made by Powers who found that the filtrate had virtually the same surface tension as pure water and that the insoluble precipitate itself was capable of promoting the entrainment of air, a finding reported also by Meilenz et al. [12]; indeed Bruere also pointed out that foam can be stabilized by the insoluble reaction product, but he apparently did not have any evidence that it could also stabilize bubbles in cement paste.

4. MECHANICS OF AIR ENTRAINMENT

4.1 Bubble Formation by Gas Generation

Quantitative data on air entrainment are hardly understandable unless the process by which air becomes entrained is kept in mind. The mechanics of air entrainment involve the action of mixers, but first, by way of orientation, we consider briefly methods of bubble formation that are not dependent on mixing action.

Gas bubbles can be produced in cement paste by various means. When powdered aluminum is mixed with cement paste, sodium hydroxide and potassium hydroxide dissolved from cement react with the metal, producing hydrogen which may remain trapped in the paste. Or, if the unstable compound hydrogen peroxide is added to the mixing water, decomposition later yields oxygen bubbles. There are some portland cements that contain small amounts of material, possibly calcium carbide, that react and produce acetylene gas. Of these sources of gas bubbles, only aluminum powder is used commercially, and it is used not for the same purpose as entrained air but for the production of a lightweight fire-resistant insulating material.

4.2 Bubble Formation by Emulsification

4.2.1 Mechanical dispersion. Entrained air is produced by a stirring or kneading process. Stirring introduces air as it draws material at the surface of a batch to the interior by a vortex action. The formation, dispersion, and stabilization of discrete air bubbles is fundamentally a process that forms an emulsion, as distinct from a foam.

To be stirred a material must be fluid under the action of the stirring device. Materials too stiff or too dilatant to be stirred are kneaded.

Such materials usually incorporate an excess of air at the start, the mixture becoming bulky (dilatant) as described in Chapter 3; there may be evidence of hydrostatic tension as described above. With continued kneading the overall air content becomes reduced, hydrostatic tension and bulkiness disappear, and the remaining air is found to be in the form of bubbles.

A concrete mixer can knead a stiff mixture or stir a fluid one. If the mixture is very stiff and dilatant, the batch may emerge from the mixer in an obviously dilatant state, its conversion to the plastic state being completed later, usually by means of vibration. If that stage of the process is not run to completion, the final product may contain portions of nonplastic, unconsolidated material containing structural voids. (Of course, some mixtures, such as those used by the block industry, are designed so that they do not become plastic, and thus structural voids are a characteristic feature.) Here we ignore unintended structural voids, considering only those that are an ingredient of the mixture—air bubbles.

In a drum-type concrete mixer stirring is produced as the fluid material follows a tortuous path around the baffles, and as it cascades from elevating buckets. The same action produces kneading when the material is too stiff to behave like a fluid. In a pan-type mixer (Eirich-type) used mostly in laboratories and factories, stirring is done by blades on a rotating vertical shaft that is mounted eccentrically with respect to the center of the pan; material is carried to them, and away, by rotation of the pan. In some such mixers, particularly those used in laboratories, perhaps half the batch is within the effective radius of the stirrer, and thus at a given instant only part of the batch is being stirred. When mixtures are too stiff for stirring, kneading is produced as the stirrers squeeze material, against the vertical wall of the pan. This action, as well as that of stirring, is aided by a scraper that cleans the walls and deflects material toward the center.

The distinction between stirring and kneading is more or less arbitrary, especially for the mixers just described. Indeed, with a given batch an initial kneading action may soon produce a fluid state that is amenable to stirring. The distinction could perhaps best be based on the effect of the action on air content when no air-entraining agent is used: If the mixing action tends to reduce overall air content from its initial level, it should be called "kneading"; if it tends to increase air content, or if it has no effect on air content, it should be called "stirring." Stirring a fluid can produce a vortex, or vortexes, and if the object is to entrain air, stirring should be done so that a vortex is produced. Vortexes in a concrete mixer may appear at the ends of baffles; where the cascading material plunges into the pool at the bottom a folding-in occurs that is not strictly a vortex, but has the same effect.

A simple device for mixing cement, water, and air is an electrically driven stirrer consisting of propellerlike blades mounted at the end of a shaft that operates at a constant speed in material held in a suitable container. If the material is fluid, such a stirrer creates and maintains a vortex from which air is fed into the body of paste as the paste circulates. Presumably, air enters as a layer between in-folding surfaces of paste, but because of mechanical instability arising from the surface tension and fluidity of water, the layer becomes quickly broken up and dispersed as separate bubbles, probably of various sizes.

If, for any reason, the paste gradually stiffens while being stirred, a point may be reached at which the action is confined to a small volume in the immediate vicinity of the blades while the rest of the batch remains quiet. At the point where the batch as a whole fails to circulate, fluidity has disappeared so far as this particular mixer is concerned. Obviously, the fluidity point for a given material will not be the same for different types of mixes.

4.2.2 Role of the air-entraining agent. Violent agitation, particularly turbulence in the flow pattern, is likely to cause collision between bubbles, giving a colliding pair of bubbles opportunity to merge into one. Such coalescence is a natural tendency because it is accompanied by a reduction of interfacial area and hence of interfacial energy, as well as by a reduction of air pressure within the bubble. In other words, the free energy of the system diminishes. The principal function of the air-entraining agent is to prevent such coalescence. It does so by forming a tough film, and it may also give the bubbles an electrostatic charge that gives them some degree of mutual repellency. Another role that can be played by a surfactant is to reduce the surface tension of water and thus enable a given shear stress to divide smaller bubbles than could otherwise be divided. With a given stirrer, reduction of surface tension increases the fraction of the circulating batch within which shear stress is able to divide bubbles larger than a given size. However, those agents that stabilize air by first producing an insoluble hydrophobic colloid (see Section 3) without reducing surface tension can have no such effect. Used in relatively high dosage, however, these agents may also be able to reduce surface tension.

4.2.3 Accumulation of air and decrease of mean bubble size while stirring. In a batch of material stirred by moving paddles or by any other means, rates of shear are not the same at different points in the batch. At some positions the shear rate is at its lowest, and in those places only relatively large bubbles can be divided. At positions of highest rates of shear, which in a paddle device would be close to the point of greatest blade velocity, division of the smallest bubbles that can be divided in that device will take place. That size is determined by the maximum rate of shear and by the maximum distorting force that can be exerted on the bubble, and probably also by the length of time such force can be maintained. The distorting force at a given shear rate depends on the viscosity of the material.

Most of the air drawn in probably enters the batch at the region of highest shear stress, but it is unlikely that at a given time and place each bubble that might be divided will be because of the briefness of its stay in that region. Circulation causes a given bubble to enter the region of high stress repeatedly, and thus the probability of its becoming divided depends partly on the length of the stirring period. Since at any given moment only a fraction of the batch is being subjected to the highest shear stress, and in successive moment different fractions of the batch are brought to this region, a certain length of time is required for say one-half the batch to have been subjected to the maximum shear rate. With material of a given viscosity, the time will be longer the larger the batch. With a given batch size and a given stirrer operating at constant speed, the rate of circulation is faster the less viscous the material; hence those factors that influence the viscosity influence the time required for a given batch to be circulated through the high shear stress region a given number of times.

Accumulation of air in paste during stirring depends also on the ability of bubbles once formed to remain discrete and on their chance to escape. The chance that one might escape is smaller the smaller the bubble. Without the aid of a stabilizing agent relatively few bubbles are able to remain discrete and become divided; the rest remain large or by coalescing become larger and escape from the batch during the period of stirring. For escape to occur during stirring, a bubble near the surface must exert enough buoyant force to move through the viscous material at a rate that will enable it to reach the surface before circulation removes it from the surface region.

The mechanics of bubble motion can be visualized in terms of a fluid being stirred in such a way as to produce a single vortex. Figure 7.2 is a sketch illustrating the components of velocity of a bubble in the surface region of a circulating body of fluid material, the bubble being drawn to indicate distortion under shear stress. Vector **A** represents instantaneous velocity in the direction of circulation; vector **B,** the upward velocity due to buoyancy of the bubble; vector **C,** the resultant velocity. The

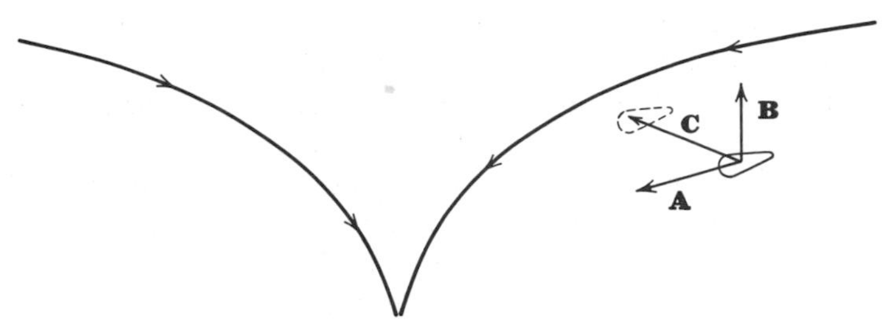

Fig. 7.2. Sketch to illustrate the velocity components for a bubble near the surface of a vortex.

dotted ellipsoid represents what the position of the bubble would be after the elapse of one unit of time if vector **A** remained unchanged during that time. Since the rate of circulation can be independent of the rate of rise due to buoyancy, vectors **A** and **B** are regarded as independent variables. At a given value of vector **A,** the magnitude of vector **B** (upward velocity) is determined by the buoyant force and the resistance to that force according to some modification of Stokes's law* such as

$$\text{upward velocity} = \frac{\bar{d}^2(\rho_f - \rho_B)g}{18\eta}\,k(\bar{d}) \tag{7.2}$$

where $\bar{d}$ is the mean bubble diameter, ρ_f is the density of the material; ρ_B is the negligible density of the bubble; g is the gravitational constant; η is the viscosity of the fluid, and k is some "correction factor" for the effect of distortion and compressibility of the bubble. It might not be independent of $\bar{d}$, as indicated. This equation is to be regarded only as a means of designating the variable factors and indicating the interrelationships approximately.

From Eq. 7.2 we see that at a given rate of circulation and a given viscosity, a bubble twice the size of another moves toward the surface perhaps four times as fast; the larger the bubble the farther from the surface it can be and still escape during the time it is in the surface region; in any case, it is probable that only those bubbles very near the surface have time enough to escape. We should expect the mean size of the bubbles retained to be smaller than that of those lost. If the emulsification process produces a constant initial bubble-size grading, we should expect the mean size of bubbles retained to diminish with

* Stokes's law is not strictly applicable since it pertains to a rigid sphere rather than to a pneumatic one; however, it will serve the present purpose.

elapsed time during stirring; experimental data shows that is what happens. As the bubbles grow smaller, their upward velocity diminishes not only because of the smaller diameter, but also because with each increment of bubbles the mixture of bubbles and fluidized paste becomes more viscous. The increase of viscosity thus tends to increase the size of bubble that can be retained at a given circulation rate, and that factor tends to reduce the rate of decrease of mean size. However, under practical conditions, increase of viscosity retards the circulation rate, which change (reduction of vector **A**) tends to offset the effect on upward velocity (reduction of vector **B**) of a given bubble. Reductions of vectors **A** and **B** are not necessarily equal.

Other aspects of the mechanics of air entrainment will become apparent as we discuss some experimental data. Air can be entrained in water alone, in wet aggregate alone, in cement paste, in plastic concrete mixtures, and in some nonplastic ones. It is with such data that we are now concerned.

5. ENTRAINMENT IN WATER

5.1 *Entrainment without an Air-entraining Agent*

A stirrer able to produce a strong vortex disperses air bubbles in water, but in terms of Fig. 7.2, vector **B** is so large relative to vector **A** that only a small volume of air can be accumulated while stirring. Escaping bubbles burst as they emerge, and, save for dissolved air, all air is lost immediately after stirring stops.

5.2 *Entrainment with an Air-entraining Agent*

With an effective air-entraining agent, such as ordinary soap, bubble formation in water under the conditions discussed above is much more copious than without. This is due, presumably, to the prevention of coalescence, or at least a reduction in the rate of coalescence, by the interfacial film produced by the agent. However, most of the stabilized bubbles escape almost as readily as before, but instead of bursting as they emerge most of them blend their surface films, forming the cells of foam structure. At the same time, some very small bubbles may remain dispersed in the water below the foam. Observed microscopically, the bubbles below the foam show Brownian motion and mutual repellency, the latter signifying electric charge. The layer of foam prevents their escape, and their Brownian activity maintains some bubble population well below the top surface, counteracting their rise due to buoyancy.

The agent used in the experiments just described functioned as a surfactant in pure water, but in limewater it produced an insoluble solid, evidently hydrophobic and colloidal, that stabilized the foam. Within the range of air-entraining agent concentrations commonly used in concrete practice, the volume of foam produced by a given agent increased with the amount of agent. For example, using different concentrations of neutralized Vinsol resin* and a high-speed electrical stirrer, volumes of air were entrained in water as is indicated in the accompanying table.

Concentration of Agent (mg/l)	Volume of Air, (cm^3/cm^3 of water)
440	0.5
900	0.8
1300	1.0
1800	1.1
2600	1.3

The lamellae of foam structure are composed of water and dual layers of stabilizing material. If enough air-entraining agent is used, practically all the water in a batch can be contained by the foam. It is therefore not far fetched to regard foam as a special case of air entrainment in water.

6. AIR ENTRAINMENT IN AGGREGATE

This subject is introduced because it leads to an understanding of the role played by the aggregate while air is becoming entrained in plastic concrete. Specifically, it demonstrates the aggregate-screen effect in its simplest form.

Experiments by H. L. Kennedy [15] demonstrated that air can be readily entrained in wet aggregate by means of an air-entraining agent and a suitable mixer. Kennedy reported an experiment in which 250 g of sand and 60 ml of water containing an air-entraining agent were stirred together. The result was a temporarily plastic mixture retaining a relatively large amount of air, 10 to 25 per cent of the overall volume. Scripture, Hornibrook, and Bryant [16] repeated the experiments and obtained similar results, as shown in Table 7.1. They reported the air contents of wet sands stirred without an air-entraining agent as well as with an air-entraining agent. Mixing was done mechanically following ASTM Method C185.

* A sodium soap of wood resin.

Table 7.1 Air Entrainment in One-size-group Sands[a]

1	2[b]	3[c]	4[d]	5[e]	6[b]	7[c]	8[d]	9[e]
	Without Agent				With Agent			
Group Number and Sieve Size (by sieve numbers)	$\frac{w}{av_a}$	$\frac{A}{V}$	$\frac{A + wv_w}{V} = \epsilon$	$\frac{wv_w/V}{\epsilon}$	$\frac{w}{av_a}$	$\frac{A}{V}$	$\frac{A + wv_w}{V} = \epsilon$	$\frac{wv_w/V}{\epsilon}$
(6) 4–8	0.464	0.083	0.374	0.78	0.440	0.142	0.404	0.65
(5) 8–14	0.485	0.089	0.387	0.77	0.464	0.194	0.449	0.57
(4) 14–28	0.509	0.076	0.388	0.80	0.485	0.232	0.483	0.52
(3) 28–48	0.551	0.060	0.394	0.85	0.509	0.259	0.509	0.49
(2) 48–100	0.620	0.014	0.391	0.96	0.551	0.181	0.472	0.62
(1) Passing 100	0.662	0.026	0.414	0.94	0.591	0.134	0.456	0.71

[a] Data are from E. W. Scripture, Jr., F. B. Hornibrook, and D. E. Bryant, "Influence of Size Grading of Sand on Air-Entrainment, *Proc. ACI*, **45,** 217–228, 1948. Figures in columns 3, 4, 7, and 8 were computed by the author from data in Tables 2 and 4 from the same article.

[b] Weight of water per unit solid volume of aggregate, g/cm^3.

[c] Volume of air per unit volume of specimen.

[d] The sum of the volumes of water and air per unit over-all volume, that is, void content, ϵ.

[e] Water content per unit void content.

6.1 *Air Entrainment without an Air-entraining Agent*

We see in column 3 of Table 7.1 that if no air-entraining agent is used the air content of compacted wet sand is generally smaller the smaller the particle size. The fact that the total voids content (water plus air) is about the same for all sands (column 4) reflects the geometric similarity of the sands, the absence of bulking due to surface tension forces (see Chapter 3), and shows that different amounts of air correspond to differences in water content. (The water content was evidently regulated to give the same apparent wetness and lubrication during mixing.)

The increase in water content with decrease in mean particle size can be explained in terms of the amount of water that can cling to particles. When the amount of water is not sufficient to inundate the aggregate, as was the case here, the water is held on grain surfaces by adhesion and by its own surface tension. During mixing, the aggregate must become dispersed beyond its compacted volume; otherwise it could not be stirred. The figures shown in column 4, however, correspond no doubt to the loosely compacted state because the dispersion of particles and enlargement of voids existing during mixing could not be maintained while the voids content was being measured. Therefore, during mixing, the percentage of water in the voids, as indicated in column 5, would be considerably smaller than the figures shown in that column. Thus the figures shown in columns 4 and 5 correspond to a larger voids content and smaller degrees of saturation than the figures show, and the differences between successive amounts of water shown in column 2 correspond to the different amounts required to lubricate the particles in the slightly dispersed state maintained during mixing.

If we compute the difference between the water per unit of solid material required for group 6 and that for group 5, the difference for groups 6 and 4, etc., we find the following ratios between differences, using the difference between No. 6 and 5 as unity: 1:2.1; 1:3.8; 1:7.4; 1:9.4. Notice that for the three largest sizes the water contents increased approximately in the same ratio as the specific surface areas, that is, 2, 4, 8. However, for the smaller sizes the water increments are considerably smaller than that required for a geometric progression. This result is due to the fact that adjacent particles share their water envelops to some degree, the degree of sharing undoubtedly being greater the smaller the particle. Strictly geometric increase of water content should be found only for isolated particles.

Thus we see that the air content of wet aggregates depends in this case upon the factors that govern water contents. It will be recalled that, in Chapter 3, air content is found to depend also on bulking effects due to cohesion arising from the surface tension of water. Such bulking effects were probably present during stirring, the voids then being dilated and less full of water. Cohesion and consequent bulking no doubt aided the stirring process and was a factor in establishing the amount of water required. However, in the quiescent, compacted state that existed when the void content of the aggregate was measured, the voids were so nearly full of water that bulking was negligible. This fact is shown by the magnitude of the figures in column 4, and especially by their similarity.

Air found in the samples as measured is probably not entrained air in the same sense that air in plastic concrete is entrained. When the interstitial space is 80 per cent full of water, or more, as shown in column 5, the air probably exists as bubbles, but under the conditions that existed while the water content was being established, that is, during stirring, the voids were less full, and air might have been a continuous phase. The significant point is that without an air-entraining agent the structure developed during mixing was not stable and the aggregate quickly settled to a compact state, the voids content (water plus air) depending on the characteristic geometry of the aggregate in its moderately compacted state.

6.2 Air Entraining with an Air-entraining Agent

When wet sand is stirred with an air-entraining agent present, the results are mostly different from those discussed above, qualitatively and quantitatively. The same dispersion of particles and dilation of voids described above is necessarily produced by stirring, but, because of the stabilizing effect of the agent, discrete bubbles formed during stirring become entities—definite components of the mixture. The degree of particle dispersion is determined mainly by the bubbles, and it is greater than that produced when no agent is present. Plasticity is achieved, and the plastic state lasts for a significant length of time after the mixing process has stopped. H. L. Kennedy was able to perform a standard slump test on concrete aggregate plasticized with water and entrained air.

When *no* air-entraining agent was used there was probably one air void per interstitial space in the sample; however, when an air-entraining agent was used there were generally more than one bubble per interstitial space, at least for the coarser aggregates, as was observed and photographed by Kennedy.

The most important difference arising from the use of an air-entraining agent pertains to the relationship between the amount of air entrained and the size of aggregate particles. Whereas without an agent air content was smaller the smaller the particles, with an agent it was greater the smaller the particles, except for aggregates composed of particles smaller than a limiting size corresponding to the 28-mesh sieve, approximately. This may be seen in the last three columns of Table 7.1. We can also see that whereas without an agent the total voids content was not influenced by air content, with an agent total void content varied with the air content. Also, in column 9, we see that while the air content became larger as aggregate size became smaller, the fraction of water in the void space also became smaller, but beyond the point of maximum air content, water occupied a progressively greater fraction of the total void space.

To understand the observed relation between particle size and amount of entrained air, we must bear in mind that the water in these mixtures was itself not able to retain much if any entrained air as bubbles. Therefore we may assume that the air contents shown in column 7 in excess of the corresponding figures shown in column 3, Table 7.1, represent the volume of air bubbles that would have escaped or that would have produced supernatant foam had they not been trapped by the aggregate. An aggregation of particles constitutes a grid or three-dimensional screen. The amount of air it can hold is determined by its void content and amount of water in the void space, governed mainly by the amount needed for lubrication.

In the compacted state each of the aggregates in Table 7.1 had about the same voids content but, as noted before, the water content of the void space was larger the smaller the particle size. Thus, in the compacted state, the net capacity for air was smaller the smaller the particle size, and the observed greater amount of entrained air in some of the smaller aggregates is necessarily to be attributed to the dispersing effect of air bubbles on aggregate particles.

Particle dispersion should be greater the smaller the mean particle size and thus the smaller mean size of the voids in the aggregate, partly because the smaller air bubbles trapped by the smaller screen are relatively less deformable than are larger bubbles and partly because the individual rock particles are lighter. Thus particles of a given small size might readily be displaced a small distance equal to say half its own diameter by means of air bubbles, whereas a particle ten times as big could hardly be displaced 10 times as far to produce the same degree of dispersion.

The very effectiveness of air bubbles in dispersing small particles seems to account also for the reversal of trend already mentioned. As particle dispersion and aggregate-void dilation increases, the effective screen size approaches the size of the largest bubble that can be accommodated in the dilated void. Therefore as dispersion increases bubbles tend to escape, and this is a factor that tends to diminish the air content. It is thus understandable that the air contents shown in column 7 reach a maximum as the aggregate size diminishes, and then fall away as the particles become still smaller.

The degree of particle dispersion produced by air bubbles can be estimated by means of the Weymouth formula discussed in Section 8.6 of Chapter 1, the convenient form for the present purpose being as follows:

$$\frac{t}{D} = \left(\frac{d_o}{d_a}\right)^{1/3} - 1 \tag{7.3}$$

where t is the clearance and D is the size of the aggregate particles; d_o is the density of the aggregate in the compacted state, and d_a is the same for the dispersed state. Thus, we let $d_o = 1 - \epsilon_4$, and $d_a = 1 - \epsilon_8$, where ϵ_4 is a value from column 4 of Table 7.1, and ϵ_8 is the corresponding value from column 8. We let $D = \bar{D}$, the average for the whole aggregate; it need not be evaluated because we are interested in the ratio $t/\bar{D}$. Using the data from Table 7.1 we obtain the results given in Table 7.2.

Table 7.2 Particle Dispersion by Stabilized Air Bubbles

Size Group	Air Content	Calculated Particle Dispersion ($t/\bar{D}$)
4–8	0.404	0.26
8–14	0.449	0.33
14–28	0.483	0.40
28–48	0.509	0.42
48–100	0.472	0.37
100–200?[a]	0.456	0.30

[a] Reported as "passing No. 100."

The figures in Table 7.2 show that air bubbles maintained particle dispersion such that the mean distance between particles was as much as 40 per cent of the mean particle diameter. Higher degrees of dispersion could not be maintained, supposedly because bubbles small enough to be trapped in the interstices of the smaller size groups are able to escape if the dispersion becomes too large a fraction of the mean particle diameter. Another possibility is that under given conditions, with a given dosage of air-entraining agent, the amount of air as bubbles small enough to be accommodated by the voids of the smaller aggregates is relatively small.

It should be kept in mind that Table 7.2, and the discussion leading to it, pertains to aggregates having different sizes, as designated, and that it does *not* indicate air-entraining characteristics of size groups when they exist as fractions of an aggregate. In any case the aggregate screen is formed by the whole aggregate, regardless of the range of sizes in the aggregate.

7. AIR ENTRAINMENT IN CEMENT PASTE

7.1 Mechanism of Air Retention

A stirrer able to fluidize cement paste can cause an accumulation of entrained air, but without an air-entraining agent only a small amount can accumulate. Presumably, unstabilized bubbles readily coalesce to form larger ones and thus maintain a relatively large average size with a consequent high rate of loss during stirring. The small volume of air that is retained without an air-entraining agent is composed mostly of relatively small bubbles. With an air-entraining agent present, bubbles accumulate rapidly and approach a maximum total volume that depends on factors yet to be discussed.

After the mixing period, vector **A** of a given bubble (see Fig. 7.2) disappears, but vector **B,** the buoyant force remains. If cement paste were a Newtonian fluid, a bubble would rise at a rate depending on the viscosity of the fluid and on the diameter of the bubble. Rising, it would experience a diminishing pressure, expand, and therefore accelerate continually until it could escape and burst or produce foam as is described above. In cement paste the bubbles in general do not rise while the paste is at rest owing to interparticle forces that give paste the properties of a weak solid, even though it seems fluid. If a bubble rises, its buoyancy must have created a shear stress in the surrounding paste that exceeded the shear strength (yield stress) of the paste. (See Chapter 10.) The fact that bubbles remain in the very top layer of samples of neat paste during the period after mixing shows that shear strength of paste is sufficient to stand the stress due to buoyancy of all but the bubbles large enough to escape immediately after stirring stops.

Retention of air by cement paste can be explained in terms of the screen effect, as for aggregate, but the screen is not a matter of aggregate geometry; instead it is a network of cohesive particles able to maintain

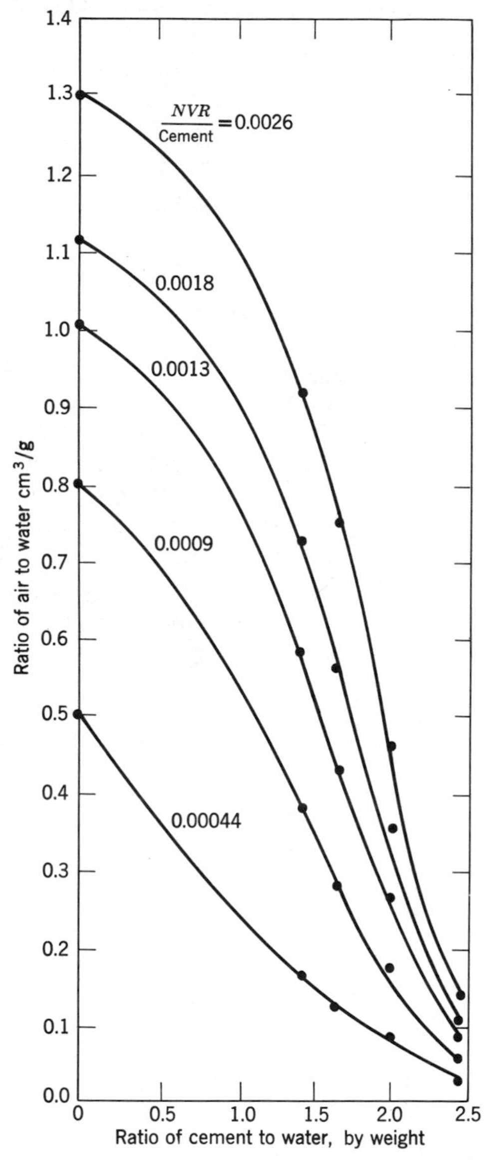

Fig. 7.3. Effect of cement content on air-entrainment in mixtures of cement and water for which a high-speed, milk-shake type of stirrer was used. *NVR* stands for neutralized Vinsol resin. (Data are from [11].)

its structural continuity because of interparticle forces of attraction. Both Bruere and Mielenz have observed that cement particles tend to adhere to air bubbles, and thus it appears that bubbles themselves are links in the cohesive structure.

7.2 *Quantity of Air Entrained*

7.2.1 Effect of cement content. As shown in Fig. 7.3, when a dilute solution of an air-entraining agent is stirred with cement, air becomes entrained, but the amount is smaller the greater the concentration of the cement. Within the range of water contents where the term "paste" or "slurry" is apt, cement particles prevent the formation of foam, all the air occurring as discretc bubbles within the body of paste. The mixture may thus be regarded as an emulsion of air in cement paste. With the particular stirrer used in these experiments it became practically impossible to produce bubbles in paste when cement concentration exceeded about 3 g/cm³ of water. A similar limit would probably be found for any device *in*capable of kneading the mixture.

The customary way of prescribing the dosage of an air-entraining agent is in terms of the amount of cement. Figure 7.4 shows a typical relation between air content and the dose of neutralized Vinsol resin

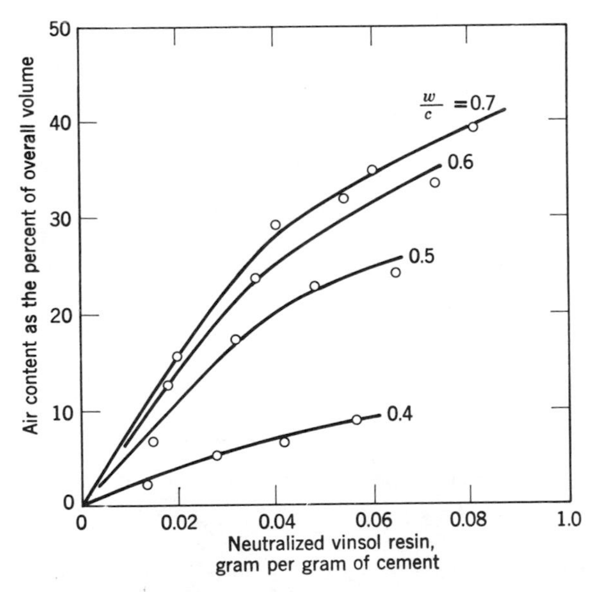

Fig. 7.4. Air entrainment in cement paste as influenced by *w/c* and dosage of air-entraining agent. (Data are from [11].)

at given water-cement ratios; the data are the same as those in Fig. 7.3. Again it is apparent that the higher the water-cement ratio the greater the amount of air entrained at a given dosage of the air-entraining agent and a given stirring procedure.

7.2.2 Stirring procedure. Data such as those in Figs. 7.3 and 7.4 show only the amounts of air entrained at the end of an arbitrary period of stirring, usually brief. Therefore the result is usually that of an incompleted process. Using a given dosage of an air-entraining agent, Bruere [18] observed that during a prolonged stirring period, with the stirrer running at a constant speed, air content increased rapidly at first and then slowly approached an upper limit. (See Fig. 7.5*a*.) Also, the rate of air-accumulation in paste increased with the rotational speed of the stirrer. (See Fig. 7.5*b*.)

The kind of stirring device, as well as its mode of operation, has a significant effect on the amount of air entrained in cement paste, and with any given device and mode of operation, the rate of air accumulation is smaller the larger the batch since only a fraction of the batch is being subjected to effective stirring or kneading at a given time, and this fraction is usually smaller the larger the batch. Some experimental results are given in Table 7.3. Air contents ranging from about 6 per cent to over 20 per cent were obtained with similar pastes.

An increase of air content with water content does not occur under all circumstances. Danielsson and Wastesson [19], using the procedure described in Table 7.3, found no increase in air content with an increase

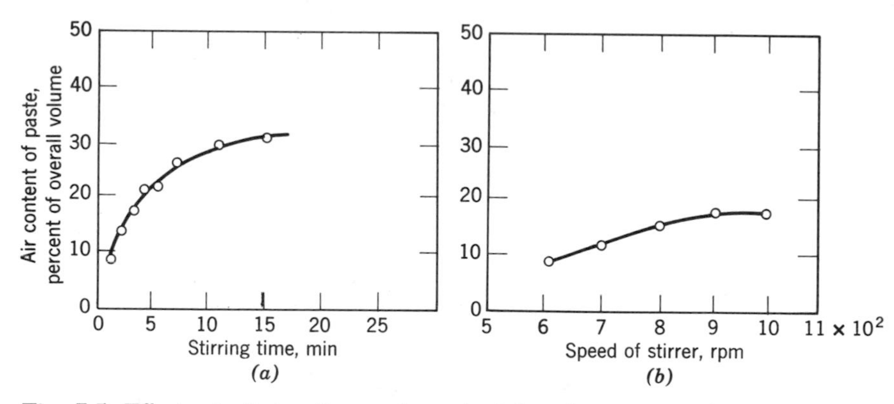

Fig. 7.5. Effects of stirring time and speed of the stirrer on the air content of neat paste. Sodium dodecyl sulfate was used as the air-entraining agent, at 0.0125 per cent by weight of the cement. (Data are from Bruere [18].) (*a*) The speed of the stirrer was 1000 rpm. (*b*) The stirring time was 3 min.

Table 7.3 Effect of Mixing Procedure on Air Content of Neat Cement Paste

Literature Reference Number	Method and Duration of Mixing of Neat Paste	$\frac{w}{c}$ (by weight)	Air Content (per cent)	Kind of Air-Entraining Agent
19	Rotating-pan mixer (Eirich type) operated 10 minutes; 1.4 cu ft (40 liters) per batch[a]	0.43	12.0	Sodium dodecyl sulfate
		0.43	7.3	Darex A.E.A.
12	Drum type concrete mixer, 3.5 ft^3 capacity, operated 4 minutes; 0.4 ft^3 (11.4 liters) per batch[b]	0.40	5.6	Neutralized Vinsol resin
		0.50	6.9	Neutralized Vinsol resin
14	Water and 500 g cement stirred by hand in 600 ml beaker for 1 minute, then for 4 minutes with electric stirrer at 1000 rpm; 0.14 ft^3 (0.4 liters) per batch[c]	0.45	17.5	Sodium dodecyl sulfate
		0.45	21.3	Sodium dodecyl sulfate
		0.45	17.4	Neutralized Vinsol resin
		0.45	22.6	Neutralized Vinsol resin

[a] U. Danielsson and A. Wastesson, "The Frost Resistance of Cement Paste as Influenced by Surface-Active Agents," *Proc. No. 30*, Swedish Cement and Concrete Research Institute at the Royal Institute of Technology, Stockholm, 1958.

[b] R. C. Mielenz, et al., "Origin, Evolution, and Effects of the Air Void System in Concrete," Part I, "Entrained Air in Unhardened Concrete," *Proc. ACI*, **55,** 95–121, 1958.

[c] After mixing and molding, molds were "tapped gently to remove any very large entrapped bubbles." Data are from G. M. Bruere, "Effect of Type of Surface-Active Agent on Spacing Factors and Surface Areas of Entrained Bubbles in Cement Paste," *Australian J. Appl. Sci.*, **8**(7), 479, 1960.

in water content; in some cases the air content diminished as water content increased. The pan-type mixer they used can produce high shear stress in a doughlike material by squeezing (kneading) action, but in initially soft or fluid paste it is relatively ineffective.

7.2.3 Water available for bubble formation. Bruere [18] observed that with a sufficient dosage of any air-entraining agent the approach toward an upper limit of air content during prolonged stirring is accompanied by an increase in apparent viscosity, as was indicated by the diminishing depth of vortex and diminishing rate of circulation. Presumably the rate of air accumulation approaches zero when the mixture of paste and bubbles approaches a consistency so stiff that the particular mixer in use is unable to produce fluidity and circulation.

Bruere's explanation of the upper limit of air content can be elaborated to throw additional light on the mechanics of air entrainment. As is shown in Chapter 10, cement paste is not fluid a short time after the end of stirring; it becomes a weak solid structure able to withstand limited shear stress. To stir this material a stress sufficiently large to exceed the shear strength of the structure must be applied to fluidize the mixture. At a low rate of shear the material shows much higher viscosity than at a high rate. This difference in viscous resistance is due to the fact that the breakdown of structure is partial; some of the interstitial fluid remains immobilized in fragments of the structure, and the amount of fluid left so immobilized is smaller the higher the rate of shear, and hence the higher the shear stress that tends to tear the structure-fragments apart.

Adopting the reasonable assumption that bubbles cannot utilize the water immobilized by paste structure, we see at once that the amount of water available for bubble formation in a given paste is, within limits, larger the higher the rate of shear maintained by the stirring or mixing device. The water that is set free by the breakdown of structure is that which gives fluidity, and it is nearly if not exactly the same as that which can be utilized by bubbles.

Formation of bubbles is itself a process that immobilizes water and increases viscous resistance to stirring. Bubbles tend to repair the structure destroyed by stirring, and they may succeed in re-forming a complete structure, of which the bubbles are a part, except in the immediate vicinity of the stirrer if there is enough power to keep the stirrer turning. In any case, when general circulation of the batch stops, the upper limit of air content has been reached.

A kneading or homogenizing device can fluidize a paste structure much too strong and stiff to be fluidized by an egg beater or propeller-type stirrer. Such a device may be designed to produce high stress in a small part of the batch and to cause the whole batch to circulate.

The pan-type concrete mixer used by Danielsson and Wastesson functions as a kneading device if the mixture is sufficiently stiff. With it and an air-entraining agent they were able to increase the air content at a water-cement ratio as low as 0.27 by weight. In such a doughy paste the mixer could give a slow kneading action able to produce a high shearing stress and temporary fluidity in the region of high stress. In the course of the long mixing period, 10 minutes, evidently most of the batch was fluidized enough times to produce an accumulation of air up to about 16 per cent, and apparently the limit might have been higher had the dosage of agent been larger or the kneading period longer. Probably this device could have entrained air even in a paste showing normal consistency under standard mixing conditions, $w/c = 0.24 \pm 0.02$.

In this connection we must keep in mind that even at normal consistency the minimum void space in portland cement is almost double that which can be accounted for by the size grading of cement particles. (See Chapter 2.) There is a considerable amount of water that would be mobile if it were not for the structure maintained by interparticle attraction. A device able to break down this bulky structure is thus able to increase the amount of water available for bubble formation. Of course, for any given particle-size grading there is a minimum water content below which no mechanical device could entrain air by emulsification. However, as this limit is approached, the mixture becomes more and more dilatant and air becomes incorporated whether or not an air-entraining agent is used.

A propeller-type stirrer operating in an open vessel is only able to fluidize a paste that has very low shear strength. For the particular conditions represented in Fig. 7.3, the indications are that the batch could not be fluidized if the amount of cement was more than about 3 g per gram of water, or $w/c = 0.33$. This composition corresponds to a water content of about 50 per cent of the overall volume. If a paste of this composition, and without an air-entraining agent, is agitated to fluidize it, most of the air it acquires will escape because, while it is being stirred, paste offers viscous resistance rather than structural resistance to the upward movement of bubbles. Thus we find that the paste that can be fluidized with a stirring device is also a paste that will contain little or no air when an air-entraining agent is not used.

On referring to Fig. 3.4 we see that the limiting paste composition indicated above, a water content of 50 per cent of the overall volume, is also the composition at which the air incorporated by the bulking effect becomes eliminated by successive additions of water. At any water content below 50 per cent of the overall volume, air is found in neat cement paste whether an air-entraining agent is used or not, and the lower the water content the higher the air content; this is true down to the water content that produces the maximum bulking effect, which is about 10 per cent of overall volume. From a water content of 50 per cent down to one of 40 per cent, the latter being the water content giving normal consistency, we are in the range of plastic pastes.

We should notice the difference between the effect of adding water to mixtures within the plastic range and that of adding water to mixtures having water contents above that range. If we begin at $w/c = 0.23$, for example, and add water to a kneaded paste containing no air-entraining agent, we reduce the air content, whereas if we add water to a stirrable paste containing no air-entraining agent, we have no effect on the air content. In either range, if the paste contains an agent, the air content increases with added water.

The water-cement ratio at which a paste can be fluidized by stirring seems to be regulated mainly by the specific surface area of the cement. For the data of Fig. 7.3, the specific surface area of the cement was 3200 cm^2/g, and thus the ratio of surface area to water at the limiting point was about

$$\frac{\sigma_c'}{w/c} = \frac{3200}{0.31 \pm .02} = 9700 \text{ to } 11{,}000 \text{ cm}^2$$

of surface per gram or cm^3 of water, where σ_c' is the specific surface area, unit area per unit weight.

7.3 Bruere's Conclusions

On the basis of many systematic experiments, Bruere arrived at an interpretation of the mechanics of air-entrainment and the role of the air-entraining agent that is somewhat different from that given above, as is mentioned in Section 3. Bruere did not mention the thixotropic structure of cement paste and thus seems not to have considered it with respect to the retention of air bubbles in cement paste. To account for such retention he postulated that an air-entraining agent must be able to make the solid particles hydrophobic so that air bubbles may adhere to them. When a cement adsorbs an agent from the solution surrounding the grains, the normally hydrophyllic state of the grain surfaces is converted to the hydrophobic. Therefore, he concluded, a satisfactory agent is one that does not become removed completely by formation and precipitation of an insoluble compound. Enough must remain in solution to maintain an adsorbed layer on the particles.

To test this explanation, Bruere made flotation tests of the kind used in ore-refining technology. If bubbles in a suspension of mineral powder capture and hold heavy mineral particles so that, when the suspension is allowed to settle, the bubbles rise forming a layer of particle-laden foam, it is concluded that the substance that stabilized the foam also made the particles hydrophobic. The experiment as conducted by Bruere consisted of shaking about 1 g of clinker particles, sieve-size (#150–#200), in 30 ml of a solution of the agent "in the presence of 10 ml of the filtrate from fresh cement paste" and observing the relative amount of clinker particles held in the foam after the sample was allowed to settle.* Similar experiments were made with silica.

Bruere observed that the majority of the agents that were able to promote entrainment of air in paste also showed "excellent flotation." He observed also that agents that are completely precipitated by the calcium ion were among those that did not entrain air in cement paste. Thus it was a reasonable conclusion that just as the bubbles and particles are held together in the flotation test, so they are in cement paste, such adhesion preventing the force due to the buoyancy of the bubbles from expelling the bubbles from the paste.

Among various items of supporting evidence, data on sodium abietate are of special interest because of the similarity of this agent to the widely used NVR (neutralized Vinsol resin). It was observed that sodium abietate could not entrain air in a mixture of pulverized silica and water, nor could it float silica particles in the flotation test; but if calcium hydroxide was added as a "conditioning agent," the results were reversed: as much as 30 per cent air was entrained in silica paste and silica particles were floated. However, this experiment provides reason also to doubt the conclusion that the hydrophobic state is essential for air entrainment.

In various parts of this book, and especially in Chapters 9, 10, and 11, the nature and significance of the flocculent state of a suspension of particles in liquid is emphasized. In the present context we are particularly concerned with the thixotropic structure which is characteristic of the flocculent state; it is the structure that is manifested by a yield stress, as is shown in Chapter 10. It is known that a mixture of silica and water is not flocculent, that is, it has no thixotropic structure and therefore cannot retain bubbles; it is known also that a small amount of calcium hydroxide, not more than enough to keep the solution saturated or even just partly so, changes a silica suspension from the non-

* It is not clear whether the 10 ml of cement solution constituted $\frac{1}{3}$ or $\frac{1}{4}$ of the total liquid. It was intended "to stimulate the conditions in cement paste, e.g., pH and the presence of electrolytes."

flocculent to the flocculent state. This alone could account for the ability of the silica paste to retain air whether the particles are hydrophobic or not.

Any doubt about this conclusion can be dispelled by the following experiment. We make a cement paste, w/c about 0.40, without an air-entraining agent and inject a few air bubbles by means of a hypodermic syringe or other means of producing small bubbles. After the specimen has become hard we break it open. The casts of the bubbles that were retained without the aid of an agent are clearly visible. The same characteristic of the flocculent state is demonstrated by the retention of bubbles formed from gas-generating material as is described in Section 4.1.

In another experiment Bruere first used dodecyl-amine hydrocloride in silica paste and found that it could float silica particles and entrain more than 30 per cent air. Then he used gelatin in silica paste and found that it could neither float silica particles nor entrain air. Finally, he used the two agents together and found that air could *not* be entrained.

Bruere interpreted these results as showing that gelatin prevented the adsorption of dodecyl-amine hydrocloride; the particles therefore did not become hydrophobic; consequently no air could be retained by the silica paste.

However, Bruere pointed out also that the gelatin made the paste "unstable," a state which he described as one that permitted segregation into three zones: "a sediment of solid particles, a layer of water, and a layer of foam mixed with fine solid particles." In our terminology this means that the gelatin converted the flocculent state maintained by dodecyl-amine hydrochloride (when it was used alone) to a nonflocculent state without thixotropic structure. It appears that the air-entraining agent might still have promoted the entrainment of air, but the paste did not have the structure to hold it when gelatin was present; apparently gelatin was a dispersing agent.

Bruere found that sodium abietate floats clinker particles and that at a concentration of 450 mg per liter it could entrain about 25 per cent air in cement paste. Accordingly he concluded that this agent made the cement particles hydrophobic and able to retain air bubbles. It would follow that some of the sodium abietate must have remained in solution to maintain a layer of adsorbed sodium abietate on the cement grains. But, the experiment to confirm the presence of soluble agent in the filtrate from the paste gave a result that was doubtful if not negative. Even after he put as much as 1000 mg per liter in the mixing water, the filtrate from the paste produced only a trace of foam that lasted a few seconds, and its "foam capacity" was reported as zero.

These considerations call attention to a significant difference between the conditions of the flotation experiment and those for air entrainment in paste. In cement paste the cement could continually maintain a high calcium ion concentration even while lime was being used to form the insoluble compound with the agent, and the reaction could run to completion, which is to say until the supply of agent was exhausted. On the other hand, the amount of calcium ion in the flotation experiment was initially only that supplied by the cement solution, which would give a concentration not more than $\frac{1}{3}$ that in the paste, the cement solution amounting to only $\frac{1}{3}$ or $\frac{1}{4}$ the total liquid. This initial concentration would be augmented by calcium ion from the clinker particles, but the specific surface area of those particles was so low that the amount of calcium ion from that source must have been very small. Thus we have reason to surmise that the residual concentration of soluble sodium abietate in the flotation test could have been higher than it was in the air-entraining test.

It is reasonable to infer that if the solution in the flotation test had been saturated with calcium ion, the results would have been different, unless the insoluble reaction product itself was able to promote flotation. If the latter is true, it means that the occurrence of flotation is not an infallible proof that the solid particles have been made hydrophobic; it might mean in this case only that a hydrophobic colloidal calcium resinate can somehow cause flotation. Some indication that this might be true can be seen in the result cited above. Sodium abietate cannot float silica particles, but in the presence of calcium hydroxide it can.

The upshot of the considerations above is that the results of flotation tests cannot be considered reliable evidence that agents which react to produce insoluble products and entrain air in paste nevertheless maintain a certain amount of agent in solution sufficient to maintain an adsorbed layer on the particles that render the particles hydrophobic. The evidence seems stronger that with such agents, some of them at least, entrainment of air is due to the insoluble reaction product, and retention is counted for in terms of the thixotropic structure of the flocculent paste.

There are, however, still some other observations that seem in line with Bruere's interpretation of the phenomena in question. Greening [25] found that the agent NVR was more efficient the higher the alkali concentration in the mixing water, up to a concentration of about 8000 mg per liter of water counting all the immediately soluble alkali in the mixture. That is to say, to produce 19 per cent air in the standard ASTM test Designation C 185, the amount of NVR required was a function of the alkali concentration. From this, following Bruere, Greening concluded that the amount of NVR remaining soluble after precipitation of the insoluble salt of Vinsol resin determined the amount of air entrained, or, in Greening's words, "The equilibrium between the soluble and insoluble portions of the vinsolate salts then seems to determine the amount of air entrained."

However, Greening did not give experimental evidence, other than the relationship with air content, that there was in any case actually a significant amount of soluble residue. On the basis of the earlier Bruere experiments on flotation it was a reasonable inference, but we have already seen that there is reason to question the applicability of the flotation experiments to the point in question.

Yet it is not easy to account for the dependence of air-entrainment efficiency on the alkali concentration. We find, however, a suggestion pertaining to the question in another of Greening's experiments. Of two lots of NVR, he found one to be more efficient than the other, and he noted that the less efficient solution had a cloudy appearance whereas the other was clear. By adding a small amount of alkali to the less efficient one, he obtained a clear solution, and he thereby equalized their efficiencies. Greening concluded that both were colloidal solutions and that the degree of dispersion of the colloidal material was a function of alkali concentration.

We have already postulated that the calcium salt of the resin is a hydrophobic colloidal material. On this basis, we see the possibility that the reaction product of the colloidal material may have a degree of dispersion related to if not identical with the degree of dispersion of the parent colloidal units. Thus on this basis we could attribute the increase in efficiency with increase of alkali concentration of this type of air-entraining agent to the increase in the number of insoluble colloidal units per gram of parent material.

In view of the consideration set forth above, the explanation of air-entrainment has proceeded and will continue to proceed on the assumption that materials such as NVR and sodium abietate produce insoluble hydrophobic colloids as air-entraining agents, and therefore do not ordinarily function as surfactants.

8. AIR ENTRAINMENT IN CONCRETE

8.1 General

Air entrainment in concrete is like air entrainment in cement paste, but with added effects due to the aggregate. It can be understood in terms of the processes already discussed separately with respect to water, paste, and aggregate. The quantity of air entrained in concrete is determined partly by the circumstance that air bubbles, cement, and water must share the void space provided by the aggregate. That amount of space is equal to the voids content of the aggregate in its compact state, plus the space added by the dispersion of aggregate particles. Since in air-entrained concrete the cement-water paste occupies most of the dilated void space of the aggregate, most of the air introduced by means of an air-entraining agent is produced in the paste fraction by an emulsifying process not different from the process of emulsifying air in cement paste alone. The shear stresses that define bubble sizes, developed during mixing, are produced primarily by the paste, the stress in the paste being induced by movement of the aggregate in the mixer.

In view of the fact that a drum-type concrete mixer rotates slowly, we might be inclined to think of it as a relatively low-speed stirring device. On the contrary, as far as the stirring of cement, water, and air are concerned, it is a high-speed device. This fact becomes apparent when we observe how the matrix material is forced through the interstices of the rolling and cascading mass of aggregate in the drum. A concrete mixer is probably more efficient than a laboratory stirrer such as that used by Bruere [14], and undoubtedly much more efficient than the pan-type (Eirich) mixer used by Danielsson and Wastesson [19] as they applied it to stirrable neat cement paste. Therefore we can regard entrainment of air in the paste component of concrete as being done by an efficient stirrer.

The various effects discussed in the preceding section for cement paste are seen also in concrete. With a given mix, and for a given consistency (slump), the air content may be increased by increasing the dosage of air-entraining agent. A given mix containing a given air-entraining agent contains more air the softer the consistency and higher the water-cement ratio.

However, there are certain important differences between entrainment in concrete and entrainment in neat paste that are to be attributed to the aggregate. When no air-entraining agent is used the air content of any cement paste that can be fluidized by stirring is very close to zero and such is the kind of paste usually found in concrete. But, in concrete the air content is not zero at any water-cement ratio or at any slump within the normal range; it usually amounts to at least 5 per cent of the volume of the paste. The air found in concrete made without an air-entraining agent might be a residue of that introduced during the bulking occurring during the initial stages of mixing, or it might be that due to the dilatation of aggregate structure required by the mixing process, particularly if the final consistency is relatively stiff, or it might be that due to temporary emulsification in the paste; or all three processes could be involved. Whatever the mode of its introduction when no air-entraining agent is used, most of the air held in concrete at the time when the concrete has been reduced to its puddled, plastic state, is held primarily by the aggregate screen.

The air content of a given mix made without an air-entraining agent is lower the higher the water content and hence the greater the slump. However, with an adequate dosage of air-entraining agent the opposite is true: the higher the slump, up to 6–7 in., the greater the amount of available water and the higher the air content. For example, Cordon [21] reported the following amounts of air for a given dosage of neutralized Vinsol resin: 5.0, 6.3, and 7.1 per cent for slumps of $1\frac{1}{2}$, 3, and $4\frac{1}{2}$ in.

The effect of prolonged mixing on the air content of concrete is usually different from the effect in neat cement paste. As we have seen, the air content of cement paste approaches an upper limit fixed by the amount of air-entraining agent or the amount of available water and perhaps by the efficiency of the stirring device. In concrete the air content may gradually approach an upper limit, but under some circumstances, perhaps most, after a certain period of rise the air content may diminish as mixing continues. Such a falling off in air content may be attributed, at least in some cases, to an increase in specific surface area of the solids in the mixture due to attrition of aggregate particles; the effect is qualitatively the same as that of adding cement without adding water or additional air-entraining agent. (See Fig. 7.3.)

In general, if for any reason a mixture gradually becomes stiffer during mixing, the amount of water available for bubble making likewise gradually diminishes. When this effect becomes great enough to offset the normal rate of accumulation of air due to continuous stirring, prolonging the process tends to diminish the air content.

Even though we must rely on a limited amount of data, it seems that the temperature of the materials in a mixer has little or no influence on the air content of fresh concrete made without an air-entraining agent *provided that the water content is adjusted with temperature so as to maintain a constant consistency.* When an air-entraining agent is used, and the amount in the batch is kept constant, the air content generally diminishes as the temperature rises, even though the water content is increased as required to maintain constant slump. In terms of the previous discussion this seems to indicate that a rise of temperature diminishes the amount of water available for bubble formation, presumably by increasing the ability of paste structure to hold water away from bubbles. The fact that air content diminishes even though the water content is increased as required to maintain constant slump indicates that consistency (as measured by the slump test) and air-entraining capacity are different functions of the water content of a given concrete mixture.

With this survey of the main phenomena in mind, let us now examine various factors on a more quantitative basis.

8.2 *Entrainment with One-size-group Aggregates*

The role of an aggregate screen in modifying the accumulation of air in cement paste, as compared with that due to emulsification in paste alone, is illustrated in Figs. 7.6 and 7.7. The uppermost curve in Fig. 7.6 represents the data in column 7 of Table 7.1; the factors governing the amount of air entrained in aggregate alone are discussed in Sections 6.1 and 6.2. Comparison of the curve for mix 1:4 with the top curve shows that the air contents of the mixtures containing cement are influenced by size of aggregate particles and hence by size of aggregate voids in generally the same way that they are when no cement

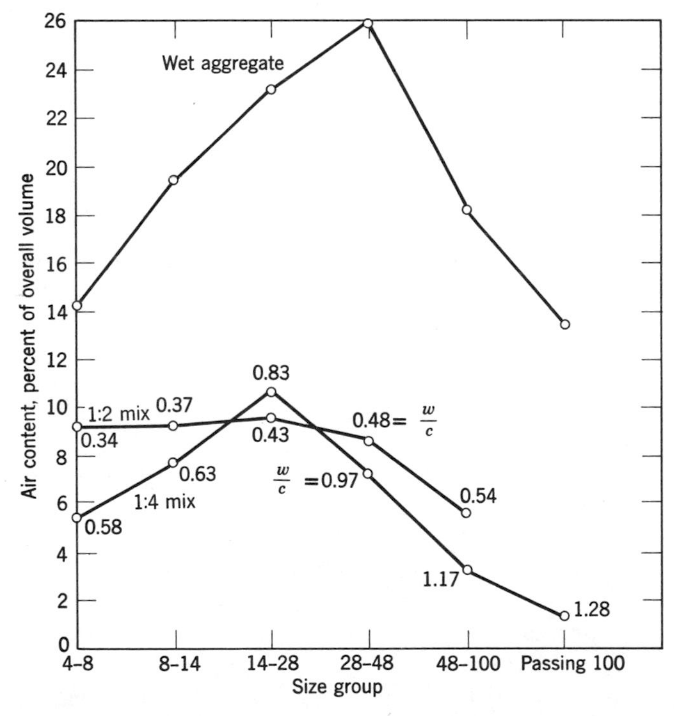

Fig. 7.6. Air entrainment in one-size-group sands with and without cement. The mortars were prepared according to ASTM Designation C 185, and neutralized Vinsol resin was used as the air-entraining agent. Numerals by points give w/c. (Data are from Scripture, Hornibrook, and Bryant [16].)

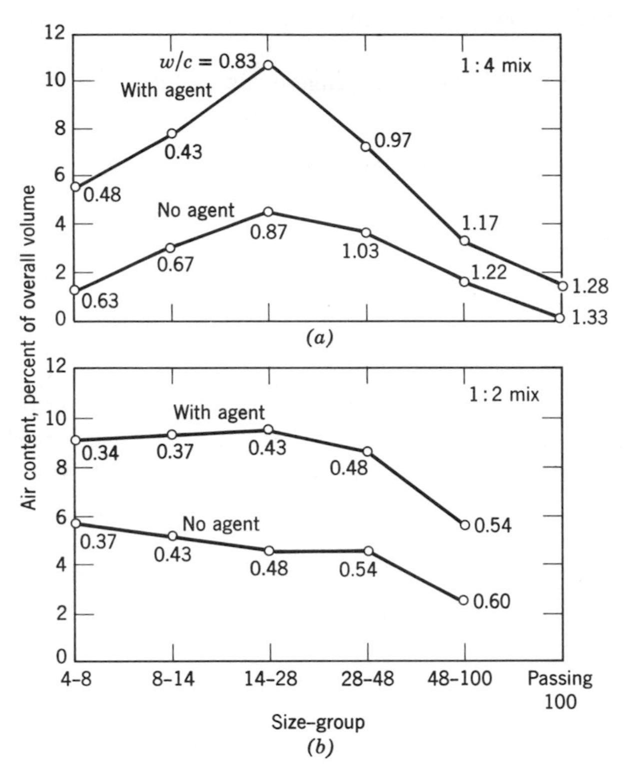

Fig. 7.7. Effect of aggregate size on entrained-air content with and without an air-entraining agent. The mortars were prepared according to ASTM Designation C 185, and neutralized Vinsol resin was used as the air-entraining agent. (Data are from [16].)

is present. In these cases in which the water-cement ratio is upwards of 0.58 by weight, the effect of the cement seems to be that of crowding out some of the air that would have been held by the aggregate screen. If the air content had been controlled by the characteristics of the paste alone, it would have increased as the water-cement ratio increased from 0.58 to 1.28 (Section 7.2.1); the factors that control air accumulation by emulsion in cement paste appear quite subordinate to the control asserted by the screen effect, which includes that of aggregate-void size.

The curve for mix 1:2 indicates that, as might be expected, accumulation of air due to emulsification in the paste becomes more apparent as the mix is made richer. If cement in the 1:4 mix crowded out air that would have been held by the aggregate alone, we should expect still more to be crowded out of the 1:2 mix. However, we see that with one exception the air contents were higher in the richer mix than in the leaner, showing that emulsification in paste was a dominant factor. The screen effect is still evident, however, as is shown by the failure of the air content to increase with the water-cement ratio as it does in neat paste.

From the relationships just discussed it can be inferred that with neat cement paste the increase in volume of entrained air with the increase in the water-cement ratio is due to the addition of relatively large bubbles. When the aggregate voids, slightly dilated, are not large enough to accommodate the voids that might be retained by neat cement paste, the paste in the mixture containing aggregate is not able to retain as much air as it would were it by itself. The effect is greater the smaller the aggregate voids.

Paste disperses aggregate, but the dispersal is not such as to make the screen less effective than it is when air is entrained in aggregate alone. From the compositions of the mixtures represented by the two lower curves in Fig. 7.6 aggregate dispersion due to air bubbles and paste was calculated by means of Eq. 7.3 with the results shown in Table 7.4; the

Table 7.4 Dispersion of Aggregate by Bubbles and Cement Paste[a]

Sieve Size	$t/\bar{D}$	
	1:2	1:4
4–8	0.082	0.014
8–14	0.081	0.021
14–28	0.096	0.059
28–48	0.103	0.059
48–100	0.106	0.068

[a] Calculated using Eq. 7.3.

degree of dispersion is indicated by the ratio $t/\bar{D}$, as it is in Table 7.2. In these mixtures the dispersal of aggregate particles by bubbles and cement paste was not more than 11 per cent of the mean particle diameter and was as little as 1.4 per cent. As we see in Table 7.2, an aggregate screen is effective at calculated dilations up to about 40 per cent of the mean particle diameter.

The total air content obtained from a given dosage of air-entraining agent in a given mix depends under most if not all circumstances on the amount of air that would be present if the agent were not used. This is clearly illustrated in Fig. 7.7 where the two lower curves of Fig. 7.6 are shown along with corresponding curves representing the same mixes without an air-entraining agent. The amount of air added by means of a given dosage tends to be independent of factors that determine the air content when no agent is used, provided that consistency is a constant factor.

In general it seems that the aggregate, through its voids content and "mesh-size," may tend to establish either an upper or a lower limit of air content in concrete. When no agent is used, bubbles that might escape from water or from paste are trapped and thus the lower limit is made greater than zero by the aggregate. When an air-entraining agent is used, the amount entrained in the paste in concrete might be less than that which would be entrained in the paste alone because the aggregate voids are too small, even though they are somewhat dilated, to accommodate the larger bubbles than can be held in a neat paste with a relatively high water content.

8.3 *Apparent Air-entraining Ability of Different Size Groups*

From the results of experiments such as those made by H. L. Kennedy [15] some investigators have concluded that in aggregate alone, or in concrete, each size group has a specific air-entraining ability. It now seems clear that no one size nor one size group in an aggregate can have an independent effect. As we have seen, the aggregate provides a screen that holds bubbles when the cement is absent or that holds bubbles in concrete when no air-entraining agent is used. Furthermore its voids, slightly dilated, provide the space where paste and bubbles may be. Therefore the relationship between the air content of a given mixture and aggregate characteristics is determined by whether or not emulsification is the major entraining process, by the voids content of the aggregate in its dilated state, and sometimes by void size. The latter two factors are functions of size range and grading; for a given size range, aggregate characteristics vary with grading. If the grading of an aggregate for a given mix is changed by increasing the proportion of a given size group, a change in air content of the mixture may be observed, and thus the change in air content *seems* to be the effect of the particular size group that was increased. But, the effect actually is due to a change in voids content of the aggregate and possibly to a change in void size.

Better understanding of how changes of grading result in changes of air content can be obtained from work by Singh [22]. Some of his data are shown in Table 7.5. The gradings are shown along with

Table 7.5 Example of the Effect of Change in the Aggregate Grading on Air Content[a]

Grading Number	Aggregate Grading (per cent in size group indicated)							Air Content (per cent)			
								1:4.5 Mix (by weight)		1:6 Mix (by weight)	
	Fine Aggregate (by sieve numbers)					Coarse		$w/c = 0.45$		$w/c = 0.55$	
	100–50	50–30	30–16	16–8	8–4	$4\text{–}\frac{3}{8}''$	$\frac{3}{8}''\text{–}\frac{3}{4}''$	No Agent	With Agent	No Agent	With Agent
1	25.4						74.6	0.3	2.2	0.3	2.9
6	24.0					38.0	38.0	1.0	2.9	0.8	3.4
2	15.4	15.4				34.6	34.6	1.0	3.9	1.9	5.0
3	12.9	12.9	12.9			30.6	30.7	1.6	4.7	2.9	5.8
4	11.7	11.7	11.7	11.7		26.6	26.6	1.8	4.9	3.0	7.0
5	11.2	11.2	11.2	11.2	11.2	22.0	22.0	2.0	5.4	3.4	6.3

[a] Data are from B. G. Singh, "Aggregate Grading Effects Air Entrainment," *Proc. ACI*, **55,** 803–810, 1959.

the air contents of two different mixtures made with and without an air-entraining agent. Whether or not an air-entraining agent was used the air contents increased in the same order as the different gradings are listed. It will be shown below that these increases in the air content were due to an increase in the voids content of the aggregate, and hence to its increased capacity for trapped air.

We have seen in Chapter 1 that when fine and coarse aggregates are mixed together, the voids content of the mixture comes out to be less than that of either aggregate alone, but at the same time the structure of the fine component becomes disturbed so that the voids content of the fine component increases. We may now add that the mean size of the voids becomes increased. In Fig. 1.8 we see that with a given percentage of coarse aggregate, dispersion of the fine aggregate was greater the larger the size ratio. Thus the effect of increasing the size ratio is to enlarge the voids and to increase the total aggregate-voids content. In Table 7.5 it is apparent that the size ratio is smallest for grading No. 1 and greatest for No. 5, with grading No. 6 falling in the second position. With the exception of grading No. 1, the grading of the coarse aggregate remains the same for each mixture. (In terms of the voids-ratio diagram for the aggregate alone this means that point C remains unchanged.)

The last two columns of Table 7.5 show that whether or not an air-entraining agent was used the amount of air retained by a given mixture was greater the greater the size ratio of fine to coarse. Bubble sizes were not reported, but it is safe to say that the maximum size of the bubble increased as the screen size increased. It should be noted also that the increase of air content due to the agent was about 3 per cent in most cases, regardless of the total air content.

8.4 Effect of Increased Air Content on Composition

When a given amount of air is intentionally introduced into a mixture already having a certain composition and consistency, the volumetric composition is altered, and the consistency usually becomes softer. To maintain the original consistency, it is necessary to reduce the water content. Increased air and reduced water may alter the content of cement and aggregate, and the relative magnitudes of these changes depend on several factors to be considered below.

8.5 Voids-ratio Diagrams

A voids-ratio diagram for the data given in Fig. 7.10 is shown in Fig. 7.8. The curve for mixtures containing no air is hypothetical, the data for it having been obtained by extrapolation as is shown in Fig.

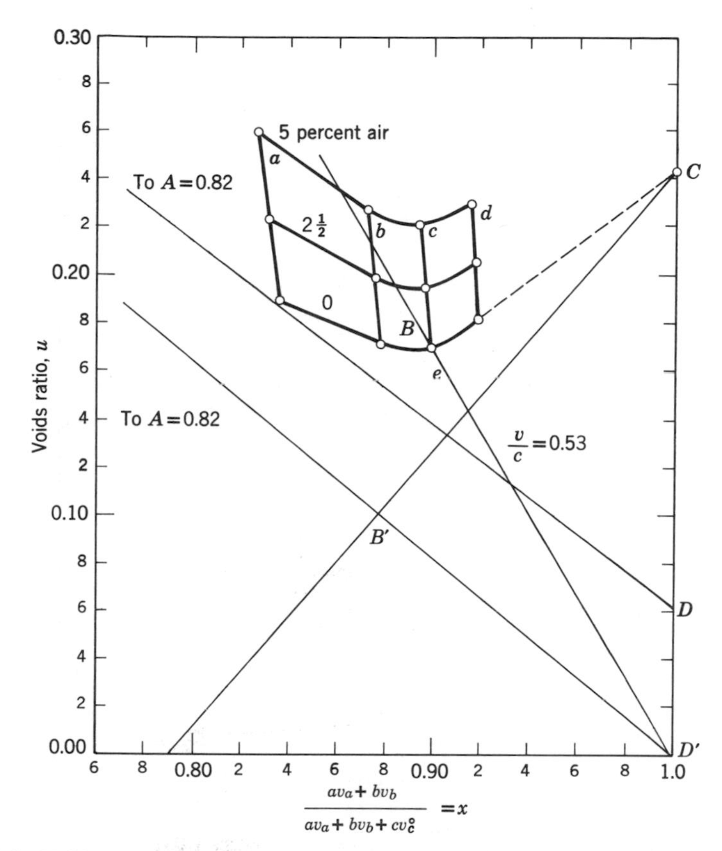

Fig. 7.8. Voids-ratio diagrams for air-entrained concretes with a slump 3 to 4 in. Values of A and D are estimated. Neutralized Vinsol resin was used as the air-entraining agent. (Data are from PCA Series 303.)

7.10. The other curves were established by calculations based on empirical relationships described further on. Because the air content is constant for each curve, the shape of each curve is determined by the water requirement only, and the water requirement is the amount that is necessary for the chosen consistency as measured by the slump test. Point B thus designates the mixture having the minimum water requirement at a given air content.

A conspicuous feature of Fig. 7.8 is that point B occurs at the same value of x (the same mix) regardless of the amount of air in the mixture. Since we have already reasoned that the paste at point B has optimum consistency, it now appears that when air is present, the consistency of the *matrix* composed of bubbles and paste is also an optimum at

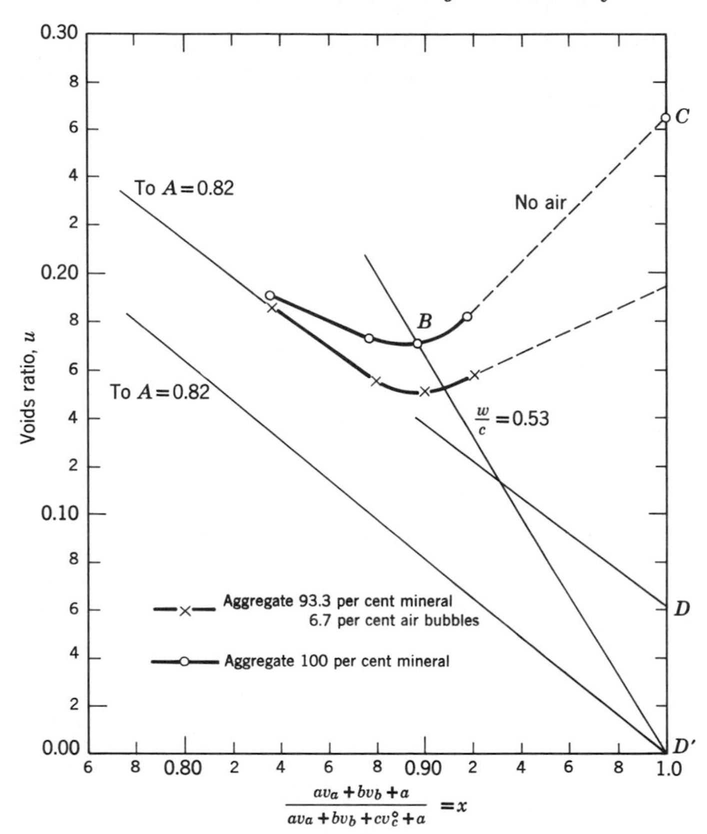

Fig. 7.9. A voids-ratio diagram in which air bubbles are regarded as a component of the aggregate.

that point. Although point B falls at the same value of x for each curve, the consistencies of the matrices are by no means the same for the different curves. With 5 per cent air, for example, the consistency of the matrix at point B must be considerably stiffer than that of the cement paste containing no air; otherwise, the unchanged slump along with the much greater dispersion of the aggregate at the upper point could not be accounted for.

It must be said that the constancy of x_B with the air content variable is not yet well understood. Apparently the virtual particle interference between aggregate and cement which causes the curve to turn away from the line from A to D toward C as x becomes greater also shapes the curve for a fixed percentage of air. (See Chapter 6, Section 3.4.)

The fluidity of the air in concrete obviously does not manifest itself in the way that the fluidity of water does. If air were substituted for water, volume for volume, the mixture would become stiffer with each increment. From this, and various facts already brought out, it is apparent that air bubbles in concrete behave to some degree like particles—pneumatic particles.

It is instructive to adopt temporarily the fiction that bubbles are solid particles and consider them to be a component of the aggregate. When the air is so regarded, the voids fraction in concrete is identical with the water content, and the voids-ratio diagram for the data in Fig. 7.8 takes the form shown in Fig. 7.9. The curve indicated by circled points is the voids-ratio diagram for no air, and that indicated by crosses is the curve for the "aggregate" composed of 6.7 per cent air bubbles and 93.3 per cent mineral aggregate (0–1½ in.).

If we regard them as a component of the aggregate, bubbles appear to improve aggregate characteristics provided that the mix is leaner than some limiting value, which in this case is indicated to be about $x = 0.82$ (or $M = 4.6$), the water-cement ratio being 0.33 by weight $(wv_w/cv_c^o) = 1.06$. This mixture is near point I, the beginning of particle interference. At higher values of x and w/c, the water-cement ratio of the paste is higher for all-mineral aggregate than for an aggregate composed partly of air bubbles.

Although the effect of entrained air thus appears to be that of giving an improved combined aggregate (bubbles and mineral matter) for mixtures showing particle interference, we shall find that it is preferable from most points of view to regard bubbles as a kind of paste-making material, comparable to cement and other mineral powders in that respect, and thus a component of the aggregate matrix. Pneumatic particles are voids, and they weaken concrete to about the same degree as does excess water; therefore treating air-bubbles as voids, as in Fig. 7.8, is obviously more realistic than treating them as aggregate, as in Fig. 7.9. We should, however, keep all aspects of them in mind: voids, pneumatic particles, and paste-making material.

8.6 Linear Relationships between Water Content and Air Content

The concrete-mix designer may choose one of several courses when adjusting a mixture to compensate in certain respects for the effects of intentionally entrained air: he may keep the cement content constant; he may keep the water-cement ratio constant; he may keep the voids-cement ratio constant; he may keep constant the proportion of cement to aggregate. In addition to these choices, he may or may not alter the grading of the aggregate to take full advantage of air bubbles.

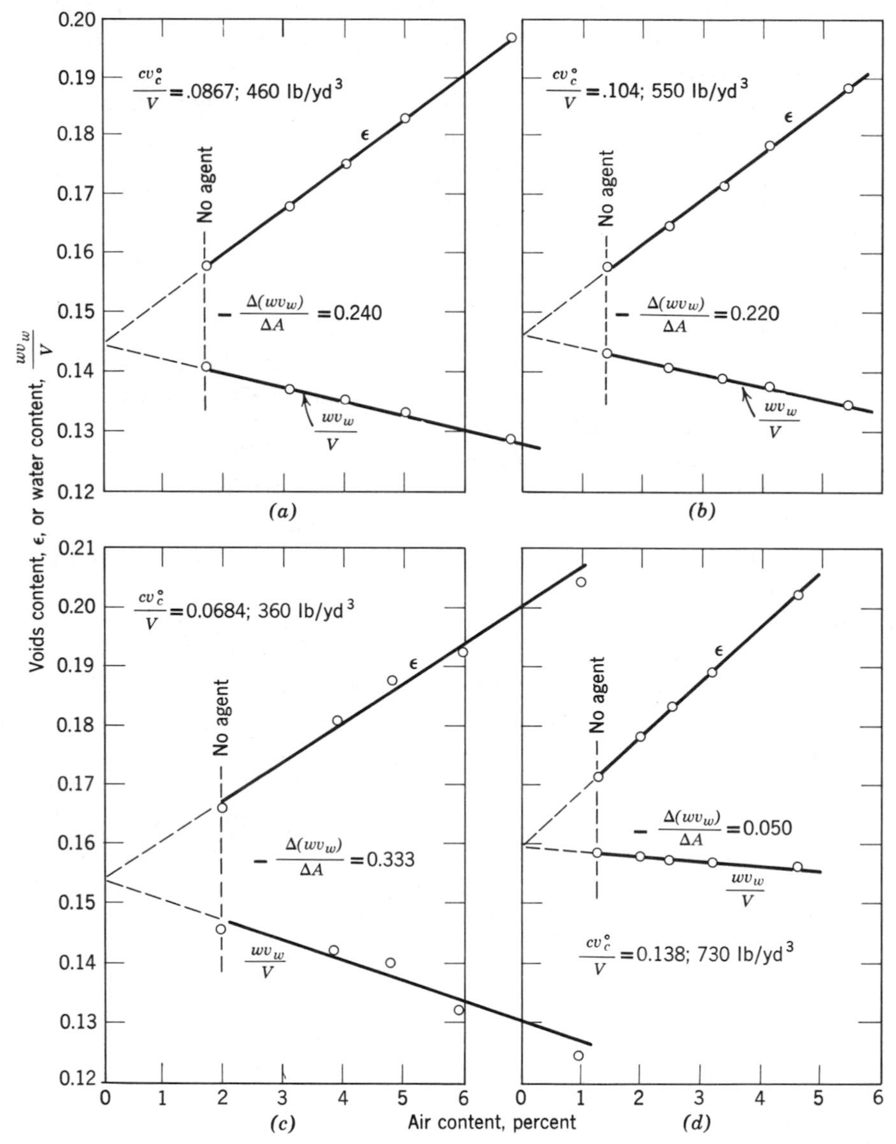

Fig. 7.10. An empirical relationship between water content and air content with cement content constant. (The same data as those in Figs. 7.8 and 7.9.)

Data reported by Gonnerman [23], Cordon [21], and others show that under certain conditions the water content of concrete at a given consistency is inversely proportional to the air content. The factors influencing this relationship need to be understood by designers of concrete mixtures.

8.6.1 Cement content constant. Figure 7.10 gives experimental data on four series of mixtures in each of which the cement content was kept constant while the air content was increased from batch to batch. As we examine these graphs, it should be kept in mind that they show how the proportions in a unit volume of concrete adjust themselves when the experimenter adjusts only the amount of water needed for a given consistency and maintains the same cement content. Cement content being constant for a given diagram, the rise of the upper line denotes a corresponding decrease of aggregate content.

As shown in Fig. 7.10, and as we have seen in other contexts, the air content of plastic concrete is substantial even when no air-entraining agent is present. Water-cement ratios of the pastes in the four mixtures made without an agent range from 0.36 to 0.55 by weight, the paste consistencies of all being well within the fluid range. Such pastes when they are made with typical cement cannot retain air without the aid of a stabilizing agent; therefore we see again that air that might have been released by the paste was held by the aggregate screen. In Fig. 7.10*a*, for example, we see from the vertical distance between the points that the air content in a mixture made without an air-entraining agent was 1.7 per cent. This air content amounts to about 7 per cent of the volume of the matrix.

The four graphs show that under the circumstances described above entrained air displaces mostly aggregate and some water, the amount being a function of the cement content of the mixture. It is apparent also that the air present when no stabilizing agent is used has the same kind of effect as the larger amounts present when an agent is used.

Voids content as a function of air content may be expressed as follows:

$$\epsilon = \frac{wv_w}{V} + \frac{\mathrm{A}}{V} = \left(\frac{wv_w}{V}\right)_0 + \left[1 + \frac{\Delta(wv_w)}{\Delta\mathrm{A}}\right]\frac{\mathrm{A}}{V} \tag{7.4}$$

The term $(wv_w/V)_0$ is the water content when air content is zero, estimated by extrapolation. The $\Delta\mathrm{A}$ is an increase in air content and $\Delta(wv_w)$, which is negative, is the corresponding change in water content, *under the condition that consistency and aggregate grading are constant.* The constant $-\Delta(wv_w)/\Delta\mathrm{A}$ is the water-reduction coefficient.

Water content as a function of air content is given by

$$\frac{wv_w}{V} = \left(\frac{wv_w}{V}\right)_0 + \frac{\Delta(wv_w)}{\Delta A} \cdot \frac{A}{V} \tag{7.5}$$

The water-cement ratio in volume units is given by

$$\frac{wv_w}{cv_c^o} = \left(\frac{wv_w}{cv_c^o}\right)_0 + \frac{\Delta(wv_w)}{\Delta A} \cdot \frac{1}{\left(\frac{cv_c^o}{V}\right)_0} \cdot \frac{A}{V} \tag{7.6}$$

or in weight units,

$$\frac{w}{c} = \left(\frac{w}{c}\right)_0 + \frac{\Delta(wv_w)}{\Delta A} \cdot \frac{1}{v_w} \cdot \frac{1}{\left(\frac{c}{V}\right)_0} \cdot \frac{A}{V} \tag{7.7}$$

Here $(wv_w/cv_c^o)_0$ or $(w/c)_0$ is the water-cement ratio when the mixture contains no air (evaluated by extrapolation) and $(cv_c^o/V)_0$ or $(c/V)_0$ is the constant cement content.

It should be understood that the linearity of the four equations above is an empirically established fact, hardly deducible *a priori;* it is conceivable that successive increments of air could have influenced consistency (as measured by the slump test) in such a way that the corresponding decrements of water would have been unequal; that is the value of $\Delta w/\Delta A$ for a given cement content would not have been independent of air content.

From data given in Fig. 7.10 it is evident that the water-reduction coefficient at a given cement content is a function of the cement content. Figure 7.11, a plot of these data, suggests that the water-reduction coefficient might be a linear function of cement content such that its value is zero when cement content is 0.157, solid volume per unit concrete volume. At this point each increment of air introduced as bubbles would replace an equal volume of aggregate without change of water content and without change of consistency as measured by slump. In practical plastic concrete the water-reduction coefficient is usually more than —0.10. If it is —0.25, for example, it means that when the cement content is held constant each 1 per cent increment of air replaces a volume of aggregate equal to 75 per cent of the volume of the added air, and a volume of water equal to 25 per cent of the added air. It means also that in order to hold the cement content constant it is necessary to increase the proportion of cement to aggregate as air content is made larger, as is shown graphically by lines *a, b, c,* and *d* in Fig. 7.8, their slopes indicating a richening of the mix as the air content is increased.

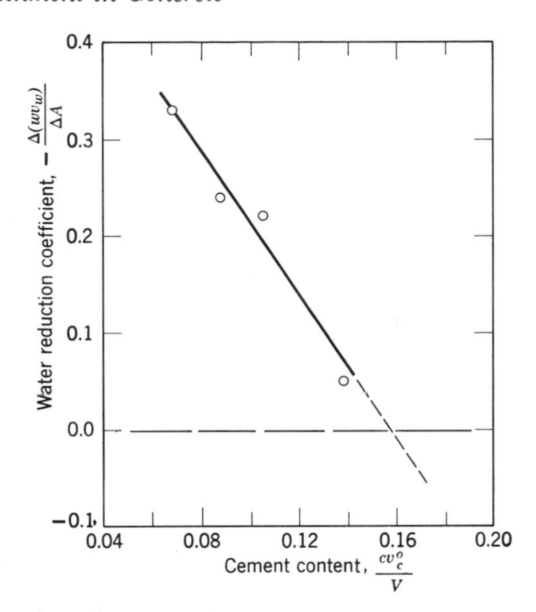

Fig. 7.11. Water-reduction coefficient as a function of cement content.

The discussion of Fig. 7.10 above pertains to results obtained when neutralized Vinsol resin was used, an agent that reacts with calcium hydroxide to produce an insoluble product and consequently is not adsorbable by cement. Since it was not adsorbed it had no effect on interparticle attraction. That is one of the reasons why air entrained by means of this agent had the same effect on water and aggregate content as the air entrained when no agent was present.

When the air-entraining agent is such that it remains soluble and some of it becomes adsorbed by the cement grains the effect of such adsorption is readily discernible, as we may see in Fig. 7.12. The solid lines represent mixtures containing an agent in which calcium lignosulfonate is the surfactant; the broken lines are for neutralized vinsol resin, those lines being drawn as in Fig. 7.10. Notice that the surface-active material gives a relationship between air content and water content or voids content that does not include the point representing the mix without an agent; total voids content and water content are both lower for a given consistency as measured by slump. Comparison with Fig. 7.10 will show that the values of $\Delta(wv_w)/\Delta A$ when the surface-active agent was used are larger than they were when neutralized Vinsol resin was used: —0.36 and —0.234 as compared with —0.258 and —0.226.

Thus we see that for a given total air content above that obtained

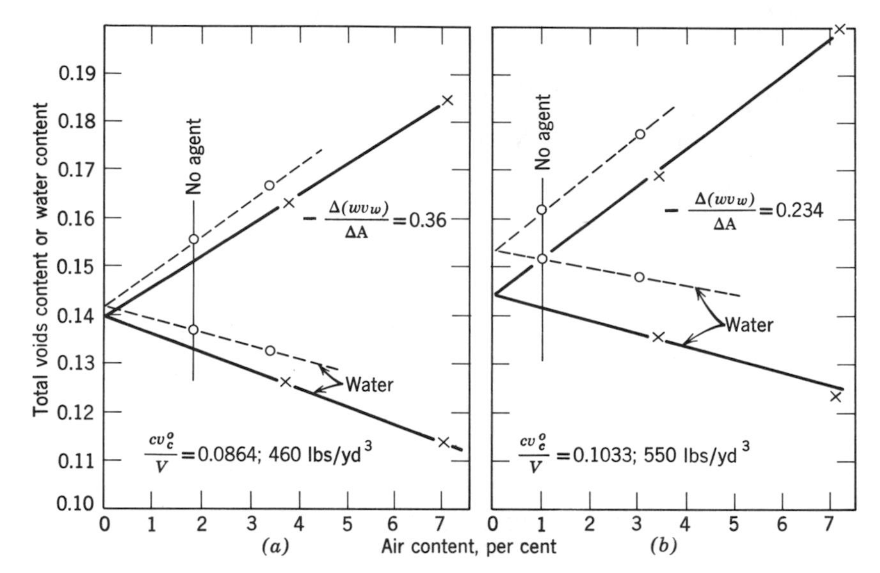

Fig. 7.12. Comparison of the effects of surface-active and insoluble air-entraining agents. The aggregate was graded 0–1½ in. The slump was 2 to 3 in. The surface-active component in the surface-active agent was calcium lignosulfonate. The insoluble agent was the calcium salt of neutralized Vinsol resin. Solid lines for mixtures with surface-active agent. Dash lines for mixtures with insoluble agent. (Data are from PCA Series 305.)

without an air-entraining agent, and for a given cement content, the water content is lower and the aggregate content is higher when we use an air-entraining agent that influences the consistency of the cement paste by being adsorbed on the cement grains, than it is when using an agent that is not adsorbed. The advantage is greater the richer the mix.

8.6.2 Constant water-cement ratio. When the designer chooses to maintain a constant water-cement ratio rather than constant cement content while increasing the air content by means of an air-entraining agent, as did Cordon [21] for example, he is of course choosing to keep the properties of the paste component of the matrix as unaffected as possible. Under this condition the mixture may be regarded as composed of three variable components: paste, air, and aggregate. An increase in the air content requires an equal and opposite change in the total volume of the other two components, but whether or not the ratio of paste to aggregate remains unchanged depends on the requirements for constant consistency.

For a constant consistency as measured by the slump test, Cordon found that if the water-cement ratio is held constant while the air content is caused to increase, the water content has to be diminished in inverse proportion to the increase of air content. Thus he found the same kind of relationship given in Eq. 7.5, but with the water-cement ratio constant instead of the cement content. This result means that successive increments of air replace equal amounts of paste, whether the composition of the paste is allowed to vary or is kept constant. However the water-reduction coefficient is not the same for the two conditions.

Under the condition that $w/c = (w/c)_0$, a constant, it follows from Eq. 7.5 and the experimental results obtained by Cordon that the cement content must be related to the air content as follows:

$$\frac{cv_c^o}{V} = \left(\frac{cv_c^o}{V}\right)_0 + \left(\frac{cv_c^o}{wv_w}\right)_0 \cdot \frac{\Delta(wv_w)}{\Delta\text{A}} \cdot \frac{\text{A}}{V} \tag{7.8}$$

or, in weight units,

$$\frac{c}{V} = \left(\frac{c}{V}\right)_0 + \left(\frac{c}{w}\right)_0 \cdot \frac{1}{v_w} \cdot \frac{\Delta(wv_w)}{\Delta\text{A}} \cdot \frac{\text{A}}{V} \tag{7.9}$$

Values of water-reduction coefficients found by Cordon are given in Table 7.6, along with corresponding values for the constant cement content, obtained from Fig. 7.11. The figures given are for aggregate com-

Table 7.6 Water-reduction Coefficients (for Entrained Air) at Different Constant Water-cement Ratios and at Constant Consistency[a]

Water-cement Ratio with Zero Air Content	Water Content with Zero Air Content		Cement Content with Zero Air Content	Water Reduction Coefficient,	See Note Below[b]
$\left(\frac{w}{c}\right)_0$	$\left(\frac{wv_w}{V}\right)_0$	(lbs/yd³)	(lbs/yd³)	$-\frac{\Delta(wv_w)}{\Delta\text{A}}$	
0.45	0.173	291	647	0.346	0.215
0.55	0.171	288	524	0.372	0.275
0.65	0.175	294	452	0.367	0.315

[a] Data are from W. A. Cordon, "Entrained Air—A Factor in the Design of Concrete Mixes," *Proc. ACL*, **42**, 605–620, 1946.

[b] Figures in this column are water-reduction coefficients for a constant cement content, the cement content in each case being that which gives the water-cement ratio in column 1, at zero per cent air. Data from Fig. 7.11.

posed of rounded particles; for angular particles, different figures were obtained. Cordon observed that for an aggregate composed of crushed limestone, both fine and coarse, water was reduced 793 lb/yd³ per per cent of air, which corresponds to a water-reduction coefficient of 0.471. This value is to be compared with 0.372, Cordon's value for rounded material.

If we keep the water-cement ratio constant, we regulate consistency by changing the paste content, whereas if we keep the cement content constant, we regulate consistency by changing the water content. In the first case the paste consistency does not change, but in the second, reducing the water content makes the paste thicker. It is to be expected, therefore, that a given change in the water content at constant w/c has less effect on the consistency of the whole mixture than it does when the cement content is kept constant. Thus the water-reduction coefficients are the larger for the condition of constant water-cement ratio, as shown in Table 7.6.

From Eq. 7.9 and data given in Table 7.6 we can calculate the change in cement content incidental to maintaining a constant water-cement ratio and constant consistency. For weight and volume in pounds and cubic yards, we have, for room temperature

$$\frac{1}{v_w} = \rho_w = 62.3 \times 27 = 1682 \text{ lb/yd}^3 \text{ of water}$$

For $(w/c)_0 = 0.45$, $(c/w)_0 = 2.22$, and with a water-reduction coefficient of -0.346, we obtain

$$\left(\frac{c}{wv_w}\right)_0 \cdot \frac{\Delta(wv_w)}{\Delta a} \cdot \frac{\text{A}}{V} = 2.22 \times 1682\left(-0.346\,\frac{\text{A}}{V}\right) = -12.92\left(100\,\frac{\text{A}}{V}\right)$$

This result shows that if the water-cement ratio is 0.45 and we wish to maintain a constant consistency, we must remove about 13 lb of cement per cubic yard for each 1 per cent of air added. Thus to increase the content of air from 1 per cent to 5 per cent, we would have to reduce the cement content 52 lb per cubic yard. Similar calculations for $w/c = 0.55$ and 0.65 give reductions in the cement content of 46 and 38 lb/yd³, respectively.

The procedure described above is questionable from the practical point of view, for it seems to imply that entrained air can perform the function of cement. It is more reasonable not to allow the cement content to diminish when we use entrained air, thus minimizing the weakening effect of the air. Indeed, to prevent any loss of strength we must increase the cement content as we increase the air content.

8.6.3 Constant voids cement ratio. If his object is to produce concrete having a specific potential strength, the designer may elect as a first approximation to maintain a constant ratio between the total voids content (water plus air) and the volume or weight of cement. The kind and degree of adjustment required may be seen in Fig. 7.8, wherein line e, the line for $v/c = 0.53$ cm^3/g of cement, is shown. For the hypothetical case of no air in the concrete, the coordinates of a mixture having a voids-cement ratio of 0.53 are $x = 0.9$ and $u = 0.17$. If we keep the cement constant while increasing the air content to 5 per cent the locus would be line c, but if we keep v/c constant, the locus is along line e, and the coordinates of the final point are $x = 0.86$ and $u = 0.236$.

Cement content in terms of coordinates u_X and x is given by Eq. 3.19:

$$\frac{c}{V} = \frac{1}{v_c^o} \cdot \frac{1 - x}{1 + u}$$

For quantities in pounds per cubic yard

$$\frac{1}{v_c^o} = 3.21 \text{ g/cm}^3 \times 62.43 \times 27 = 5411 \text{ lb/yd}^3$$

If no air is present,

$$\frac{c}{V} = 5411 \frac{0.10}{1.17} = 462 \text{ lb/yd}^3$$

and if 5 per cent is air,

$$\frac{c}{V} = 5411 \frac{0.14}{1.236} = 613 \text{ lb/yd}^3$$

indicating an increase of 151 lb/yd^3.

Using the values of x given above, we obtain $M = 1{:}9.00$ if no air is present and $M = 1{:}6.14$ if 5 per cent of the mixture is air, which indicates the change required to maintain the given voids-cement ratio at a constant consistency and constant aggregate grading.

8.7 *Substitution of Air Bubbles for Sand*

The large increase in the cement content required to prevent an increase in the voids-cement ratio in the example given above is usually not required. Usually some loss of strength is tolerable in view of the other benefits obtained. Also the increase in voids-cement ratio can be minimized by substituting air bubbles for some of the sand, thereby increasing the water-reduction factor.

Cordon [21] found that if the percentage of sand in the aggregate was reduced by 1 per cent for each 1 per cent increase in air content while the water-cement ratio was held constant, the consistency also remained constant; he found also that the water-reduction due to displacement of paste by air was about 40 per cent greater than when he left the grading unchanged.

Later, Committee 613 of the American Concrete Institute [24] included in its general recommendations a procedure that amounts to using air bubbles as the equivalent of sand. The volume fraction of cement and that of coarse aggregate are held constant, the water content is determined so that at the desired air content the mixture has the chosen consistency. The rest of the volume is made up of air-bubbles and sand.

The volume fraction of sand is made such as to satisfy the following equation:

$$\frac{av_a}{V} = \left(\frac{av_a}{V}\right)_0 - \frac{A}{V}$$

where $(av_a/V)_0$ is the sand content when no air-entraining agent is used, presumably corresponding to the optimum percentage of sand in the aggregate. This method of adjusting a mixture amounts to reducing the sand content approximately by Cordon's rule, but at the same time preventing a reduction of cement content.

With some mixtures little or no increase in the cement content is required to maintain strength if the factors discussed above are correctly taken into account.

REFERENCES

[1] Pearson, J. C., "Volumetric Method for Determining the Air Content of Freshly Mixed Concrete," *Proc. ASTM,* **44,** 343–350, 1944.

[2] Menzel, C. A., "Development and Study of Apparatus and Methods for the Determination of the Air Content of Fresh Concrete," *Proc. ACI,* **43,** 1053–1072, 1947.

[3] Klein, W. H., and S. Walker, "A Method of Direct Measurement of Entrained Air in Concrete," *Proc. ACI,* **42,** 657–668, 1946.

[4] Verbeck, G. J., "The Camera Lucida Method for Measuring Air Voids in Hardened Concrete," *Proc. ACI,* **43,** 1025–1039, 1947.

[5] Rosiwal, A. *Verhandl, K.-k. geol. Reichanstalt,* 1898, p. 143. See also Johannsen and Stephenson, *J. Geol.,* **27,** 212–220, 1919.

[6] Rexford, E. P., "Discussion of The Camera Lucida Method for Measuring Air Voids in Hardened Concrete," *Proc. ACI,* **43,** 1025–1039, 1947.

[7] Brown, L. S., and C. U. Pierson, "Linear Traverse Technique for Measurement of Air in Hardened Concrete," *Proc. ACI,* **47,** 117–123, 1950.

[8] Lindsay, J. D., "Illinois Develops High Pressure Air Meter for Determining Air Content of Hardened Concrete," *Proc. Highway Research Board,* **35,** 424–435, 1956.

[9] Erlin, B., "Air Content of Hardened Concrete by a High-Pressure Method," *J. PCA Research and Development Laboratories,* **4,** 24–29, 1962.

[10] Backstrom, J. E., R. W. Burrows, R. C. Mielenz, and V. E. Wolkodoff, "Origin, Evolution, and Effects of the Air Void System in Concrete," Part II, "Influence of Type and Amount of Air-Entraining Agent," *Proc. ACI,* **55,** 261–272, 1958.

[11] Powers, T. C., "Void Spacing as a Basis for Producing Air-Entrained Concrete," *Proc. ACI,* **50,** 741–760, 1954.

[12] Mielenz, R. C., V. E. Wolkodoff, J. E. Backstrom, and H. L. Flack, "Origin, Evolution, and Effects of the Air Void System in Concrete," Part I, "Entrained Air in Unhardened Concrete," *Proc. ACI,* **55,** 95–121, 1958.

[13] Lauer, K. R., *The Mechanisms of Air Entrainment in Mortars,* Purdue Univ., doctoral dissertation, 1960. Abstract in *Dissert. Abs.,* **21**(7), 1889, 1961.

[14] Bruere, G. M., "Effect of Type of Surface-Active Agent on Spacing Factors and Surface Areas of Entrained Bubbles in Cement Pastes," *Australian J. Appl. Sci.,* **11**(2), 289–294, 1960. Australia, C.S.I.R.O. Reprint. Abstract in *C.S.I.R.O. Abst.,* **8**(7), 479, 1960.

[15] Kennedy, H. L., "The Function of Entrained Air in Portland Cement," *Proc. ACI,* **40,** 515, 1944.

[16] Scripture, E. W., Jr., F. B. Hornibrook, and D. E. Bryant, "Influence of Size Grading of Sand on Air-Entrainment," *Proc. ACI,* **45,** 217–228, 1948.

[17] Weymouth, C. A. G., "A Study of Fine Aggregate in Freshly Mixed Mortars and Concretes," *Proc. ASTM,* **38,** Part II, 354–372, 1938.

[18] Bruere, G. M., "Air Entrainment in Cement and Silica Pastes," *Proc. ACI,* **51,** 905–919, 1955.

[19] Danielsson, U., and A. Wastesson, "The Frost Resistance of Cement Paste as Influenced by Surface-Active Agents," *Proc. No. 30,* Swedish Cement and Concrete Research Institute at the Royal Institute of Technology, Stockholm, 1958.

[20] Ish-Shalom, M., and S. A. Greenberg, "The Rheology of Fresh Portland Cement Paste," *Portland Cement Association Project B74.* Published in part in "Proc. Fourth International Symposium on the Chemistry of Cement," *Nat. Bur. Std. (U.S.)* Monograph 43, **2,** Session V, paper V-54, pp. 731–744, 1960.

[21] Cordon, W. A., "Entrained Air—A Factor in the Design of Concrete Mixes," *Proc. ACI,* **426,** 605–620, 1946.

[22] Singh, B. G., "Aggregate Grading Affects Air Entrainment," *Proc. ACI,* **55,** 803–810, 1959.

[23] Gonnerman, H. F., "Tests of Concretes Containing Air-Entraining Portland Cements or Air-Entraining Materials Added to Batch at Mixer," *Proc. ACI,* **40,** 477–507, 1944.

[24] ACI Comm. 613, "Recommended Practices for Selecting Proportions for Concrete (ACI 613-54)," *Proc. ACI,* **51,** 49–64, 1954.

[25] Greening, N. R., "Some Causes for Variation in Required Amount of Air-Entraining Agent in Portland Cement Mortars," *J. PCA Research and Development Laboratories,* **9**(2), 22–36, 1967.

[26] Bruere, G. M., "Mechanism by which Air-Entraining Agents Affect the Viscosities and Bleeding Properties of Cement Pastes," *Australian J. Appl. Sci.,* **9,** 349–359, 1958.

8

Characteristics of Air-void Systems

1. INTRODUCTION

The principal purpose of entraining air in concrete, that of protecting hardened cement paste from destruction by frost, can be accomplished by providing a sufficiently large number of bubbles per unit volume of paste to produce a cellular structure in which the cell walls, composed of hardened paste, are only a few thousandths of an inch thick. For a given percentage of air, the average thickness of the walls between bubbles is smaller the smaller the mean bubble diameter and hence the larger the number of bubbles. Therefore we are concerned not only with the quantity of entrained air, but also with the characteristics of the air-void system, particularly the bubble-size distribution. The size distribution of bubbles in concrete can be measured, but the method is tedious, of limited accuracy, and relatively costly.

Although such data as have been reported on bubble-size distribution are considered in Section 6, we now are concerned mainly with a function of size distribution that can be measured with relative ease, namely the specific surface area of the bubbles, which is defined as the boundary area of the air voids divided by the volume of the voids.

The specific surface area of bubbles in fresh cement paste or concrete has never been measured, and it seems unlikely that it ever shall be. Therefore it is a practical necessity to assume that the casts of bubbles found in hardened paste or concrete are representative of the bubbles as they existed in the fresh state. There are reasons to believe that the system of bubbles found in a specimen of hardened concrete is not exactly the same as that which existed when the sample was in the plastic state, but it seems unlikely that in the present context any important error will result from assuming identity. The present purpose is to use such data as a means of studying the role played by entrained air as an ingredient of freshly mixed concrete. Of course, the principal ultimate purpose of entrained air is served to one degree or another by the system of bubbles as it exists in hardened concrete.

2. MEASUREMENT OF SPECIFIC SURFACE AREA

2.1 General Procedure

As was demonstrated by T. F. Willis [1], regardless of what the size distribution of the bubbles may be their specific surface area is simply

$$\alpha = \frac{4n}{\text{A}/V} = \frac{4}{\bar{l}} \tag{8.1}$$

where α is the specific surface area, which is boundary area of bubbles divided by total volume of bubbles, $\bar{l}$ is the mean length of chords found by the traverse line across bubble sections we describe in Chapter 7, n is the average number of bubble sections encountered along a unit length of traverse line, A is the volume of air, and V is the volume of the specimen.

Willis's derivation of Eq. 8.1 was based on a system containing randomly distributed spheres. Smith and Guttman [2] showed that the model need not be restricted as to shape; voids might have any shape, irregular or not, or various shapes, and yet in principle the correct value of specific surface area is obtainable. The necessary restriction is that the voids must be in random array.

The usual procedure for evaluating α is to determine A$/V$ by the linear traverse method (see Chapter 7) and at the same time count the number of bubbles intercepted by the line of traverse. The latter when divided by the total length of traverse, gives $\bar{l}$.

2.2 Accuracy

The accuracy of values of the specific surface area α as it currently is measured is not definitely known. The indications are that if the operator is experienced and the air content is within the normal range the air content, A$/V$, generally agrees within about half a percentage point with the value obtained when the fresh concrete is tested by the pressure method (Chapter 7); but even the counting of the number of interceptions n seems to be subject to considerable uncertainty, for different operators frequently report different counts on the same specimen. Some degree of uncertainty seems inescapable, for each bubble should be represented by chords ranging from zero to a length equal to the diameter of the bubble, and accuracy is limited by the resolving power of the microscope and the ability of the operator to distinguish shallow bubble sections from the plane surface of the specimen.

It has been found that in some cases operators in different laboratories examining the same specimens obtain values of α and A/V that are in good agreement. In other similar comparisons this has not been true.

The data used here were obtained mostly from three sources. For the most part, data from a given source seemed internally consistent, and data from different sources could be compared, provided that conclusions based on small differences, say less than 15 per cent, were avoided.

3. DETERMINANTS OF α IN NEAT PASTES

3.1 Range of Values

Bubbles found in different mixtures of cement and water show a considerable range in their specific surface areas. Mielenz [3] reported α's ranging from about 770 to 1330 in.^{2}/in.^{3}; Danielsson and Wastesson [4] reported values ranging from about 860 to 3900 in.^{2}/in.^{3}; Bruere [5] found values ranging from about 700 to 2100 in.^{2}/in.^{3} (340–524, 511–1535, and 276–827 cm^2/cm^3, respectively). Differences found by each investigator were due to the same factors that control quantity: method and duration of stirring, consistency of the paste, or the water-cement ratio; the kind and amount of air-entraining agent used. There are also effects, relatively minor, due to differences in kinds and amounts of solutes in the mixing water and to differences of temperature. Differences in extremes found by different investigators might have been due in part to differences in accuracy of measurement.

We shall now consider experimental data pertaining to the variable factors influencing air entrainment in neat paste. The mechanics of air entrainment described in Chapter 7 should be kept in mind.

3.2 Duration of Stirring

Since the smallest bubbles probably are produceed in regions of highest shear stress, since only part of the batch is being subjected to such stress at any given time, and since, during stirring, losses if any are predominantly of the larger bubbles (see Chapter 7), it follows that the specific surface area of bubbles should progressively increase with the length of stirring time with a given dosage of air-entraining agent.

Experimental data given in Table 8.1 show that to be true. For the particular conditions represented it appears that a specific surface area of about 2000 $in.^2/in.^3$ (787 cm^2/cm^3) is the highest specific surface area that could be produced. (We may note in passing that the largest values are about one-twelfth the typical specific surface area of cement.)

Table 8.1 Effect of Duration of Stirring on the Air Content and Specific Surface Area of Bubbles in Neat Cement Paste[a]

Stirring Period (min)	Air Content (per cent)	Specific Surface Area α	
		$in.^2/in.^3$	cm^2/cm^3
1	13.6	1470	579
3	21.4	1760	693
4	21.4	1830	720
7	26.4	1940	764
15	26.7	2000	787

[a] Data are from G. M. Bruere, "The Relative Importance of Various Physical and Chemical Factors on Bubble Characteristics in Cement Paste," *Australian J. Appl. Sci.*, **12,** 78–86, 1961.
The value of w/c was 0.45. The agent used was sodium dodecyl sulfate; the dosage was 0.025 per cent of the cement.

3.3 *Speed of Stirrer*

With a given water-cement ratio, dosage, and period of stirring, the specific surface area of bubbles in cement paste is larger the higher the speed of the stirrer, as is shown in Table 8.2. Since the shear stress is higher the higher the rate of stirring, the result is in line with explanations of the mechanics of bubble production in Chapter 7.

3.4 *Effect of Dosage and Kind of Air-entraining Agent*

Under given stirring conditions, and with a given kind of air-entraining agent, Bruere [5] found that the amount of entrained air increased with the dosage, but the specific surface area of the bubbles was independent of dosage, except for the smallest amounts of some agents, in which cases the specific surface areas were relatively low at low dosages. The exceptions mentioned can be ignored for they occur at dosages below those normally required in application.

Table 8.2 Influence of Speed of Stirrer on the Specific Surface Area of Bubbles Produced in Neat Cement Paste[a]

Speed of Stirrer (rpm)	Air Content (per cent)	Specific Surface Area α		Mean Specific Surface Diameter (μ)
		in.²/in.³	cm²/cm³	
750	13.7	1230	484	124
1000	21.4	1760	693	87
1200	20.3	2150	846	71

[a] Data are from G. M. Bruere, "The Relative Importance of Various Physical and Chemical Factors on Bubble Characteristics in Cement Paste," *Australian J. Appl. Sci.*, **12,** 78–86, 1961.

The value of w/c was 0.45. The agent used was sodium dodecyl sulfate constituting 0.025 per cent of the cement. The mixing time was 3 min at 20–22°C.

The mean specific surface area of entrained bubbles was not the same for different agents, as is shown in Table 8.3. Why these differences occur cannot be explained definitely from data now available; they probably involve differences in lowering of surface tension, differences in strength and elasticity of the stabilizing films produced, and possibly

Table 8.3 Characteristic Specific Surface Areas of Bubbles Produced in Cement Paste by Different Air-entraining Agents[a]

Name of Agent	Type	Specific Surface Area α		Mean Specific Surface Diameter (μ)
		in.²/in.³	cm²/cm³	
"Igepon T"	Anionic	2150	846	71
Sodium dodecyl sulfate	Anionic	1880	740	81
"Darex AEA"	Anionic	1475	581	103
Neutralized Vinsol resin	Anionic	1360	535	112
Dodecyl trimethyl ammonium bromide	Cationic	1190	468	128

[a] Data are from G. M. Bruere, "The Relative Importance of Various Physical and Chemical Factors on Bubble Characteristics in Cement Paste," *Australian J. Appl. Sci.*, **12,** 78–86, 1961.

The value of w/c was 0.45. The stirrer was operated at 1000 rpm for 4 min at 20–22°C.

differences in effects on the viscosities of bubble-containing pastes. The relatively large bubble size shown for neutralized Vinsol resin may reflect the fact that, as used in cement paste, it does not lower surface tension. As is discussed in Chapter 7, this material reacts with lime released from cement during mixing and forms an insoluble product which becomes the stabilizing agent. (Incidentally, it should be noted that although neutralized Vinsol resin is an anionic surfactant when dissolved

Table 8.4 Influence of the Water Content of the Paste on the Specific Surface Area of Bubbles

w/c	Air Content (per cent)	Specific Surface Area α in.^{2}/in.^{3}	Specific Surface Area α cm^2/cm^3	Mean Specific Surface Diameter (μ)
Sodium Dodecyl Sulfate, 0.020 mg/g Cement[a]				
0.27	16.8	2540	1000	60
0.35	9.5	1600	630	95
0.43	12.8	1500	590	102
0.51	8.5	864	340	176
Darex AEA, 0.180 mg/g Cement[a]				
0.27	7.5	2820	1110	54
0.35	4.7	2490	980	61
0.43	7.3	1650	650	92
Sodium Dodecyl Sulfate, 0.025 mg/g Cement[b]				
0.40	16.7	2050	807	74
0.45	21.4	1760	693	87
0.50	25.8	1420	559	107

[a] Data are from U. Danielsson and A. Wastesson, "The Frost Resistance of Cement Paste as Influenced by Surface-active Agents," Proc. No. 30, Swedish Cement and Concrete Research Institute at the Royal Institute of Technology, Stockholm, 1958.

[b] Data are from G. M. Bruere, "The Relative Importance of Various Physical and Chemical Factors on Bubble Characteristics in Cement Paste," *Australian J. Appl. Sci.*, **12**, 78–86, 1961.

in water, as shown in Table 8.3, that fact is irrelevant to its use in cement paste or concrete.)

As is shown further on, the differences between different agents demonstrated in Table 8.3 might not be conspicuous in concrete mixtures, for the screen effect tends to limit the maximum size of the voids and thus it modifies the specific surface area of the voids. Presumably the screen effect is independent of the kind of agent used.

3.5 Paste Viscosity or Water-cement Ratio

The specific surface area of bubbles produced in neat cement paste is lower and the mean size larger the higher the water-cement ratio, as is shown in Table 8.4. This, too, is what we would expect from arguments already made: the lower the viscosity of the mixture the lower the shear stress and the larger the bubble that can withstand shear stress.

Data for mixtures containing sodium dodecyl sulfate are shown in Fig. 8.1 in terms of the mean specific surface diameters. Although the data are few and not entirely consistent, they suggest that for a given means of kneading or stirring, the mean diameter increases linearly with the water-cement ratio, at least for most of the practical range of the water-cement ratio. We see again that the device which was capable of producing the higher rate of shear produced the smaller bubble size at a given water-cement ratio.

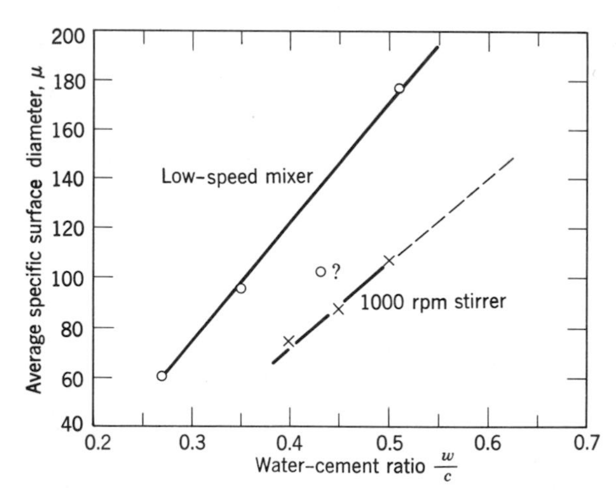

Fig. 8.1. The relation between mean bubble size and the water-cement ratio of neat paste. Sodium dodecyl sulfate was used as the air-entraining agent. (⊙ —Data from Danielsson and Wastesson [4]. × — data from Bruere [6].)

The size of the largest bubbles that can be retained by neat cement paste has not been reported, but it is common to find bubbles in sections of hardened paste as large as 2000 μ (0.08 in.). Thixotropic structure, which becomes evident as soon as stirring stops, thus seems to have sufficient shear strength to arrest the rise of such bubbles, but this ability diminishes as the water-cement ratio is made higher. For that reason we should not expect the lines in Fig. 8.1 to remain linear over an indefinite range; at the limit, which is pure water, discrete bubbles cannot be retained.

4. VALUES OF α IN CONCRETE: NO AIR-ENTRAINING AGENT

4.1 *Range of Values*

In Chapter 7 it is shown that when no air-entraining agent is used in an ordinary concrete mixture, the paste alone is generally too thin to retain air, but the aggregate traps a certain amount depending mostly on its void content and grading, and on the content of cement and other similarly fine material. The mean size of voids so trapped is generally, but not always, larger than that of bubbles found in air-entrained paste. Whereas the specific surface area of bubbles found in cement paste may range from about 800 to 2000 $in.^2/in.^3$, depending on the kind of agent, Mielenz [7] reported that samples from 22 structures having an average air content of 2 per cent contained bubbles having an average specific surface of 443 $in.^2/in.^3$ (174 cm^2/cm^3). (See Table 8.5.) In six of the 22 cases the specific surface areas range from 600 to 1100 $in.^2/in.^3$, the average air content for these being 1.1 per cent. For the largest specific surface value the air content was 0.3 per cent. At the other extreme, six of the 22 had a specific surface of not over 200 $in.^2/in.^3$ (79 cm^2/cm^3) and the average air content was 2.7 per cent. These figures reflect the fact that factors that tend to make the air content of non-air-entrained concrete low also tend to make the specific surface of the bubbles retained relatively high.

4.2 *Determinants*

The factors that determine the specific surface area of voids in concrete when no air-entraining agent is used are generally the same as those controlling the amount of air, as is discussed in Chapter 7. In general, a relatively large air content signifies relatively coarse voids—low specific surface area. For a given aggregate, and given concrete consistency, specific surface area is usually smaller the higher the cement content,

Table 8.5 Characteristics of Voids Found in Nonair-entrained Concrete from Structures[a]

Structure	Part of Structure Represented	Air Content (per cent)	Specific Surface Area in.^{2}/in.^{3}	Spacing Factor $\bar{L}$ (in.)
La Prele Dam, Wyoming	Mass concrete	0.8	816	0.012
Stewart Mountain Dam, Arizona	Mass concrete	1.4	127	0.051
Shasta Dam, California	Mass concrete	0.3	1111	0.011
Black Canyon Dam, Idaho	Mass concrete	0.7	440	0.020
Grand Coulee Dam, Wash.	Mass concrete	0.6	315	0.030
Bonneville Dam, Oregon-Washington	Mass concrete	2.9	645	0.009
American Falls Dam, Idaho	Mass concrete	1.5	976	0.009
Wu Sheh Dam, Formosa	Mass concrete	3.1	152	0.036
Grassy Lake Dam, Wyoming	Mass concrete	1.3	280	0.028
Rosa Dam, Washington	Mass concrete	1.1	444	0.019
Parker Dam, California-Arizona	Mass concrete	0.3	500	0.025
Buck Dam, Virginia	Mass concrete	0.2	800	0.021
All American Canal, California	Canal lining	1.6	548	0.014
Tornillo Canal, Texas	Canal lining	5.7	400	0.011
Hoover Dam, Arizona-Nevada	Spillway	0.9	482	0.019
	Underbed below terrazzo floor of monument	3.8	247	0.023
Flume and wasteway, Owyhee Project, Oregon	Lining	1.0	460	0.021
Air Force Base, Colorado Springs, Colorado	Roof cornice	3.3	444	0.014
Street, Kimball, Nebraska	Pavement	3.7	184	0.030
Parking apron, Casper, Wyoming	Pavement A	2.2	107	0.061
	Pavement B	1.9	150	0.047
Residence, Wheatridge, Colorado	Patio slab	3.7	117	0.047
Average		2.0	443	0.025

[a] See Section 6 concerning the spacing factor. Data are from R. C. Mielenz et al., "Origin, Evolution, and Effects of the Air Void System in Concrete, Part IV, The Air Void System in Job Concrete," *Proc. ACI*, **55,** 507–517, 1958.

at least up to the cement content corresponding to point B in the voids-ratio diagram. At a given point to the right of point B, the specific surface area is generally lower the coarser the sand. Use of mineral powder in such mixtures (see Chapter 3) gives a relatively low air content and a relatively high specific surface area. Stiff mixtures are likely to contain voids of low specific surface area unless thoroughly vibrated or otherwise well manipulated. Effectiveness of puddling or vibrating would be a variable factor in any case.

5. DETERMINANTS OF α IN CONCRETE WITH AN AIR-ENTRAINING AGENT

5.1 *Range of Values*

When normal amounts of air-entraining agents are used, the specific surface areas generally are upwards of 600 in.^{2}/in.^{3}, (236 cm^2/cm^3), as is shown in Table 8.6. Some concrete mixtures have relatively high air contents even when no air-entraining agent is used, usually because of lack of sufficient fine material in the aggregate. When an air-entraining agent is used with such a mixture the voids are likely to have a relatively low specific surface area.

5.2 *Water-cement Ratio and Dosage of Air-entraining Agent*

We have already seen that under given conditions the specific surface area of bubbles formed in neat paste is nearly independent of the dosage of the air-entraining agent, and thus it is independent of the air content. We have that basis for expecting the specific surface area of bubbles found in concrete also to be independent of the dosage of the air-entraining agent. *But, this is not at all true for concrete.* For example, a certain concrete mixture in which the specific surface area of the bubbles was 226 in.^{2}/in.^{3} and for which no air-entraining agent was used, showed the following specific surface areas with successively increasing dosages of neutralized Vinsol resin: 615, 744, and 776 in.^{2}/in.^{3}. The air content increased from 2.1 to 8.5 per cent.* To control entrained air in concrete effectively, we must understand why such variation occurs.

The first question to be considered is why at a given water-cement ratio the air-entraining agent usually fails to produce as high a specific surface area as it does in neat cement paste. There is no reason to suppose that small bubbles fail to be produced in a concrete mixer, for measurements show that the full range of small sizes is present.

* P.C.A. Series 367, References 336–340.

Table 8.6 Characteristics of Voids Found in Air-entrained Concrete from Structures[a]

Structure	Part of Structure Represented	Air Content (per cent)	Specific Surface Area in.^{2}/in.^{3}	Spacing Factor $\bar{L}$ (in.)
Hungry Horse Dam, Montana	Mass concrete, Exterior	3.6	909	0.005
	Mass concrete, Interior	5.6	851	0.003
	Mass concrete, Exterior	3.3	870	0.005
	Mass concrete, Interior	3.5	755	0.005
Canyon Ferry Dam, Montana	Mass concrete, Exterior	2.9	816	0.006
	Mass concrete, Interior	4.8	702	0.004
Davis Dam, Nevada-Arizona	Mass concrete	3.1	1111	0.004
Angostura Dam, South Dakota	Mass concrete	3.7	1111	0.004
Friant-Kern Canal, California	Canal lining	4.2	615	0.008
Poudre Supply Canal, Colorado	Canal lining	10.6	1111	0.002
Tecolote Tunnel, California	Tunnel lining	5.0	1081	0.004
Carlsbad Caverns Project, New Mexico	Highway curb and gutter[b]	4.9	909	0.005
South St. Vrain Project, Colorado	Highway bridge footing[b]	5.3	1600	0.003
Big Bend Nat'l Park Project, Texas	Highway culvert[b]	2.1	1250	0.006
Hugo Wild Horse Project, Colorado	Highway pavement[c]	4.3	930	0.005
Evans-LaSalle Project, Colorado	Highway pavement[c]	4.7	909	0.005
Denver-Bloomfield Project, Colorado	Highway pavement[c]	4.5	1212	0.004
Paonia North Project, Colorado	Highway pavement[c]	4.3	1290	0.004
Limon-Genoa Project, Colorado	Highway pavement[c]	3.9	1143	0.004
Breed Underpass, Breed, Colorado	Highway underpass[c]	2.4	1111	0.006
Building, Buffalo, N.Y.	Structural member	5.6	678	0.006
Railroad bridge	Prestressed deck slab[d]	3.5	727	0.007
Average		4.4	986	0.005

[a] Data are from R. C. Mielenz et al., "Origin, Evolution, and Effects of the Air Void System in Concrete, Part IV, The Air Void System in Job Concrete," *Proc. ACI*, **55**, 507–517, 1958.

[b] Control cylinders compacted by hand rodding, supplied by the Bureau of Public Roads, U.S. Department of Commerce.

[c] Control cylinders compacted by hand rodding, supplied by the Colorado Department of Highways.

[d] Slab cast and prestressed at a plant in Denver, Colorado.

[e] See Section 6 concerning the spacing factor.

Therefore the low specific surface area at low dosage must be assumed to be due to a disproportionate amount of air occurring as large bubbles, relative to the proportions of large bubbles found in neat cement paste.

Bubbles in concrete have a larger average size than those in comparable neat paste apparently for reasons that are not exactly the same in different circumstances. In some cases, especially the lean mixtures containing pastes of low cement content, no air-entraining agent, and no supplementary fine material, the aggregate structure provides spaces large enough, and a screen small enough, to prevent the escape of bubbles too large to be retained by the thin, watery paste alone. However, this explanation does not apply generally when an air-entraining agent is used. As we shall see, the most usual case seems understandable in terms of the mean size of the air voids that are present in a given mixture when no air-entraining agent is used; there is good experimental evidence that these same voids are present even when an air-entraining agent is used and that their mean size is usually relatively large.

When an air-entraining agent is used, stabilized fine bubbles effect a dispersal of aggregate beyond the limit that can be reached by increasing the water content. The aggregate content of the concrete is reduced by such a dispersal almost as much as the air content is increased, if the cement content is held constant. (See Fig. 7.8.)

We may imagine the first increments of small stabilized bubbles forming, in effect, a layer of pneumatic particles around each aggregate particle, or rather a layer of bubble-stiffened paste. The situation can be represented diagramatically as it is in Fig. 8.2*a*, *b*, and *c*. Figure 8.2*a* represents the foundation of the structure, that is, the structure of the aggregate alone. Figure 8.2*b* represents the slightly dispersed aggregate structure in a matrix composed of paste and unstabilized, trapped air bubbles, the air content being represented by one large bubble held along with some paste in the "cage." The slightly dispersed particles, each particle augmented by a layer of adhering paste having thickness equal to half the distance of dispersion, form a virtual aggregate geometrically similar to the foundation structure represented in Fig. 8.2*a*.

Finally, in Fig. 8.2*c* we see the further dilatation (exaggerated) produced by fine bubbles which, in effect, thicken the layer of adhering paste. Experimental evidence showing that such dispersion actually occurs, given in Chapter 7, should be kept in mind. Again, if the particles are thought of as having virtual boundaries beyond their solid surfaces at a distance equal to half the distance of dispersal, the virtual aggregate is geometrically similar to the foundation structure in Fig. 8.2*a*, and the fraction of total volume available to the large bubble (the original air content) is the same as represented in Fig. 8.2*b*. Thus,

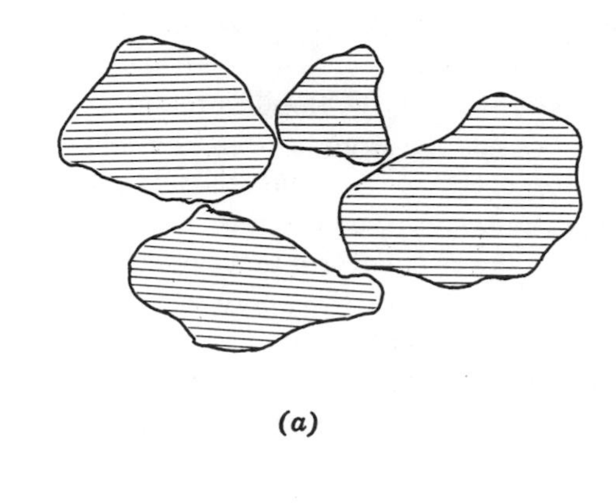

(a)

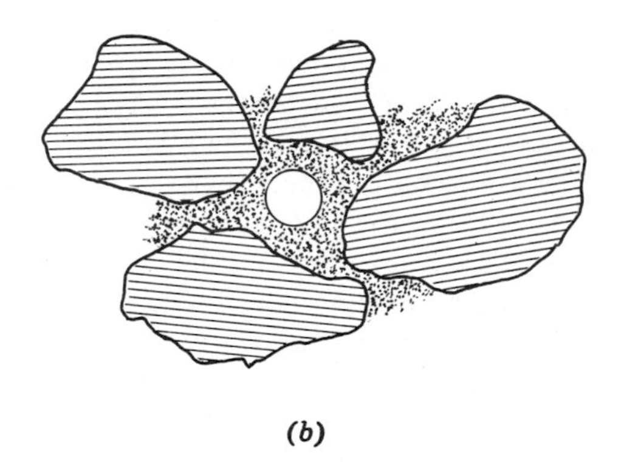

(b)

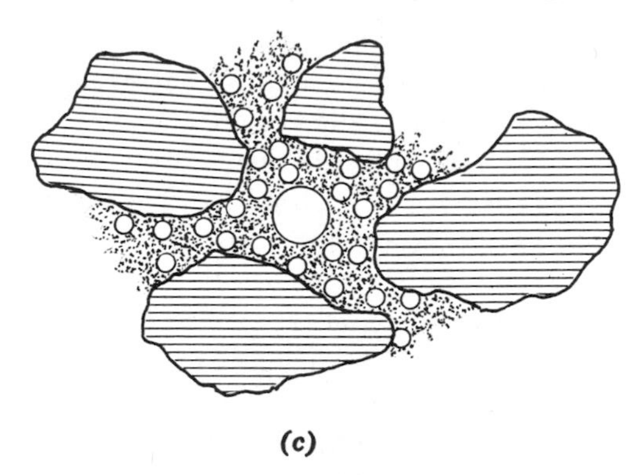

(c)

Fig. 8.2. A sketch indicating how the content of entrapped air tends to remain unaffected by entrained air.

on the basis of this model, it seems possible that the original system of coarse bubbles, represented here by one large bubble, could persist in the presence of additional air introduced by means of an air-entraining agent. The same air, so to speak, seen in Fig. 8.2*b* appears also in Fig. 8.2*c*, and although it has become a slightly lower percentage of overall volume, it remains a significant percentage of total air content.

The picture just imagined is oversimplified, but it serves to rationalize the supposition that, for at least a first approximation of the truth, the bubble content of concrete can be regarded as a combined aggregate of fine and coarse bubbles, each having a characteristic specific surface area. Under given mixing and handling conditions, the characteristics of the fine bubbles are determined by the kind of air-entraining agent and especially by the water-cement ratio of the paste; the characteristics of the coarse bubbles are determined by the aggregate structure and by the specific position of the mix with respect to point B of the voids ratio diagram, and by the amount and kind of fine material in the mixture, other than portland cement. If this is true, or approximately true, it means that the properties of one class of bubbles remains unaffected, or approximately unaffected, by the presence of the other.

On the basis of an assumption of *complete* independence we can consider the specific surface area of the air in a given specimen to be made up as follows:

$$\alpha = \alpha_o(1 - r) + \alpha_f r \tag{8.2}$$

where α_o and α_f are the specific surface areas for the coarse and fine volume fractions of the bubbles and where

$$r = \frac{\mathrm{A} - \mathrm{A}_o}{\mathrm{A}} \tag{8.3}$$

A is the actual volume of air, and A_o the volume present when no air-entraining agent is used.

It is convenient to transform Eq. 8.2 as follows:

$$\frac{\alpha}{\alpha_o} = (1 - r) + \frac{\alpha_f}{\alpha_o} r$$

Subtracting 1 from both sides, we obtain

$$\left(\frac{\alpha}{\alpha_o} - 1\right) = \left(\frac{\alpha_f}{\alpha_o} - 1\right) r \tag{8.4}$$

According to this equation, experimental values of the parameter on the left plotted against corresponding values of r should be a straight

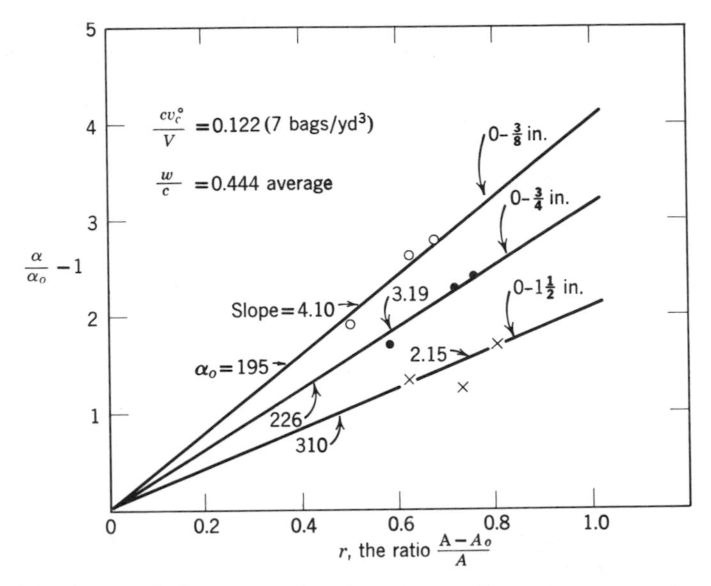

Fig. 8.3. A test of the assumption that the specific surface areas of entrapped air and entrained air are independent.

line having a slope equal to the ratio of the specific surface area of the fine bubbles to that of the coarse, minus 1.0. This relationship could be expected to hold only when different values of r are obtained from the same mix by using different dosages of air-entraining agent from batch to batch. Because the specific surface area of the fine bubbles, although nearly independent of dosage, is a function of the water-cement ratio of the paste, another necessary condition is that *the water-cement ratio should be the same for each batch.* Data meeting these requirements approximately are plotted in Fig. 8.3.

The continuously graded aggregate was composed of sand and crushed limestone in each case, one aggregate comprising nine size groups (0–1½ in.), the second eight size groups, and the third seven. Specimens were taken from $3 \times 3 \times 11\frac{1}{4}$ in. prisms which had been compacted by a standard hand-rodding procedure. It is apparent that Eq. 8.2 is at least a good first approximation of the bubble systems in this group of specimens.

Solving Eq. 8.2 for α_f, as shown in Eq. 8.5, we see that the value of α_f can be calculated from the slopes of the lines in Fig. 8.2.

$$\alpha_f = [1 + (\text{slope})]\alpha_o \tag{8.5}$$

The values in in.²/in.³ indicated by the three groups of data from specimens containing 658 lb/yd³ ($cv_c^o/V = 0.12$) are

Seven size groups; $w/c = 0.45$: $\alpha_f = 990$; Std. Dev. = 64
Eight size groups; $w/c = 0.44$: $\alpha_f = 946$; Std. Dev. = 20
Nine size groups; $w/c = 0.43$: $\alpha_f = 940$; Std. Dev. = 79

There is no significant difference statistically between the three figures for α_f. The mean value of α_f, for the three groups is 960 in.²/in.³, with a standard deviation of 62. The differences of slope in Fig. 8.2 are thus indicated to be due to differences in α_o.

Data from the same source for two lower cement contents and the same three aggregates gave the same indication as Fig. 8.3, but the data appeared to be much more erratic, partly because the individual values of the water-cement ratio deviated rather widely from the mean and apparently because of difficulties with the measuring technique. Instead of the graphical method, a value of α_f was calculated for each datum, using the following form of Eq. 8.2:

$$\alpha_f = \alpha_o + \frac{\alpha - \alpha_o}{r} \tag{8.6}$$

The results of these calculations for all three cement contents are given in Table 8.7. It is apparent from comparison of average values for each of the three groups that the specific surface area of the fine component of the bubbles in concrete is smaller the higher the water-cement ratio of the paste, qualitatively the same as for entrainment in neat cement paste.

The relationship between calculated specific surface diameters of the fine components ($6/\alpha_f$) and water-cement ratios of the pastes in these concrete mixtures is shown in Fig. 8.4, along with the lines from Fig. 8.1, for comparison; also, other data from Backstrom [8] are shown as crosses.*

Although there is leeway for other interpretations, it seems that the rate of increase of bubble size with water-cement ratio at water-cement ratios below about 0.47 is similar to that for neat cement phase, indicated

* The data from Backstrom plotted in Fig. 8.4 are not strictly comparable with the others because they are not calculated from Eq. 8.6 but are the values directly observed in the respective concrete mixtures. The specimens were small and vibrated intensely, and thus α_o was made relatively high, having small effect on α. Therefore, it is believed that Backstrom's data, especially the point for $w/c = 0.35$ serve to show approximately what the trend of the PCA data would have been if there had been such data from very rich mixtures.

Table 8.7 Calculation of the Specific Surface Area of Fine Components of Bubbles in Concrete by Eq. 8.6

Aggregate							
Number of Size Groups	Maximum Size (in.)	Water-cement Ratio (Avg)	r	α_v	α_f	Std. Dev.	Specific Surface Diameter μ
Cement: 658 lb/yd³; $\frac{cv_c^o}{V} = 0.122$							
7	$\frac{3}{8}$	0.45	0.62	195	991[a]	64	60
8	$\frac{3}{4}$	0.44	0.76	226	946[a]	20	63
9	$1\frac{1}{2}$	0.43	0.80	310	886[a]	79	68
Cement: 517 lb/yd³; $\frac{cv_c^o}{V} = 0.096$							
7	$\frac{3}{8}$	0.53	0.58	189	732[b]	158	82
8	$\frac{3}{4}$	0.51	0.71	296	750[a]	132	80
9	$1\frac{1}{2}$	0.49	0.75	260	754[a]	38	80
Cement: 376 lb/yd³; $\frac{cv_c^o}{V} = 0.069$							
7	$\frac{3}{8}$	0.71	0.52	205	583[c]	119	103
8	$\frac{3}{4}$	0.68	0.62	184	676[a]	84	89
9	$1\frac{1}{2}$	0.62	0.67	199	669[a]	68	90

[a] Average of 3.
[b] Average of 4.
[c] Average of 2.

by the lines from Fig. 8.1, but in the higher range of w/c, the rate of increase is not as great as would be expected in neat paste. This aspect of the relationship seems to indicate that in neat paste the decrease in specific surface area, or increase in mean size, accompanying an increase of water-cement ratio involves an increase in maximum as well as average bubble size, as would be expected from the decrease of shear stress under given stirring conditions. In concrete the bubbles cannot be larger than the spaces defined by the aggregate structure, and if these spaces are too small to accommodate the largest bubbles that can be generated in neat paste, the average size of the bubble in cement cannot increase with w/c as much as it would in neat paste.

With some aggregates and mixtures, we might expect the change of slope to occur farther to the right, particularly if the mixture contains supplementary powdered material. Also, we might expect aggregates very deficient in size groups 1 and 2 (deficient in —50 mesh material) to give a curve somewhat different from that in Fig. 8.4, because it would be able to accommodate larger bubbles than in a normally graded aggregate.

As is indicated in Fig. 8.4, the inflection of the curve occurs near the water-cement ratio corresponding to the mixture at point *B* of the voids-ratio diagram. Because we have already learned that to the left of this point the influence of aggregate structure on the properties of fresh concrete rapidly diminishes, and to the right particle interference is dominant, we find it understandable that the relationships should be the same as it is for neat cement in one range, and not the same in the other.

It should be kept in mind that the foregoing discussion pertains to the calculated specific surface area of the fine component of the bubbles in concrete. We have learned about factors that influence α_f in Eq. 8.2; in Section 4 some of the factors that control bubble size of the coarse component in terms of α_0 were considered. It is how clear that the specific surface area of the voids in any given concrete mixture may, at least for a first approximation, be considered to be a function of all these variable factors. It is not surprising that even with an officially accepted air-entraining agent, a wide variety of bubble systems can be found among different concretes, especially if the total volume of air is the same in each case. The characteristics of the aggregate

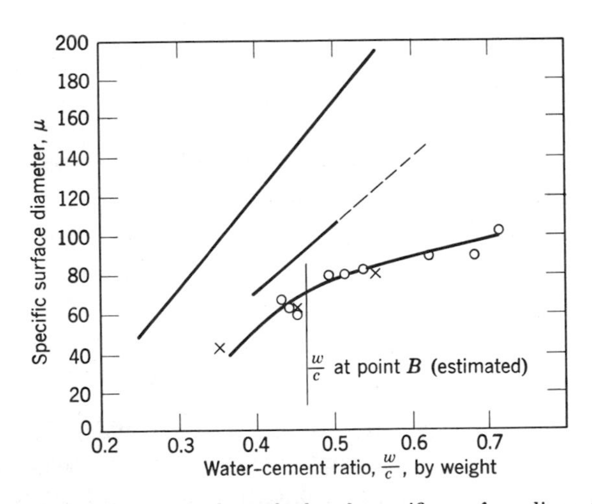

Fig. 8.4. The relation between the calculated specific surface diameter of the fine component of bubbles in concrete mixtures and the water-cement ratio. (⊙ —Data from Table 8.7; × —data from Backstrom [8].)

should be scrutinized just as carefully as those of the air-entraining agent in order to obtain a sufficiently high value of α.

5.3 Surface Texture of Sand

Experiments by Backstrom, et al. [8] showed that in concrete mixtures the specific surface area of the entrained air is influenced by the surface texture of the sand grains. When Backstrom used sands composed of grains that seemed smooth under a microscope, the average specific surface area of the bubbles was, for four different sands, 742 in.^{2}/in.^{3}, whereas for seven sands having rough grain surfaces, it was 1037 in.^{2}/in.^{3}

5.4 Manipulation

Up to this point we have considered air voids found in concrete after the manipulations required for producing specimens have been carried out. During such manipulation the air content of the batch diminishes, and the specific surface area of the retained bubbles increases. The void systems found in concrete specimens therefore depend to an important degree on the way the material is handled before the measurements are made. Test results from two mixtures handled in various ways are given in Table 8.8. The mixtures were alike except for dosage of air-entraining agent.

The data show that ordinary rodding eliminated quantities of air up to about 5 per cent of the volume of the batch, or up to about 70 per cent of the amount of air originally present, and that the effectiveness of rodding depended on the size and shape of the container.

Vibration (under laboratory conditions) eliminated even larger quantities of air. In studies of the effects of prolonged vibration Higginson [9] found that regardless of the amount of air originally present and regardless of the consistency of the mixture, all but about 1 per cent can be removed from laboratory specimens. (Similar results were reported by Crawley [16].) With a given mixture but with various water contents Higginson found that the average rate of loss during vibration was larger the larger the water content and hence the more the slump. For a given mix having different air contents but the same slump, the average rate of loss was the same regardless of differences in initial air content.

The rate of loss of air from laboratory specimens during vibration is influenced by the size of the container and thus by the size of sample. For example, starting with an air content of 9 per cent, Higginson found that one-half the air could be removed from a 0.5 ft^3 sample in a standard unit-weight measure in 50 sec, whereas one-half the air in a 0.2 ft^3

Table 8.8 Effect of Ordinary Manipulations on the Amount and Specific Surface Area of Entrained Air[a]

State of Sample[b]	Air Content (per cent)		Specific Surface Area α in.²/in.³	
	Low Dosage	Normal Dosage	Low Dosage	Normal Dosage
State a	7.8	10.3	296	656
State b	3.4	7.8	421	755
State c	2.9	7.1	444	784
State d	2.3	6.3	563	870
State e	2.2	5.7	615	1030

[a] Data are from J. E. Backstrom, et al., "Origin, Evolution, and Effects of the Air Void System in Concrete, Part III, Influence of Water-Cement Ratio and Compaction," *Proc. ACI*, **55**, 359–375, 1958.

[b] State a: The batch was discharged from the mixer, with no manipulation.
State b: The sample was rodded in a $\frac{1}{2}$ ft³ measure.
State c: The sample was rodded in a slump cone.
State d: The sample was rodded in a prism mold (bar mold).
State e: The sample was vibrated in a cylinder mold.

sample (6 × 12 in. cylinder) required about 12 sec. However, in tests made during construction of two concrete dams, where concrete was handled in large batches and placed in massive sections by vibration, Higginson found practically no loss of air during transport and final vibration. This observation seems compatible with laboratory observations just mentioned. Since in dam construction concrete is handled in batches containing several cubic yards, and since the concrete is deposited in relatively thick layers, it is not surprising to find the rate of loss reduced to a negligible level. Rate of loss from pavement slabs and the like might not be negligible, but even under such circumstances the rate is not likely to be as high as that observed in the laboratory. Higginson concluded that the large rates of loss under laboratory conditions were due to intensification of vibration by reflection from the walls of the container.

The effect of vibration is to break down to one degree or another, and temporarily, the floc structure of cement paste, and thus to fluidize it. While the paste is in the fluidized state, maintained by continual vibration, buoyant force causes bubbles to rise to the surface. If the vibration is sufficiently intense to produce circulation of the material in the mold, conditions for the escape of air are enhanced, but at the same time new air may become emulsified, or air already present may become further dispersed into smaller bubbles, factors that tend to offset the rate of loss. In any case, the final air content would be expected to level off at a relatively low air content, all but the smallest bubbles eventually being lost.

Disproportionate loss of the larger bubbles during vibration in the laboratory is clearly indicated in the data on the change of specific surface area shown in Table 8.8 and by various other data reported by Backstrom. There are strong indications that the voids first eliminated are those that are determined primarily by the structure of the aggregate, whereas those that are retained are those produced primarily by emulsification.

Referring to Eqs. 8.6 and 8.3, let us redefine α_0 as the specific surface area of the voids in concrete in state a, identify $(A/V)_0$ as the corresponding air content, and let (A/V) stand for the air content after further manipulation. Using these quantities we can calculate the specific surface area of the air not expelled α_f. The results of such calculations for the batch with the larger dosage of air-entraining agent are as follows:

For change from state a to state b, α_f = 1063 in.^{2}/in.^{3}
For change from state a to state c, α_f = 1068 in.^{2}/in.^{3}
For change from state a to state d, α_f = 1208 in.^{2}/in.^{3}
For change from state a to state e, α_f = 1493 in.^{2}/in.^{3}

Now if we calculate α_f from the increase of air content and specific surface area when the air content is increased by means of an increase in the dosage of air-entraining agent, we find the following:

For state a, α_f due to increased dosage = 1784 in.^{2}/in.^{3}
For state b, α_f due to increased dosage = 1013 in.^{2}/in.^{3}
For state c, α_f due to increased dosage = 1019 in.^{2}/in.^{3}
For state d, α_f due to increased dosage = 1046 in.^{2}/in.^{3}
For state e, α_f due to increased dosage = 1291 in.^{2}/in.^{3}

For states b, c, d, and e, we see that α_f, calculated from the change in α due to the decrease in the air content by manipulation, and α_f, calculated from the increase in α due to the accompanying increase in the air content due to increased dosage, are similar. The high value for state a, 1784, is probably too high because of the relatively high nonuniformity of unhandled material and the consequent sampling error; this calculation involved two different samples.

On comparing the two sets of calculated values we see that for conditions *b*, *c*, and *d* the air not expelled by rodding is the same as that present because of the use of an air-entraining agent, the part that has a characteristic specific surface area, whereas that expelled is that present primarily because of aggregate structure.

In both sets of calculations given above there is an indication that intense vibration increases the value of α_f, that is, that vibration itself produces small bubbles, probably by dividing larger ones, a conclusion reached before by Backstrom et al. [8].

The experimental findings reported above have important practical implications; they indicate that losses of air from air-entrained concrete during handling have desirable consequences. Since the air that is lost consists mostly of large bubbles, the number of bubbles lost is by no means commensurate with the volume of air lost. Backstrom et al. [8] give the following example. When the air content of a certain concrete mixture was reduced from 6.7 to 1.2 per cent by vibration, the number of bubbles per cubic centimeter of matrix was reduced from 905,000 to 434,000. Although reduced, the remaining number apparently was adequate: Freezing and thawing tests indicated a high degree of protection with either number. The elimination of excess air reduced the void-cement ratio from 0.77 to 0.58 cm^3 of water per gram of cement which is enough to increase 28-day compressive strength more than 1000 $lb/in.^2$

5.5 Temperature

The specific surface area of the bubbles in air-entrained concrete is influenced by the temperature of the batch at the time of mixing, as is also the quantity of air, as discussed in Chapter 7. If when temperature changes the water content of the batch is adjusted to maintain constant consistency as indicated by the slump test, the specific surface area appears to be smaller the higher the temperature if the air contents are moderate; but, if the air contents are high, the specific surface area appears to be influenced little if any. It seems that temperature has little effect on α_f, the specific surface of the fine component of the bubbles, but at a higher temperature a smaller *quantity* of the fine component is added to the original quantity of coarse bubbles, and thus the composite system shows a relatively lower specific surface area.

The data from which these indications were derived pertain to specimens compacted by hand-rodding in the standard manner. Probably if the specimens had been compacted by vibration the specific surface area of the bubbles would be relatively independent of temperature. However, no data have been seen that specifically settle this point.

6. BUBBLE SIZE DISTRIBUTION

6.1 General Considerations

The theory of protecting concrete from the effects of frost by means of entrained air, together with experience, leads to the conclusion that every bit of cement paste must be within a few thousandths of an inch, 75 μ perhaps, of an air space. In general, for a given percentage of air, the average distance between bubbles is smaller the greater the specific surface area of the bubbles.

As a means of estimating the average spacing of bubbles in cement paste, Powers [10] proposed a spacing factor calculated from air content, paste content, and specific surface area of bubbles, by one of the following two formulas, whichever gives the smaller value:

$$\bar{L} = \frac{p}{\alpha \mathrm{A}/V} \tag{8.7}$$

$$\bar{L} = \frac{3}{\alpha}\left[1.4\left(\frac{p}{\mathrm{A}/V} - 1\right)^{1/3} - 1\right] \tag{8.8}$$

where $\bar{L}$ is a number related to the distance between bubbles, called the *spacing factor;* p is the paste content (volumes of water and cement per unit volume of sample). Examples of values obtained with and without air-entraining agents are given in Tables 8.5 and 8.6.

In most cases Eq. 8.8 gives the smaller value of $\bar{L}$ and hence is the one most used. It is based on a model composed of a system of equal spheres, each having a diameter equal to the average specific surface diameter of the actual bubble system. Each sphere is supposed to be at the center of a cube of paste, the total volume of such cubes equalling the total volume of paste plus air in the concrete. The spacing factor $\bar{L}$ is the difference between the radius of a sphere circumscribing such a cube and the radius of the enclosed sphere (air bubble), and thus it is, in the model, the half-distance between bubbles along the diagonals of the cubes.

The factor so obtained is obviously not the actual average spacing in a system of graded sizes, nor is it necessarily related to the actual spacing always in the same way. The actual center-to-center spacing depends on the number of bubbles per unit volume of paste, and for a given specific surface area there might be different size distributions and correspondingly different numbers of bubbles. It is thus possible for a given spacing factor based on specific surface area to correspond to various actual spacings.

Equation 8.7 is simply the volume of paste divided by the boundary area of the bubbles. The quotient is thus a "hydraulic radius" of the paste-filled region occupied by the bubbles, the actual mean distance bubble to bubble being some unknown function of the unknown shapes of these spaces.

If exact knowledge of the distances between bubbles should prove to be necessary, it is obvious that either of the factors described above will not be satisfactory. A forward step toward solution of the analytical aspect of the problem has been made by Philleo [11], who developed a basis for calculating the probability that a given fraction of the total volume of paste is within a specified distance from the nearest void. Application of the analytical procedure proposed by Philleo requires knowledge of the number of bubbles per unit volume of paste. To obtain this figure, the size distribution of the bubbles must be known.

6.2 The Analysis by Lord and Willis

Lord and Willis [12] showed how bubble-size distribution can be obtained by modifying the linear-traverse procedure described in Chapter 7. The procedure for determining air content only requires a record of the cumulative total of the lengths of the individual chords made by the line of traverse and the boundaries of bubble sections. The modification requires that each chord length be recorded separately, and sorted into relatively narrow length intervals. The task is made easier by the auxiliary electronic equipment shown in Fig. 7.1.

6.3 Limitations

The accuracy of estimating the number of voids and the void-size distribution has not yet been established, nor has the precision, that is, reproducibility. However it is known that satisfactory results can be obtained only at the expense of extreme care and by operators with ability to withstand prolonged tedium. A single specimen may require measurement of more than 1000 chords. Even under the best of conditions, accuracy is limited by the resolving power of the microscope and the ability of the operator to distinguish shallow bubble sections from the flat surface in cases where most of the bubble was cut away.

It remains to be learned from future studies whether data having adequate accuracy can be obtained. It also remains to be seen whether data on void-size distribution are any more definitive of the effectiveness of entrained air than are data on specific surface area and whether the advantage, if any, is sufficient to justify the difference in cost of the information. The few data that have been published are believed to represent careful and patient work with adequate equipment. Most of such data that have been analyzed by the Lord and Willis procedure are given below.

6.4 *Effect of Cement Content of Concrete*

Void-size distributions of entrained air in rich, medium, and lean concretes are given in Table 8.9, the richness of the mixes being indicated

Table 8.9 Effect of the Water-cement Ratio on Bubble-size Distribution[a]

Diameter of Bubbles[b] (μ)	Number of Bubbles per cc of Concrete		
	$w/c = 0.35$	$w/c = 0.55$	$w/c = 0.75$
20	255,234	73,240	4000
40	128,386	39,685	20,661
60	26,035	16,285	7878
80	13,183	8191	3234
100	7348	5215	1589
120	1847	4002	1103
140	1354	889	694
160	812	832	531
180	639	523	779
200	1271	1381	1672
250	118	129	196
300	33.7	120	136
350	15.3	61.1	118
400	59.8	94.2	57.9
450	47.7	54.1	31.8
500	68.4	104.0	79.5
600	4.3	14.9	18.2
700	3.1	17.8	17.8
800	20.3	1.6	1.7
900	—	1.3	1.3
1000	—	9.7	8.7
2000	—	1.0	2.0
Total	436,480	150,852	42,811
Specific surface area α in.^{2}/in.^{3}	1380	800	500
Air content, per cent	4.8	5.2	5.3
Paste content, per cent	41.2	26.9	24.1
Number of bubbles per cc of paste	106.0×10^4	56.0×10^4	1.8×10^4

[a] Data are from J. E. Backstrom, et al., "Origin, Evolution, and Effects of the Air Void System in Concrete, Part III Influence of Water-Cement Ratio and Compaction," *Proc. ACI*, **55**, 359–375, 1958.

[b] The figures give upper limits of length intervals; for example 40 indicates 20 to 40 μ.

Table 8.10 Comparison of Bubble Systems Produced with Two Different Air-entraining Agents[a]

Diameter of Bubbles[b] (μ)	Number of Bubbles per cc of Concrete	
	Agent D	Agent L
20	210,470	605
40	91,646	3403
60	30,761	2494
80	11,501	1608
100	7319	1023
120	4125	554
140	1056	548
160	1344	272
180	539	278
200	254	247
250	245	173
300	87.7	77.6
350	129.0	52.8
400	61.1	28.0
450	54.3	3.8
500	18.1	1.8
600	11.0	14.0
700	8.0	3.2
800	10.6	5.6
900	0.4	4.0
1000	1.5	4.6
2000	1.0	0.8
3000	1.5	0.2
4000	—	0.2
Total	359,645	11,401
Air content, per cent[c]	5.1	2.3
Specific surface area α in.^{2}/in.^{3}	1143	476
Paste content, per cent	25.0	25.0
Number of bubbles per unit volume of paste	144×10^4	4.56×10^4

[a] Data are from J. E. Backstrom, et al., "Origin, Evolution, and Effects of the Air Void System in Concrete, Part II, Influence of Type and Amount of Air-entraining Agent," *Proc. ACI,* **55,** 261–272, 1958.
[b] The figures give upper limits of diameter intervals; for example, 40 corresponds to 20 to 40 μ.
[c] Values were obtained by the linear traverse method. The corresponding values by the pressure method were 5.4 and 4.8 per cent.

Table 8.11 Effect of Vibration on Bubble-size Distribution[a]

Diameter of Bubbles[b] (μ)	Number of Bubbles per cc of Concrete	
	Vibrated 2 sec	Vibrated 50 sec
20	155,839	68,869
40	96,180	29,887
60	19,830	15,641
80	14,666	2981
100	6014	1836
120	5847	838
140	1855	491
160	884	306
180	772	265
200	418	180
250	525	70.1
300	139	19.1
350	65.6	2.0
400	45.0	20.4
450	23.9	23.0
500	18.2	5.7
600	4.1	0.6
700	12.7	3.6
800	7.4	—
900	1.8	—
1000	13.9	—
2000	0.6	—
3000	0.7	—
4000	0.2	—
Total	303,163	121,438
Air content, per cent	6.7	1.2
Specific surface area α in.^{2}/in.^{3}	870	1540
Paste content, per cent	26.8	28.1
Number of bubbles per unit volume of paste	113×10^4	43.2×10^4

[a] Data are from J. E. Backstrom, et al., "Origin, Evolution, and Effects of the Air Void System in Concrete, Part III, The Air Void System in Job Concrete, *Proc. ACI,* **55,** 507–517, 1958.

[b] The figures give upper limits of diameter intervals; for example 40 corresponds to 20 to 40 μ.

by their respective water-cement ratios. As would be expected from earlier discussion, the observed number of bubbles is higher the higher the specific surface area, and calculations show that the number of bubbles calculated from the specific surface diameter is considerably smaller than that calculated from size distribution: 16.4×10^4, 5.3×10^4, and 1.5×10^4, to be compared with 106×10^4, 56×10^4, and 1.8×10^4. Only for the leanest mix are the figures alike, and in this case it seems questionable that the figures indicate no more difference than they do.

6.5 *Effect of Characteristics of Air-entraining Agent*

Some air-entraining agents produce relatively unstable systems of large bubbles. A size analysis of one such system is shown in Table 8.10, along with data from a satisfactory agent. Agent *L* was obviously the inferior material.

6.6 *Effect of Vibration*

When air bubbles escape from fresh concrete while the concrete is being vibrated, bubbles of all sizes are lost, and the largest sizes tend to be eliminated completely. As may be seen in Table 8.11, 50 sec of vibration under laboratory conditions eliminated practically all bubbles larger than 500 μ. In this case the air content was reduced from 6.7 to 1.2 per cent, a reduction of 82 per cent, and the number per unit volume of paste was reduced 62 per cent. At the same time the specific surface area was nearly doubled. Freezing tests indicated that the specimens remained well protected with only 1.2 per cent air in the form of very small bubbles.

7. INSTABILITY OF AIR BUBBLES

7.1 *Theoretical Considerations*

We are already aware that an air-entraining agent stabilizes the bubbles formed during mixing and makes possible a relatively large accumulation of air in the form of bubbles of various sizes. Mielenz et al. [3] pointed out that such a system of bubbles, although stable in the sense already discussed, is intrisically unstable thermodynamically. To maintain thermodynamic equilibrium at a given uniform temperature, the pressure of air against water must be everywhere the same, and the water must be saturated with air, the amount dissolved depending on temperature. It is easy to show that such conditions cannot prevail in fresh concrete.

A 1-ft cube of fresh plastic concrete with one face exposed will present

1 ft^2 of water surface to the atmosphere; if the unit cube contains 5 per cent air in the form of bubbles having a specific surface area of 1000 $in.^2/in.^3$, it presents about 600 ft^2 of water surface to air internally. Thus it is that in air-entrained concrete the principal areas of contact between water and air are at the boundaries of bubbles, and the amount of air required to saturate the water depends principally on pressures at the bubble boundaries.

At a given level below the top surface of a mass of fresh concrete, the air pressure in a bubble is equal to the hydrostatic pressure at that level plus the pressure due to surface tension; thus

$$P = P_h + P_\gamma = P_h + \frac{4\gamma}{d} \qquad (8.9)$$

where P is the air pressure in the bubble, P_h is the hydrostatic pressure, P_γ is the pressure due to surface tension, γ is the coefficient of surface tension, and d is the diameter of the bubble. From Eq. 8.9 it is clear first of all that internal air pressure is generally higher than external air pressure and that the water in concrete must become supersaturated with air, with respect to the air in the atmosphere. Consequently, we must suppose that at any exposed surface air passes continually from the water in air-entrained concrete to the atmosphere. However, such a possible loss of air should be inconsequential in comparison with the possible rate of transfer of air from one bubble to another.

Since bubble diameters in a freshly formed system may range initially from perhaps 2 or 3 μ up to 2000 μ or more, it is evident from Table 8.12 that pressures in adjacent bubbles may differ by as much as 1 atm or even more. If the water surrounding a small bubble should become saturated with respect to the pressure in the small bubble, it will have become supersaturated with respect to the water surrounding a larger bubble. Hence we should expect air to diffuse through the water from a smaller to the larger bubble, diminishing the smaller and enlarging the larger. The average pressure on the air will have diminished; hence the total volume of air will have increased.

On theoretical grounds, Mielenz et al. [3] who were the first to discuss the subject, stated that the approximate rate of transfer from one bubble to another may be expressed as follows:

$$\frac{dB}{dt} = \frac{aD\gamma}{\theta}\left[\frac{1}{d_1} - \frac{1}{d_2}\right] \qquad (8.10)$$

where dB/dt is the rate of mass transfer of air; a is the area of the smaller bubble; D is a diffusion constant, which is larger the greater the solubility of the gas and which depends also on the permeability to air of the stabilizing film at the bubble boundary; θ is the average distance between bubble boundaries; and d_1 is the diameter at a given time of the smaller bubble and d_2 that of the larger.

It should be noticed that Eq. 8.10 can represent the rate of diffusion to an outside surface if d_2 is taken as infinite, provided that the bubble is very near the surface; for such a bubble, rate of loss at any given moment is inversely proportional to the bubble diameter at that moment. We may see also in Eq. 8.10 reason to believe that there is some lower limit to the size of a bubble that can last long enough to leave a cast

Table 8.12 Air Pressure in Bubbles Due to Surface Tension Only[a]

Bubble Diameter (μ)	P_γ	
	Atm	psi
640	0.0045	0.066
320	0.009	0.132
160	0.018	0.264
80	0.036	0.528
40	0.072	1.06
20	0.144	2.11
10	0.288	4.22
5	0.576	8.45
2.5	1.15	16.9

[a] Calculations are based on Eq. 8.9 with $\gamma = 72.75$ dyn/cm.

in hardened cement paste. Mielenz expressed the view that it is "improbable that bubbles originally less than 10 μ will be preserved."

Under given conditions of diffusion, rate of transfer from a small bubble to a large one should continually increase, for the transfer causes the difference between the two diameters to increase. In a system containing bubbles of many sizes, bubbles of intermediate size may at first increase in diameter as they receive air from smaller bubbles, but after the supply of air from smaller bubbles becomes depleted, they may then begin to shrink by loss of air to larger bubbles; in general, at any given time bubbles smaller than the average should be diminishing while those larger than the average should be increasing in size. Mielenz pointed out that the final result of transfer of air from smaller to larger bubbles is not only an increase in total volume of air, but also a decrease in the number of bubbles, and a reduction in the specific surface area of the air void system. The extent to which such changes can take place in concrete should be limited by the length of the period during which the cement paste remains weak enough to allow the larger bubbles to expand. Mielenz concluded that all factors that increase the time of setting will increase the extent to which the process described above progresses.

7.2 Experimental Observations

The principal deductions from Eqs. 8.9 and 8.10 were verified very well by Mielenz and his coworkers, mostly by semiquantitative observations of the behavior of bubbles produced in water, in limewater, and in thin cement slurries. They were able to see the growth of large bubbles at the expense of small ones. They had observed also that the bubbles found in hardened paste of high water-cement ratio have a lower specific surface area than those in pastes of lower water cement ratio, and they considered this to be further confirmation of their deductions; it was logical to suppose that the difference in specific surface area is due to the relatively favorable conditions for diffusion of air from bubble to bubble in the thinner paste during the period when the paste was quasifluid.

However, as we have already seen, the difference in specific surface area in question is to be expected from considerations pertaining to the mechanism of air-entrainment in cement paste. That is to say, the difference in specific surface area of the bubbles can be accounted for in terms of the viscosity of the paste, duration of stirring, and speed of the stirrer. In the present case we may assume that the duration and speed of stirring are constant factors and that the relatively low specific surface area of the bubbles in the thinner paste is due to the lower viscosity of that paste and consequently lower shearing stresses.

It seems doubtful, therefore, that an explanation of differences in specific surface area based solely on factors controlling the interchange of air between bubbles can be sufficient.

Mielenz et al. [3] using neat cement pastes mixed 4 min in a drum-type concrete mixer, found the air content measured by the linear traverse method to be substantially higher than that indicated by the pressure method applied to the fresh paste. For example, at $w/c = 0.60$ the figure for the fresh state was 3.3 per cent, whereas that for the hard state was 8.8 per cent; 13 such observations were reported, all showing more air in the hardened paste than in the fresh. These observations were in line with expectations based on Mielenz's theoretical considerations already given. However, there were data on concrete mixtures paralleling some of those on neat cement showing in each of four cases substantially *more* air in the fresh state than in the hardened, just the opposite of the observations on neat cement.

Bruere [14] attempted to verify Mielenz' deductions by experiments carried out on cement pastes within the normal range of compositions. By using an accelerator (ammonium carbonate) he caused air-entrained pastes to become rigid within 3 min and, from later measurements on the hardened material, found that both the amount and the specific surface area to be the same as when the paste was allowed to remain plastic for the normal period, in this case about 2 hr. Also, he lengthened the period of plasticity to about 4 hr (by adding citric acid) and again found the amount and specific surface of the casts found in the hardened specimen to be the same as for the other two conditions. He entrained nitrous oxide, which is about 50 times as soluble as air and found a relatively low specific surface area as would be predicted by Mielenz, but the area did not change after the period of stirring was over; this conclusion was indicated by the similarity of values obtained from retarded and accelerated pastes. Bruere concluded that all the measurable loss from small bubbles occurred while the material was still in the mixer and that none occurred after the period of mixing.

In contrast, Larson et al. [15] obtained results from experiments on concrete that seem to give some support to Mielenz's deductions. To different batches of a certain concrete mixture Larson added different dosages of a surfactant, the purpose being to retard the setting of the mixture various degrees. Along with each dose of surfactant he added a dose of neutralized Vinsol resin, the dosage being adjusted to give an air content of about $5\frac{1}{2}$ per cent.* The surfactants alone entrained some air, but at the dosages required for the desired degrees of retardation they generally produced less than the $5\frac{1}{2}$ per cent wanted.

The same program was carried out with three different surfactants. With two of the three surfactants the specific surface area of the bubbles was smaller the greater the dosage of the surfactant; with the third the specific surface areas were relatively low, but there was no downward trend with increased retardation.

Thus the results with two of the three surfactants seem to be in line with Mielenz's theory, and those from the third might be reconciled with it. Nevertheless we must consider these results to be inconclusive

* It should be recalled that when neutralized Vinsol resin is used with cement paste it does not function as a surfactant because of its reaction with $Ca(OH)_2$ to form an insoluble air-entraining agent.

because the kind of air entraining agent was a variable factor. That is to say, we know that different agents produce bubbles having different specific surface areas, and although we may consider a certain mixture of a surfactant and Vinsol resin to function as one agent, we cannot expect the properties of such a composite agent to be independent of the proportions of its ingredients. We see in Table 8.3 that there is no evidence that retarding surfactants of the kind used by Larson produce low specific surface areas, even though there was as much opportunity for diffusion of air from small bubbles to larger ones as there was in Larson's tests.

Yet we cannot dismiss Mielenz's data from cement paste showing that the air content does increase, presumably because of the enlargement of the larger bubbles by air received from the smaller ones; the amount of indicated growth is too large, and the observations too numerous, to be ascribed to systematic error. Nonetheless they seem in conflict with Bruere's results.

Seeking an explanation of the discrepancy between Bruere's findings and those of Mielenz, we note that at $w/c = 0.45$, Bruere obtained a specific surface area of about 1600, whereas Mielenz, with the same kind of agent, obtained about 1000 $\text{in.}^2/\text{in.}^3$ The difference is probably due to different mixing procedures, Bruere's method apparently producing higher shear stress and thus smaller bubbles. If we assume on this basis that Mielenz's sample contained many large bubbles, whereas Bruere's did not, we can assume also that in Eq. 8.10 the differences between d_1 and d_2 were correspondingly different and thus find a basis for finding a larger rate of transfer in the Mielenz specimens.

On the whole, it now seems doubtful that the thermodynamic instability of air-bubble systems is able to bring about spontaneous changes in the air-bubble systems in concrete. Probably the cohesive strength of fresh paste or concrete at any composition within the normal range is such to offer effective resistance to the enlargement of large bubbles, and the structure of the paste component is such that it greatly reduces if it does not eliminate occurrences of direct contact between bubbles, thus maintaining long diffusion paths as compared with the short distances that prevailed in Mielenz's microscopical experiments.

REFERENCES

[1] Willis, T. F., Discussion of "The Air Requirement of Frost Resistant Concrete" by T. C. Powers, *Proc. Highway Research Board,* **29,** 203–211, 1949.

[2] Smith, C. S., and L. Guttman, "Measurement of Internal Boundaries in 3-Dimensional Structures by Randon Sectioning," *J. Metals* (*Trans. AIME*), **5,** 81–87, 1953.

[3] Mielenz, R. C., V. E. Wolkodoff, J. E. Backstrom, and H. L. Flack, "Origin, Evolution, and Effects of the Air Void System in Concrete," Part I, Entrained Air in Unhardened Concrete," *Proc. ACI,* **55,** 95–121, 1958.

[4] Danielsson, U., and A. Wastesson, "The Frost Resistance of Cement Paste as Influenced by Surface-active Agents," Proc. No. 30, Swedish Cement and Concrete Research Institute at the Royal Institute of Technology, Stockholm, 1958.

[5] Bruere, G. M., "Effect of Type of Surface-Active Agent on Spacing Factors and Surface Areas of Entrained Bubbles in Cement Pastes," *Australian J. Appl. Sci.,* **11**(2), 289–294, 1960. Australia, C.S.I.R.O. Reprint. Abstract in *C.S.I.R.O. Abst.,* **8**(7), 479, 1960.

[6] Bruere, G. M., "The Relative Importance of Various Physical and Chemical Factors on Bubble Characteristics in Cement Paste," *Australian J. Appl. Sci.,* **12,** 78–86, 1961.

[7] Mielenz, R. C., V. E. Wolkodoff, J. E. Backstrom, and R. W. Burrows, "Origin, Evolution, and Effects of the Air Void System in Concrete, Part IV, The Air Void System in Job Concrete," *Proc. ACI,* **55,** 507–517, 1958.

[8] Backstrom, J. E., R. W. Burrows, R. C. Mielenz, and V. E. Wolkodoff, "Origin, Evolution, and Effects of the Air Void System in Concrete, Part III, Influence of Water-Cement Ratio and Compaction," *Proc. ACI,* **55,** 359–375, 1958.

[9] Higginson, E. C., "Some Effects of Vibration and Handling on Concrete Containing Entrained Air," *Proc. ACI,* **49,** 1–12, 1952.

[10] Powers, T. C., "Void Spacing as a Basis for Producing Air-Entrained Concrete," *Proc. ACI,* **50,** 741–760, 1954.

[11] Philleo, R. E., "A Method of Analyzing Void Distribution in Air-Entrained Concrete," unpublished paper, Portland Cement Association Research and Development Division, Skokie, Illinois, 1955.

[12] Lord, G. W., and T. F. Willis, "Calculation of Air Bubble Size Distribution from Results of a Rosiwal Traverse of Aerated Concrete," *ASTM Bull.,* 56–61, October 1951.

[13] Backstrom, J. E., R. W. Burrows, R. C. Mielenz, and V. E. Wolkodoff, "Origin, Evolution, and Effects of the Air Void System in Concrete," Part II, "Influence of Type and Amount of Air-Entraining Agent," *Proc. ACI,* **55,** 261–272, 1958.

[14] Bruere, G. M., "Rearrangement of Bubble Sizes in Air-Entrained Cement Pastes during Setting," *Australian J. Appl. Sci.,* **13**(3), 222–227, 1962.

[15] Larson, T. D., J. L. Mangusi, and R. R. Radomski, "Preliminary Study of the Effects of Water-Reducing Retarders on the Strength, Air Void Characteristics, and Durability of Concrete," *Proc. ACI,* **60**(12), 1739–1753, 1963.

[16] Crawley, W. O., "Effect of Vibration on Air Content of Mass Concrete," *Proc. ACI,* **49,** 909–920, 1953.

9

Intermolecular Forces, Adsorption, and the Zeta Potential

1. INTRODUCTION

We have already made various references to interparticle forces, but without explaining them. It is time now to discuss these forces in greater detail, for it is not possible to comprehend basic concrete technology without becoming familiar with their nature and their several manifestations in concrete. Our immediate purpose is to become familiar with forces among particles in fresh paste and concrete.

We are concerned with solids composed of countless ions or molecules. Such bodies attract nearby molecules or ions and at the same time repel them in a way that is not fundamentally different from the action between individual molecules in the gaseous state. These forces of attraction and repulsion are universal and relatively nonspecific; any particle attracts any other particle whether they are alike or different and whether or not chemical reaction between them is possible. The universal, nonspecific interactions just mentioned comprise several mechanisms giving rise to forces of attraction known collectively as *van der Waals forces*.

We are concerned also with electrostatic forces between solids because they have much to do with the physical properties of cement paste. They are controlled by the nature of the solid surfaces and the composition of the aqueous solution with which the surfaces are in contact.

Van der Waals forces have some degree of specificity; that is, attraction is not quantitatively equal between various pairs of molecules or between solid surfaces and different molecules or ions. Indeed where water molecules are involved, cohesion usually involves a definite though weak chemical bond—the hydrogen bond—as well as the less specific van der Walls attraction common to all molecules.

One consequence of universal attraction, with its secondary degree of specificity, is the phenomenon called *adsorption*. A solid surface attracts and holds momentarily the molecules of a fluid with which it is in contact, strongly or weakly according to the degree of attraction. Water is strongly adsorbed by the solid material in concrete, and any mechanical interference with the normal process of adsorption is likely to meet with powerful resistance.

Adsorption involves not only the molecules of a fluid, but also it may involve ions, particularly the ions of a substance in solution. When ions are adsorbed by a solid surface, those of one sign are usually attracted, while those of the opposite sign are repelled. Such *selective adsorption* gives rise to various important effects, as is seen further on.

Thus, we have ample reason to consider a study of interparticle forces as prerequisite to the study of basic concrete technology. However since the subject is almost as broad and deep as the whole of physical chemistry, we shall have to limit our discussion here to the minimum necessary to develop terminology and principal concepts.

We begin by discussing the nature of van der Waals forces between pairs of molecules.

2. VAN DER WAALS FORCES

The complex subject of van der Waals forces is treated fully in such textbooks as *Physical Chemistry* by Glasstone [1], *Physical Chemistry of Surfaces* by Adamson [2], or *Physical Adsorption* by Brunauer [3]. The following condensed and simplified discussion is derived mostly from these sources.

There is of course the law of universal gravitation, which says that there is a force of attraction between all things proportional to the product of the masses divided by the square of the distance between them. However this law is unable to account for experimental observations of action between atoms and molecules. There are chemical bonds of various kinds, but they are different both in intensity and kind from those that account for the adsorbed state or liquid state. Thus we start our discussion of van der Waals forces with the assumption that those forces are different from gravitation and different from chemical bonds.

It should be understood that the forces of attraction, as well as forces of repulsion, are matters of observation for which no explanation is offered; they are regarded as fundamental aspects of nature as we know it.

2.1 The Orientation Effect

A water molecule is made up of one large oxygen atom with two hydrogen atoms attached asymmetrically. Each hydrogen atom has one unit of positive electrical charge, and the oxygen, two negative charges. The hydrogen-oxygen-hydrogen centers are so located that the center of the negative charge does not fall directly between the two positive charges. Because of this, when a molecule of water is placed between electrically charged plates it tends to become oriented with respect to the plates, the necessary force-couple arising from the separated centers of positive and negative charges in the molecule. This type of molecule is called a *permanent dipole,* or, a *polar molecule.* The tendency of a polar molecule to become oriented is due to its *dipole moment,* and the degree of polarity of a dipole is expressed in terms of the value of its dipole moment. The dipole moment of water molecules can be determined from the dielectric constant of the fluid.

It might seem that an aggregation of permanent magnets would result in no net magnetic force, but because of thermal activity, which interferes with the establishment of the completely random orientation that would result from magnetic forces alone, molecules in a liquid or gaseous state become so oriented that mutual attraction exceeds mutual repulsion, leaving a net cohesive force which is sufficient to account for most of the cohesion in so strongly polar a material as water. The cohesion due to permanent dipoles is known as the *orientation effect.*

It has been deduced that the potential energy of two molecules due to the orientation effect (dipole-dipole interaction) should be inversely proportional to the sixth power of the distance between them.

2.2 Nonpolar Forces—The Dispersion Effect

There are various kinds of molecules that have no permanent dipole moment, and yet they are able to cohere. Such cohesion is attributed mostly to fluctuating, or instantaneous, dipoles in the constituent atoms. Any given atom appears to be a dipole at a given instant because its electrons are never distributed symmetrically. An extreme example is the hydrogen atom in which the centers of positive and negative charge can never coincide because it has only one electron. Glasstone explains the effect of these fluctuations as follows:

". . . it may be imagined that an *instantaneous* picture of a molecule would show various arrangements of nuclei and electrons having dipole moments. In a molecule with no permanent moment these rapidly varying dipoles when averaged over a large number of configurations give a resultant of zero. Nevertheless, the temporary dipoles are able to induce in other molecules dipoles in phase with themselves, so that as a result there is a net attraction between molecules" [1, p. 291].

The force just described goes by the seemingly improbable name of *dispersion effect;* it is so called because the cohesive force arising from it depends on the electromagnetic radiation which is characteristic of the molecule, and that characteristic is also a factor determining the refractive index (light-dispersion) of the material.

Deryaguin [4] explains electromagnetic interaction as follows:

"Molecules are like radio antennas in which oscillating electric charges emit a train of electric and magnetic vibrations. When the vibrations from one antenna reach a second, they set its charges in oscillation. These oscillations become in turn a source of waves that reach back to the first antenna, exerting a force on its moving charges. . . . The wavelength of most radio waves is measured in meters or centimeters. But the radiation sent out and absorbed by molecules are light waves only ten-thousandths of a millimeter long."

The interaction is strong when the distance between radiators is of the same order as the wavelength of the radiation.

The potential energy between two molecules or atoms separated by distances substantially greater than their own dimensions was calculated to be inversely proportional to the sixth power of the distance between them.

2.3 The Induction Effect

The dispersion effect just described is essentially an effect arising from polarity induced by electromagnetic interaction. Permanent dipoles also are able to induce polarity in neighboring molecules, according to the degree to which the centers of opposite charge become separated, or farther separated, by the action of a neighoring permanent dipole. The characteristic dipole moment created in this way depends on *polarizability* of the molecule. The effect of induced polarity is qualitatively the same as permanent polarity; it contributes to intermolecular cohesion, and its contribution is called the *induction effect.*

2.4 Combined Effects

The relative contributions to total cohesion by the principal mechanisms described above depend on the permanent dipole moment of the molecule in question. When the permanent dipole moment is large, the dominant factor is the orientation effect; when the permanent dipole moment is small, the dispersion effect is dominant; in all cases the induction effect is relatively small. There are still other effects not mentioned above, but they are for our purposes unimportant.

Each kind of attraction varies inversely as the distance separating the molecules, the proportionality constant depending on electromagnetic properties of the molecules; the attractions are different for each mechanism. For any given kind of material, the three different proportionality constants can be combined into one because all three vary as the same function of distance of separation.

The force of attraction varies as the seventh power of the distance of separation; the potential energy of a pair of molecules is inversely proportional to the sixth power of the distance between them.

3. FORCE OF REPULSION

In addition to the forces of attraction just discussed, there is a force of intermolecular repulsion. The force of repulsion, which becomes effective when the electron clouds of adjacent atoms begin to merge, is a very complicated function of electromagnetic characteristics. However, as pointed out by Glasstone [1], at least for the simpler molecules a good approximation can be given on a simple basis: *The force of repulsion varies approximately as the thirteenth power of intermolecular distance; the potential energy due to repulsion is inversely proportional to about the twelfth power of the intermolecular distance.*

4. COMBINED ATTRACTION AND REPULSION

The relationships described above can be expressed as Eq. 9.1:

$$F = \frac{\alpha}{r^7} - \frac{\beta}{r^{13}} \tag{9.1}$$

where F is the force in dynes and α and β are constants characteristic of the kind of molecule and the existing temperature and pressure. The first term gives approximately the van der Waals force of attraction, and the second the force of repulsion.

The potential energy U between two molecules separated by distance r is given by the integral of $F\,dr$, which is

$$U = -\frac{\alpha}{6r^6} + \frac{\beta}{12r^{12}} \tag{9.2}$$

The negative sign of the first term, which term is due to attractive force, means that work has to be done on the molecules to separate them.

It is apparent that the force of attraction diminishes very rapidly as the distance between molecules increases; it falls away practically to nothing when the distance of separation exceeds a few molecular diameters. The force of repulsion diminishes much more rapidly. Thus, although both forces have exceedingly short effective ranges, attraction has a much longer range than repulsion.

From Eq. 9.2 it is apparent also that there is a certain value of r for a given kind of molecule and hence for given values of α and β, at which the sum of forces acting on the molecules is zero and the potential energy of the pair is at a minimum.

The above discussion of forces of attraction pertains to pairs of molecules. For aggregated molecules, particularly those in a liquid, the same considerations apply but with quantitative differences. For example, the orientation effect is complicated by the competition among different molecules for control of the orientation of a particular molecule within their reach. Thus in Eq. 9.2 the value of α for water molecules in the gaseous state should be expected to be different from the value for water molecules in the liquid state.

Keeping in mind the reservations mentioned above concerning quantitative exactness, we can represent the potential energy of the molecules in a fluid on the basis of Eq. 9.2. The potential energy would be a mean of the various potential energies that exist among the molecules making up the fluid, and the distance between the molecules would also be an average, $\bar{r}$. On this basis we obtain such a graphical representation as is shown in Fig. 9.1. In diagram *b* the top and bottom curves represent, schematically, the second and first right-hand terms of Eq. 9.2, respectively; the solid curve represents the total energy and is of course the sum of the two terms in Eq. 9.2. Notice that both components approach zero energy asymptotically and that the energy due to repulsion approaches it much more rapidly than does the energy due to attraction. Thus beyond a certain small distance of separation the effect of repulsion is negligible whereas that of attraction remains effective. At $r = r_o$ the particles have minimum potential energy; in that position a particle is said to be at the bottom of its *potential trough*.

Diagram *a* of Fig. 9.1 represents Eq. 9.1, the solid line curve giving the net force. At $r = r_o$ the net force is zero. To make r smaller than r_o, a compressive force must be applied; this is required by the opposing repulsion shown by the extension of the curve into the lower quadrant. To make r exceed r_o, traction must be applied, increasing with r up to $r = r'$. At r' forces would be in unstable equilibrium, for at any larger value of r a given tractive force meets with decreasing resistance. It is evident that such intermolecular stress-strain relationships are fun-

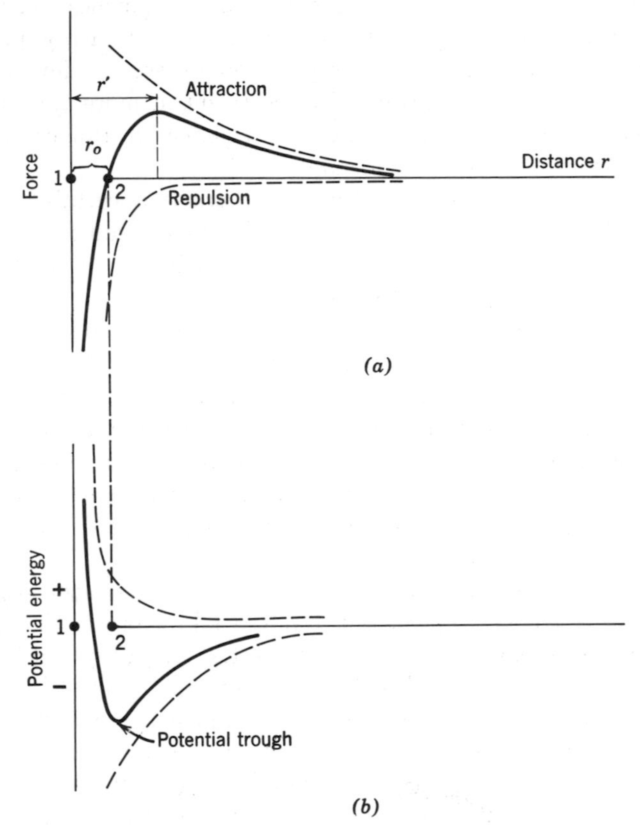

Fig. 9.1. Schematic representation of forces between two atoms, 1 and 2, and corresponding potential energies, as functions of distance. (After Houwinck, by permission of Cambridge University Press.)

damentally like the observed stress-strain relationships for macroscopic crystalline bodies.

5. HYDROGEN BONDS

The principal force of attraction in water is the orientation effect due to the permanent-dipole nature of the molecules. An aggregation brought about by orientation forces alone would be characterized by a rather random orientation of molecules and a relatively dense packing. However, in the liquid state the majority of water molecules, as seen on a molecular scale, are not randomly oriented, and they are far from being densely packed. Moreover, the total energy of bonding is larger than that accounted for by the van der Waals forces described above.

The additional effect which accounts for the actual density and other important characteristics of water and of ice is due to the formation of hydrogen bonds. When two water molecules are near each other, a hydrogen atom of one molecule is likely to form a weak chemical bond with the oxygen of the other. In fact, the oxygen atom of a given water molecule is able to form two hydrogen bonds, one with one water molecule and the other with another water molecule, and each of the two hydrogen atoms is able to bond with two other water molecules. Thus each water molecule is able to form hydrogen bonds with four other molecules. These bonds are so located and directed that the four molecules tend to group together in a tetrahedral structure.

6. ADSORPTION

The van der Waals forces of attraction operate not only between pairs of molecules of the same kind, but also between unlike molecules, or between molecules and ions. Moreover, they operate between the molecules or ions making up the surface of a solid body and individual molecules or ions.

Individual ions or molecules in a gas move at high velocity. When one strikes a solid surface it may rebound without any net loss of energy, or it may be caught by the forces of mutual attraction and be held for a time. During the time that a molecule is held, it is said to be *adsorbed,* and the aggregation of adsorbed molecules is said to be in the *adsorbed state*. This state cannot be said to be identical with either the liquid or solid state. Any adsorbed material is called the *adsorbate,* and the solid that adsorbs it is called the *adsorbent.*

The adsorbed state is a condensed state, and therefore the process of adsorption may be thought of as being like the process of compressing a vapor. Just as the compression of water vapor to the liquid state at a constant temperature entails the work of compression and the loss of latent heat of liquefaction, so isothermal adsorption entails doing the work of compression and the loss of the latent heat of conversion from the gaseous state to the adsorbed state. The latent heat of adsorption of water vapor on the materials of concrete exceeds the latent heat of liquefaction.

When a solid surface is in contact with a body of liquid water, there is a layer of adsorbed water between the liquid phase and the solid phase. Similarly, when a solid surface is in contact with ice, there is a layer of adsorbed water between the two solid phases.

6.1 Dynamical Aspects of Adsorption

When a bare surface is exposed to an atmosphere containing water vapor, some of the molecules will be caught and held for a time, as is mentioned above. The length of time a given molecule will linger depends upon the amount of energy it loses on impact. On the average this amount is greatest for those molecules that come in direct contact with the surface, and it therefore depends on to what degree the surface is covered with molecules at the time of impact. According to data by Powers and Brownyard [5], the average loss of energy by the molecules forming the first monomolecular layer on the bare surface is about 15,600 cal/mole for a surface of hydrated-cement gel. For molecules forming the second layer the loss of energy is about 12,530 cal/mole, and for amounts exceeding 5 layers, the energy loss appears to be not significantly different from the normal heat of condensation which is 10,540 cal/mole at 20°C.*

The amounts of energy given above indicate the minimum amount that a molecule in one or another of the designated situations must acquire from the milieu of thermal activity before it can escape (evaporate) from the layer. The average length of time required from the moment of adsorption to that of escape—*the lingering time*—can be calculated if the total heat of adsorption is known [6]. On the basis of the figures given above, the lingering times at 20° are as follows:

For molecules most strongly bound, 1.8 sec
Average time for the first layer, 0.04 sec
Average time for second layer, 220×10^{-6} sec
Average time for third to fifth layers, 7×10^{-6} sec

These figures show that the adsorbed state is a very dynamic one. Even the most strongly attached molecules remain adsorbed for less than 2 sec before changing to some other situation. In the second layer the average lingering time is conveniently expressed in millionths of a second; it is only about 30 times as long as the lingering time of a water molecule entering the surface of a body of water.

6.2 The Adsorbed Film

The quantity of water adsorbed by a given adsorbent at a constant temperature depends on the concentration of adsorbate molecules in the

* Heat of condensation is given here as the enthalpy plus the work of compression RT. At 20°C the sum is about 9058 + 582 = 10,540 cal/mole.

adjacent space. Since the concentration of vapor molecules is very nearly proportional to their pressure and since pressure can be measured more easily than concentration, the amount of adsorption is usually dealt with as a function of water vapor pressure. Moreover, it is convenient to express vapor pressure as a fraction, or percentage, of the maximum possible vapor pressure at the existing temperature and atmospheric pressure, which is to say the relative vapor pressure, or the humidity.

When it is in contact with air at a humidity of about 17 per cent an adsorbent such as rock or cement gel adsorbs one molecular layer; at 60 per cent humidity the amount is enough for two molecular layers; at 100 per cent humidity the thickness of the adsorbed film reaches a maximum, the number of layers being uncertain. One thing is certain: Only the first two layers have physical properties markedly different from water in bulk. We return to this point later.

In fresh cement paste, the adsorbed film is not formed by capturing water molecules from the vapor phase, for there is no vapor phase to speak of within the mass; it is formed from the liquid water with which the solids are in contact. The adsorbed film on a plane surface in such a situation has its maximum thickness, presumably the same as it would have if it had formed in contact with saturated vapor.

The relation of the film formed directly from a liquid to one formed from vapor can be seen by imagining that the process was that of first isothermally compressing saturated vapor to the liquid state, losing the isothermal heat of condensation in the process, then bringing the liquid into contact with the adsorbent, whereupon a further change of state takes place, that is, the change from the liquid state to the adsorbed state, with loss of the latent heat of that process.

The dynamic characteristic of the adsorbed film cannot manifest itself by evaporation and condensation as described above, but it may be supposed that there is a continual interchange of molecules between the film and the contiguous liquid, adsorbed molecules being continually "dissolved" by the liquid while others are being taken from the liquid into the film. Within the film we should expect continual interchange between regions of strong and weak adsorption.

The mobility of adsorbed molecules gives mobility to the adsorbed film. If by a mechanical means, such as pressing two surfaces together, some of the film is displaced, the displaced molecules are able to maintain a pressure opposing the displacement. This counter pressure has been called the *disjoining pressure* [7]. It is with this property of adsorbed water that we are concerned further on.

7. VAN DER WAALS ATTRACTION BETWEEN SOLID BODIES

The foregoing discussions amounted to a preparation for a discussion of forces between solid bodies composed of many molecules or atoms. Such bodies may be called colloids since they have dimensions less than about 5×10^{-5} cm, but they are larger than inorganic molecules. Or they can be larger bodies having sizes in the microscopic or macroscopic range. Whether colloidal, microscopic, or macroscopic, all these solid bodies are subject to forces of mutual attraction.

The forces of gravitation are always present among discrete bodies, but among small particles the gravitational force is negligible in comparison with forces arising from electromagnetic and electrostatic interaction. In other words, gravitation is negligible in comparison with van der Waals forces of attraction and with certain forces of repulsion yet to be discussed.

Although we need not make quantitative calculations of interparticle forces when discussing phenomena in which the forces discussed above play a prominent role, we shall examine results of one attempt to evaluate van der Waals attraction experimentally, and describe the theoretical relationships between force of attraction and distance between bodies, our purpose being to gain some idea of the magnitude of the attraction and how it varies with distance between the bodies.

7.1 Theoretical Considerations

Theoretical analyses of the problem of attraction between macroscopic bodies have generally dealt with spheres or flat parallel plates. Most theories have been based on the assumption that each molecule in one body is attracted by each and every molecule in the other nearby body and that the total attraction is the sum of all the interactions between pairs. The dispersion effect discussed in Section 2.2 is considered to be the principal source of attraction between particles. On this basis, the force of attraction between parallel plates turns out to be inversely proportional to the third power of the distance between them. Thus

$$P_s = \frac{A}{H^3} \tag{9.3}$$

P_s is the force of attraction per unit area (dyn/cm²), the subscript s indicating that the third-power law for force is theoretically valid only for *short* distances of separation, as discussed below; H is the distance between the surfaces in cm; A is a parameter comprising characteristics of the material.

On the basis of the first term of Eq. 9.1 we see that the force of attraction between two molecules separated by 2 units of distance is $(\frac{1}{2})^7$, or $\frac{1}{128}$ of the attraction between them when they are separated by 1 unit; from Eq. 9.3 we see that the force of attraction between parallel plates at 2 units of distance is $\frac{1}{8}$ of that at a distance of 1 unit. Thus we see why the forces of attraction between multimolecular bodies are considered to be relatively long-range forces.

The potential energy of two parallel plates due to their mutual force of attraction is, for small values of H, the integral of $P_s dH$, given in Eq. 9.4.

$$U_s = -\frac{A}{2H^2} \tag{9.4}$$

Equations 9.3 and 9.4 are simplified expressions of the basic theory and are considered applicable for only a limited range of distance. The distance of separation H must be greater than one molecular diameter and not greater than about 10^{-5} cm.

To obtain a law valid for all values of H, Overbeek [8] used Eq. 9.4 and applied a correction factor derived from theoretical considerations. Thus he used the relationships

$$P = \frac{A}{H^3} f(p) \tag{9.5}$$

and

$$U = -\frac{A}{2H^2} f(p) \tag{9.6}$$

where $f(p)$ is a reduction factor, p being inversely proportional to the characteristic wavelength of the electromagnetic interaction between the surfaces. For a wavelength of 10^{-5} cm, for example, the reduction factor is 0.98 at a distance of 8 Å; at 32 Å it is 0.90; at 1600 Å it is 0.10, which is to say the attraction is only 10 per cent of that given in Eq. 9.3.

Thus theoretical studies give us a basis for estimating van der Waals attraction between parallel plates of material except that the theory mentioned does not provide a value for the constant A. However we now have recourse to experimental work that enables us to estimate A in a fairly satisfactory way, well enough at least for purposes of illustration.

7.2 *Estimation of Attractive Force*

We have already brought out the point that forces of attraction between multimolecular bodies are considered to be long-range forces. However the expression "long-range forces" in the present context should not be misunderstood: a long-range van der Waals force is perceptible over a distance of about 1 μ at most, or about 40 millionths of an inch. Also, we should be aware of the magnitude of forces involved: at a separation of 10^{-4} cm, the attractive force between parallel plates amounts to about 0.01 dyn/cm^2. It is easy to see that to measure such forces and distances is not a simple matter. Nevertheless, it has been done. Experiments were completed successfully in the Soviet Union by Deryaguin and Abrikossova [4], and in England by Prosser and Kitchener [9]. The Russians succeeded in obtaining data on the force between bodies made of several materials, including quartz.

By using the results of one of the experiments just mentioned with theoretical Eq. 9.5, it was possible to find the corresponding value of the constant A as shown below:

$$P_A = \frac{6.25 \times 10^{-13}}{H^3} f(p) \tag{9.7}$$

where P_A is the approximate force of attraction between a pair of flat parallel plates of quartz in air, dynes/cm^2. Solutions can be obtained for given values of H by using appropriate values for $f(p)$ given by Overbeek [8].

Equation 9.7 is theoretically applicable to all values of H, provided that the smallest is substantially greater than two molecular diameters. Of course, the constant is not the same for different materials, and it is not the same for attraction between unlike materials as it is for attraction between bodies made of the same material.

Solutions of Eq. 9.7 are given in Table 9.1. The indicated attraction between plates 8 Å apart is about 1,200 bars, or 17,260 psi. At a distance of 16 Å the indicated attraction is 144 bars, or 2,088 psi. It is evident that if the surfaces are separated as much as 300 Å, the attraction is very small.

Incidentally, it is also evident that a structure in which the solid elements are separated no more than about 30 Å van der Waals attraction could account for tensile strengths of the same order of magnitude as those observed in hardened cement pastes and the like. However this is not to say that strengths are actually quantitatively accounted for on that basis. In the first place, there may be bonds other than van der Waals forces to be considered (but not for fresh cement paste), and in the second place there may be and usually is a disjoining force that tends to offset the force of attraction. Moreover, there is probably an effect due to the presence of material other than air between the solid surfaces, as discussed below.

Table 9.1 Solutions of the Equation

$$P_A = \frac{6.25 \times 10^{-13}}{H^3} f(p)$$

for Parallel Plates of Silica in Air at Various Distances[a]

H (Å)	$f(p)$[b]	P_A dyn/cm²	Bars[c]	psi
6	0.98	2.83×10^9	2830	41046
8	0.975	1.19×10^9	1190	17260
16	0.95	1.44×10^8	144	2088
32	0.91	1.73×10^7	17.3	250
80	0.77	940,000	0.94	13.6
160	0.60	90,000	0.09	1.3
320	0.39	7,400	—	—
480	0.28	1,600	—	—
800	0.18	180	—	—
1600	0.10	15.2	—	—

[a] P_A = force of attraction, dynes/cm².
H = distance between plates, cm.
$p = 2\,H/\lambda$, where λ is the wavelength of the electromagnetic interaction between plates, taken as 10^{-5} cm.
[b] $f(p)$ = the value of p at a given value of H as computed by Overbeek, and as given in column 2.
For large values of H, $f(p) = 0.98/p$.
[c] 1 bar = 0.987 atm = 14.504 psi.

7.3 Effect of Intervening Material

The discussion so far pertains only to solid parallel plates in air. When material of appreciable density fills the space between the plates, the effect of that intervening material must be considered. Brunauer [3, pp. 98 and 193] reported electromagnetic van der Waals forces to be independent of intervening material, a conclusion defended by Polanyi [10]. However, this point of view has been contested by Overbeek [8] and by Deryaguin [11]. Overbeek presented an analysis leading to the conclusion that if parameter A of Eq. 9.5 pertains to action between plates in a vacuum (or in air), it would have to be reduced by a factor of $1/D$ for any solid or liquid material, D being the dielectric constant of the intervening material at a specific frequency of electromagnetic interaction. If this view is correct and the surrounding medium is other than air, we must use the following relationship:

$$P_A = \frac{A/D}{H^3} f(p) \tag{9.8}$$

For a given material, the value of D is a function of the wavelength of the electromagnetic interaction between surfaces; it is actually taken as equal to the square of the optical index of refraction at the wavelength in question. In the present case the wavelength is perhaps 2000 Å, which for water corresponds to about $n = 1.4$. Hence D would be about 2. However, specific data on the values of A and D for a given material in water are lacking, and we are able only to illustrate the points yet to be discussed on the basis of assumed values of A/D.

8. REPULSION DUE TO DISJOINING PRESSURE

There are two kinds of forces that tend to disperse the particles in a dense aggregation of small particles in water. One kind is electrostatic repulsion, which is taken up later. The other is repulsion due to film pressure, which has been referred to above as disjoining pressure. It would be logical to let the term disjoining pressure comprise both kinds of dispersive force, as was done recently [12], but we shall continue to let it stand for the effect due to the adsorbed molecular film.

In general, we are concerned with aggregations in which the smallest interparticle distance is much smaller than the particles. Therefore, as was done in Section 7, we use a model of parallel plates separated by a small distance H. We shall consider the plates to be immersed in water and later on immersed in an aqueous solution of electrolytes.

At this point we are especially interested in the thickness of adsorbed films on bodies in contact with liquid water. Such knowledge will enable us to estimate that distance between solid surfaces which cannot be reduced without developing a counter disjoining pressure. Furthermore it will make it possible for us to estimate the magnitude of disjoining pressure as a function of the distance between surfaces.

8.1 Estimation of Thickness of the Adsorbed Film

It is theoretically possible for adsorption on a plane surface to build a "film" of unlimited thickness (Brunauer [3]; de Boer [6]). However all but a few molecular layers could be displaced with negligible effort, and therefore the practical maximum thickness is much less than the theoretical maximum.

The practical maximum thickness of a water film on mineral oxides cannot be given with certainty. It seems reasonable to base an estimate on the assumption that that part of the adsorbed water which shows a heat of adsorption greater than the normal heat of condensation is certainly held strongly enough to offer significant resistance to displacement. Data by Harkins and Jura [13] pertaining to adsorption of water on particles of titanium dioxide indicate a limit of about five molecular layers. Accordingly, we assume that the practical maximum thickness of a water film on silica and the like is five molecular layers.

The next step is to estimate how much pressure is required to displace portions of the five-layer film.

8.2 Estimation of Disjoining Pressure

The pressure required to displace a film is also the pressure a film might develop if it were to penetrate a space less than about 10 molecular diameters wide; that is to say, it is the disjoining pressure. Something of the magnitude of that pressure can be given, as is developed below.

When two parallel flat plates, immersed in water and being moved toward each other, come to within a certain distance of separation H_m, their adsorbed films meet and any further reduction of distance between the plates involves displacement of adsorbed water and an increase in the applied force. It can be shown from considerations of conditions for thermodynamic equilibrium between the water in the film and the water surrounding the plates that the pressure on the films required to maintain the distance $H < H_m$ is given approximately by the following equation:

$$P_R = 2.303 \frac{RT}{M(v_w)_a} \log h \tag{9.9}$$

where P_R is the force per unit area on the water between the plates in excess of that on the water that is not between the plates; h stands for the relative vapor pressure or humidity, and its value is that of an atmosphere with which the *reduced* film between the plates would be in equilibrium without the pressure P_R, which is to say that if the adsorbed layers were reduced to a thickness $H/2$ by immersing the plates in humid air instead of doing so mechanically with the plates under water, h is the humidity of the air corresponding to that thickness. The other values are constant for a given absorbate and conditions: R is the gas constant, T is absolute temperature, M is the molecular weight of water, $(v_w)_a$ is the specific volume of adsorbed water. The specific volume of adsorbed water cannot be independent of pressure; however its volume change is so small in comparison with the change in pressure it can safely be assumed to remain constant for present purposes.*

* The specific volume of adsorbed water is probably also a function of the amount of adsorbed water per unit area of surface, but it is nevertheless usually treated as a constant.

Obviously the calculated pressure on the plates under the conditions described above is the same as the excess pressure in the adsorbed film; it is thus identical with the disjoining pressure.

The following values may be used for the constant factors.

$$R = 8.3 \times 10^7 \text{ ergs/mole °K}$$

$$T = 294\text{°K (21°C; 70°F)}$$

$$M = 18.02 \text{ g/mole}$$

$$(v_w)_a = 1.0 \text{ cm}^3\text{/g (assumed)}$$

These values give

$$P_R = 3.12 \times 10^9 \log h, \text{ in dyn/cm}^2 \tag{9.10}$$

$$P_R = 3120 \log h, \text{ in bars} \tag{9.11}$$

$$P_R = 45{,}200 \log h, \text{ in psi} \tag{9.12}$$

Although it is natural to think of disjoining pressure as a positive force, it must actually be treated as a negative quantity because it opposes the force of attraction which is considered positive. This situation arises from the convention that an increase in the distance between two bodies is considered a positive change and from another convention that the potential energy of two bodies due to their mutual attraction is zero when they are separated by an infinite distance. It then comes out that the potential energy due to attraction at any finite distance of separation is negative, and the force of attraction is positive. Since h is 1.0 or a fraction, the equations above give zero or negative numbers for disjoining pressure. Nevertheless, we may continue to think of disjoining action as being due to pressure. The negative number can be thought of as indicating what the pressure in the reduced film would be without the added pressure.

8.3 Disjoining Pressure as a Function of H

Because we are here considering the relationship between disjoining pressure and van der Waals attraction and because van der Waals attraction is a function of H, we need to know the relationship between disjoining pressure and H. This relationship can be stimated from the relationship between the thickness of the adsorbed layer and the humidity h. A reasonable estimate can be based on adsorption data obtained by Powers and Brownyard [5] from samples of hardened cement paste. Although other data used above were for silica, or titanium dioxide, the heats of adsorption for these materials and cement paste are nearly the same, and on that basis we are justified in assuming that the thicknesses of absorbed films are nearly the same; this is an example of the relatively low degree of specificity of van der Waals adsorption, mentioned before.

If w_a is the weight of water adsorbed by a unit of weight adsorbent and S is the surface area of the adsorbent, the thickness of the layer t is given by

$$t = \frac{w_a(v_w)_a}{S} \tag{9.13}$$

Also

$$S = a_1 \frac{V_m N}{M} \tag{9.14}$$

where V_m is the weight of water required to form a monomolecular layer of water on a unit weight of the adsorbent, in grams; $N = 6.024 \times 10^{23}$, which is Avogadro's number; a_1 is the area covered by one adsorbed molecule, assumed to be 11.4×10^{-16} cm^2 (Brunauer, Kantro, and Copeland [14]); and other symbols have the same significance as above. Combining Eqs. 9.13 and 9.14 we find

$$t = \frac{M(v_w)_a}{a_1 N} \frac{w_a}{V_m} \tag{9.15}$$

Using the numerical values given for Eq. 9.10, we obtain

$$t = 2.63 \frac{w_a}{V_m} \text{ Å} \tag{9.16}$$

which indicates that the effective diameter of a water molecule in the adsorbed state is about 2.63 Å.

Because V_m is the weight required for a monomolecular layer, Eq. 9.16 gives the thickness of the layer in terms of the number of molecular layers. Values of w_a/V_m up to about 2 can be evaluated accurately by means of the Brunauer, Emmett, Teller (BET) equation [3, 5], which for the present purpose may be written as follows:

$$\frac{w_a}{V_m} = \frac{Ch}{(1 - h)[1 - (1 - C)h]} \tag{9.17}$$

where C is a constant for a given adsorbent-adsorbate system, related to the heat of adsorption. On the basis of data by Powers and Brownyard, we may assume it to be between $C = 20$ and $C = 50$ for the kind of materials with which we are concerned.

Values of w_a/V_m are given in Table 9.2. For values up to $N = 0.5$ (about two molecular layers) the values given are from Eq. 9.17. For higher values of h the values were estimated graphically by a smooth-curve extension to the arbitrary terminal point $w_a/V_m = 5$ at $h = 1.0$; that is, it was arbitrarily assumed that the limit for adsorption on a plane surface is five molecular layers, as discussed before.

Table 9.2 Calculated Thickness of the Adsorbed Film at Various Humidities[a]

Humidity	$C = 20$		$C = 50$	
(h)	w_a/V_m	H	w_a/V_m	H
0.05	0.54	2.84	0.76	4.00
0.10	0.77	4.04	0.94	4.94
0.15	0.92	4.84	1.06	5.58
0.20	1.04	5.46	1.16	6.10
0.30	1.28	6.74	1.36	7.16
0.40	1.55	8.16	1.62	8.52
0.50	1.90	10.00	1.96	10.3
0.60			2.43	12.8
0.70			3.00	15.8
0.80			3.57	18.8
0.90			4.30	22.6
1.00			5.00	26.3

[a] w_a = weight of adsorbed water.
V_m = weight of adsorbed layer one molecule thick.
H = distance between surfaces in Å = $5.26\ w_a/V_m$.
C = a constant (BET parameter). The values for $C = 20$ are practically the same as for $C = 50$ when the humidity > 0.50.

Values for H for $h > 0.50$ were obtained graphically using the assumption that the limit of adsorption on a plane surface is five molecular layers.

From the relationship between h and H given in Table 9.2 we may now, by Eqs. 9.11 and 9.12, calculate the corresponding values of H and the disjoining pressure as shown in Table 9.3. Estimations of film thickness are given on two different bases to show that the assumption about the value of C does not make much difference except at small values of h.

We see that the disjoining pressure that exists when plane solid plates are very near to each other can be very large. For example at a spacing

Table 9.3 Calculated Disjoining Pressure between Parallel Plates at Different Distances[a]

		P_R		*H* from Table 9.2 (Å)		
h	log *h*	bars	psi	$C = 20$	$C = 50$	Average
0.05	−1.301	4060	58,900	2.84	4.00	3.42
0.10	−1.000	3120	45,200	4.04	4.94	4.49
0.15	−0.8234	2570	37,300	4.84	5.58	5.21
0.20	−0.6990	2180	31,600	5.46	6.10	5.78
0.30	−0.5299	1650	23,900	6.74	7.16	6.95
0.40	−0.3979	1240	18,000	8.16	8.52	8.34
0.50	−0.3010	939	13,600	10.00	10.30	10.15
0.60	−0.2218	692	10,000		12.8	
0.70	−0.1549	483	7,000		15.8	
0.80	−0.0969	302	4,400		18.8	
0.90	−0.0458	143	2.100		22.6	
1.00	0	0	0		26.3	

[a] h = relative humidity.
P_R = repulsive force between parallel plates due to disjoining pressure of adsorbed water = 3120 log h bars at 21°C.
H = distance between parallel plates in Å.
C = a constant (BET parameter).
1 bar = 10^6 dyn/cm² = 14.502 psi.
The last line is an arbitrary assumption; see text.

of about 8 Å units it is about 18,000 lb/in.². It should be kept in mind that this is the force required to keep the water surrounding the plates from entering the space between them and forcing the plates apart. As mentioned above, the estimate that the force required would be virtually zero at a spacing of 26 Å is uncertain, but the estimate has a reasonable basis.*

* Actually the above estimates are reasonable but probably somewhat inaccurate. The method of evaluating t by Eq. 9.13 and the use of Eq. 9.17 involve the tacit assumption that the adsorption of water molecules in hardened cement paste occurs uniformly over all the interior surfaces without hindrance. However, there is good experimental evidence that an unknown but significant fraction of the internal surface area bounds spaces that are too narrow to permit unhindered adsorption. Ignoring this fact, as we do above, leads us to underestimate what the film thickness should be at a given humidity. In the present context the resulting distortion is believed to be not serious. Possibly the maximum film thickness is somewhat greater than 30 Å.

9. POTENTIAL ENERGY DUE TO VAN DER WAALS ATTRACTION AND DISJOINING PRESSURE

When parallel plates (or particles) are under water, they have a certain potential energy amounting to the algebraic sum of the potential energy due to forces of repulsion and the potential energy due to van der Waals attraction plus whatever other force is needed to maintain the existing distance between surfaces. The force of repulsion may include electrostatic forces as well as the film pressure discussed above, but for the present we consider disjoining pressure only, that is to say, the pressure arising from partial displacement of the adsorbed layer.

9.1 Potential Energy Due to Attraction

It follows from Eq. 9.8 that the potential energy of two bodies due to the van der Waals force of attraction between them is

$$U_A = -\frac{A/D}{2H^2} f(p) \tag{9.18}$$

where U_A is the potential energy due to attraction, and the other symbols have meanings already given. For silica in air, we have already estimated that $A = 6 \times 10^{-13}$. If $D = 2$ as indicated in Section 7.3, we find $A/2D = 2.5 \times 10^{-13}$. However, since we might expect the value of A for hydrated cement (the material that coats cement particles in cement paste) to be somewhat higher than that for silica, we shall make further estimates on the basis of 2.5×10^{-13} and 5×10^{-13}: there are obvious uncertainties.

9.2 Potential Energy Due to Disjoining Pressure

Potential energy due to disjoining pressure can be obtained by integration of the $P_R(H)$ relationship as follows:

$$U_R = \int_{26}^{H} P_R(H)\, dH \tag{9.19}$$

where U_R is the potential energy due to the repulsion force from the disjoining pressure. Values of P_R are those given by Eq. 9.9, and corresponding values for H are obtainable from Table 9.3.

9.3 Potential Energy Due to Combined Attraction and Repulsion

The total potential energy of two bodies under forces of attraction and repulsion is of course the algebraic sum of the energies due to the two forces taken separately; that is

$$U = U_R + U_A \tag{9.20}$$

For Eq. 9.18 two values of $A/2D$ were used: 2.5×10^{-13} and 5×10^{-13}. For the energy due to repulsion, a plot of data in Table 9.3, $-P_R$ versus H, was integrated graphically, using the average value of H from the two values of C. The sums of the two energies are plotted as functions of H in Fig. 9.2.

We note that either curve in Fig. 9.2 is analogous to the corresponding curve in the lower diagram of Fig. 9.1. Just as the forces of attraction and repulsion between individual molecules define a certain intermolecular distance at which the pair has minimum potential energy, so the forces of van der Waals attraction and disjoining pressure between two multimolecular bodies define a position of minimum potential energy. A potential trough is indicated at about $H = 25$ Å by either of the curves.

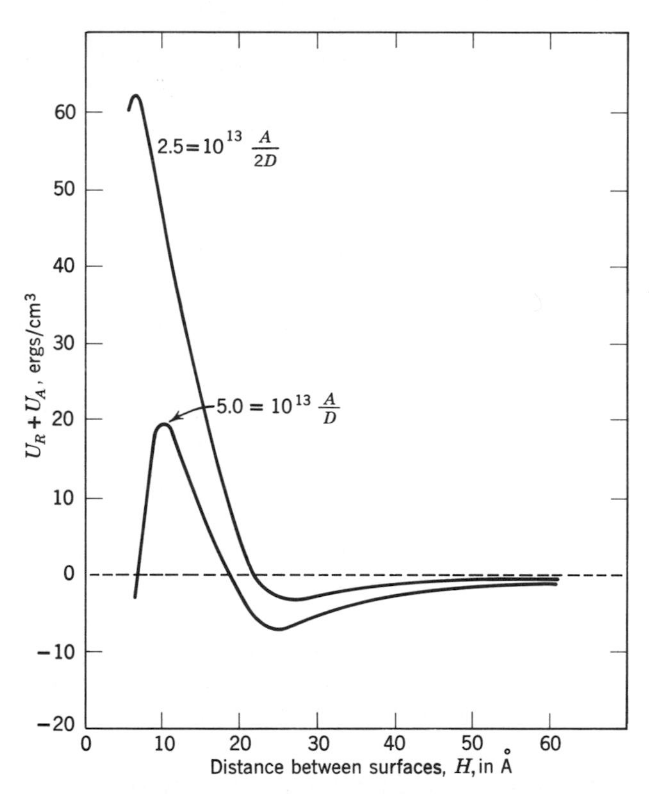

Fig. 9.2. Calculated potential energy of two parallel silica plates in water as a function of the distance between them. The energy due to van der Waals attraction was calculated using Eq. 9.18. Energy due to repulsion was calculated from data in Table 9.2 and Eq. 9.19. The repulsion due to electrostatic charge was neglected. The curves represent different assumptions of the value of $A/f(D)$, as indicated.

Figure 9.2 illustrates the point that within the range of distances indicated the forces of attraction and repulsion are of the same order of magnitude, but they vary with H at different rates. As is shown in the next section, the force of repulsion due to disjoining pressure may be supplemented by electrostatic repulsion in such a way as to modify the potential trough, or to destroy it.

Instead of two plates, we may consider Fig. 9.2 to represent the average state in a suspension of mineral particles. The graph then indicates that the particles would tend to cohere but at the same time remain separated by a distance of perhaps 25 Å. Thus we would expect the suspension to appear flocculated if dilute and flocculent if sufficiently concentrated. If the condition of the suspension corresponded to the lower curve, the mixture (paste) would appear more strongly flocculent than it would if it were represented by the upper curve; it would have a stiffer consistency, for example. Actually, a suspension of silica particles in water appears to be flocculent when the net electrostatic charge on the particles is zero, which indicates that Fig. 9.2, based on data from parallel silica plates, is at least qualitatively correct.

To the left of the potential trough the curve rises to a maximum and then falls, indicating the possibility if not probability of another trough at a very small value of H. However it cannot be said that this indication is real because the accuracy of Eqs. 9.9 and 9.18 is especially questionable at the small values of H where the inner trough might be. However that may be, it seems clear that if the particles are initially separate, van der Waals attraction alone cannot surmount the barrier due to disjoining pressure, and they will remain separate even while showing cohesion, which is to say, while showing evidence of floc structure.

We may note also that the right-hand side of the potential trough is not steep. This indicates that relatively little force is required to increase the distance between a pair of particles and that the floc structure can be destroyed or modified with relatively little effort, or with relatively little change in the force of interparticle repulsion. Thus, mechanical agitation of a slurry changes its consistency, as is brought out in Chapter 10. Changes in the electrolytic environment of the particles can do the same, sometimes to a much greater degree.

10. ELECTROSTATIC FORCES

As indicated above, repulsion force may involve electrostatic repulsion as well as repulsion due to the disjoining pressure of adsorbed water. Indeed, in some systems electrostatic charge may be the only source of repulsion; the effect of the charge on dry cement in air is an example. We are now concerned with electrostatic effects in aqueous suspensions, the effects being small relative to those due to film pressure.

The phenomena to be considered can be introduced by referring to some of the many experiments that have been reported.

10.1 Buzagh's Experiments

Buzagh [15] completely filled a vessel that had parallel sides and a bottom made of a quartz plate with a dilute suspension of quartz particles in water and allowed the particles to settle to the bottom. He then sealed the top and carefully inverted the vessel. Using a microscope, he found particles adhering to the inverted bottom, which indicated that the van der Waals attraction between them and the quartz plate was greater than the force of gravity. Of the various particle sizes in suspension, only the intermediate sizes were found clinging to the plate. The large particles had fallen because they were too heavy to be held by the van der Waals attraction; the small ones had fallen because their thermal agitation together with their weight was sufficient to overcome attraction. The number of adhering small particles diminished with time, as would be expected from the randomness of thermal activity.

The experiment just described demonstrated the effect of van der Waals attraction, with which we are already familiar. The presence of a force of repulsion was demonstrated by a variation of the same kind of experiment. After he had allowed the particles to settle to the bottom, Buzagh examined the thin deposit of particles at the bottom microscopically. Although van der Waals attraction and gravitation were acting in the same direction, it was evident that many of the particles were not actually in contact with the quartz plate. Such particles seemed to be held in place; they would not slide when the vessel was tilted until a certain angle was exceeded. Yet the smaller ones could be seen to "whirl and dance" in a two-dimensional thermal (Brownian) movement. Buzagh concluded that such particles, although immobilized from sliding, were separated from the surface by a liquid film having a thickness of several hundred molecular diameters. Such a phenomenon seems not to be accounted for by disjoining pressure, for the thickness of the effective adsorbed layer is not believed to be that thick. (See Section 8.1.) As shown below, far-reaching electrostatic repulsion is the most likely explanation of this action at such a long distance.

Buzagh found from repeated experiments of the first kind that the percentage of the total number of particles able to adhere after the vessel had been inverted depended markedly on the amount and kind of electrolyte dissolved in the water. Since such variations of conditions have little or no effect on van der Waals forces of attraction, it was apparent that the force of repulsion is sensitive to the kinds and amounts of ions in solution. It depends in fact on the composition of an electrical double layer that exists around each particle, and, in Buzagh's experiment, also on the flat surface near which particles were held. Some of Buzagh's results, as tabulated by Overbeek [8], are given in Table 9.4. The figures show that a few multivalent ions have the same effect as a large number of univalent ions.

Table 9.4 Concentration of Electrolyte Necessary to Increase the Adherence Number of Quartz Particles from 50 to 95 Per Cent[a]

Kind of Electrolyte	Concentration (millimoles per liter)
LiCl	1.4
NaCl	1.0
KCl	0.4
$CaCl_2$	0.04
$SrCl_2$	0.03
$BaCl_2$	0.02
$AlCl_3$	0.04
$Th(NO_3)_4$	<0.002

[a] The value in pure water is 50 per cent.

It is generally true that electrostatic repulsion at a given electrolyte concentration depends mostly on the valence of the ion having a charge opposite to that of the surface (or particle). We see in Table 9.4 that a tervalent ion may have an effect several hundred times as great as that of a univalent ion. A multivalent ion may be so effective that it can reduce the surface charge to zero, or beyond; in the latter case the particle charge becomes reversed.

10.2 The Electrical Double Layer

When a solid bearing a negative charge, for example, is placed in contact with an aqueous solution of an electrolyte, the concentration of the positive and negative ions in the vicinity of the solid surface becomes modified, as is shown schematically in Fig. 9.3. In this example

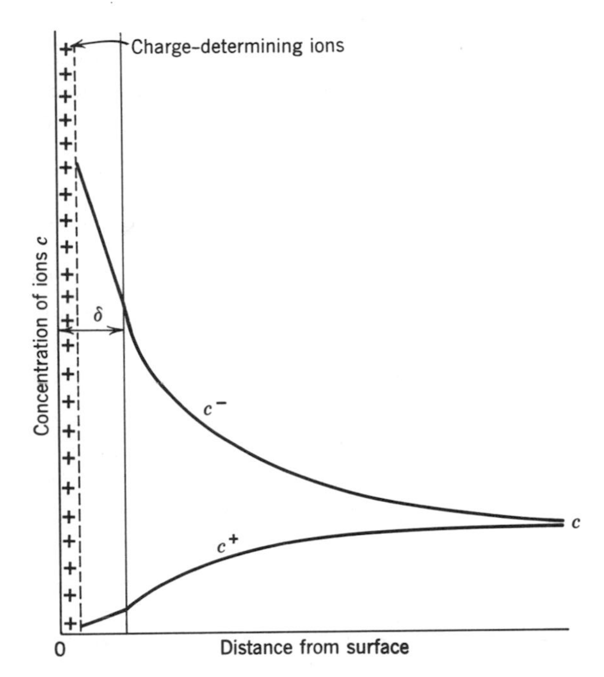

Fig. 9.3. Illustration of the presumed distribution of ion concentration (charge density) in the vicinity of a particle surface. The symbol c is the total ion concentration per unit volume of solution, c^- is the concentration of negative ions, c^+ is the concentration of positive ions; δ indicates the thickness of the Stern layer; space to the right of delta is the Guoy diffuse layer.

the dry solid may be supposed to bear a negative charge because its surface is composed predominantly of negative ions. In a dilute solution of an electrolyte positive ions are preferentially adsorbed by the solid surface, and if there is chemical affinity, the ions may be strongly adsorbed, even chemisorbed, and thus made hardly distinguishable from the adsorbent. These strongly adsorbed positive ions give the surface a net positive charge and are called the *charge determining ions.* Since the number of positive and negative charges in the solution are equal, the system as a whole is electrically neutral, and if localized concentrations of one kind of ion are produced by selective adsorption, a counteracting concentration of the opposite kind of ion must occur.

At an earlier time it was assumed that if a positive ion is adsorbed, a negative ion would appear beside it to maintain electrical neutrality, thus producing a double layer of charges, one positive and one negative. It is now realized that such a simple double layer cannot exist. Although, in our example, the positive surface attracts the negative ions according to Coulomb's law, the electrostatic force is opposed by a tendency for the negative ion to diffuse away from the surface as would be required to equalize its concentration. Similarly positive ions in solution are electrostatically repelled by the positive surface, but they tend to diffuse toward the surface. The result is that instead of a simple layer of counter ions, there is a diffuse cloud of ions near the surface, the predominant charge in the cloud being such as to tend to neutralize the surface charge. This is indicated in Fig. 9.3 by the shapes of the two concentration curves, one for positive and the other for negative ions. Theoretically, the spatial distribution follows an exponential law.

As indicated in Fig. 9.3, the concentration curve is not believed to be exponential all the way to the surface. A principal cause for such a departure from a mathematically regular change in the concentration gradient is that the charges are not mathematical points; they are carried by ions having finite volumes. Another factor is that each ion is "solvated," which means that it is surrounded by a certain number of water molecules to each of which it is electrostatically bonded. The distance marked δ in Fig. 9.3 designates the thickness of the surface region where charge concentration is modified by the factors just mentioned. This region is called the *Stern layer;* it is obviously closely related to the adsorbed layer.*

Postulation of the Stern layer amounted to an embellishment of a previous theory known as the *Gouy theory of the diffuse double layer,* or sometimes as the *Gouy-Chapman theory.* According to that theory, the ion concentration was supposed to vary exponentially all the way to the surface; the double layer of charges was made up of the charged surface for one side, and the other side was considered to be at the center of gravity of the net charge of the diffuse cloud of counterions. When the concept of the Stern layer was introduced, the Gouy-Chapman theory was retained, but the outside boundary of the Stern layer was regarded as the inside boundary of the diffuse layer.

From the foregoing discussion of Fig. 9.3, it is apparent that a solid surface with its adsorbed ions and outer diffuse cloud of ions, is, in effect, a charged electrical condenser; actually, by the Stern theory, it constitutes two condensers in series. Electrostatic capacity depends on the potential difference between charged surfaces, or virtual surfaces, and on the static dielectric constant of the medium, about 80 for water.

Corresponding to the charge distribution illustrated in Fig. 9.3, there are electrostatic potentials as shown in Fig. 9.4. At the surface, the selec-

* In colloid chemistry, the term *lyosphere* is used where I have used adsorbed layer.

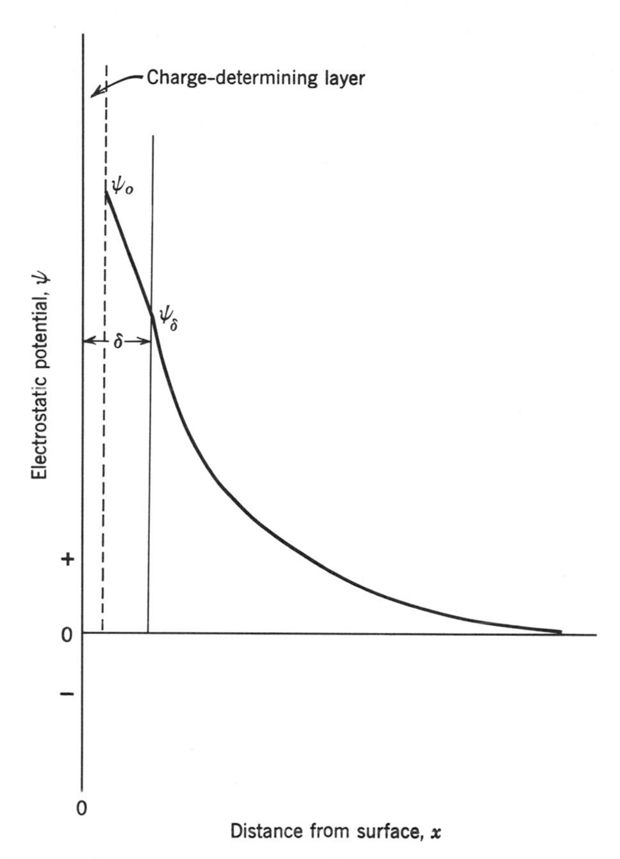

Fig. 9.4. Electrostatic potentials corresponding to charge densities illustrated in Fig. 9.3. The symbol ψ is the electrostatic potential at distance x from the surface; ψ_o is the surface potential; ψ_δ is the potential of the Stern layer; δ is the thickness of the Stern layer.

tively adsorbed positive ions give a net positive potential indicated by ψ_o relative to a remote point in the surrounding solution where positive and negative charges are equally concentrated. A smaller net positive potential, ψ_δ, is at the outside of the Stern layer. The relationship between ψ_o and ψ_δ is subject to wide variations, as discussed further on.

10.3 Electrostatic Repulsion

When two identical solid bodies immersed in an electrolyte solution are moved toward each other, a force of electrostatic repulsion begins to develop when their ion clouds begin to merge. The ion clouds (or diffuse layers) around each body have the same effect qualitatively as the electron clouds around atomic nuclei. Such repulsion augments that due to the disjoining pressure above (Section 8).

The force of repulsion between ion clouds is basically that given by Coulomb's law, but there are complications due to the diffuse distribution of the charge. The subject is much too complex to warrant full discussion here, and we need not attempt it because we do not need to deal with the quantitative aspects of it. (Actually, a rigorous quantitative treatment has not been achieved even by specialists in the physical chemistry of surfaces; simplifying assumptions are required in all but the simplest situations.)

It turns out that the distance at which repulsion may become effective can be considerably greater than the effective range of disjoining pressure, suggested in Table 9.3. Electrostatic repulsion at such relatively great distances is very small, but it may be the same order of magnitude as the van der Waals attraction at that distance; consequently, a potential trough may occur at relatively great distances when electrostatic effects are present. Evidently it was such a relatively remote potential trough that held the oscillating silica particles observed by Buzagh. (See Section 10.1 above.)

We may recall also that Buzagh observed that the interparticle force varied greatly with the kind and amount of ions in solution. We see now that the net force felt by particles a certain distance apart depends, at relatively large values of H, mainly on the electrostatic potential that exists at that distance. As mentioned above, the potential at a given distance x from a surface in the surrounding solution is given by exponential functions. The simplest possible relationship, valid for very restricted conditions, is the following:

$$\psi = \psi_\delta e^{-kx} \qquad (9.21)$$

where ψ is the potential at distance x from the Stern layer. Parameter k is composed as follows:

$$k = (\text{constant}) \left(\frac{n_o z^2}{DT} \right)^{1/2} \qquad (9.22)$$

where n_o is the total number of ions, positive and negative, per unit volume of the solution at a remote point where the net charge is zero; z is the number of charges per ion, that is, the valence; D is the static dielectric constant of water; T is the absolute temperature.

Equation 9.21 shows that the potential at a distance $x = 1/k$ is $1/e$ of the Stern potential. Thus $1/k$ marks the center of gravity of the space charge. Accordingly $1/k$ is taken as the effective thickness of the diffuse layer, that is, of the ion cloud.

Considering the composition of k as it is given in Eq. 9.22, we see that the effective thickness of the diffuse layer is smaller the higher the ion concentration n_o other factors being equal. Also, at a given concentration it is smaller the higher the valence of the ion. Since it is raised to the second power, valence is a relatively sensitive factor.

The relationship seen in Eq. 9.21 is an oversimplification, for the magnitude of the Stern potential is also a function of the type of electrolyte and electrolyte concentration. In fact, it can be reduced to zero, or even reversed in sign, as pointed out above.

Although general, complex equations have been worked out, they all contain the exponential term, and parameter k has about the same significance discussed above. For our present purpose, it is sufficient to identify the controlling variables, as has been done.

11. DISPERSE, THIXOTROPIC, AND FLOCCULATED STATES

The foregoing discussion of interparticle forces and potential energies indicates that in suspensions of particles any of a wide range of states can exist and that in a given suspension different states will exist under different internal conditions. It is convenient to consider two extreme states, and an intermediate one, as are designated in the title above.

11.1 The Disperse State

The term "disperse" may take one of several meanings when used in different contexts, and therefore we had better discuss some of the meanings other than the particular one it has in this discussion, to avoid possible confusion. In some contexts "to disperse" means to comminute either artificially or naturally. The resulting collection of particles may be called "a dispersion." A system of graded particles is sometimes called "polydisperse," in contradistinction to a *monodisperse system* in which all the particles have the same size. In the concrete industry "disperse" is seldom given such meanings.

"Disperse" is also used in the sense of "scattered," "deflocculated"; "mutually repellent"; this is its meaning in this chapter. A *disperse system*, or *disperse suspension*, is a mixture in which interparticle potential troughs are absent or if they are present, they are negligible.

When a potential trough does not exist, particles have no tendency to remain near each other; instead, when two particles approach each other, they repell each other. In other words, a high electrostatic potential, particularly a high Stern-layer potential, denotes the disperse state. If the particles are very small, they exhibit spontaneous random motions, the Brownian motion previously mentioned, but they avoid each other because of their mutual repulsion.

Brownian motion is due to momentary translatory forces generated by random bombardment from the molecules of the surrounding fluid, and it is thus regarded as thermal motion. We note this phenomenon because it takes place only when there is practically no net interparticle attraction, and thus it can be used as a sensitive index to the state of the fine particles in a mixture. Under certain conditions small cement particles show Brownian motion, but usually they do not.

11.2 The Thixotropic State

The word "thixotropic" was originally used to describe a property of an aqueous suspension of colloidal iron oxide, that is, an iron oxide "sol." If we take an iron oxide sol of suitable concentration and add a certain small amount of sodium chloride to it, we find that the sol soon "sets" to form a gel. Then when we stir the gel, or shake it vigorously, it becomes fluid. However, after a time of rest the solid state returns. This transformation can be carried out repeatedly on the same sample. The setting of the sol does not generate heat, and therefore the phenomenon was at first called "the isothermal sol-gel transformation."*

The thixotropic state was believed to be characterized by a moderately deep potential trough at a considerable distance of particle separation. At rest the particles were thought to tend to remain in their respective positions of minimum potential energy. Usually, this means that they are not uniformly distributed in space, but form a structure that could be described as a three-dimensional net, a reticulum. When such a structure is agitated by stirring, shaking, or vibrating, the particles are subjected to forces that tend to "lift" them from their potential troughs and free them from interparticle bonds, thus destroying the firmness of the structure. After such a disturbance, the particles by thermal movement (Brownian motion) seek and find their stable positions, and thus the structure is recreated.

Such a limited concept of thixotropy has not proved adequate. With the passage of time and accumulation of observations on a wide variety of materials, the definition of thixotropy has become less clear, and authorities differ about certain areas of its use. As implied above, there is an interval between time of destruction and time of repair of a thixotropic structure. As originally observed, the interval was a matter of perhaps minutes or longer, but it is possible to create a series of systems

* "Sol" is the term for a colloidal solution.

in which the setting time is progressively shorter from one to the next. At the limit it seems that setting time might be almost instantaneous. That is, a suspension may appear to be fluid while actually in motion but to have some degree of firmness immediately after motion stops.

Such behavior as that just described is characteristic of thick, flocculent suspensions described as *paste*. Here we shall regard a paste as being a thixotropic system, even though the thixotropic setting time may be too brief to be observed directly. It appears that the characteristic attributes of paste are always associated with a moderate degree of flocculence.

11.3 The Strongly Flocculated State

When the particles in a fluid suspension are without electrostatic charge and when there is no barrier due to an adsorbed layer of water molecules, van der Waals attraction is able to draw the particles together as closely as surface topography permits. The result is the state of strongest possible flocculation. A state approximating this is found when cement is mixed with toluene, a nonpolar liquid that is weakly adsorbed and which is unable to dissolve electrolytes. In this case, the repulsion force is supposedly due mainly to the intrinsic charge on the solid surface. The potential trough is very close to the surface and relatively deep.

12. ELECTROKINETIC PHENOMENA

A charged particle in an electric field is subject to a force directed toward the side of the field bearing a charge opposite to that of the particle. When the force produces particle movement, the process is called *electrophoresis*. If a porous body composed of charged particles is placed in a conduit filled with an aqueous solution and oppositely charged electrodes are then placed in the solution on each side of the porous body, the fluid will flow through the body in the direction opposite to that of the electrophoresis of free particles bearing the same charge; this phenomenon is called *electroosmosis*. If there is no charge on the electrodes and fluid is caused to flow through the body mechanically, a potential difference between the electrodes develops, called the *streaming potential*.

The phenomena described above involve relative motion between solid and fluid. Since some of the fluid adheres to the surface, a plane of shear develops in the liquid near the surface, a plane that may or may not coincide with the outside of the Stern layer (Adamson [2]). Actually, the shear "plane" is very likely a zone of variable viscosity.

If we could observe successive "layers" of molecules and ions along a line from a point in the liquid to the surface, we would at first find a uniform coefficient of viscosity from layer to layer, but near the surface, the viscosity would be found to increase rapidly. The strongly adsorbed material, with its comparatively long term of molecular lingering, shows a response to shear stress corresponding to a viscosity several hundred thousand times the viscosity of the same material when it is not adsorbed. The difference in mobility under stress is so great that it is quite reasonable to consider the strongly adsorbed material to be immovable under the conditions considered here, and to use the concept of shear plane.

12.1 The Zeta Potential

The potential at the plane of shear is called the *zeta potential;* it is also called the *electrokinetic potential.* The motive force acting on a particle in a given electric field is directly proportional to the zeta potential, and the direction of movement depends, of course, on the sign of the zeta potential. Likewise, the rate of electroosmosis is directly proportional to the zeta potential. At a given rate of flow in a given system in which the flow is induced mechanically, the streaming potential is also directly proportional to the zeta potential.

Thus, for calculation of electrokinetic effects, the zeta potential is precisely the potential needed. It can be measured by methods based on electrophoresis or electroosmosis. However, for calculation of electrostatic repulsion, we need the Stern layer potential rather than the zeta-potential, assuming they are different. It seems, however, that these two potentials are not likely to be much different. Anyhow, the zeta potential has long been used in connection with studies of interparticle forces, especially as manifested by flocculation and dispersion. In the rest of this chapter we consider the zeta potential to be virtually the same as the Stern layer potential.

13. SOME PROPERTIES OF CEMENT SUSPENSIONS

In connection with various subjects in other chapters we frequently encountered manifestations of the forces discussed in this chapter. Now we shall consider some aspects of cement suspensions that show rather clearly how van der Waals forces, electrostatic potential, and disjoining pressure influence interparticle attraction.

13.1 The Zeta Potential in Cement Pastes

Not much on the electrokinetics of cement paste has been reported. However Steinour [16] measured electroosmosis in various cement pastes and in pastes made with one of the principal components of portland cement, beta-dicalcium silicate. The results showed that the particles of cement normally bear a positive charge. Also, beta-dicalcium silicate gave results substantially the same as those given by cement. The data on beta-dicalcium silicate being the more extensive, we shall discuss those.

Some of Steinour's data are given in Table 9.5. Flow rates are shown for two different test periods. The fourth and fifth columns, added by the present author, give relative zeta potential expressed as per cent of the potential observed in a paste made with distilled water; the figures are simply the relative flow rates, for in a given system flow rate is proportional to zeta potential.

Table 9.5 Effect of the Concentration of Sodium Chromate on Electroosmosis in Beta-dicalcium Silicate Paste[a]

Na_2CrO_4 (equivalents per liter)	Rate of Flow in Arbitrary Units		Relative Zeta Potential (per cent)	
	3rd period	6th period	3rd period	6th period
0	229	253	100	100
0.041[b]	102	100	44	40
0.122	1	38	0	15
0.244	−40	−24	−17	−10
0.487	−64	−63	−28	−25

[a] Data are from H. H. Steinour [17].
A positive charge is indicated by a positive flow rate.
[b] Data on this line are averages of two tests.

The first item in Table 9.5 gives the flow rates when the paste was made with distilled water. The flow of the liquid was toward the positive electrode, showing that the solid particles were positively charged. Calculation of the zeta potential on an approximate basis gave 11 mV. In this experiment the nearly pure beta-dicalcium silicate was in an aqueous solution produced by its own hydrolysis during the initial period of reaction with water. (See Chapter 11.) The principal ions so produced were the divalent calcium ion (positive) and the monovalent hydroxyl ion (negative). (In pastes made with cement the solution would contain also the positive monovalent sodium and potassium ions and negative divalent sulfate ions.)

Presumably, calcium ions were strongly adsorbed, giving the beta-dicalcium silicate particles the observed positive charge and creating some unknown electrostatic potential ψ_o. Owing to partial neutralization by the negative counter ions, the potential at the outside of the Stern layer would be a value ψ_δ, which is lower than ψ_o, as previously discussed. We consider the Stern layer potential to be virtually the same as the zeta potential, 11 mV in this case.

The other items in Table 9.5 show how supplementary ions influence the zeta potential. Positive monovalent sodium ions might add to the total positive surface charge, or they might displace some divalent calcium ions from the charge-determining adsorbed layer. The actual situation is not known, but the net effect on surface charge should be small. On the other hand, supplementing the monovalent hydroxyl ions in the Stern layer with divalent chromate ions, or even a replacement of hydroxyl by chromate ions, should cause an increase in the potential drop in the Stern layer or, in other words, should lower the zeta potential.

Experimental results in Table 9.5 show just such an effect as that described above. The zeta potential was not only lowered, but also, as the amount of added electrolyte was increased from test to test, the zeta potential passed through zero and became negative. This indicates that at a sufficiently high concentration the negative ion becomes the charge-determining ion, presumably forming a compact dual layer of positive and negative strongly adsorbed ions, with the negative ions outside.

Table 9.6 contains other data by Steinour showing how ion valence influences the zeta potential at a given mole concentration of added electrolyte. The first item is the same as the first in Table 9.5. In the following group of nine items, the electrolytes were so chosen that the positive ions were mostly monovalent, and the valence of the negative ions ranged from 1 to 4. Looking at the final column, we see that the relative zeta potential was the lower the higher the valence of the negative ion, or, more generally, the higher the valence of the ion having a charge opposite to that of the surface; or, in still other words, the higher the valence of the counter ion. In the last five items of this group, the counter ions were sufficiently effective to give negative zeta potentials. In terms of Fig. 9.4 this means that the Stern-layer potential passed through zero into the negative region. Of course, the dominant ion of the diffuse layer was reversed also.

Some of the surface-active agents now being used in concrete produce the effect to be expected from polyvalent counter ions. Calcium lignosulfonate, for example, produces very large negative ions that presumably crowd into the Stern layer. Ernsberger and France [17] found by

Table 9.6 Effect of the Kind of Electrolyte on Electroosmosis in Beta-dicalcium Silicate Paste[a]

Chemical Formula	Ion valence +	Ion valence −	Rate of Flow in Arbitrary Units	Relative Zeta Potential
None[b]			229	100
$NaNO_3$	1	1	190	83
$NaCl$	1	1	164	72
$NaOH$	1	1	73	32
$Na_2S_2O_6$	1	2	42	18
Na_2SO_4	1	2	−38	−17
[c]Na_2CrO_4	1	2	−64	−28
$K_3Fe(CN)_6$	1	3	−121[d]	−53
$Ca_2Fe(CN)_6$	2	4	−187	−82
$Na_4Fe(CN)_6$	1	4	−312[e]	−136
$Ca(NO_3)_2$	2	1	322	140
$CaCl_2$	2	1	348	152
[f]$Ca(OH)_2$	2	1	398	174
$Ba(NO_3)_2$	2	1	344	150
$BaCl_2$	2	1	357	156
$Ba(OH)_2$	2	1	283	124
$Co(NH_4)_6Cl_3$	3	1	380	166

[a] Data are from H. H. Steinour [17].

Each solution with which beta-dicalcium silicate paste was prepared contained 0.33 hydrogen equivalents of solute per 1000 g water, except as noted. Each preparation contained 35 g beta-dicalcium silicate, 18 cm^3 of distilled water, and the appropriate weight of electrolyte.

[b] No electrolyte was added, but $Ca(OH)_2$ was produced by electrolysis of some beta-dicalcium silicate.

[c] By mistake almost $\frac{1}{2}N$ instead of $\frac{1}{3}N$ was used.

[d] The fourth half-hr result was reported. Evidently foaming prevented a good result for the third half hour since no data were recorded for that period.

[e] Difficulty in reading (foaming?) prevented good readings except during 15 min of this interval. The value given is twice the 15 min flow.

[f] Saturated solution, hence, only $\frac{1}{25}N$.

electrophoresis measurements that adsorption of the lignosulfonate ions gave cement partices a pronounced negative charge.

It is apparent also from the first group in Table 9.6 that valence is not the only factor determining the effectiveness of ion. For example, of the three compounds of monovalent positive and negative ions, the hydroxyl ion is clearly more effective than the nitrate or chloride.

The last seven lines in Table 9.6 pertain to pastes made with aqueous solutions of the compounds indicated. In each of these solutions the negative ion was monovalent and the positive ion was di- or trivalent. In every case the zeta potential was substantially increased above the potential that existed in the pastes made with distilled water.

There is evidence in these data that the hydrolysis of beta-dicalcium silicate was not sufficient to saturate the solution with respect to the calcium ion. This is shown by the fact that when a saturated solution of calcium hydroxide was used instead of distilled water there was an increase in the potential. It seems clear, therefore, that in these experiments the surface potential was generally not as high as it might have been because the adsorption of the charge-determining calcium ion was not at the maximum, and because the surface potential was not at its maximum, the zeta potential was also lower than it might have been. Thus, when Steinour added more calcium hydroxide, or any other electrolyte of the same type with regard to positive-ion valence, the zeta potential was increased.

All these results are in agreement with Gaskin's view [18] that in cement paste the charge-determining ion is principally the calcium ion, whereas the negative ions are those that tend to neutralize the charge, and to produce the flocculent state. As we have already seen, in freshly mixed cement paste negative ions include not only hydroxyl but also sulfate ions.

Changes in the zeta potential discussed above are accompanied by changes in the interparticle forces in the pastes. At one extreme the forces are balanced in such a way that they maintain a moderately flocculent state; at the other extreme they maintain a state of dispersion. One way of observing such changes in interparticle forces is to measure the volumes of sediment after suspensions have settled, as discussed in the following section.

13.2 Observed Changes in Interparticle Forces

A simple method of observing the relative strengths of cohesive forces among particles due to changes in the zeta potential in a suspension is to measure the volume of sediment after a given quantity of suspended material has settled out; the stronger the interparticle bond, the greater the specific volume of the sediment, or as is commonly said, the greater the *sedimentation volume*. The explanation of this relationship is essentially the same as the explanation of the bulking of sand, given in Chapter 3, but the cohesive force is made up of the van der Waals attraction minus the sum of the electrostatic repulsion and disjoining pressure, instead of forces arising from tension at air-water interfaces. In a suspension, the only air-water interface is that at the top of the surface of the mixture, and liquid surface tension therefore has no effect on submerged particles.

In a typical experiment 40 ml of fluid was placed in a 1×6 in. test tube and to this was added 20 g (about 6.3 cm^3) of cement. The mixture was shaken briefly and then allowed to stand for 5 min in a stoppered tube. Then it was shaken by hand vigorously for 1 min and placed in a rack. After 48 hr of settlement the volume of sediment was measured and recorded as the volume of sediment per unit solid volume of solids.

In one group of experiments suspensions were made in various mixtures of ethyl alcohol and water, the mixtures ranging from pure water to pure alcohol. The results are shown in Fig. 9.5. Certain of their features are not easy to explain and may be due to extraneous factors, but the principal indications are clear in terms of factors discussed in Sections 6 through 10.

At ordinate zero the sedimentation volume is that of cement in water. The value plotted corresponds to the volume observed 48 hr after the

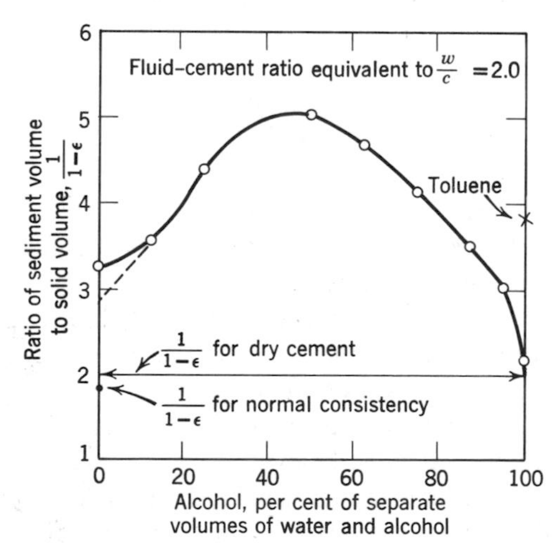

Fig. 9.5. Sedimentation volumes of cement in various mixtures of water and alcohol.

beginning of settlement, but because of the effect of the setting or congealing of paste, settlement could not have continued for more than about 3 hr, and possibly less than that. Usually within 2 hr inundated cement paste stops settling and begins to *expand* slightly. Therefore the observed sedimentation volume of cement in water is somewhat ambiguous with respect to its relation to interparticle forces in freshly mixed cement paste. The broken-line extension of the curve terminating below the plotted point is probably a good estimate of what the volume would have been without the interference of setting; it indicates a volume significantly greater than that of dry, settled cement.

In view of the indicated volume of settled dry cement, it appears that the net interparticle attraction is greater between cement particles in water than it is between cement particles in air. But this may not be true: observe that the force due to gravity, which acts against cohesive force during the building of a sediment structure, is smaller for particles in water than for particles in air, on account of the buoyant force. For example, cement particles that weigh 3.15 g/cm^3 in air, weigh about 2.15 g/cm^3 in water, and thus the net downward force on cement particles in water is the fraction $2.15/3.15 = 0.68$ of the force in air. Thus even if net interparticle attractions were equal, we should expect a larger sedimentation volume in water than in air, and if the net interparticle attraction in water is smaller than in air, it is still possible for the sedimentation volume to be larger. Since the quantitative relationship between interparticle attraction and sedimentation volume is unknown, we cannot pursue this point to a definite conclusion. It seems, however, that although the force of attraction between cement particles in water is not greatly different from that in air, it would be expected to be smaller.

Net interparticle attraction between particles in water is indicated not only by sedimentation volume but also by the fact that during and after settlement the water above the sediment remains clear, that is, free from the cloudy appearance given by small particles. This means that the small and large particles stick together while they fall.

The properties of a suspension of cement in water are due mainly to the electrolytically controlled zeta potential, as is discussed above, and to the strongly adhering film of water, the latter characteristic placing it in the category called *lyophilic suspension* (or hydrophilic suspension). When either the zeta potential or the film is altered, interparticle force is also altered, as discussed below.

When ethyl alcohol (denatured) was substituted for some of the water, sedimentation volume increased as is shown by the left half of the curve in Fig. 9.5, signifying an increase in cohesive strength. Explanation of this feature of the curve is necessarily speculative and tentative because of a lack of quantitative data on changes of zeta potential and changes in the repulsion force due to disjoining pressure; however, the data help us to reach a reasonable understanding of principal phenomena.

Alcohol tends to dissolve the layer of adsorbed water molecules, but because an alcohol molecule is polar, and in that way is like water, alcohol itself can be adsorbed and thus provide an effective barrier such as that discussed in connection with Fig. 9.2. The effective repulsion force due to disjoining pressure does not necessarily decrease when alcohol is substituted for water, although a decrease actually is indicated by Fig. 9.5, the increase in sedimentation volume having that meaning. The decrease of repulsion may be accounted for in terms of the change in electrostatic repulsion.

Since calcium hydroxide is less soluble in ethyl alcohol than in water, the calcium ion concentration diminished as alcohol was substituted for water, and the positive electrostatic charge due to adsorbed calcium ions correspondingly diminished. After the concentration of adsorbed positive ions decreased to a certain level, the positive charge on the surface just offset the intrinsic negative charge of the solid surface, and thus gave a net surface charge of $\psi_o = 0.0$. At this point which presumably is represented by the apex of the curve in Fig. 9.5, only the disjoining pressure remains to limit the force of interparticle cohesion due to van der Waals attraction. Thus the rise of the curve up to the point representing about 50 per cent alcohol is indicative of the effect of removing electrostatic repulsion between particles, without much change, presumably, in repulsion due to disjoining pressure.

With a further increase in the per cent of alcohol, there was a corresponding further reduction in the calcium ion concentration, and the surface charge of the solid became progressively more negative. The charge finally reached a maximum negative value when the point representing pure alcohol was reached. This transition is indicated by the right-hand half of the curve in Fig. 9.5.

With 100 per cent alcohol, we apparently got a sedimentation volume that was determined by a relatively strong repulsion force comprising the disjoining pressure due to adsorbed alcohol molecules and the electrostatic repulsion due to a net negative surface charge on the particles. If there was any zeta potential at all, it was probably negligible, which means that electrostatic repulsion was due to direct action between surface charges rather than to mutual repulsion of ion clouds.

In pure alcohol the smallest particles of cement were observed to remain suspended for a relatively long time after the coarse ones had settled, and they exhibited Brownian motion. The state is called *dis-*

persed, although the sedimentation volume is similar to the settled volume of dry cement, a fact suggesting that van der Waals attraction becomes effective after gravity brings dispersed particles close together.

The alcohol-cement system just described may be called a lyophilic suspension with little or no electrolyte, and presumably without zeta potential, but with an intrinsic surface charge.

In Fig. 9.5 we see examples of a lyophilic suspension with a zeta potential (100 per cent water), a lyophilic suspension with apparently no surface potential (40 to 50 per cent alcohol), and finally a lyophilic suspension with probably no zeta potential. In suspensions of the first category weak flocculence is observed; those in the second category are strongly flocculent; those in the third are nominally disperse, and actually are in that state before sedimentation brings the particles close together.

One point in Fig. 9.5 represents a suspension of cement in the organic, nonpolar liquid, toluene. This item is an example of a lyophobic suspension with no zeta potential but supposedly with a full intrinsic negative charge on the particles. Molecules of toluene are not strongly adsorbed, and thus cannot present a high potential barrier to keep the particles apart. It seems safe to assume therefore that in this case van der Waals attraction is opposed mainly by electrostatic repulsion due to the intrinsic charge on the surface of the particles. The position of the point in Fig. 9.5 indicates that interparticle cohesion between cement particles in toluene is greater than it is in water or in alcohol, and the appearance of the suspension suggested relatively strong flocculation.

It is possible to convert a lyophobic, strongly flocculent suspension of cement to a lyophilic, weakly flocculated suspension, or to a disperse suspension, by means of a suitable surfactant. For example, cement and kerosene, a lyophobic mixture, can be converted to a lyophilic state by means of oleic acid. If we add oleic acid drop by drop while we stir the mixture, the properties of the mixture change gradually until finally a stage is reached at which the cement-kerosene paste exhibits properties similar to that of a cement-water paste of the same proportion of solid to liquid. However, a few additional drops of oleic acid converts the cement-kerosene paste to a state resembling that described for mixtures of cement and ethyl alcohol. In other words, it is possible to carry the transition all the way to the disperse state.

14. THE FLOCCULENT STATE OF CEMENT IN WATER

In the foregoing discussion we have dealt with various factors that determine the physical characteristics of mixture of cement with different fluids, bringing out how the several factors may be combined in different ways to produce mixtures having widely different characteristics. Generally we infer that the relationships pertain to any suspension, thick or thin. In this section we shall consider thick suspensions, cement pastes in particular.

We have already distinguished between the terms floccu*lated* state and floccu*lent* state. The flocculated state is one in which separate clusters of particles exist. The grains making up each cluster are held together, and apart, by the combined interparticle forces of attraction and repulsion discussed earlier in this chapter. The separate clusters, called *floccules,* are easily observed in a dilute suspension for they settle out as separate units. The flocculent state may be thought of as a suspension composed of floccules so concentrated that the floccules merge into a continuous mass, thus giving the whole system the aspect of a single large floc. However this is an oversimplified concept, as we shall see when the following matters are considered.

In the flocculent state the individual particles cannot be evenly distributed in space. To see that this is so, we must consider some numbers pertaining to average interparticle distance.

In a typical cement paste, $w/c = 0.50$, the cement having a surface area of 10^4 cm^2/cm^3, the quotient of water volume by solid-surface area is about 1.6×10^{-4} cm. This quotient is the hydraulic radius m. The average distance between solid surfaces is between $2m$ and $4m$. Thus we deduce that in the paste just described, the average interparticle distance is between 3 and 6 μ or upwards of 30,000 Å. Such a distance is far greater than any reasonable estimate of the distance between particles when they are in their positions of minimum potential energy with respect to interparticle attraction and repulsion, and since the particles tend to occupy positions of minimum potential energy, we may visualize them as forming masses within which the particle concentration is greater than the average concentration for the whole mixture. Such a structure would result from bringing about a concentration of floccules, thus creating a continuous floc structure but with floccule interstices, as already suggested.

However, it would be extending the concept too far to assume that the variation of paste density with water content is due only to differences in the packing of uniformly dense floccules. In Chapter 11, Section 4.4.6, it is shown that the water-cement ratio of a floccule formed of ordinary cement in a dilute suspension is about 1.25 by volume, or 0.4 by weight. Hence any paste having that water-cement ratio or a lower one could not have a structure composed of floccules each of which has a water cement ratio of 0.4. Indeed it is doubtful that pastes having relatively high water-cement ratios should be regarded as aggregations of floccules, for there is good evidence that the density of floc structure is different for different water-cement ratios.

One must keep in mind that pastes are not made by concentrating previously formed floccules, but by stirring cement and water together in the desired proportions. The mechanical act of mixing cement and water produces hydraulic forces that tend to keep each particle as far as possible from its neighbors while the mixing process is under way. The action thus tends to create a uniform dispersion from which the thixotropic floc structure presumably develops as soon as stirring ceases. It appears that this structure forms quickly and that the particles do not have much time to move into positions of greatest stability. Therefore, the floc structure (corresponding to the structure of individual floccules formed in a dilute suspension) tends to enclose more water the higher the original water-cement ratio.

Another factor that influences the structure of flocculent paste is the efficiency of the mixer. A mixer that produces a high shear stress tends to give a floc structure having a "fine" texture; on the other hand, a low-energy method of mixing, such as stirring cement and water by hand, produces paste having a relatively "coarse" texture.

It should be mentioned that the conclusions above about paste structure are difficult if not impossible to confirm by direct observation, but they are well supported by indirect evidence.

15. SOME REMARKS ON TERMINOLOGY

The purpose in writing this chapter was to give an understanding of the subject matter of surface physics and colloid chemistry as manifested in one aspect or another of the properties of concrete. Many difficulties were encountered, not the least being the state of flux of the science and consequent differences of opinion among experts.

The terminology of colloid science is often bewildering. According to some authors, materials fall into one of two classes: *lyophobic* (liquid hating) and *lyophilic* (liquid loving); in the former class are the insoluble materials, particularly metals in a colloidal state, and in the latter class are the organic materials: gelatin, and the like. But, it turns out that the materials of interest here do not belong clearly in either category. Silica, most components of hydrated cement, and many nonmetallic minerals, are relatively insoluble in water, but they are by no means as insoluble as the metals and therefore cannot be considered hydrophobic to the same degree that metals are hydrophobic. Moreover, these materials are different from organic colloids, but again the difference in some respects seems to be a matter of degree. Gelatin, for example, is so strongly hydrophilic that it may be reduced to a molecularly dispersed state by water, whereas silica gel, or hydrated cement, shows only a tendency in that direction, although a definite one: strong adsorption of the surrounding liquid may be regarded as the initial step toward dissolution, and indeed strong adsorption implies some degree of solubility. Glasstone speaks of "surface solution" as an aspect of liquid adsorption.

Since, as we have seen above, a given nonmetallic inorganic material may appear lyophobic toward one kind of liquid and lyophilic toward another, and a given material can be made to change from lyophobic to lyophilic in a given liquid, the dichotomous definition given above is not very useful here.

The terms lyophobic and lyophilic as used in this chapter reflect the following observations and concepts. If the liquid is strongly adsorbed in a given suspension of solids in a liquid, the solid may be said to be lyophilic; if it is not, it may be called lyophobic. Cement is lyophilic toward water (hydrophilic) but it is lyophobic toward organic liquids such as kerosene or toluene.

The two terms apply also in connection with the phenomenon called *spreading-wetting* or just *wetting*. If when placed in contact with a flat solid surface a drop of liquid spreads to form a thin film, the surface is said to be lyophilic toward that liquid; but if the drop does not spread, the surface is called lyophobic. If the liquid is water, the solid is called *hydrophobic*, or *water-repellent*.

When water spreads on a solid, it does so because the force that causes spreading, actually van der Waals attraction between water molecules and the surface, exceeds the maximum force that can be developed by the surface tension of the water drop, a force that resists the creation of more water surface by spreading. The force due to the surface tension of the liquid increases as the drop spreads, and with some combinations of solid and liquid, it may become equal to the spreading force while the drop is still intact, but flattened. In such a case the edge of the drop forms a certain angle with the surface, called the *contact angle*. The contact angle can take on any value between zero and nearly 180 degrees, zero denoting complete spreading-wetting and 180 degrees complete nonwetting; in other words, a zero contact angle indicates the solid to be completely lyophilic, and a 180 degree contact angle indicates it to be completely lyophobic toward the particular liquid in question.

It is correct to infer that there are innumerable intermediate degrees of "wettability," and even the same material, seemingly, may show various degrees of wettability toward a given liquid. A clean metal, for example, is hydrophobic in the sense just discussed, but as the surface of the metal becomes oxydized it becomes progressively more hydrophilic.

REFERENCES

[1] Glasstone, S., *Text Book of Physical Chemistry,* 2nd ed., van Nostrand, Princeton, N.J., 1946.
[2] Adamson, A. W., *Physical Chemistry of Surfaces,* Interscience, New York, 1960, 320–323.
[3] Brunauer, S., *The Adsorption of Gases and Vapors: Physical Adsorption,* Princeton University Press, Princeton, N.J., 1945.
[4] Deryaguin, B. V., and I. I. Abrikossova, "Direct Measurements of Molecular Attraction," *J. Phys. Chem. Solids,* **5,** 1–10, 1958.
[5] Powers, T. C., and T. L. Brownyard, "Studies of the Physical Properties of Hardened Portland Cement Paste," *Proc. ACI,* **43,** 549–602, 1947; Bulletin 22, Part 4, Portland Cement Association, 1948.
[6] de Boer, J. H., *The Dynamical Character of Adsorption,* Oxford University Press, New York, 1953.
[7] Deryaguin, B. V., "A Theory of Capillary Condensation on the Pores of Sorbents and of Other Capillary Phenomena Taking into Account the Disjoining Action of Poly-molecular Liquid Films," *Acta Physichochim. (USSR),* **12**(2), 181–200, 1940.
[8] Overbeek, J. T. G., "Interaction Between Colloidal Particles," in H. R. Kruyt, ed., *Colloidal Science,* Vol. 1, Elsevier, New York, 1952, p. 257.
[9] Prosser, A. P., and J. A. Kitchener, "Direct Measurement of the Long-Range van der Waals Forces," *Nature* (London), **178,** 1339, 1956; *Proc. Royal Soc.,* **A242,** 403, 1957.
[10] Polanyi, M., "The Potential Theory of Adsorption," *Science,* **141,** 1010–1013, 1963.
[11] Deryaguin, B. V., "The Force Between Molecules," *Sci. Am.,* **203**(1), 47–53, 1960.
[12] Deryaguin, B., S. Nerpin, and N. Tchurayev, "Effect of Film Transfer Upon Evaporation of Liquids from Capillaries," *RILEM Bulletin, New Series No. 29,* 93–97, 1965.
[13] Harkins, W. D., and G. Jura, *J. Am. Chem. Soc.,* **66,** 919, 1944.
[14] Brunauer, S., D. L. Kantro, and L. E. Copeland, "The Stoichiometry of the Hydration of Beta-dicalcium Silicate and Tricalcium Silicate at Room Temperature," *J. Am. Chem. Soc.,* **80,** 761–767, 1958.
[15] Von Buzagh, A., *Colloid Systems,* Technical Press Ltd., London, 1937.
[16] Steinour, H. H., Private communication about unpublished work. Also, "The Setting of Portland Cement," Portland Cement Association Research Bulletin No. 98, 1958, pp. 86, 122.
[17] Ernsberger, F. M., and W. G. France, "Portland Cement Dispersion by Adsorption of Lignosulfonate." *Ind. Eng. Chem.,* **37,** 598–602, 1945.
[18] Gaskin, A. J., Discussion, "Fourth International Symposium on the Chemistry of Cement," *Nat. Bur. Std. (U.S.)* Monograph 43, **2,** 746, 1960. Also, private communication, March 14, 1961.

10

Rheology of Freshly Mixed Concrete

1. INTRODUCTION

Rheology is the science dealing with deformation and flow of matter. In this chapter we discuss its applications to fresh concrete. We shall study the process of shaping and molding in terms of the internal strains and corresponding applied forces incidental to that process; we shall therefore be dealing with intrinsic properties that can be measured quantitatively and expressed in fundamental units. Our subject is not workability, as most people understand that term, although the rheological characteristics of a concrete mixture constitute the most important factor in workability, and perhaps the only aspect that can be evaluated quantitatively.

The following discussion is restricted to an outline of some elementary rheological concepts and to a presentation of what seems to be the most informative data on paste and concrete. In some areas questions will be raised concerning the relevance of conventional rheology to the study of workability. First, however, we review some of the rheological characteristics of concrete that can be seen by observation of normal operations.

1.1 Behavior of Freshly Mixed Concrete

A concrete mixture may be so designed that when it is discharged from the mixer it is bulky and nonplastic and remains so during handling. When so designed, it can be molded only by pressure, tamping, or vibration, and in the process it does not become plastic. (See Chapter 3,

Section 3.) Compacting and molding such a mixture involves overcoming internal resistance to movement of one part of the material with respect to another part, and thus it involves internal strain and stress. Apparently such operations have not been studied by methods of rheology, and we can say no more about them. The kind of concrete most widely used emerges from the mixer in a plastic state or in a state such that it becomes plastic before the process of transporting and placing it is complete. (See Chapter 3, Section 4.)

Let us observe a plastic mixture as it emerges from a mixer and moves down a chute into a wheeled conveyor, a concrete buggy. If the mixture has a stiff consistency, we see that it does not flow, it slides in the chute; if it is of a softer consistency we may observe a tendency for the top and middle part to move a little faster than the rest, but sliding is still the principal mode of motion. We follow the loaded buggy along the runway at the top of a wall form to the point of placement; the load is dumped into the form, and we note that it tends to stand as a heap until a vibrator or some other mechanical means causes it to flow and to flatten.

While the buggy was being trundled along the somewhat uneven runway, we noticed that the top of the load flattened and the whole mass became fluidlike, tending to remain level as the container became momentarily tilted. While the buggy was being emptied, we saw that the contents seemed to pour somewhat as a fluid, but at the same time we might see a succession of fractures appearing as the outflowing material was "bending" over the lip of the buggy. While the pouring was in process, the concrete at the point of impact tended to flow as molten lava does, but when the last of the load had fallen, the whole mass instantly "froze" in whatever shape it had at that final instant. The behavior described cannot properly be called that of a fluid, but fluidity may exist momentarily.

The characteristics just discussed are shown vividly in Fig. 10.1, which is a photograph taken during the final stages of construction of Hoover Dam on the Colorado River. Concrete was transported from a mixing plant in a large bottom-dump bucket via cableway to the point of deposit. At the instant the photograph was made the last of the batch was just entering the "pool" of concrete, the recoil of the suddenly unloaded cableway having snatched the bucket from view. The concentric waves visible on the pool resemble the waves formed by dropping a pebble in a pond, but there is a significant difference: The waves freeze and would become permanent if the concrete were not subject to further manipulation.

On the Hoover Dam project, carried out in the early 1930's, the final

Fig. 10.1. The momentarily fluidized state of concrete. (The photograph was taken at Hoover Dam, Arizona, by F. R. McMillan.)

shaping of the concrete was done mostly by means of shovels and boots. In more recent times the placing of mass concrete of stiff consistency is accomplished by means of vibrators. The mixture is deliberately made so stiff that when first deposited a batch stands in a heap in a more or less dilatant state (Chapter 3, Section 4.1). Workmen complete the placement by inserting vibrators into the heap repeatedly. Vibration causes the heap to flow and flatten.

The effect of a vibrator on a stiff concrete mixture becomes apparent to a man standing on the fresh concrete. He tends to sink when the vibrators are turned on, but when they are shut off, he can walk almost dryshod over the surface, as he could before vibration was applied.

Although a mixture may appear dilatant and thus nonplastic before it is vibrated, proper vibration converts it to the plastic state; actually, during vibration a fluid state prevails, as is shown below.

After vibration has been stopped fluidity disappears but plasticity persists. Under the circumstances described above, the persistence of plasticity can be detected in a simple way: If a man stands on a surface in a straddled stance and rhythmically shifts his weight from one foot to the other, he can cause a considerable area of the surface to "teeter," that is, he can induce wave motion.

The initial dilatant state described above is perhaps exceptional, observable only when the consistency is unusually stiff or the mixture unusually deficient in cement or other fine material. However, other characteristics are observable to one degree or another in all plastic mixtures. Some of the phenomena are manifestations of interparticle forces discussed in Chapter 9.

1.2 *Behavior of Freshly Mixed Paste*

We shall find it advantageous to consider the rheological characteristics of cement paste before attempting to discuss the rheological characteristics of concrete. Such characteristics of paste depend partly on the nature and extent of certain chemical reactions with water that occur during the period of mixing. Although it would be beyond the scope of this book to discuss these reactions in detail, we must become familiar with their general aspects.

1.2.1 Initial chemical reaction. During a short period beginning at the time of first contact between cement and water, a relatively rapid chemical reaction occurs with evolution of heat. An example is shown in Fig. 10.2, which gives rate of heat liberation as a function of time. Normally, there are four stages of reaction: Immediately after contact is made between cement and water a short period of dissolution and rapid exothermic chemical reaction takes place. During this interval, the rate of heat evolution increases rapidly and reaches a peak rate

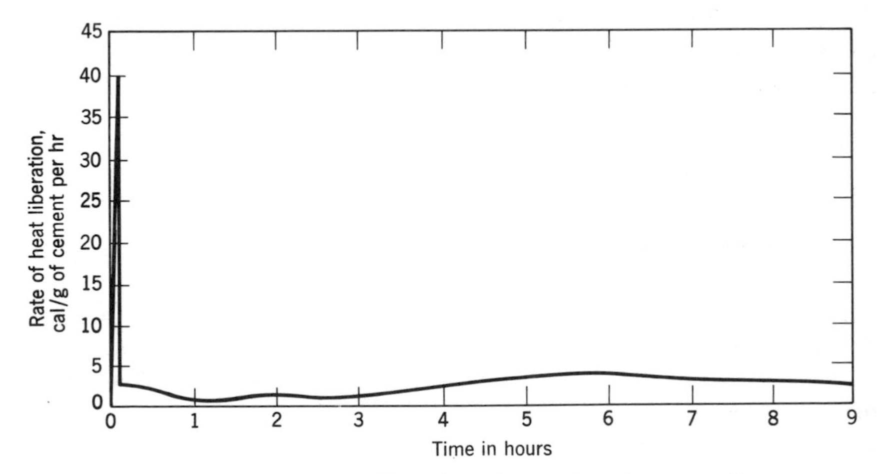

Fig. 10.2. Rate of heat liberation of cement paste versus time.

of 40 or more cal/g/hr. Then the rate drops rapidly, marking the end of the first stage. During the next hour, more or less, the rate remains low, decreasing to about 1 cal/g/hr. Then it begins to increase, marking the end of the second stage. During the third stage, the rate of heat evolution increases and reaches a peak at about the sixth hour after mixing; this is about the time the final set is reached in a paste having normal consistency. The fourth stage begins as the rate again begins to fall, dropping back to less than 1 cal/g/hr within 24 hr, and to still lower rates after that. We are here concerned only with the first two stages.

During second stage the paste remains plastic; hence, the second stage may be called the *dormant period*. At room temperature the dormant period normally lasts 40 to 120 min, depending on the characteristics of the cement and on whether or not a chemical accelerator or retarder has been used; indeed, retarders are able to prolong the dormant period much beyond 120 min.

The characteristics just described may not be manifested experimentally unless the paste has been properly handled.

1.2.2 The role of mechanical mixing. It should be understood that merely adding water to cement does not produce a plastic mixture. That is to say, if dry cement were allowed to absorb water by capillary action, it would become a firm though weak solid within a few minutes. The plastic state is achieved by a mechanical action that disperses the cement grains in an aqueous solution which provides the electrolytic environment necessary for plasticity.

The rheological characteristics of a batch of fresh paste depend on how the batch is mixed. When cement and water are mixed by gentle stirring or kneading, the mixture remains stiff relative to the consistency that can be produced by vigorous mechanical action [1]. To produce a homogeneous paste having the lowest possible stiffness, a laboratory mixer able to produce a high rate of shear is necessary.

The length of the mixing period is important; if it is of the order of 30 sec or less, the paste soon acquires *"brief-mix-set"* [2], a phenomenon attributed to the grains becoming stuck together by gel in the process of forming on grain surfaces. This kind of set, which seems to be common to all portland cements, may be prevented by continuing the mixing for about 1 min.

However even when the mixing period is long enough to prevent brief-mix-set, which is usually the case, it may yet be too short to eliminate another phenomenon called *false set.* False set is usually due to the presence of plaster of paris ($CaSO_4 \cdot \frac{1}{2}H_2O$), which is present in most

Table 10.1 Mixing Schedules Required by Different Cements to Eliminate False Set[a]

Cement			Condition after 1-min. Mix			Final Condition	
ASTM Type	Lot Number	w/c (by weight)	R[d]	Slump (in.)	Required Mixing (min)	R[d]	Slump (in.)
1	15753	0.36	27	1.05	2-0-0	25	1.25
1	15760[b]	0.35	41	0.85	2-2-2	33	1.10
2	15755[c]	0.34	32	1.10	2-2-2	26	1.10
3	15757[c]	0.34	33	0.95	2-3-2	27	1.05
4	15762[b]	0.34	42	0.85	2-5-2	22	1.40

[a] Specific surface of each cement was 1500 cm^2/g (Wagner Turbidimeter). The tests were made on 1:2 (by weight) mortars made with Elgin sand; the w/c was that required for (1 $\pm$ 0.05)-in. slump, with a 6-in. cone, after a 1 min. mix. The first number of the mixing schedule is the length of the first period of mixing; the second is the period of rest; the last is the final mixing period.
[b] This cement showed marked false set.
[c] This cement showed evidence of moderate false set.
[d] Remolding test, number of jigs.

portland cements.* The plaster forms a weak solid structure that can be destroyed by remixing, and if the plaster-of-paris reaction has been completed at the time of remixing, false set will not occur again.

The required length of time for precluding false set is not the same for all cements, and thus in laboratory experiments it is advisable to allow a period of rest between an initial and final mixing period, the necessary length of the rest period having been determined by trial for each cement to be used. Examples of mixing schedules required to eliminate false set are given in Table 10.1.

Most laboratories are not equipped with mechanical equipment to knead pastes of normal consistency. For example, it has been observed that at $w/c = 0.24$, an inadequate laboratory mixer operated for 5 min produced a paste that would slump 2 in. in the standard test, but when the size of batch was doubled, mixing for the same length of time in the same mixer resulted in a relatively stiff consistency; the slump was only 1 in.

* During the manufacture of cement, gypsum ($CaSO_4 2H_2O$) is added during the final grinding stage. Under the combined action of the heat generated in the mill and the drying action of the anhydrous components of the cement, the gypsum becomes partially dehydrated; that is, it becomes plaster of paris.

With the exception of matters pertaining to false set, time effects and mixing procedures are principally laboratory problems. Under normal field conditions, the time of mixing can hardly be so short as to permit brief-mix-set, and the rolling mass of aggregate in the concrete mixer homogenizes the paste as effectively as the most vigorous laboratory stirrer.

In the sections that follow we examine rheological data on pastes from various laboratories, and we find some differences where it seems there should be similarity. In some cases the differences may be explained in terms of known differences in methods of handling the pastes; in other cases descriptions of handling methods are omitted, and we must resort to conjecture or perhaps leave the discrepancies unexplained.

2. SOME CONCEPTS AND TERMS OF RHEOLOGY

Rheological terms have been used above and in earlier chapters with the expectation that they would be at least partly understood and not be misleading. Now we proceed to define them as explicitly as possible and to discuss the relevant concepts.

2.1 Consistency

Rheology in the present context can be regarded as the science pertaining to that difficult to define term, "consistency." Although it pertains to consistency, rheology does not involve the measurement of consistency, for, like "workability," "consistency" cannot be defined quantitatively.

Consistency has been defined as "the relative mobility or ability of fresh concrete or mortar to flow" [3]. It has also been defined as a degree of firmness, density, viscosity or resistance to movement or separation of constituent particles (the consistency of syrup); the manifestation of mutual attraction of particles at different moisture content (the consistency of soil) [4]. In ASTM Designation E24-58T consistency is defined as "the resistance of a non-Newtonian material to deformation."

As the term is sometimes used, we can speak of an increase or decrease in consistency, or we can say that one substance has more consistency than another. However, because a unit quantity of consistency is not defined, the meaning of such statements remains vague. With reference to plastic concrete mixtures, the term "consistency" is usually used in the way we use the term "temperature"; just as we speak of a high or low temperature, implying a high or low atomic kinetic energy, so we speak of the dry or wet consistency, implying the observed degree of the thinning effect of water. Similarly, we may call a material "stiff" or "soft," implying differences in shearing resistance. By this usage, the word "consistency" needs an adjective to indicate differences or changes.

From the standpoint of the rheologist, consistency is related to resistance to deformation, and deformation involves stress, strain, and time. Thus the quantitative aspects of consistency can only be expressed in such units. It can be said that the consistency of concrete is the result of those rheological characteristics that change when the water content of a given mix is changed, or when the temperature is changed, or when a surfactant is added.

Regardless of the possible differences of interpretation of the word, the standard slump and flow tests are often referred to as measures of consistency.

2.2 Elasticity

The capability of a strained body to recover its size and shape after it has become deformed by external forces is due to the property of the material called *elasticity*. Accordingly, an elastic deformation is defined as a deformation that disappears upon removal of the external forces causing the deformation and the stresses associated with it. When they are stressed in ordinary ways materials are not perfectly elastic; however many materials approximate ideal elastic behavior closely.

In the present context we are mostly concerned with forces that produce distortion rather than volume change [5]. This distortion may be described as shear strain, the orientation being such as to coincide with the distortion. The corresponding stress producing the shear strain is a shear stress and in an ideally elastic body is proportional to it. That is,

$$\tau = G\gamma \tag{10.1}$$

where τ is the shear stress, γ is the shear strain, and G is a proportionality constant called the *shear modulus*.

2.3 Viscosity

Viscosity is the property of a substance that enables it to yield continually under a constant shearing stress. It may be thought of as internal friction, for the work done in maintaining continual shearing is dissipated as heat. Maxwell defined the *coefficient of viscosity* as follows:

"The coefficient of viscosity of a substance is measured by the tangential force on unit area of either of two horizontal planes at unit distance apart, one of which is fixed while the other moves with unit velocity, the space between being filled with the viscous substance."

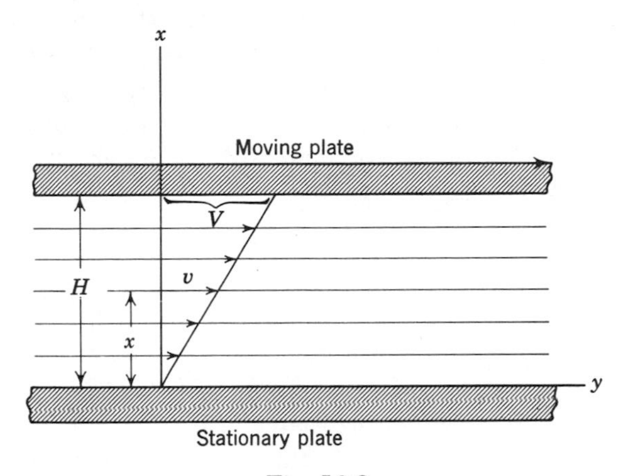

Fig. 10.3.

Consider Fig. 10.3 [5, p. 9]: It represents two parallel plates separated by a distance H, with fluid between them, the top plate moving at a constant rate such that it moves distance V in one second; v is the velocity at a point at distance x from the bottom place. It is apparent that the velocity gradient in the fluid is

$$\frac{dv}{dx} = \frac{V}{H} \tag{10.2}$$

Maintaining such motion requires a certain tangential force creating shearing stress τ between unit areas of any two of the imaginary layers of fluid as one layer slides over the other. Such shear stress is proportional to the velocity gradient; that is,

$$\tau = \eta \frac{dv}{dx} \tag{10.3}$$

where η is the coefficient of viscosity.

Equation 10.3 pertains to an ideal fluid, that is, a fluid that flows at a constant rate under any tangential force, however small the force, and in which the velocity is strictly proportional to the force. A liquid showing such behavior, or approximating it closely, is called a *Newtonian liquid,* and the laminar mode of flow is called *Newtonian flow,* in contradistinction to *turbulent flow* which develops at a higher velocity.

2.3.1 Effect of suspended particles on viscosity. A mixture of liquid and rigid particles has a higher coefficient of viscosity than that of the liquid alone. If the forces acting between the suspended particles are negligible, which is to say if the particles are in the dispersed state, the effect is a hydrodynamic one. Each rigid particle generates a perturbation of flow lines that requires a certain expenditure of energy not required when flow lines are undisturbed. Albert Einstein (1905) deduced the following relationship:

$$\eta = \eta_o(1 + 2.5\varphi) \tag{10.4}$$

where η is the viscosity of the mixture and η_o is that of the liquid alone; φ is the solid-volume concentration of spherical particles.

Equation 10.4 was derived for one sphere in an infinite volume of liquid; it rests on the assumption that the disturbance propagated by the particle can spread without reflection and without interference from other disturbances. Practically, the effects of interferences not considered in the derivation proved to be negligible for particle concentrations up to about 3 per cent by volume. At higher concentrations, the viscosity increases with concentration much faster than is indicated by Eq. 10.4. Numerous more or less successful attempts to extend the analysis to take into account mutual effects between particles have been made, but we are not concerned with them.

2.4 Plasticity

There is as much vagueness about the meaning of plasticity as there is about consistency. However, it is generally agreed that a material capable of being molded or shaped is plastic, and that the term "plasticity" implies the ability of a soft material to retain a given shape. A soft plastic material is not devoid of elasticity, but of the usual amount of strain produced in molding, the elastic part is negligible.

Although the above approach to the subject of plasticity leads to a fairly clear concept, some difficulties appear on further consideration. The stipulation that the term implies the ability to hold a given shape raises the question, What shape? A truncated cone of freshly mixed concrete having an 8-in. base, 12-in. altitude, and 4-in. top diameter (standard slump cone) ordinarily cannot withstand as much shear stress as the force of gravity can produce in a body of concrete having that shape unless it is supported by the mold. This is shown by the slumping of the mass when the mold (slump cone) is removed. However the same material can withstand the shear stresses produced by gravity after it has slumped and thus acquired a new shape. Also, a seemingly fluid paste that can be poured cannot stand unsupported in the shape of a cylinder, but if allowed to flow, it will find a shape in which shear stresses do not exceed its shearing strength, and it can then retain that shape against the force of gravity which tends to level it completely.

Closer examination of paste in a beaker reveals that its mobility does not necessarily indicate that it is a liquid before pouring. At $w/c = 0.40$, for example, the specific gravity of cement paste is about 1.97. It would seem, from the principle of Archimedes, that a wooden object with a specific gravity of about 0.5 could not remain immersed in cement paste. Yet, when we thrust a toothpick partly under the surface of cement paste, we find that it remains there despite the buoyant force that tends to expel it.

In noting the stationary state of the toothpick we are observing a body partly embedded in a plastic solid material, that material being made solid by interparticle forces already discussed (Chapter 9). However it is a soft solid too weak to support its own weight under all circumstances, and a sustained shearing stress of sufficient magnitude may be able to convert it from a solid state to a fluid state. The minimum stress necessary to maintain a liquid state is a characteristic of the material (in the present case, paste) and is called the yield value or yield stress.

The general nature of the conversion from a plastic state to a fluid state is easily visualized in terms of Fig. 10.4, which gives the relationship between the rate of forced flow through a small cylindrical tube (capillary) and the corresponding motive force.

Curve 1 is the relationship for a Newtonian fluid. The mechanics of Newtonian flow in a capillary is like that shown in Fig. 10.3, except that in a capillary the flow is visualized in terms of concentric circular laminae, instead of parallel layers.

Curve 2 represents one of the various kinds of non-Newtonian flow, the kind depicted being that of a material which is at first a soft plastic solid, usually called a *Bingham body*. The mechanics of this kind of non-Newtonian flow can also be given in terms of Fig. 10.3. It will be discussed in terms first used by Bingham [6], although for application to cement paste and concrete it will be necessary later to modify the interpretation.

Let us suppose that the material between the plates (Fig. 10.3) is at first an elastic solid. When a tangenital shearing force is applied, the material at first deforms elastically,* but it does not begin to flow until a stress just exceeding the elastic limit of the material is reached. After such a stress has been created, and after a steady rate of flow has been established, the flow rate is supposed to be proportional to

* It is shown further on that for pastes the initial deformation is mostly nonelastic.

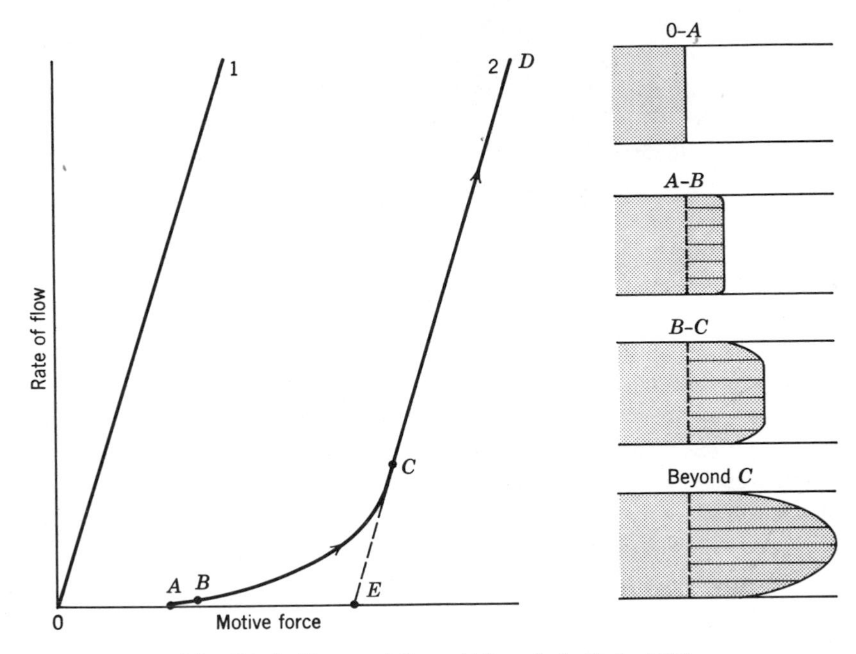

Fig. 10.4. Types of flow. (After de la Peña [8].)

the amount by which the shearing stress exceeds the yield stress, and in this case

$$\tau - f = U \frac{dv}{dx} \tag{10.5}$$

where f is the elastic resistance that must be overcome. The symbol U is the viscosity of the material in the fluidized state. For soft materials, it is usually called the *plastic viscosity*, and Bingham called its reciprocal the *mobility*.

According to Bingham's reasoning, we should expect no flow at any shearing stress up to $\tau = f$. Then at any stress $\tau > f$, there should be a steady flow, the rate of flow being proportional to $\tau - f$. Therefore Bingham assumed that point E in Fig. 10.4, when converted to terms of stress, is the yield stress f.

Bingham explained the segment ABC of the curve in Fig. 10.4 as follows. When a force just exceeding A is applied, steady movement of the material begins, the initial movement being due to structural failure, or fluidization, at the periphery only. The nature of the initial movement is indicated in sketch A-B; it is called the *plug flow*. With

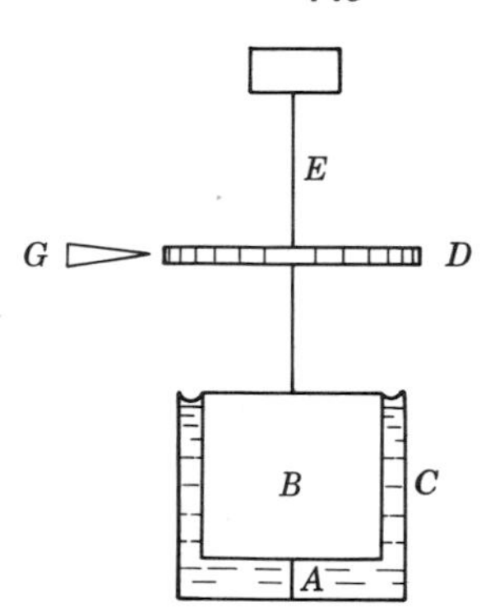

Fig. 10.5. Elements of a coaxial-cylinder viscometer. *A* indicates the sample, *B* the bob, *C* the cup, *E* the torsion spring, *D* and *G* the scale and reference point.

increasing shearing stress, the diameter of the plug diminishes while the thickness of the fluidized peripheral region increases (sketch *B-C*), and at a sufficiently high velocity the velocity gradient approaches the parabolic form indicated in the sketch at the bottom, all the material apparently (not actually) having become fluidized.

Bingham concluded that rheological diagram such as curve 2 in Fig. 10.4 is characteristic of a plastic material and therefore that plasticity can be evaluated quantitatively in terms of the yield stress and the plastic viscosity or its reciprocal, the mobility.

From data obtained with a capillary viscometer, Bingham evaluated U and f, assuming that the slope of the apparently straight line part of the diagram is proportional to $1/U$, and that point E, expressed in terms of stress, is identical with f.

However, many studies have been made since Bingham carried out his pioneer work, and it is now realized that the Bingham constants cannot be evaluated accurately from such data. The Bingham constants can be evaluated accurately by means of a coaxial cylinder viscometer, illustrated diagrammatically in Fig. 10.5. It consists of a cylindrical container (cup) that can be rotated around its vertical axis, and an inner cylinder (bob) restrained from rotating by a spring. The material to be tested of course occupies the space between the bob and the cup. When the cup is rotated, torque is transmitted to the bob by the intervening material. The theoretical rheological diagram obtained for a Bingham body is like that shown in Fig. 10.6.

Reiner and Riwlin [in 5, p. 119], starting with Eq. 10.3, the basic Bingham equation, showed that for a coaxial-cylinder viscometer the rheological diagram can be represented by Eq. 10.6:

$$\Omega = \frac{K_1}{U} T - K_2 \frac{f}{U} \tag{10.6}$$

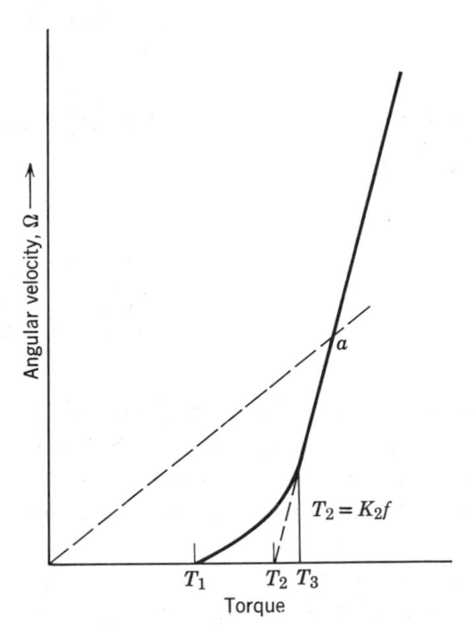

Fig. 10.6.

where Ω is the angular velocity of the cup in radians per second, T is the observed torque, K_1 is an instrument constant, and

$$K_2 = \ln \frac{R_o}{R_i} \tag{10.7}$$

where R_o and R_i are the outer and inner radii of the sample, and ln is the symbol for the natural logarithm. Thus, K_2 is also an instrument factor.

According to Eq. 10.6, there is a certain value of torque T that corresponds to zero angular velocity of the cup. Designating this value as T_1 we see that T_1 is related to the yield stress f as follows:

$$T_1 = \frac{K_2}{K_1} f \tag{10.8}$$

Since the constants K_1 and K_2 are those needed to convert observed torque (moment) to stress, it follows that T_1 indicates the yield stress directly. However under the usual testing procedures T_1 is not easily observed, and it becomes more convenient and accurate to determine the yield stress from the second term of Eq. 10.6, that term being represented by T_2 in Fig. 10.5.

The curved part of the rheological diagram between T_1 and T_2 is analogous to plug flow as illustrated in Fig. 10.4. The geometry of the system is such that at a given speed of rotation the shearing stress in the sample is greatest at the surface of the bob, radius R_i, and least at the inner surface of the cup, radius R_o. Therefore at any speed of rotation producing a torque between T_1 and T_3 a certain thickness next to the bob will have become fluidized while the rest of the material remains a solid shell rotating with the cup. As the speed of rotation is increased, the fluidized part increases at the expense of the solid part, and when the torque reaches T_3 the shearing stress next to the wall of the cup will have just exceeded the yield value, and the entire sample will have become fluidized.

2.5 Apparent Viscosity

A material giving a flow curve like that shown in Fig. 10.6 would not be distinguishable from a Newtonian fluid at any single value of torque. For example, a Newtonian fluid represented by the dashed line in Fig. 10.6 would show the same ratio of torque to angular velocity as does the non-Newtonian material at the torque represented at point a, where the lines intersect.

The equivalent Newtonian viscosity of a non-Newtonian material at a specific shearing rate is the *apparent viscosity* of the non-Newtonian material. Obviously, one material may show a wide range of apparent viscosities according to the arbitrary choice of shearing rate. We may note also that a false indication of apparent viscosity can be obtained if the chosen shearing rate does not exceed one giving a torque greater than T_3, which is to say, if the shearing rate is insufficient to fluidize the entire sample.

2.6 Thixotropy

The discussion above might seem to indicate that for a given material the Bingham constants are single valued, and thus stand for definite characteristics of the material. For some pastes this indication is substantially valid, but for others it is not. For some mixtures plastic viscosity seems to increase as the shearing rate increases, and for others it decreases. For the latter class the relationship between shearing stress and shearing rate as shearing rate increases may be curved all the way to the highest shearing rate used. Moreover when we attempt to retrace the curve by reductions in the shearing rate, a pronounced hysteresis may appear, the "down-curve" being straight and indicating a value of T_2 smaller than that indicated by the "up-curve." Furthermore, the degree to which the hysteresis develops depends on the manner in which the experiment is carried out; specifically, it depends on how long each shearing rate is maintained constant before changing to the next one.

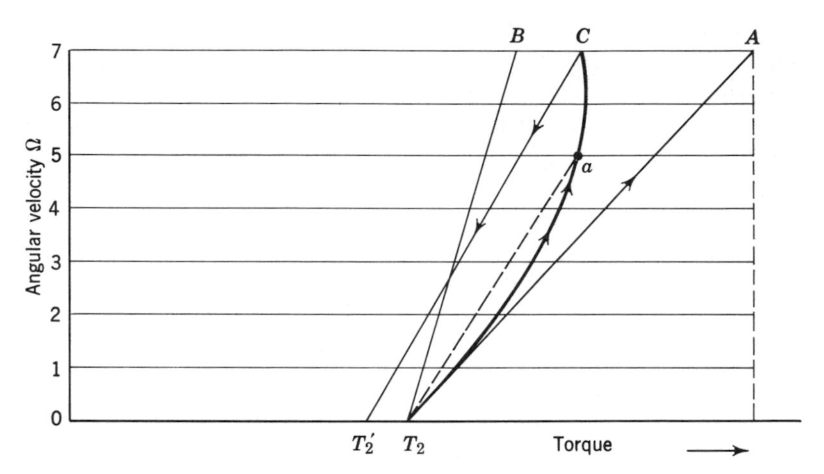

Fig. 10.7. Types of curves obtainable from a thixotropic material. (After Dellyes [7].)

The rheological characteristics just described are said to be due to the progressive breakdown of the structure as shearing rate increases and to the tendency of the structure to re-form. A material having such characteristics is called *thixotropic,* or, less committally, it is called a *shear-thinning material.* (Thixotropy is discussed in Chapter 9, Section 11.2; it is a phenomenon primarily due to interparticle forces.)

Dellyes [7] discussed the many different diagrams that can be obtained from tests of one thixotropic material in terms of Fig. 10.7. In Fig. 10.7, T_2 corresponds to the yield value, as it does in Fig. 10.6. The diagram actually represents a hypothetical case. It is assumed that a coaxial-cylinder viscometer is operated at various constant speeds Ω_1, Ω_2, . . . , Ω_7, the elapsed time at each speed, Δt, being different for different series of tests. At any given speed, the torque appears highest at the earliest possible reading and diminishes with time, approaching a minimum value. If Δt at each speed is made long enough for the torque to reach its minimum value, the flow curve is a straight line such as T_2B; it represents complete thixotropic breakdown, or at least the greatest degrees of breakdown that can be achieved at each of the several rates of shear.

On the other hand, if the time allowed for thixotropic breakdown at a given speed is kept as short as possible so that Δt is as near zero as possible, a straight line such as T_2A is obtained; it represents a minimum of thixotropic breakdown.

Thus lines T_2B and T_2A are two extremes for a given material; the first represents $\Delta t = \infty$ and the maximum thixotropic breakdown, and the second represents $\Delta t \doteq 0$ and minimum thixotropic breakdown. Practically, Δt cannot be made infinite, but it is feasible to make it large enough to obtain virtually maximum breakdown for a given speed.

If Δt is established by arbitrary choice and is such that maximum thixotropic breakdown does not occur, a curve such as T_2aC is obtained. Any number of such curves can be obtained, within the limits already discussed, by making different choices of Δt. For each up-curve there is a corresponding down-curve, as indicated by CT'_2; down-curves for a given Δt are always straight lines.

When an up-curve such as T_2aC is obtained, a straight line from T_2 to any point on the curve indicates an apparent plastic viscosity analogous to the apparent viscosity discussed above. It is represented in this example by the broken line from T_2 to a.

One interpretation of the behavior described above begins with the assumption that the particles in a paste at rest are linked by "bonds." When the paste is subjected to continuous shearing, a certain number of bonds are broken each second while a smaller number of new bonds are produced until, after a given period of shearing, the rate of bond breaking becomes equal to the rate of bond formation, the equilibrium having been reached with a smaller number of effective bonds than existed at the start. The higher the shearing rate, the smaller the net number of bonds and thus the smaller the apparent viscosity. The stress indicated at T_2 indicates the shearing strength when the maximum number of bonds are intact.

On the basis of concepts that are discussed in Chapter 9, we would be inclined to explain thixotropic breakdown in terms of reduced interparticle attraction rather than in terms of rates of bond breaking and restoration. We would suppose that the breakdown is due to an increase in the smallest distance between particles. Thus, when particles are caused to move with respect to each other through the liquid in which they are suspended, as in continuous shearing, hydraulic forces are generated which tend to separate and reorient the particles to minimize the resistance to movement. Supposedly, the general effect is to reduce the greatest distances between particles and to increase the smallest distances. In other words the effect is to decrease the deviation from the average interparticle distance, the deviation being normal to the flocculent state of the paste.

Since interparticle attraction varies with the third or fourth power of interparticle distance (Chapter 9), it is apparent that the increase in attractive force due to the reduction of the largest interparticle distances could be negligible as compared with the decrease in the attractive force due to an increase in the smallest distances. The effect could be greater the greater the velocity and duration of shearing because the magnitude of the hydraulic forces bringing about the dispersal and re-orientation depends on velocity and because time is required to produce the ultimate effect of any given rate of shearing.

However, in any case, there must be a limit to the amount of change possible. Once this limit has been reached, as a result of increasing the shearing rate, any further increase in the shearing rate cannot produce more structural change; the paste will act like an ideal Bingham body, and the rheological diagram will be like T_2B in Fig. 10.6.

When the experiment is conducted in such a way to produce less than the ultimate structural effect, the curve T_2aC is produced. The linearity of the down-curve indicates that the forces of attraction, weakened by the shear-oriented positions of the particles, are not able to alter the structure appreciably during the short time allowed for establishing the curve. Therefore, T_2' indicates what the shear strength the structure would be if the particle orientation produced at the highest rate of shear persisted in the static state.

If we conduct the experiment that produced the graph T_2aC in Fig. 10.7, and then at point C increase Δt to the time required for maximum breakdown, the torque would decrease from C to B. Then, still using the increased Δt, we could obtain the down curve BT_2, which represents the behavior of an ideal Bingham body. In general, a thixotropic material should behave as an ideal Bingham body if the experiment is made to achieve a steady state at each and every speed of rotation.

However, viscometer tests are usually carried out according to an arbitrary schedule, and if a curve such as T_2aC is obtained, the material is called thixotropic. Although thixotropy is not a property expressible in fundamental units, the area between the up-curve and the down-curve for an arbitrary Δt has been used to indicate the relative degree of thixotropy, the assumption being that the area will be greater the greater the degree of thixotropic breakdown existing at the top point of the up-curve. The difference between T_2' and T_2 has also been used as such an index. From this point of view, it is natural to say that a material is not thixotropic when the up-curve and the down-curve coincide.

The deduction just mentioned seems to be a questionable one. If the particles in a paste are only weakly flocculent, it is possible that the arbitrarily chosen Δt may be long enough to give a substantially complete breakdown at each speed above that giving torque T_3. In such cases the rheological diagram would be like T_2B and reversible. Yet the material would not be qualitatively different from other materials that are considered to be thixotropic; if a smaller value of Δt had been chosen, a hysteresis loop would have been observed.

The questioned interpretation amounts to defining thixotropy in terms of Δt. If the material does not break down completely in an arbitrary time, hysteresis is observed and the sample is called thixotropic. If a complete breakdown *does* occur at each speed, it would be called non-thixotropic. However, rheologists do not appear to have adopted such a definition explicitly, the choice of Δt being a matter of individual choice. It seems, therefore, that we could correctly call a material thixotropic if, under arbitrary test conditions, it gives any one of the kinds of flow-curves discussed in connection with Fig. 10.7.

In closing this section it should be said that the subject of thixotropy is not as important to us as the amount of space given to it might indicate. The discussion seems to be necessary because published papers on the rheology of cement paste, or on other material rheologically like cement paste, nearly always involve the term "thixotropy." The meaning attached to the term is not always clear, and when it does seem clear, the meaning is often found in conflict with another author's interpretation.

2.7 *Dilatancy*

The term *dilatancy* has already been used in earlier chapters where its meaning in certain contexts was probably clear, but some further discussion is appropriate here. As remarked by de la Peña, in his extensive review of rheology as applied to concrete [8], the characteristic called *dilatancy* is not always easy to recognize. Furthermore, a material may be dilatant under some conditions and not under others. In its simplest form, however, as first discussed by Reynolds in 1875, it is unmistakable; some of Reynolds's observations, as summarized by de la Peña, are given below.

2.7.1 Coarse aggregations. If we *partly* fill an unstretchable bag with an aggregate composed of strong rigid balls, we find we can change the shape of the aggregate with negligible effort. However, if we fill the bag completely, we find that the aggregate cannot be deformed to a visible degree without bursting the bag. This observation means that to produce a macroscopic shear strain under the conditions just described, we must produce an increase in overall volume to permit one "layer" of balls to slide over another.

If we completely fill a rubber bag with a compacted aggregate, the aggregate can be deformed because the rubber will stretch to allow it. However, if we fill the interstices in the aggregate with water, expelling all the air, and then close the bag, the aggregation will become rigid because any deformation requires expansion against atmospheric pressure.

If the rubber bag contains a small excess of water over the amount necessary to fill the voids of the compacted aggregate, a limited deformation can be effected without changing the volume, and with relatively little effort. However the aggregate becomes rigid at the moment the volume of solids in the dilatant aggregate becomes equal to the total volume of water; any further deformation requires dilation against atmospheric pressure. In such cases a dilatant aggregation undergoes a change in voids content (a change of interstitial space) from that corresponding to its maximum density to that corresponding to its minimum density. If the initial degree of compactness is not that which gives the maximum density, the dilation required is correspondingly smaller.

In all experiments such as those described above not all parts of an aggregation undergo dilatation simultaneously, and at the end of the experiment various degrees of dilatation may exist in different parts of the aggregation. This phenomenon is related to that observed when one walks on the wet sand of an ocean beach: The sand surrounding each footprint seems to become dry because the water, which in the undisturbed sand is just enough to fill the voids, becomes unable to do so when the sand is dilated as the result of displacement.

2.7.2 Fine aggregations. Another aspect of dilatancy is manifested by certain concentrated dispersions of quartz powder in water. Such a dispersion will flow like a viscous liquid when it is being spread slowly with a spatula; however, the same mixture takes on a dry appearance and offers much resistance if rapid spreading is attempted.

A suspension of quartz powder can easily be penetrated with a glass rod if pressure is applied slowly. However if the rod is dropped perpendicularly from a certain height, it is almost completely stopped at the surface at the instant of contact. Thereafter it penetrates slowly until it reaches the bottom of the container.

As mentioned in Section 2.6, some mixtures, when tested in a coaxial-cylinder viscometer, appear to increase in viscosity as the shearing rate is increased. According to Eq. 10.5, such an effect cannot be accounted for in terms of viscous resistance alone. It seems to be another manifestation of dilatancy. Examples of dilatancy manifested in this way are shown in Fig. 10.8. (The data are due to Greenberg, Jarnutowski, and Chang [9].) Each curve is a flow diagram obtained with a coaxial-cylinder viscometer from suspensions of pulverized quartz in water at solid concentrations ranging upwards from 43 per cent. At the lowest solid-volume concentration only normal viscous resistance is shown; the curve

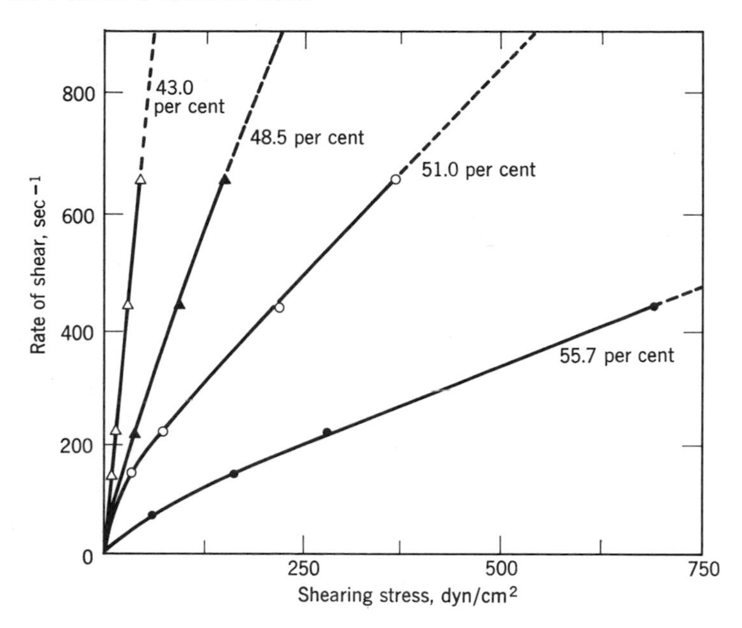

Fig. 10.8. Effect of the concentration of solids on the rheological properties of quartz powder suspensions in water [9].

is linear and its slope indicates a viscosity of 0.068 poises, which is about 7.5 times that of water alone. But, at higher concentrations the flow curves are not linear, each indicating a disproportionate increase in shearing stress as the shearing rate increases. As indicated above, the increase in viscosity with increase in shearing rate may be called an effect of dilatancy, although no actual increase of volume is involved. The increased resistance to shearing seems to be a particle-interference effect.

It should not be overlooked that none of the suspensions discussed above exhibits a yield stress, for they do not have a structure due to interparticle forces; therefore they cannot exhibit thixotropy. We shall see further on, however, that some materials are able to show thixotropy and dilatancy at the same time; they have a yield stress, but at the same time they tend to become more rather than less viscous as the shear rate increases above the yield stress.

3. PASTES IN THE FLUIDIZED STATE

Having established in the preceding section certain pertinent concepts and terms of rheology, let us now examine the characteristics of pastes

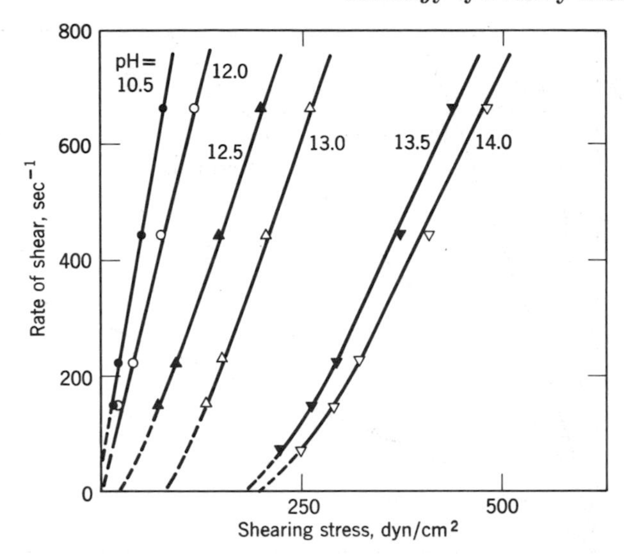

Fig. 10.9. Effect of pH on the rheological properties of 43 per cent quartz powder suspensions [9].

in the fluidized state, especially cement paste. Choosing to deal with the fluidized state of paste suggests that the consistency of concrete is considered to be due, at least in part, to the rheological characteristics of its paste component in the fluidized state. Such an assumption is valid for concrete while it is being placed by means of vibration.

3.1 Silica Pastes

As is indicated above, the rheological characteristics of a given suspension depend mainly on interparticle forces. Pastes made of pulverized quartz and water are useful for study for their interparticle forces are easily changed. The changes are brought about by changing the kind and concentration of solute and by regulating the hydrogen ion concentration, pH [10]. By raising the pH sufficiently, we can change a quartz-powder suspension from a Newtonian liquid to a thixotropic fluid, as shown in Fig. 10.9. Raising the pH from 10.5 to 14.0 by adding sodium hydroxide increased the yield value from zero to 230 dyn/cm^2 and the apparent viscosity at 658 rad/sec from 0.11 to 0.86 poises.

Other tests showed that at a pH of 10.5 the sedimentation volume was at a minimum, and that it increased rapidly as the pH increased above that value.

As is discussed in Chapter 9, Section 13.2, relative sedimentation volume can be taken as a measure of relative interparticle cohesion, the greater the sedimentation volume the greater the force of cohesion. In the viscometer test an increase in interparticle attraction is manifested by an increase in the yield stress and in the coefficient of plastic viscosity U. The latter factor shows up in Fig. 10.9 as a decrease in the slope of the curve as the pH increases.

3.2 *Cement Paste*

3.2.1 Various characteristics. The rheological characteristics of cement paste in the fluidized state depend on many factors, including the method of paste preparation, and the elapsed time between the cessation of mixing and the beginning of the rheological test. If the time that has elapsed before the test is begun does not exceed the dormant period (Section 1.2.1), the rheological characteristics are not greatly influenced by time, that is, by the age of the sample. However, if the time that has elapsed falls within the period of setting, usually upwards of 2 hr, age has a pronounced effect. As shown in Section 1.2.2, much depends on whether the mixing period extends beyond the period of brief-mix-set and especially on whether or not the final mixing occurs after the period of false set, if any.

Working with pastes mixed in a standardized manner, and mixed in such a manner as to eliminate the effects of false set, Ish-Shalom and Greenberg [11] observed three different types of flow curves, as illustrated in Fig. 10.10. Each diagram represents two consecutive tests on

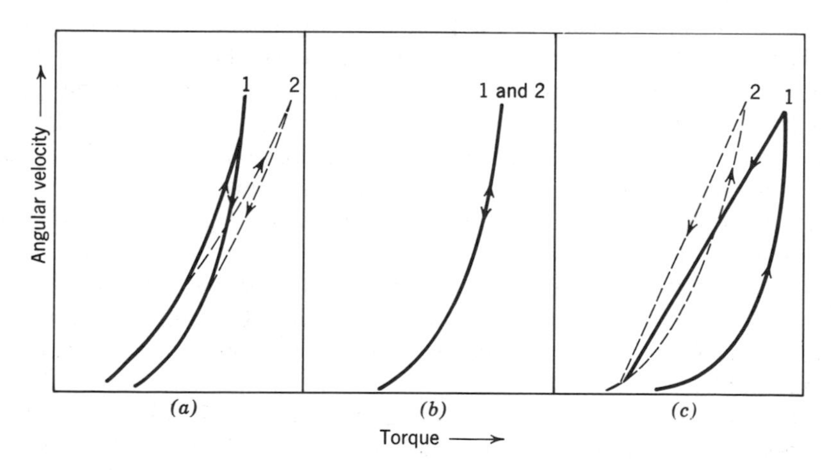

Fig. 10.10. Types of diagrams obtained from cement pastes [11].

one sample, one test giving an up-curve and a down-curve in a total time of about 3 min, and with an arbitrary Δt, not stated.

In diagram *c* the curves indicate partial thixotropic breakdown, the second test showing the residual effect of the partial breakdown occurring during the first test. The down-curves are straight and fall to the left of the up-curve.

In diagram *b* the up-curve and down-curve coincide, and the first and second tests give the same result. As is discussed above, the type of curve shown in diagram *b* has been interpreted to indicate nonthixotropy, but it seems more reasonable to assume that it represents the behavior of paste in a weakly flocculent state such that thixotropic breakdown is complete within the interval Δt and at a relatively low rate of shearing. The ability of such a paste to "set" soon if not immediately after having been fluidized, definitive with respect to thixotropy, is shown by the shape of the lower end of the down-curve; in that part of the down-curve the equivalent of plug flow seems to begin as soon as the lowest shearing stress (that at the wall of the cup) falls below a certain limit indicated by the beginning of the curved part of the down-curve.

The rheological characteristics of the sample represented in diagram *a* appear to be just the opposite of those of the sample represented by diagram *c*. That is to say, diagram *a* shows the maximum shear resistance that developed during the second test to be greater than that developed during the first test, whereas the opposite is true for the other sample. The type of behavior shown in diagram *a* was called antithixotropic by Dellyes [7]. The same term was adopted by Ish-Shalom and Greenberg [11].

As brought out by Gaskin [12] it seems reasonable to conclude that the type of behavior illustrated in diagram *a* can be accounted for in terms of thixotropy and dilatancy appearing simultaneously in the same sample. The existence of yield stress is accounted for by interparticle attraction, relatively weak, giving rise to correspondingly weak thixotropy. At low rates of shear thixotropic effects are dominant, but at higher rates of shear, dilatant characteristics become apparent. According to Gaskin [12], "particles temporarily lock together in bridging arrays when forced into contact with one another during shear." Considering the strong disjoining effect of the adsorbed water (Chapter 9), we would qualify the word "contact," but the explanation nevertheless seems to be reasonable.

The behavior illustrated in diagram *c* is invariably observed when the test is made after the beginning of the period of set, that is, at an age of 2 to 3 hr. It may appear also when a paste is only 15 min old, but at that early age the pattern illustrated in diagram *a* is frequently obtained; that is, the so-called antithixotropic stage is observed.

When a cement paste shows antithixotropy at the start, its rheological characteristics change with time; it passes through the state represented in diagram *b* and eventually reaches the strongly thixotropic state indicated in diagram *c*. According to the interpretation given above, such a paste is one in which dilatancy is dominant over thixotropy at the start, but eventually thixotropy is dominant. As we shall see further on, it is a simple matter to demonstrate the dilatant state of a thick cement paste and at the same time recognize the interparticle cohesion that gives it thixotropic characteristics.

The type of rheological diagram obtained from a given paste depends not only on age, but also on various other factors not yet well understood. For example, Ish-Shalom and Greenberg found that a paste mixed for 1 min and tested at the age of 15 min had predominantly thixotropic characteristics, whereas the same kind of paste mixed for 10 min and tested at the age of 15 min was strongly antithixotropic; that is, dilatancy was dominant. The same was true when the tests were carried out on samples 45 min old, except that for the sample that had been mixed for 10 min the reversible state exemplified by diagram *b* rather than definite dilatancy was indicated.

In another set of tests in which the water-cement ratio of the paste was the variable factor, dilatancy seemed dominant during the first test (one exception occurred at $w/c = 0.45$), and thixotropy was dominant in every case during the second test.

Since we shall not become deeply involved in the characteristics of paste in its fluidized state, perhaps the principal point to be noted here is that, as we have seen in another connection, the rheological properties of cement paste are not simple, single-valued properties. The measured factors are no doubt the resultant of several underlying properties, both chemical and physical. Generally, however, results showing orderly relationships to variable factors can be obtained when the routine of testing procedures is properly established and rigorously maintained.

3.2.2 Effect of cement concentration. Several investigators have obtained rheological parameters for pastes made with various proportions of cement. We shall examine the results obtained by Papadakis [10] and by Ish-Shalom and Greenberg [11].

The apparatus, methods, and materials employed by Papadakis were such as to give reversible flow curves of the type shown in Fig. 10.10*b*. Such curves were obtained for pastes made with four different portland cements having similar specific surface areas. The values of plastic

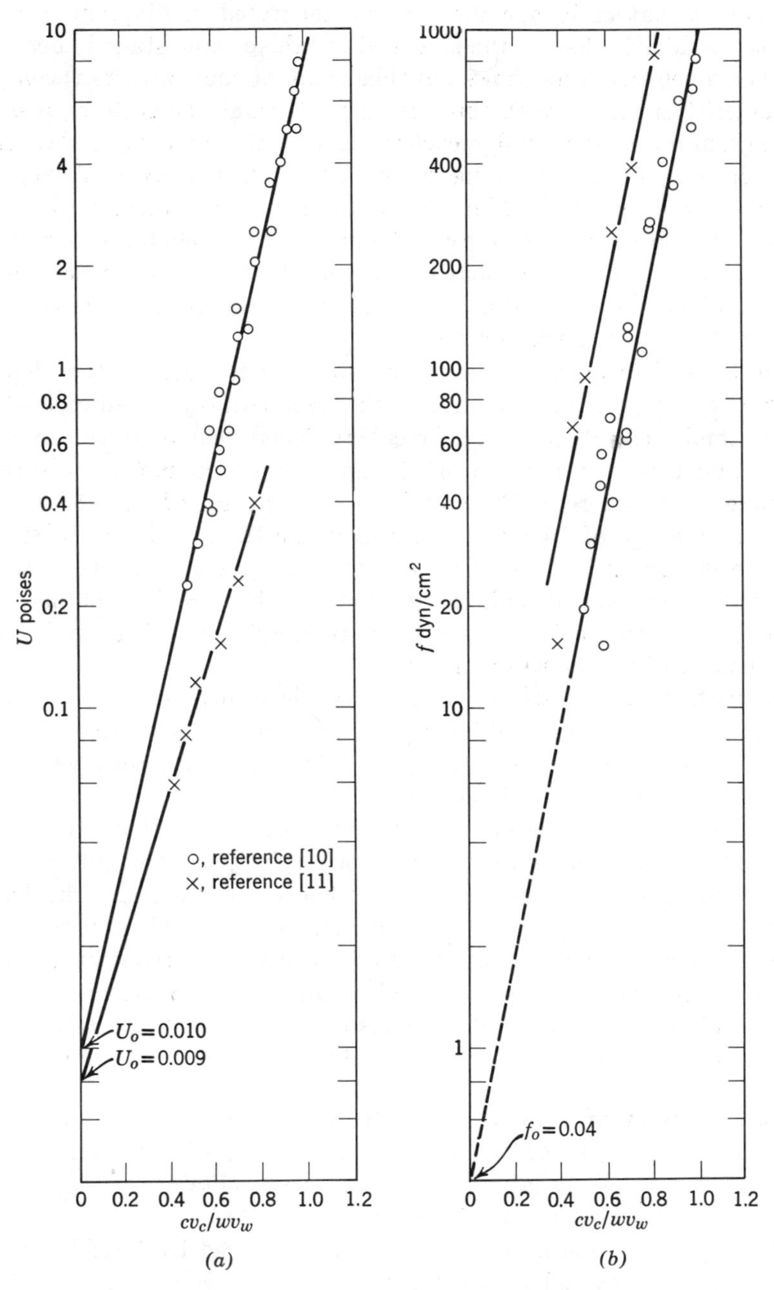

Fig. 10.11. Effect of the cement content of paste on plastic viscosity U and yield stress f [10, 11].

Table 10.2 Rheological Characteristics of Portland Cement Pastes of Various Water-cement Ratios[a]

Water-cement Ratio		Volume Fraction of Cement[b]	Age of Paste[c] (min)	First Test		Second Test	
$\frac{w}{c}$	$\frac{cv_c^o}{wv_w}$			f (dyn/cm²)	U (poises)	f (dyn/cm²)	U (poises)
0.4	0.78	0.439	15	810	0.39	—	—
			40	970	0.52	970	0.48
0.45	0.70	0.410	15	390	0.22	390	0.26
			45	405	0.26	390	0.26
0.5	0.62	0.384	8	250	0.16	260	0.17
			30	250	0.16	268	0.15
0.6	0.50	0.342	15	90	0.12	100	0.12
0.7	0.44	0.307	15	64	0.085	71	0.085
			45	52	0.074	52	0.074
0.8	0.39	0.280	15	16	0.058	16	0.061

[a] Data are from M. Ish-Shalom and S. E. Greenberg, "The Rheology of Fresh Portland Cement Pastes," Proceedings of the Fourth International Symposium on the Chemistry of Cement," National Bureau of Standards Monograph 43, U.S. Department of Commerce, 1962, Volume II, pp. 731–743. The specific surface area of the cement was 3800 cm²/g (2265 Wagner).

[b] $\frac{cv_c^o}{V}$

[c] At time of start of test.

viscosity and yield value were obtained from each flow-curve and the results were plotted as shown in Figs. 10.11*a* and 10.11*b*. Also shown are points from Table 10.2 representing data obtained by Ish-Shalom and Greenberg added by the present author.

The data on plastic viscosity conform fairly well to an equation having the form

$$U = \eta_o e^{k(cv_c^o/wv_w)} \tag{10.9}$$

where η_o is the viscosity of pure water at the temperature of the experiment; k is a constant, probably also a function of temperature. For the Papadakis data, $\eta_o = 0.01$ and $k = 6.7$. For the Ish-Shalom and Greenberg data $\eta = 0.009$ and $k = 4.86$.

Equation 10.9 has the form proposed by Arrhenius in 1887 [13] for Newtonian viscosity rather than plastic viscosity, except that Arrhenius found a function corresponding to $\exp k(cv_c^o/V)$ rather than $\exp k(cv_c^o/wv_w)$.

In Fig. 10.11*b* we see that yield stress is related to the amount of cement per unit of water in the paste by the same type of relationship as that expressed in Eq. 10.9, but of course with different constants. A constant f_o, evaluated by extrapolation of the line as drawn by Papadakis, appears to be about 0.04 dyn/cm^2.

The data from Ish-Shalom and Greenberg plotted in Fig. 10.11 indicate relatively low values for U and high values for f_o, compared with the other data. The specific surface area of the cement they used was about the same as the average of those used by Papadakis, but the cements might have otherwise been different. It is not possible to account for the differences in rheology from information at hand. However it is well to bear in mind remarks above about the influence of differences in testing procedure on the results obtained from one kind of paste.

3.2.3 Effect of surfactants. Surfactants are likely to affect the consistency of paste. In this connection Bruere [14] reported test results in terms of apparent viscosity,* as shown in Fig. 10.12. His interpretation of these data was substantially as follows. Agents 4 and 7 reduced apparent paste viscosity at all concentrations. Agent 4, calcium lignosulfonate, produces lignosulfonate anions which become adsorbed with the polar ends oriented toward the water. (It does not necessarily decrease the adsorption of water, and it increases electrostatic repulsion; it is a dispersing surfactant.)

Agent 7, saponin, is not molecularly adsorbed; instead, according to Rehbinder et al. [in 14], it is a colloidal material that forms a protective coating around the grains, thus reducing interparticle attraction.

Agent 6, "Lissapol N300," had no effect on viscosity. (Bruere offered no explanation.)

Agent 5, hexadecyl trimethyl ammonium bromide, *increased* apparent viscosity for concentrations up to a certain value, beyond which further additions decreased viscosity. The elongated molecules of this agent are adsorbed with their nonpolar ends oriented toward the water, thus making the cement grains hydrophobic and increasing interparticle attraction. However, at higher concentrations a second layer begins to form on the first, the molecules of the second layer being oppositely oriented. Thus as the second layer builds up, the grains again become hydrophilic and interparticle attraction is accordingly reduced with corresponding effects on apparent viscosity. Such a material seems to act either as a flocculating, neutral, or dispersing surfactant, depending on concentration.

* The values were not strictly apparent viscosity, but rates of shearing under constant torque.

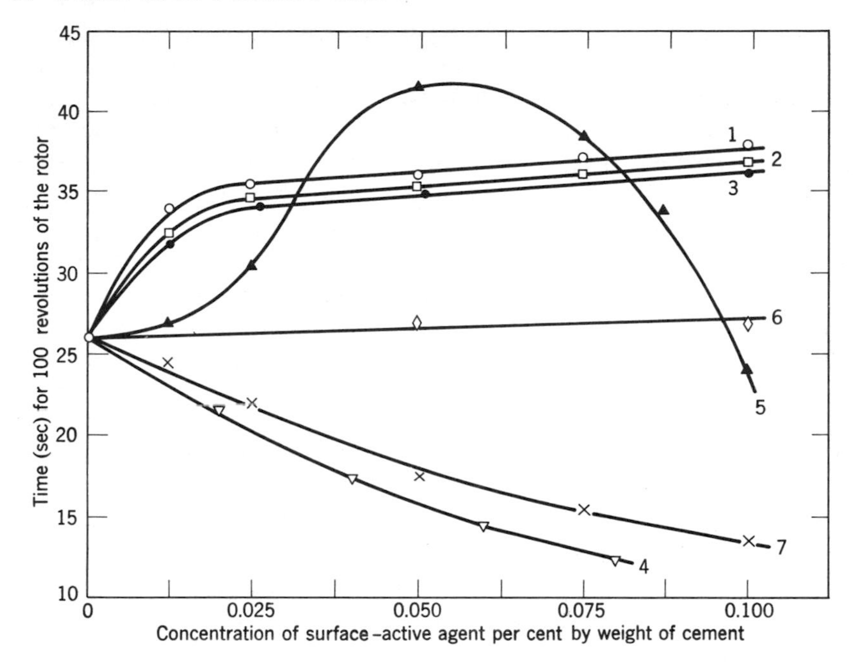

Fig. 10.12. Effect of different types of surfactants on apparent viscosity. The value of w/c was 0.40 by weight. The temperature of the pastes was $20 \pm 1°C$. The weight of the driving rotor in the viscometer was 300 g. The surface-active agents used are indicates by numerals: 1, sodium dodecyl sulfate; 2, sodium abietate; 3, "Darex AEA"; 4, calcium ligonsulfonate; 5, hexadecyl trimethyl ammonium bromide; 6, "Lissapol N300"; 7, saponin. (Data are from Bruere [14].)

Agents 1, 2, and 3 are anionic and, in amounts greater than about 0.01 per cent by weight of the cement, each increases paste viscosity. This, according to Bruere, seems to be due to an increase in interparticle attraction, and that in turn is due to the coating of adsorbed surfactant, which renders the particle hydrophobic. Bruere decided that

"The anionic agents do not show . . . reversal of effect on paste viscosity because they are precipitated by lime in cement paste and exist in the mixing water as sparingly soluble calcium salts. Hence, the concentration of the anionic agents in solution cannot become high enough to form a second adsorbed layer of molecules on cement grains which is necessary to reverse the nature of the surface" [14].

The interpretation just given with respect to agents 1, 2, and 3 is based on evidence that such materials (anionic surfactants) become adsorbed on cement grains, make the grains hydrophobic, and *increase* interparticle attraction. But, there is evidence that sodium abietate at least does not act that way; indeed, it does not function as a surfactant in cement paste. In Chapter 7, Section 7.3, and in Figure 7.12, we saw direct evidence that calcium lignosulfonate has an effect on the consistency of concrete that is independent of the effect of entrained air, whereas in Fig. 7.10 we saw evidence that neutralized Vinsol resin has no such effect on concrete consistency. (Vinsol resin resembles sodium abietate.) Adsorption of calcium lignosulfonate by the cement grains evidently increases the fluidity of the paste, thus reducing the water requirement of the concrete for a given consistency; but with Vinsol resin (Fig. 7.10) the same kind of evidence is entirely absent: there certainly is no evidence of an effect independent of that of the entrained and entrapped air. (See Chapter 7.)

However the lack of evidence that the fluidity of the paste was reduced is not definitive, for the concrete mixtures were in the *BF* category and thus a stiffening of the paste would not necessarily increase the water requirement of the concrete mixture. Definitive data on this point *were* given by Bruere; he found that sodium abietate had little or no effect on the bleeding capacity of cement paste. In fact a slight reduction was indicated, whereas an *increase* in interparticle attraction would surely have reduced bleeding capacity. (See Chapter 11.)

It seems, therefore, that the observed effects of sodium abietate and the like should be attributed to the solid material produced by the reaction between calcium hydroxide and the agent, referred to by Bruere. As brought out in Chapter 7, agents of this kind as used in concrete do not seem to become adsorbed on the cement grains to an appreciable extent, and therefore we are on shaky ground if we attribute observed physical effects directly or indirectly to adsorption. However for those agents that are adsorbed and do not form insoluble compounds, the interpretation given by Bruere seems well founded.

Papadakis [10] gave data on one surfactant in terms of plastic viscosity and yield stress, as shown in Fig. 10.13. The material was identified as a "deflocculating plastifier"; apparently it was not an air-entraining agent. As shown in diagrams *a* and *b*, it slightly reduced plastic viscosity at each water-cement ratio, and the yield stress was strongly affected, becoming almost zero at the highest water-cement ratio used. A principal point of interest is that the decrease in *apparent* viscosity was evidently due much more to reduction in yield stress than to reduction in plastic viscosity. It seems likely that for those agents that increased the apparent viscosity, the effect was due mainly to an increase of yield stress.

Bruere and McGowan [38] experimented with certain polyelectrolytes, particularly one called "Krilium," and found that some of them were

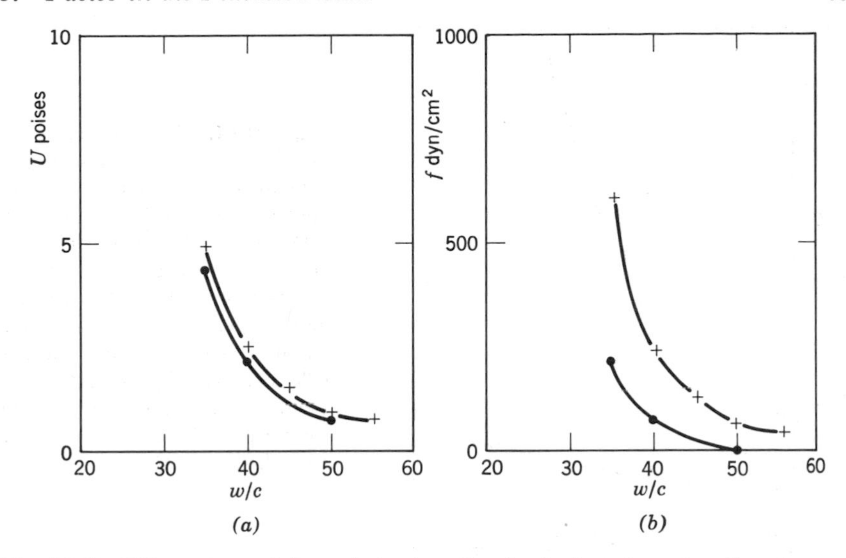

Fig. 10.13. Effect of a certain surfactant on the rheological characteristics of cement pastes. The cross indicates mixtures without a surfactant; the dot, mixtures with a surfactant. (Papadakis [10].)

capable of stiffening cement paste to about any degree desired. The same effect was readily observed in concrete mixtures.

It might be supposed that a dispersing surfactant is the preferred kind of air-entraining agent. However, as is shown in Chapter 3, and especially in Chapter 7, Fig. 7.8, the cement-water paste should have a certain consistency which is optimum for the particular aggregate being used. For a given cement, and in the absence of a surfactant, the optimum consistency is likely to be that of a paste having a water-cement ratio somewhere near 0.5 by weight. If the mixture in question is a rich one, that is, one to the left of point B in the voids-ratio diagram, it should be advantageous to use a surfactant of the dispersing type. On the other hand, if the mixture is one to the right of point B, an agent which stiffens the paste might be advantageous. In any case, the final choice should be based on experimental results.

3.2.4 Effect of entrained air. In Chapter 7 it is shown that entrained air increases the stiffness of cement paste. Bruere [14] investigated the effect by means of the coaxial-cylinder viscometer and reported results in terms of apparent viscosity. An example of his results is shown in Fig. 10.14. The paste was mixed in a closed system which made it possible to control the amount of entrained air with a fixed amount of air-entraining agent present. Two of the agents represented in Fig.

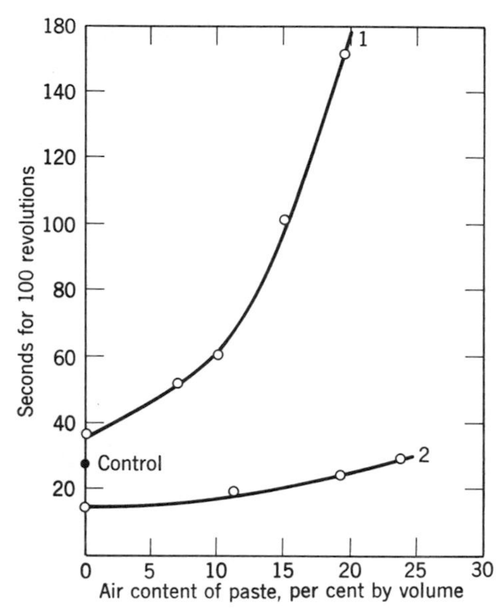

Fig. 10.14. Effect of entrained air on the apparent viscosity of cement paste, amount of agent constant. The value of w/c was 0.44 by weight. The weight of the driving rotor in the viscometer was 190 g. The surface-active agents used are indicated by numerals: 1, sodium abietate, 0.05 per cent by weight of cement; 2, saponin 0.10 per cent by weight of cement. (Data are from Bruere [14].)

10.12 were used, one of them being sodium abietate and the other a protective colloid, saponin, which, as is shown in Fig. 10.12 is a dispersing agent.

For either material, introduction of air increased apparent viscosity, but the effect was much the greater for the material producing an insoluble precipitate. In separate experiments Bruere [14] observed that when agents producing an insoluble product were used, air bubbles tended to adhere to cement grains, whereas with agents producing soluble products, they did not. This difference seems to be an adequate basis for explaining the difference between the two curves.

4. FRESHLY MIXED PASTE IN THE PLASTIC STATE

4.1 *Behavior under static stress*

V. V. Stolnikov [16] investigated the rheology of cement paste from a point of view quite different from that leading to the conventional use of viscometers. He measured the strain produced by a constant shearing stress as a function of time, the method being one attributed to Rehbinder.

The paste to be tested was placed in a rectangular container of square cross section, 3×3 cm $\times 5$ cm high. A zinc plate, 2×2 cm was embedded in the paste in a vertical position. The surfaces of the zinc plate were grooved horizontally (riffled) to a depth of 0.75 mm to prevent slipping. An upward force (traction) was applied to the plate, presumably by means of a thin wire or fiber, the force being measured by a dynamometer of some sort. The resulting movement of the submerged plate relative to the position of the container was measured by means of a micrometer microscope. The procedure was to apply a given force suddenly, holding it constant for a time while observing the movement, and then removing the force suddenly and observing the recovery.

Figure 10.15 is one of the diagrams obtained at relatively low shear stress. A force producing a shear stress of 1230 dyn/cm² (0.0178 psi) was applied suddenly and maintained for 105 secs. An "instantaneous" shearing-strain of about 0.004 was followed by further strain at a diminished rate until the strain increased to 0.011, which seemed to be close to a limit. Then the stress was suddenly reduced to zero. This gave an instantaneous recovery equal to the initial elastic response, followed by a gradual return to zero shear strain.

Such behavior as that just described indicates the paste that Stolnikov used to be a Kelvin solid, which is to say, a solid that is essentially elastic but whose deformation under constant stress is retarded by internal friction. Its stress-strain-time characteristics can be represented by a model composed of a spring and dash pot connected in parallel, as

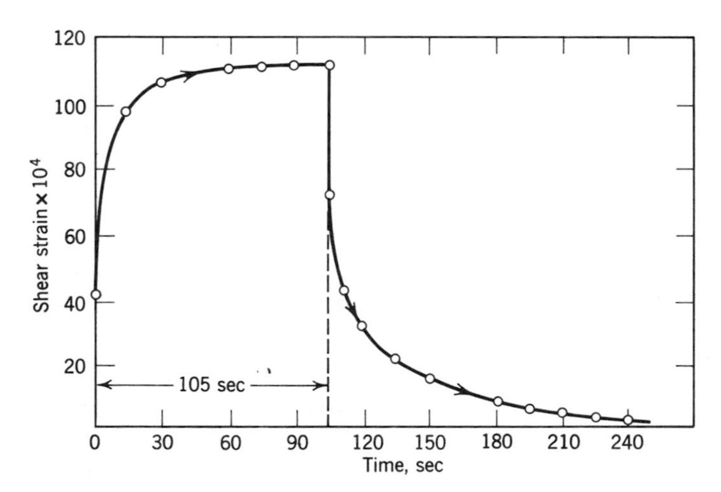

Fig. 10.15. Strain versus time for cement paste of normal consistency. The shear stress τ_o was 1230 dyn/cm² or 0.0178 psi during the first 105 sec; thereafter it was zero. (Data are from Stolnikov [16].)

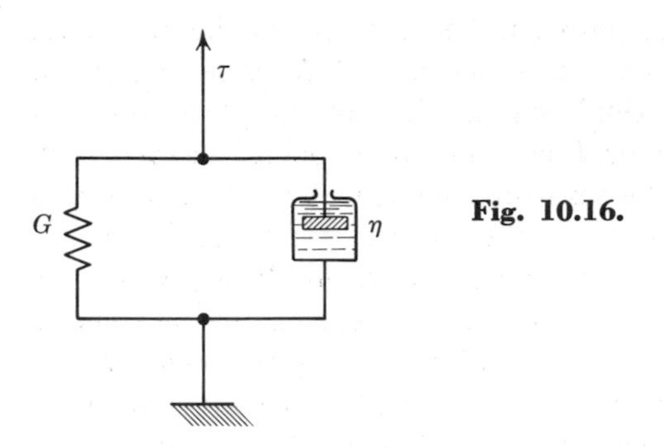

Fig. 10.16.

is shown in Fig. 10.16. The time-dependent parts of the deformation are called reversible creep.

When Stolnikov applied a stress 50 per cent higher than that which he used in the experiment just described, he obtained the results shown in Fig. 10.17. At this stress the same kind of deformation shown in Fig. 10.15 began, but at the same time viscous flow of the material occurred as is indicated by the broken-line extension of the straight part of the diagram. The curve therefore is the resultant of the delayed elastic response characteristic of a Kelvin solid, plus viscous response, that is, plastic deformation.

The occurrence of plastic deformation indicates that the applied stress exceeded the yield stress of the material. Under these conditions the deformation was not limited by the elastic elements, as represented in

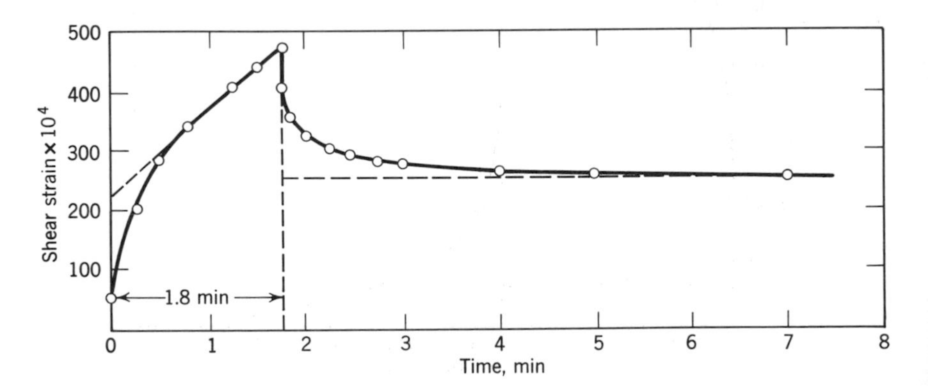

Fig. 10.17. Behavior of the same material represented in Fig. 10.15 but under higher shearing stress. The shear stress τ_o was 1840 dyn/cm^2 or 0.0267 psi during the first 1.8 min; thereafter it was zero. (Data are from Stolnikov [16].)

Fig. 10.15, and shear-strain presumably could have continued indefinitely at a constant rate.

However, the stress was removed after an arbitrary length of time, whereupon there was elastic recovery, instantaneous and delayed. The strain due to viscous flow was of course not recovered because the energy that produced it had been dissipated as heat.

As mentioned above, the plastic deformation occurred because the applied stress exceeded the elastic limit of the material, which Stolnikov gives as 1240 dyn/cm^2 (0.018 psi) for the given material. At stresses above this level, the paste behaves as would a model represented in Fig. 10.18. In series with the model representing a Kelvin solid (below the dash line) is a combination representing what Stolnikov refers to as a *Maxwell-Schvedov body:* a spring in series with a device representing the elastic limit, followed by a dash pot.

According to Fig. 10.18, a stress gives rise to resistance that can be calculated in terms of two shear moduli and two coefficients of viscosity (solid viscosity), together with an elastic limit. Stolnikov evaluated each of these constants for the particular material used in his experiments, obtaining results given in Table 10.3. It appears that, as Stolnikov tested it, cement paste of normal consistency behaves according to Bingham's concept of a plastic solid. At stresses below the elastic limit, it behaves elastically. At higher stresses, it flows viscously. The elastic limit appears to be identical with the yield stress.

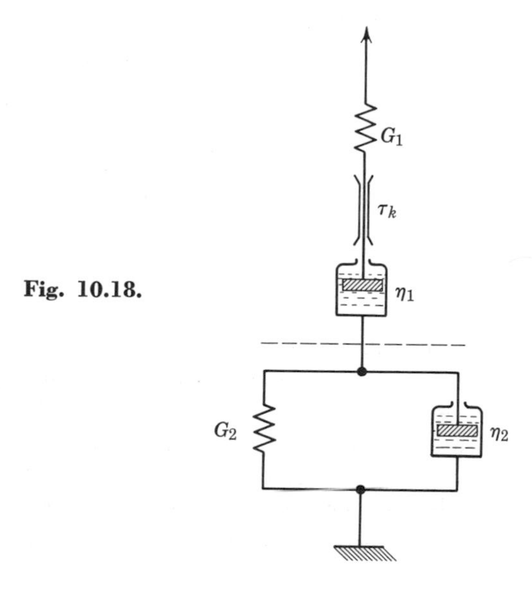

Fig. 10.18.

Table 10.3 Rheological Constants for a Neat Cement of Normal Consistency as Evaluated by Stolnikov

Shear modulus, G_1	4.45×10^6 dyn/cm²	64.5 psi
Shear modulus, G_2	1.93×10^6 dyn/cm²	28.0 psi
Viscosity, η_1	3.8×10^6 poises	
Viscosity, η_2	1.8×10^6 poises	
Elastic limit, τ_k	1240 dyn/cm²	0.018 psi
Relaxation time	8.5 sec	

4.2 *Relation to Consistency of Concrete*

The question now arises as to the connection between the factors given in Table 10.3 for a certain cement paste and the consistency of concrete. First of all, it is clear that such paste cannot normally be found in a plastic concrete mixture. (See Chapter 3.) Its consistency is much too stiff. Therefore, the values for elasticity and viscosity shown in Table 10.3 are not those of any pastes found in plastic concrete mixtures.

Although Stolnikov reported data on only one paste and only on that at normal consistency, his method is applicable in principle to any thixotropic paste. However, to apply the method to pastes having compositions within the practical range would probably require exceedingly delicate measurements. This conclusion is based on Fig. 10.11*b* and an assumption that the elastic limit as measured by Stolnikov is proportional to the yield stress as measured by Papadakis. On this basis, the relative elastic limits would diminish with an increase in w/c as is indicated in the accompanying table.

w/c	Relative Elastic Limit
0.240	1.0
0.310	0.1
0.437	0.01
0.730	0.001

The indication is that for pastes such as those used in concrete, with w/c generally exceeding 0.45 by weight, the elastic limit usually would be less than 1 per cent of the value for neat cement at normal consistency, about 0.018 psi (Table 10.3). In the practical range of water-cement ratios it would range from about 0.0002 to 0.00002 psi.

It thus seems that the pastes produced in concrete are so weak that they could exhibit viscous behavior only; that is to say, a concrete mixture should show a constant rate of flow under any stress exceeding its very low elastic limit. But, direct observation shows that concrete does not flow under a constant shearing stress, although under some circumstances it may slide in a chute, giving superficially the appearance of flow. Under a given shearing stress, it deforms a certain amount and no more.

The behavior just described becomes more understandable as we examine data presented below, but at this point it is appropriate to point out that the missing factor in Stolnikov's experiments is dilatancy (Section 2.7). A constant stress of the same order of magnitude as the elastic limit produces small strains and does not develop particle interference* and consequent dilatancy. Under different circumstances, with stress greatly exceeding the yield stress, the deformations are relatively large, they develop more rapidly, and cumulative resistance due to dilatancy precludes continuous flow under constant stress.

Dilatancy in cement paste of normal consistency, paste such as that used by Stolnikov, was described by Powers and Wiler [17] as follows:

> "It may be observed easily by forming a ball of neat cement of about normal consistency and then twisting the ball so that one hemisphere tends to rotate with respect to the other. The twisting is accomplished with little effort up to a certain angle of distortion at which the required effort for further distortion increases sharply. Any distortion beyond that point where the required force increases is accompanied first by a drying of the surface in the zones where the strain is greatest, and then by rupture."

It may be recalled from Section 2.7.2 that resistance due to particle interference can be observed even where actual volumetric dilatation of the sample does not occur, dilatation being a manifestation of extreme particle interference.

We may note in passing that the tendency toward dilatancy in cement paste increases as interparticle attraction is reduced. This can be demonstrated in the following manner. In performing the test for normal consistency of hydraulic cement (ASTM Designation: C187-58) the necessary amount of mixing water is gauged according to a standard degree of penetration of the needle of the Vicat apparatus. Since

* The term "particle interference" in the present context does not have exactly the same connotations as when it is used in connection with the Weymouth theory as we discuss it in Chapters 1 and 6. However, it is apparent that virtual particle interference in the sense of the modified Weymouth concept discussed in Chapter 6 would be conducive to the development of particle interference as visualized here—that developing as the result of shear strain.

the use of a surfactant having a dispersing effect decreases interparticle attraction and increases the fluidity of a paste, we might expect such a surfactant to reduce the amount of water required to obtain the standard penetration of the Vicat needle. However, experiments showed little or no actual reduction in the water requirement because the paste became dilatant under the action of the rapidly penetrating needle.

In the following section we shall examine the rheological characteristics of cement paste and concrete alluded to briefly above, which are more closely related to the consistency of concrete than any characteristics examined thus far.

5. FRESHLY MIXED MORTAR OR PASTE IN THE PLASTIC STATE

In Section 4 we learned that the response of paste to a small static shearing force has no obvious relation to consistency of concrete. We shall now consider work reported by Powers and Wiler [17], and some later unpublished work in which Pickett collaborated, that demonstrate rheological characteristics different from those already discussed and that have a close relationship to the consistency of concrete.

As indicated in the title above, paste and mortar will be considered together. A tandem treatment is preferred, but not possible because of the limited scope of the data available; the experimental project referred to above was never completed. It is feasible to consider the two together because it turned out that there are no qualitative differences between the data obtained from mortar and those from paste.

Although the term "mortar" is often convenient, mortar is not qualitatively different from concrete. Therefore, in the present context, mortar can be considered concrete containing aggregate having a maximum size limited by the No. 4 sieve.

5.1 A Recording Plastometer

Figure 10.19 shows a recording plastometer developed by Powers and Wiler [17]. Its dimensions were such that it could accommodate mixtures containing aggregate. Its design was based on that of a coaxial-cylinder viscometer, yet it has some unique features described further on.

The sample was placed in the cylindrical container mounted on the turntable, the container being 12 in. in diameter and 8 in. deep. A hollow drum 3 in. deep and 8 in. in diameter was mounted concentrically, with its bottom 4 in. above the bottom of the container by means of the vertical shaft seen in the photograph; it corresponds to the bob in a viscometer. The sample to be tested filled the space under and around the drum.

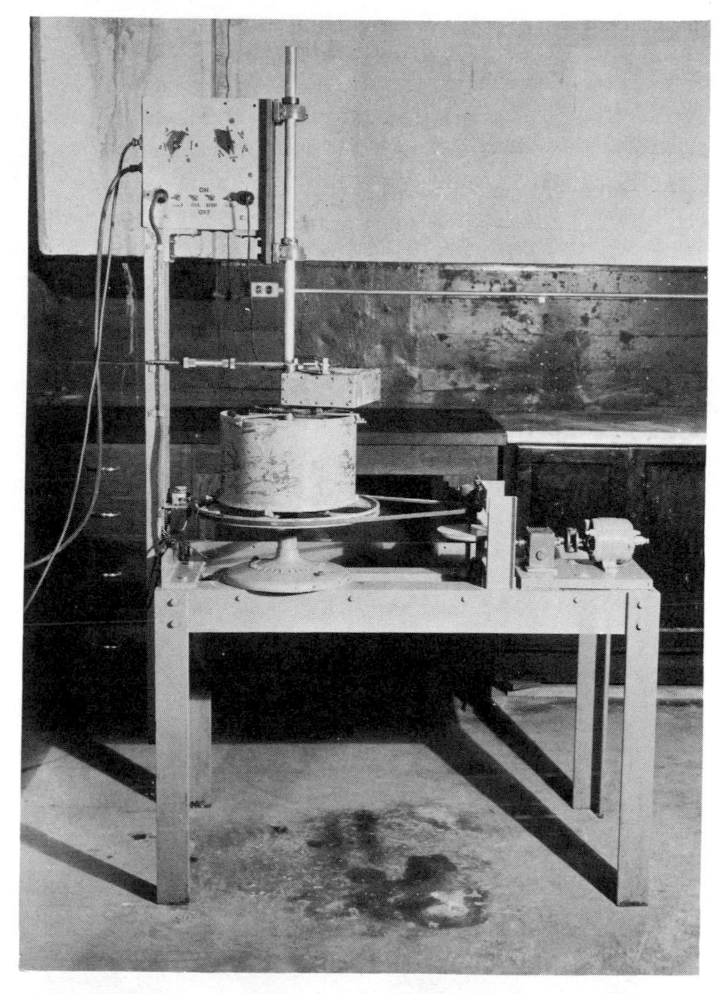

Fig. 10.19. The original model of the recording plastometer of Powers and Wiler [17].

When the container was caused to rotate, the tendency of the drum to rotate with the sample was opposed by a spring-couple that held the drum nearly static; the container wall and the vertical surface of the drum was covered with ribbed rubber to prevent slip. The rotation of the cylinder and the torque required to prevent rotation of the drum were automatically recorded. Two pens made traces on graph paper moving at a known rate to record the stress, strain, and elapsed time.

A sample could be tested in various ways. The container could be rotated continuously at any of several rates, as when measuring the viscosity of a fluid, or it could be caused to oscillate.

After the first phase of the project was reported [19], the sample container and other features of the apparatus were redesigned by Professor Pickett to reduce the difference between the maximum and minimum shearing stress in the sample. Figure 10.20, top, is a close-up view of the revised container ready to receive the sample, and Fig. 10.20, bottom, shows the two parts separated. The container was made in the shape of a ring, and the drum of the original apparatus was replaced by a ring called the *stator*. The stator divided the sample into two parts which were referred to as *annuli*, and was so positioned that the stresses at corresponding points in the two annuli were nearly equal. The ring, which dipped 2 in. into the sample, corresponded to the bob in a conven-

Fig. 10.20. The main parts of the revised model of the recording plastometer.

tional viscometer. It was subject to maximum shearing stress on its outside surface and to minimum shearing stress on its inside; consequently results were reported in terms of an average stress. Maximum stress exceeded the minimum by 32 per cent in the inner annulus and by 30 per cent in the outer. Maximum stress was about 12 per cent higher than the average value reported.

The dimensions of the redesigned apparatus were such that $\gamma = 7.62\Omega$, where γ = shearing strain, and Ω the rotation of the container in excess of the rotation of the stator, in radians. The stator was subject to some rotation due to the stretch of the springs that restrained it.

By observation of a radial crease made on the surface of a sample of paste or mortar, it was confirmed that during distortion the strain in the sample was linear for limited distortions, or practically so. A typical rate of distortion was about 0.05 rad/sec or 2.86°/sec. Such a rate is considered to be in line with the sluggish distortions that concrete undergoes during placement. Doubling the rate of distortion had very little effect on the results.

Work with the improved device described above was interrupted by World War II and the incomplete results remained unpublished. The program was never resumed. Since most of the data have not previously been published, some experimental details are given here.

The machine described above was not perfected, and the data obtained are not wholly reliable in all respects. Particularly, it cannot be assumed that every detail of shape of the stress-strain curves reflects characteristics of the material rather than of the machine. At the beginning of a test, and especially when the machine was continuously oscillating, there were certainly some transient effects due to changes in the kinetic energy of the sample as well as to changes in the kinetic energy of certain parts of the mechanism. However, because the plastometer was operated at relatively low speeds, these effects were not large, and the observed data were used without any attempt to correct them.

As we shall see further on, nonlinear segments of stress-strain curves are usually represented by an average stress-strain coefficient for a given amount of distortion.

The sample in the plastometer was 4 in. deep; the stator ring dipped into it 2 in.; therefore the total torque observed included an end effect that was not taken into account when the machine constants were derived. Experiments indicated that about 12 per cent of the observed torque was due to end effect, but no corrections were applied.

5.2 *Typical Test Results*

5.2.1 Continuous rotation. The most obvious way to test a sample by means of the apparatus described above is to operate the apparatus

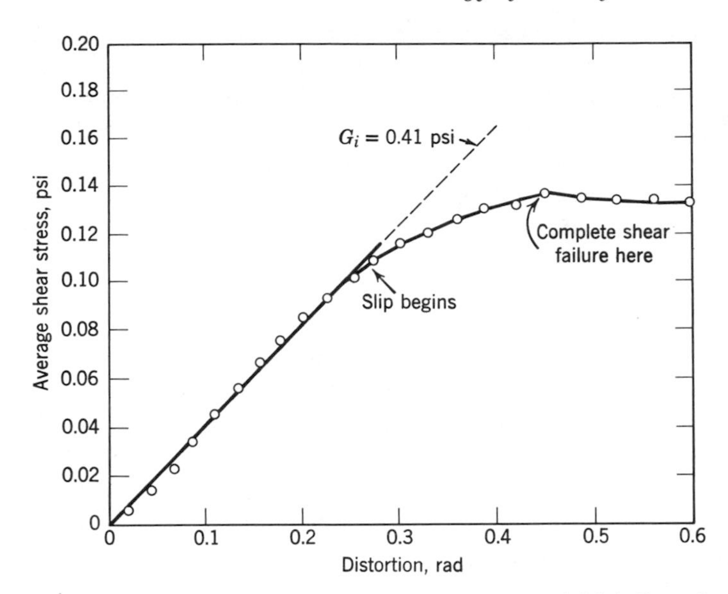

Fig. 10.21. A typical stress-strain diagram for a mortar—initial distortion. A 1:2 mortar was used, with $w/c = 0.376$ and a slump of 8 in. The rate of distortion was 0.05 rad/sec.

in the same way as its prototype is operated, that is, to cause the container to rotate continually in one direction while the torque required to hold the stator is observed. Tested in this manner, a sample of mortar and another of paste gave results shown in Figs. 10.21 and 10.22. Each curve gives the average shearing stress and the corresponding shearing strain.

After the motor was started the sample began at once to undergo shear strain. The indicated shear stress increased approximately in proportion to strain. The deviations from linearity were of such a nature as to be accounted for by kinetic effects.

As the failure stress was approached, the strain *seemed* to increase faster than stress; however, the plastometer recorded only the motion of the sample container relative to the stator rather than actual strain, and it thus could indicate the actual strain in the sample only as long as there was no slip at the boundaries. When slip occurred, it was visible. There was no slip at low stress, but (see Fig. 10.21) the falling away of the curve from the extended tangent at higher stress was due at least partly to slip.

Therefore, although the stress at failure is known, the indicated strain at failure may be somewhat too high. In these cases the average shear stress at failure was about 0.127 psi for the mortar, and about 0.051 psi for the paste.

Failure was localized at the two places where the shear stress was highest: at the surface of the inner wall of the container and at the outside surface of the stator. Because of the fact that the stress in those places is greater than the average stress by a factor of 1.12, as mentioned above, the shear strength is indicated to be about $1.12 \times 0.127 = 0.142$ psi for the mortar, and correspondingly 0.057 for the paste.

We may recall in this connection Bingham's reasoning that plastic flow begins when stress just exceeds the elastic limit of the material and that plastic flow is viscous flow. This seems to be true for ductile metals, but it is true of cement paste only under very restricted conditions. As shown in Sections 4.1 and 4.2, an elastic limit of the order 2×10^{-2} psi can be observed in a very stiff paste, and viscous flow can be seen when the stress exceeds the elastic limit moderately. But, when the elastic limit is of the order 10^{-4} or 10^{-5} psi, as it is estimated to be for the pastes in ordinary concretes, the relationships are different.

It seems that the stress produced in deforming a mass of paste or concrete greatly exceeds the elastic limit of the paste, and, as already noted, the deformation is almost entirely a plastic one, but not primarily viscous. It might be called *visco-dilatant deformation,* with dilatancy dominant, but not dominant to the extent of producing a volume-increase.

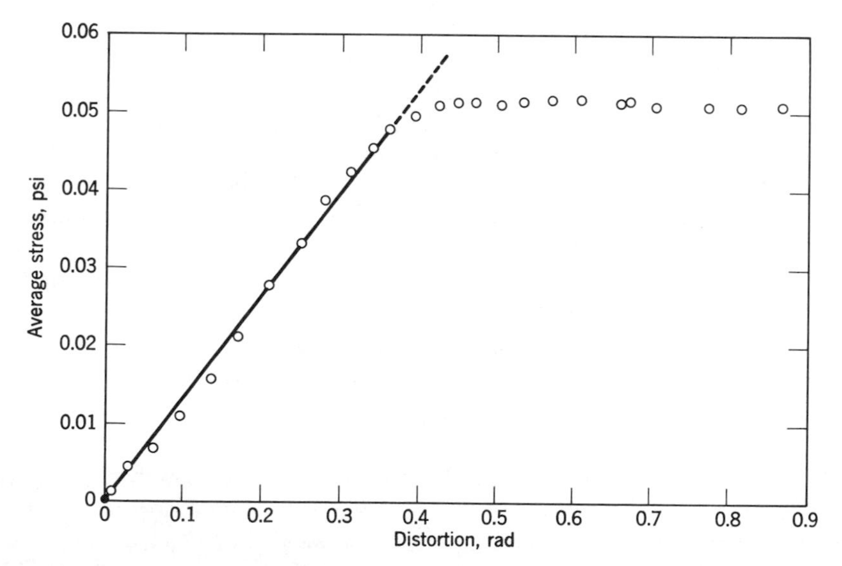

Fig. 10.22. A typical stress-strain diagram for neat cement paste—initial distortion. $w/c = 0.28$; the slump was 6 in. The rate of distortion was 0.162 rad/sec.

(See Section 2.7.2.) Dilatancy dominates to the extent of causing the resistance to shear-strain to be proportional to the strain, up to the stress that causes localized shear failure under conditions where the stress is not uniform.

On the basis of such observations we may say that one of the rheological attributes of a visco-dilatant plastic solid is the characteristic increase of stress required to produce a unit increase of strain. It seems appropriate to call this coefficient the *modulus of stiffness;* it is analogous to the conventional modulus of shear, except that it pertains to strains that are almost entirely nonelastic. Occasionally, it is advantageous to use the reciprocal, which may be called the *modulus of softness.*

To explain the rheological characteristics denoted by the modulus of stiffness, it seems necessary to attribute to dilatancy the buildup of resistance that takes place as the strain increases, as has already been suggested. A concrete mixture is intrinsically dilatant, and the dilatancy is of the kind illustrated by the experiment with quartz powder in water rather than of the kind of dilatancy shown by a compact rigid aggregation. (See Section 2.7.1.) Yet the concrete is able to undergo deformation plastically, and its capacity for plastic deformation depends on the state of dispersion of the solid material in the fluid-filled space, as has been emphasized in previous chapters. However, owing to the relatively high concentration of particles, and to their uneven arrangement, effects of particle interference appear as soon as distortion starts.

A tendency toward dilatation does not inevitably produce a volume increase; in the tests discussed here, none was observed. In the plastometer the sample is confined except at the top, and volume expansion is effectively opposed. It seems likely that the interfering particles are forced to become rearranged so that they can be accommodated while undergoing the required displacement. During the initial distortion of the sample, and as shearing strain approaches the limit for fracture at the nominal surface of the ribbed-rubber cover on the metal, some particles in the fracture zone must be pushed back into the body of the sample to make continuous shearing possible.

Continuing rotation after shear failure is analogous to maintaining plug flow of the kind illustrated in sketch *A-B*, Fig. 10.4. It is also like the plug-flow of concrete in a pipeline when concrete is transported by pumping.

Even if enough friction could be developed in the fracture zone (a thin fluidized layer) it would not be possible to fluidize a normal concrete mixture. The above-mentioned particle displacement, which makes local failure in shear possible without dilatation, that is, the pushing aside of interfering particles, could not possibly occur simultaneously throughout a sample of concrete. Instead, at some limit of strain complete interlock would occur and deformation could not be further increased unless enough stress could be developed at the boundaries to produce the necessary dilatation of the sample as a whole.

The resistance due to dilatancy can be thought of as a propagation of the displacements of particles, or groups of particles, as they are forced to move to reduce the water-filled spaces between them. Actual contact need not be produced. Indeed, as has already been noted, the adsorbed water film could prevent that, but the mechanical effect would be essentially the same if contact did occur.

In concrete, particle interference is not to be ascribed to this or that group of particles in the aggregate but to the particulate pattern consisting of all the particles present. Thus, if a solid sphere were embedded in fresh concrete and caused to move by pulling on it with a thin wire, the resistance to any movement involving more than a very small strain in the surrounding material would be due mainly to particle displacements propagated from the point of disturbance to the walls of the container, all the particles in the directions of propagation being involved, cement as well as aggregate. The magnitude of the resistance should be a function not only of the composition of the mixture but also of the dimensions of the mold or container holding the concrete, just as the rate of fall of a particle through a viscous liquid depends not only on the viscosity of the liquid but also on the size of the container.

The contrast between the behavior in the plastometer of visco-dilatant plastic material (that is, thick paste or concrete) and that of a viscous material was demonstrated experimentally. The annulus was filled with a liquid having high viscosity, an aqueous solution of glucose. With the sample container rotating continuously there was no localized failure in shear, and the instrument registered the viscous nature of the material correctly. Then increments of standard Ottawa sand were introduced up to a volume concentration approaching 50 per cent. This increased the apparent viscosity, but each mixture still behaved like a fluid, showing shearing strains throughout the sample at all speeds.

Other experiments were made in which pulverized silica was suspended in pure water. This nonflocculent mixture behaved like the mixture of glucose and sand. However when a small percentage of calcium hydroxide was added the suspension became flocculent and its rheological behavior became substantially the same as that of cement paste.

Thus it seems clear that the rheological properties of thick paste and concrete are not due alone to relatively high particle concentration. They depend also on interparticle attraction which interferes with ready modification of the existing particulate pattern while shear strains are being produced. It seems that thixotropy promotes dilatancy under these circumstances.

We may note in passing that the action of a concrete mixer produces large and rapid shear strains that fluidize the paste and at the same time promote dilatation. Consequently, the behavior of a given mixture in a mixer may be quite different from that which it exhibits while being transported and placed. If the mixture is made to have a stiff consistency, dilatation in the mixer may be so severe that the material emerges in a crumbly state, the shearing strains in the mixer having far exceeded the limit that could be accommodated plastically. Yet, the same mixture after being compacted may be quite plastic while undergoing limited strain slowly.

5.3 Oscillation at Constant Amplitude

The recording plastometer could be caused to oscillate automatically within a small arc, rather than to rotate continuously. When it was oscillating the instrument revealed rheological characteristics of paste and mortar other than those discussed above. Usually the arc of oscillation was chosen so that the maximum stress developed would not exceed about one-half the shearing strength of the material.

Operated in the manner just described, the machine produced a graphical record such as the section of strip-chart shown in Fig. 10.23. On this chart the line of reference is at the center. Displacements above

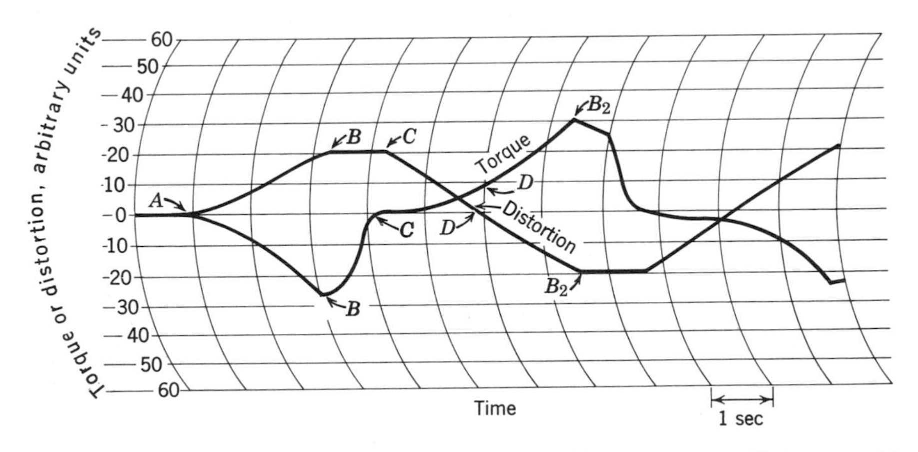

Fig. 10.23. Typical record obtained when the machine is set to oscillate about 10 degrees on either side of the original position. The same pattern is obtained from neat cement paste or mortar.

or below that line indicate the amount of distortion and the amount of torque required to prevent the rotation of the stator. The equally spaced curved lines represent time, one line per second. At the time marked A on the diagram the pens registered zero torque and zero distortion; then in the interval AB distortion took place in a clockwise direction as is indicated by the upper trace; the corresponding torque acting on the stator is indicated by the lower trace. (In the original these traces are distinguished by different colors.)

As soon as the deformation reached the desired point B the driving motor automatically stopped and began rotating slowly in the opposite direction while the stress in the sample dropped to zero, as is indicated by the registered torque. The distortion remained unchanged during this interval. At the instant the stress reached zero, the driving motor automatically resumed full speed causing the sample container to rotate counterclockwise until the distortion with respect to the initial state was the same as it was during the first clockwise half-cycle.

If, after the first half-cycle, the distortion is measured from the position of zero torque, it follows that the total amount of clockwise or counterclockwise distortion in subsequent cycles is twice that of the initial distortion.

Tests at various speeds of distortion indicated that the effect of rate was small enough to be disregarded, a result consonant with other indications of the minor role of viscosity and the dominance of dilatancy. Stress and strain data were transcribed by reading corresponding points from the two traces on the tape, and applying appropriate machine constants.

A typical stress-strain diagram for cyclic distortions is shown in Fig. 10.24, which represents test results from neat cement at normal consistency. The letters A, B, and C correspond to the same designations in Fig. 10.23. The curve from A to B represents the initial clockwise distortion; B to C covers the period during which the shear stress that had developed in producing the initial distortion was allowed to disappear; from C to D the machine rotated counterclockwise to the original position of the sample container; without changing speed, the motion continued counterclockwise from D to B_2; during the period from B_2 to C_2 the shear stress disappeared; then the machine returned to clockwise rotation from C_2 to D_2 and on to E where the first complete cycle was finished. The curve from D_2 to E is a repetition of the initial quarter-cycle but it begins with some stress already in the sample for there was no pause at D_2 for stress to return to zero.

The rather complicated stress-strain diagram given in Fig. 10.24 contains most of the characteristics found in all other diagrams of its kind.

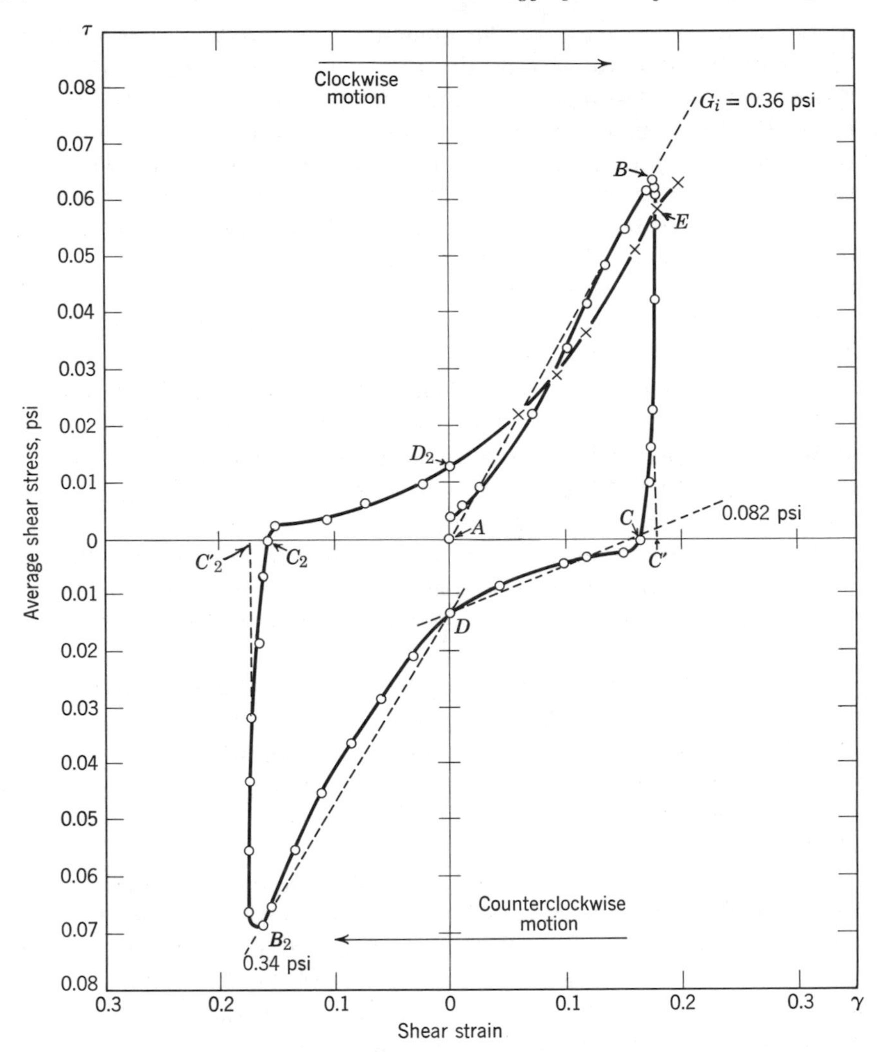

Fig. 10.24. A stress-strain diagram for cyclic distortions. A neat cement paste was used with $w/c = 0.24$. The rate of distortion was 0.08 rad/sec.

Therefore by examining it carefully we shall find the most important rheological characteristics developed by this kind of handling.

The initial distortion, *A* to *B*, shows at the start an apparent development of some stress without observable strain, but this indication is no doubt due to maladjustment of the machine at the start of this particular run, for under the circumstances stress is the consequence of strain. The principle indication is that an increase in stress is practically proportional to the increase in strain, as is discussed in Section 5.2.1. While the stress is returning to zero, interval B to C (about 1.17 sec), the strain diminishes a little. This is believed to represent the recovery of a small amount of elastic energy that had become stored in the sample, although the indicated amount of recovery is probably not an accurate indication of elastic strain. A few experiments were made by disconnecting the driving mechanism and applying a small distorting force by hand; on releasing the force, a small amount of instantaneous elastic recovery was easy to observe.

At point D, the sample container had returned exactly to its initial position, and the cycle may be considered half complete. However it is convenient to regard point B to be the beginning of the first complete cycle, and to call the segment A to B the *initial* distortion.

The amount of work done on the sample by the machine for a given cycle may be evaluated by means of a planimeter. However for some purposes it is convenient to calculate the work for certain segments of the diagram by an approximate method, as follows.

Letting W stand for energy expended, that is, work, we have

$$dW = \tau \, d\gamma \tag{10.10}$$

For any segment of the diagram where it is feasible to assume stress to be proportional to strain, or to use the average stiffness for that segment, we may write

$$\tau = G\gamma \tag{10.11}$$

where G is $d\tau/d\gamma$, the modulus of stiffness. Then, substituting from Eq. 10.11 into Eq. 10.10, and writing the integral expression, we have

$$W = G \int_{\gamma=0}^{\gamma} \gamma \, d\gamma = \frac{G}{2} \gamma^2 \tag{10.12}$$

For example, in Fig. 10.24 $G = G_i = 0.36$ psi, where G_i is the modulus of stiffness for the initial distortion, and the corresponding distortion is $\gamma_i = 0.175$ rad. The work of initial distortion is

$$W_i = \frac{0.36}{2} (0.175)^2 = 0.055 \frac{\text{in.} \cdot \text{lb}}{\text{in.}^3}$$

Thus it can be said that for the initial distortion of this material, the modulus of stiffness is 0.36 psi, and the work required for a distortion of 0.175 radians is 0.055 in. · lb/in.^{3}

The change from state C to state D is a complete reversal of the initial distortion, but, as is indicated on the graph, it is accomplished with much less effort than was required for the initial distortion. The average stiffness is indicated by the slope of a straight line from C to D, which in this case is 0.082 psi. Hence the work required for this segment of the cycle was

$$W_{CD} = \frac{0.082}{2}(0.175)^2 = 0.0123 \frac{\text{in.} \cdot \text{lb}}{\text{in.}^3}$$

This result shows that in reversing the initial distortion, the machine encountered stiffness amounting only to $(0.082/0.36)100 = 23$ per cent of that encountered during the initial distortion, and of course the work required was smaller by the same factor. It was observed that a counter-distortion can always be produced with much less effort than the antecedent opposite distortion, which means that the first distortion has a softening effect. However the effect is limited by the direction and magnitude of the previous distortion, there being no softening effect in a direction normal to the direction of shearing.

Let us now consider the segment D to B_2 of the counterclockwise part of the cycle. The average modulus of stiffness from D to B was 0.34 psi for a distortion of 0.16 rad, which is a little lower than the modulus for the initial clockwise distortion; an extrapolation of the curve indicates that if the distortion had been continued to 0.175 rad, the average stiffness would have been 0.35 psi, only a little less than the value for the initial distortion, 0.36 psi. This similarity shows that the softening effect of a given shear strain on the resistance to the reversal of that strain does not extend beyond the limits of the original strain.

The sector B_2 to C_2, during which stress falls to zero, shows the same evidence of a slight elastic recovery that was seen in the first leg of the cycle. From C_2 to D_2, the first segment of the second half of the first complete cycle, the average stiffness was the same as it was from C to D, 0.082 psi. The last-mentioned point shows that the deformation from D to B_2 had practically the same softening effect for its reversal as was shown for the initial distortion, A to B, and C to D.

Finally, for the sector D_2 to E, we find that the distortion was repeated with considerably less effort than was needed to create the initial distortion. The average stiffness encountered in going from D_2 to E was 0.24 as compared with 0.36 psi, and the work required was 0.037 in. · lb/in.^{3}, which is 67 per cent of that for the first distortion of the same kind.

The feature of the stress-strain diagram just discussed is quite in harmony with the fact that workmen find it advantageous to use a puddling or quaking action in causing concrete to flow. By repeating a motion in the direction of the desired flow, the resistance to flow in that direction is reduced.

5.3.1 Behavior of a viscous material. Figure 10.25 shows a cycle of operation with an aqueous solution of glucose in the sample container. As described above, reversal of the machine starts at point (1) and stress drops to zero, as is indicated at (2). Then counterclockwise motion begins and reaches a constant rate of 0.136 rad/sec in the segment indicated on the graph. In the second half of the cycle, (3) to (1), the machine was operating at a lower rate, 0.059 rad/sec, as is indicated on the graph. We see that during each half-cycle, the stress rises to a limit and remains constant, the level being higher the higher the rate of distortion.

In the example above, the two levels of shear stress in relation to

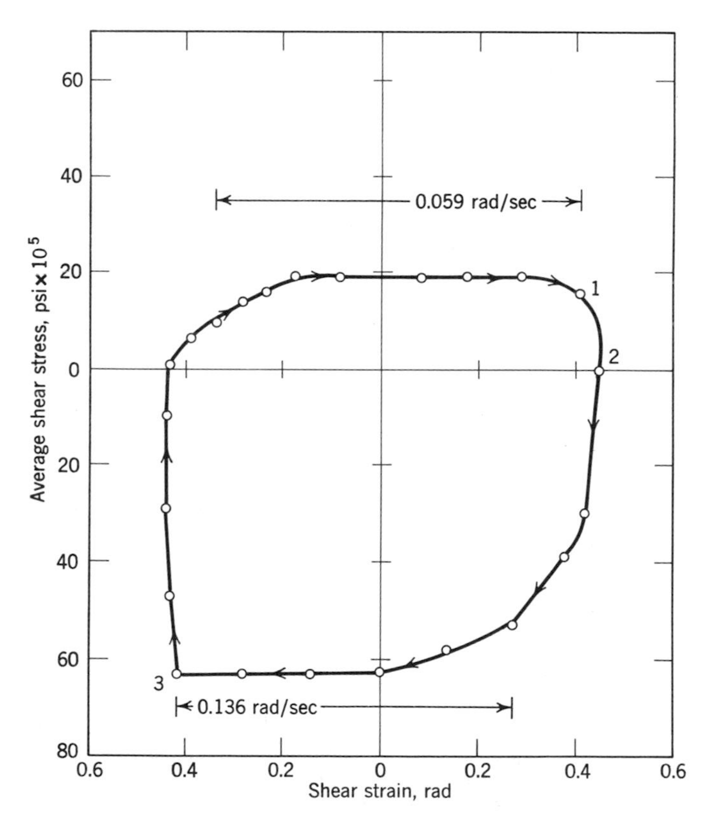

Fig. 10.25. A stress-strain diagram for cyclic distortion of a viscous liquid, in this example a glucose solution.

the respective shear rates do not indicate the same coefficient of viscosity, which may mean that the liquid is not strictly Newtonian or that inertial effects in the instrument were giving false indications. However the main point to be considered is the marked contrast between the pattern obtained from a viscous material and that obtained from cement paste and concrete mixtures.

5.3.2 Behavior of mortar. As indicated above, the features found in Fig. 10.24 are found also in the stress-strain diagrams for mixtures containing aggregate. An example is shown in Fig. 10.26, which represents a mixture made with (0–#4) aggregate and having a consistency such as to give a slump of about $5\frac{1}{2}$ in. Only the initial deformation and the clockwise halves of the complete cycles are shown, and of the complete cycles only the second and sixth are represented. For each of the cycles up to the sixth the deformation was kept within 0.22 rad of the zero position, but during the sixth and final cycle, deformation was allowed to continue. The resistance increased up to $G = 0.42$ psi, and it

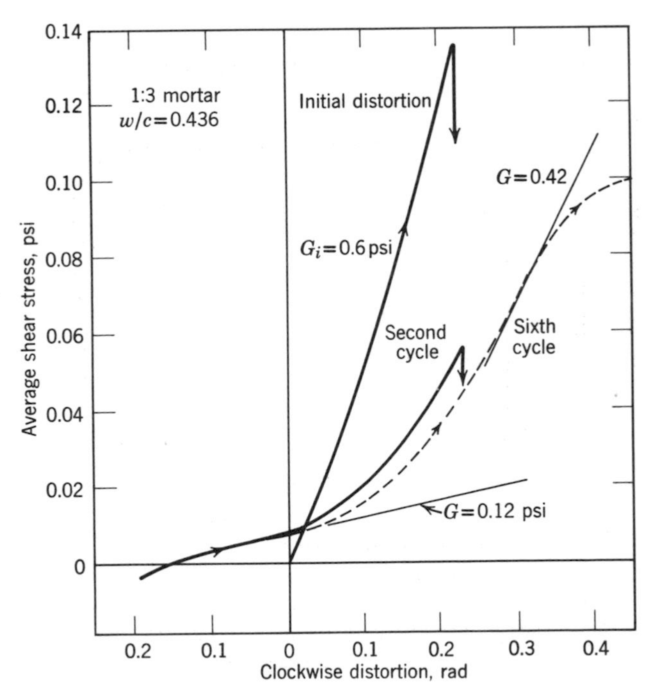

Fig. 10.26. A partial stress-strain diagram for the cyclic distortion of mortar. A 1:3 mix was used with $w/c = 0.436$.

have gone higher if the sample had not begun to fail in shear at its boundary. The falling away of the curve from the extended tangent indicates the beginning of such failure.

Figure 10.26 illustrates very well the many different consistencies a given sample of concrete is capable of showing. During a limited oscillatory deformation, this sample showed stiffness as low as 0.12 psi, only one-fifth of the initial stiffness.

Table 10.4 Reduction of Shearing Resistance During Repeated Distortions of the Same Kind[a]

Cycle[b]	Maximum Distortion: Radians	Maximum Distortion: Degrees	Maximum Stress $10^3\ \tau$ (psi)	Rate of Distortion (rad/sec)
i	0.1375	7.88	8.01	0.022
1	0.1375	7.88	5.82	0.028
2	0.1375	7.88	4.37	0.110
3	0.1375	7.88	3.64	0.110
4	0.1375	7.88	3.64	0.034
5	0.1320	7.56	3.28	0.041
$5\frac{1}{2}$[c]	0.1870	10.72	11.3	0.021[d]
6	0.1320	7.56	5.46	0.042
7	0.1430	8.19	3.28	0.116
8	0.1430	8.19	2.91	0.116

[a] The mix was 1:3 mortar with (0–#4) sand; $w/c = 0.44$.
[b] Cycle *i* is the initial clockwise quarter cycle. Cycle 1 denotes the first complete cycle, the stress being read at the end of the clockwise movement, except as explained in note *c* below.
[c] At the end of the second half of the fifth cycle, the machine failed to reverse at the proper point, giving the larger distortion shown.
[d] Average value for an irregular rate.

We have seen above that as distortions of exactly the same kind are repeated by the process of oscillation through a given arc, the limits of which are symmetrical with the initial position, the resistance grows progressively smaller. This is shown again by the data in Table 10.4 and in Fig. 10.27. These data pertain to a mixture containing aggregate. The values tabulated and plotted are the stresses at points E, as in Fig. 10.24, marking the end of each clockwise motion. At the end of the fifth cycle following the initial distortion, the stress was only about 40 per cent of that required for the initial distortion.

The letters S and F denote relative rates of distortion, slow and fast,

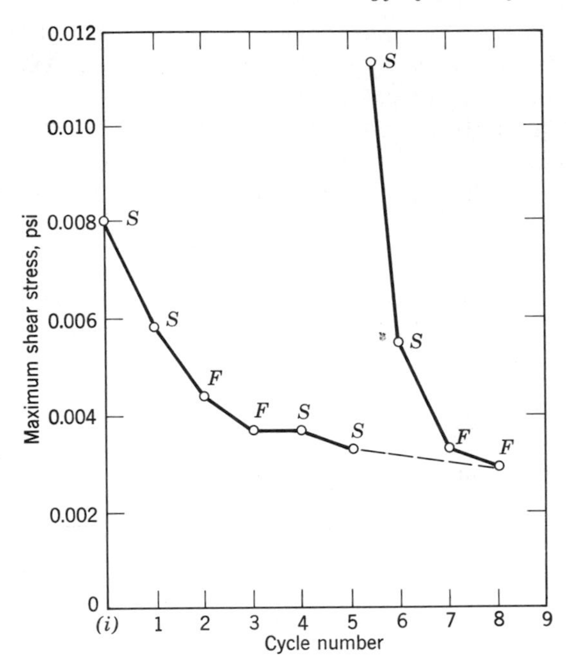

Fig. 10.27. Maximum shearing stress versus cycle number. A 1:3 mortar was used, with $w/c = 0.44$ and a slump of 6 in. The letter i designates the initial distortion. S and F designate slow and fast distortions.

the actual rates being given in the last column of Table 10.4. There is no indication that changing the rate has a significant effect on the stress required to produce a given deformation; the indication is that if there is a fraction of the total resistance that can be ascribed to viscosity, that fraction is small.

Between the fifth and sixth cycle, there was a temporary failure of control that allowed about twice as much distortion as was intended; the effect is shown by the discontinuity of the curve. On resuming normal oscillation, maximum resistance quickly returned to what appears to be a continuation of the initial progressive softening effect of repeated action.

5.4 *Oscillation at Increasing Amplitude*

When the amplitude of oscillation is caused to increase with each cycle, the effect is that shown in Fig. 10.28. This diagram shows the clockwise halves of the first four complete cycles after the initial distortion; the terminal points of each curve are marked with the number of the cycle. Notice that the increase in maximum resistance rises in proportion to the increase in amplitude of oscillation, as is shown by the broken line; the slope of this line may be regarded as a modulus of stiffness when increasing the total deformation in the manner described. In this case it has a value of $G = 0.53$ psi. The average stiffness for the initial distortion, not shown, was about $G_i = 1.0$.

5.5 Relation of Stiffness to Water Content

We have just seen that a given sample of paste or concrete may show many different degrees of stiffness in the plastometer depending on the immediate history of the sample with respect to previous deformations. Maximum stiffness is observed during the initial deformation, and the minimum during the reversal of the same deformation. It would

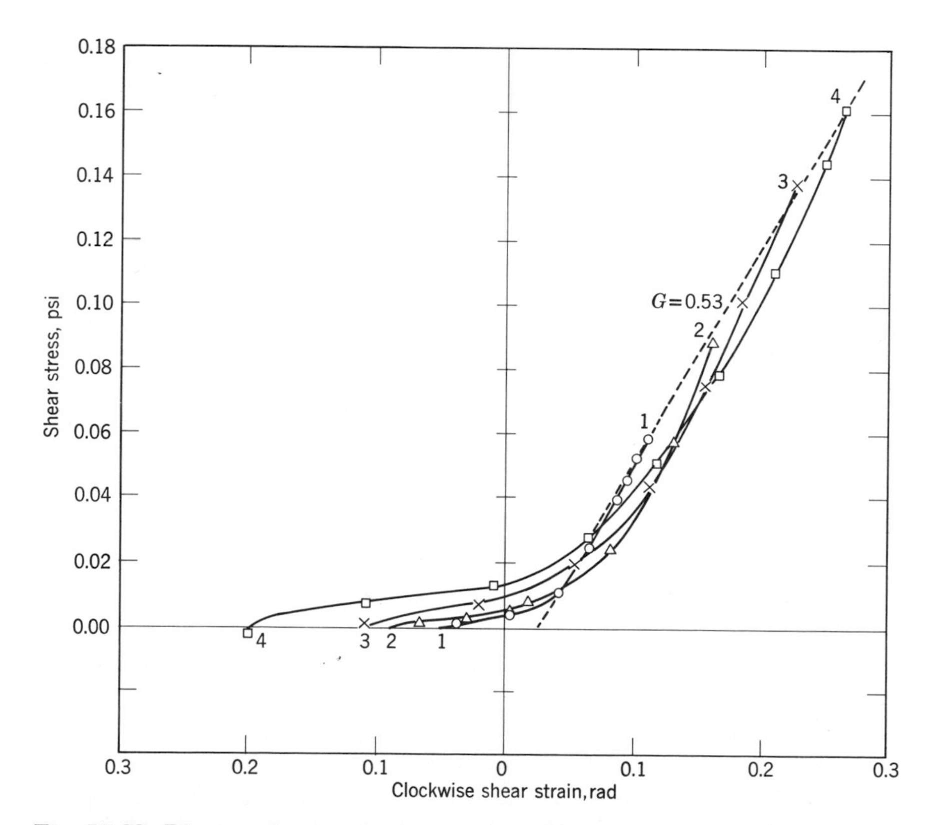

Fig. 10.28. Diagram showing the increase in maximum stress as a function of the increase in amplitude during cyclic distortions. A 1:2 mix was used with $w/c = 0.341$ and a slump of $3\frac{3}{4}$ in. G_i was estimated to be 1.0 psi.

might seem, however, that any given mixture can be adequately characterized in terms of its maximum modulus of stiffness G_i. Therefore, as we now consider the effect of a change of water content (or solids content) on the modulus of stiffness, we shall consider its effect on the initial modulus.

The subject of stiffness is introduced in Chapter 4, Section 2.2 where it is shown that

$$G_i = G_o e^{k_1(V_s/wv_w)} \tag{10.13}$$

where G_o is an empirical constant having the dimension of stress; k_1 is another empirical constant depending on the characteristics of the solid material in the mixture; V_s/wv_w is the volume of solid material per unit volume of water.

It will be recalled that Eq. 10.13 is the same kind of relationship found by Papadakis for cement paste in the fluidized state, and we shall see further on that it holds also for concrete in the fluidized state. In Chapter 4 we found it to hold for paste and concrete in the plastic state, the data on stiffness being that obtained by the remolding test or the slump test. (In the latter case the constant S_o, corresponding to G_o, was not the same for neat cement as for mixtures containing aggregate.)

As mentioned above, such an exponential relationship was proposed by Arrhenius in 1887 for the effect of solute molecules on the viscosity of solutions and for the effect of colloidal particles on the viscosity of suspensions, Newtonian viscosity being indicated in both cases. He found the value of the exponent to be a function of the amount of added material per unit volume of the mixture, or to the complementary amount of solvent or fluid. This difference is possibly a manifestation of the difference between materials in which the shearing resistance is viscous and those in which it is predominantly dilatant. More is said about this further on.

5.5.1 The volume effect. As we have already seen, if a body of neat cement paste is subjected to shear strain, the amount of force required is greater the greater the amount of strain. Now, if, say, one-half the paste in a unit volume is replaced by rigid particles dispersed at random in the paste, a unit of overall strain produces more than a unit of strain in the paste. Since the force necessary to shear the paste is proportional to the strain in the paste, it follows that a greater shearing force is required to distort the mixture than is required to distort pure paste by the same amount. Such a mechanical effect is analogous to the hydrodynamic effect discussed in 2.3.1.

The point of principal interest is that the mechanical effect of rigid particles in a viscous matrix is a function of the volume concentration only, and thus it is independent of the size of the particles.

5.5.2 The size effect. If the volume effect discussed above were the only one, we should expect the magnitude of the effect to be a function of the fraction of paste displaced. In Chapter 3, Sections 5.3.2 and 5.8, evidence of such a volume effect was pointed out in terms of the amounts of water required to maintain basic consistency, but at the same time it was shown that there is another effect clearly related to the size of the particles: the smaller the average size, or the higher the specific surface area, the greater the amount of added water per unit volume of added material. It is apparently for this reason that various attempts to develop laws based on Einstein's equation (Eq. 10.4) for the present case have not been successful. There is a size effect apparently related to the probability of particle interference during deformation of the plastic mass.

Thus the value of k_1 in Eq. 10.13 depends on the composition of the material making up the volume of solids V_s as well as on volume per se. With given materials, this is a matter of the relative proportions of aggregate and cement.

5.5.3 The water-requirement factor. The volume of solids may be made up of cement or of mixtures of cement and aggregate. In the latter case it is convenient, as in Chapter 4, to let

$$V_s = cv_c^o(1 + M) \tag{10.14}$$

and thus

$$G_i = G_o e^{k_1(1+M)(cv_c^o/wv_w)} \tag{10.15}$$

k_1 may be called the water-requirement factor; as shown below, it is a function of M and thus is a constant only for a given mix of given materials.

5.5.4 Evaluation of the water-requirement factor. It was shown in Chapter 4 that data on the modulus of stiffness conform to the following relationship:

$$\log\left(\frac{G}{G_o}\right) = k_2(1 + M)\,\frac{cv_c^o}{wv_w} \tag{10.16}$$

where $k_2 = k_1 \log e = 0.4343k_1$.

Plottings of $\log G_i$ versus cv_c^o/wv_w are given in Fig. 10.29 for the few data on G_i now available. In constructing this graph, the value of G_o was estimated by extrapolating the line through the points for the neat cement specimens, the value obtained being $G_o = 3 \times 10^4$. Then, using that

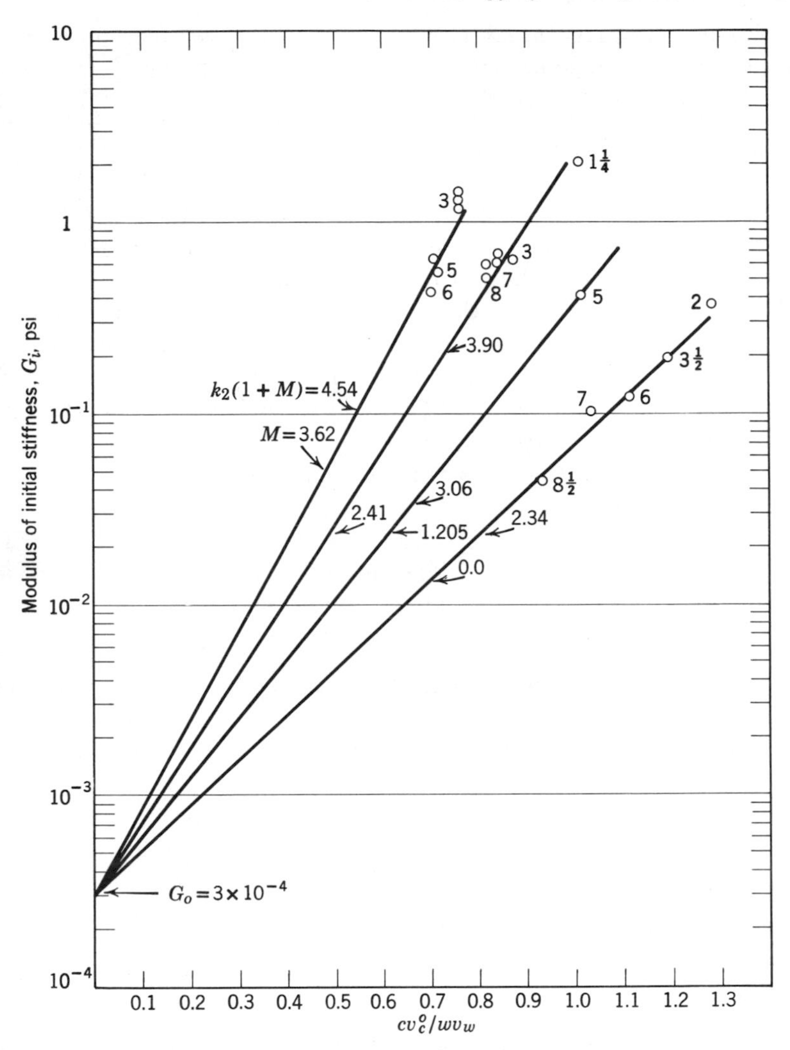

Fig. 10.29. Modulus of initial stiffness versus the ratio of cement volume to water volume for different mixes. Aggregate No. 265-11 was used. (See Table 3.1.) Numerals near the points indicate slumps.

value, the slopes of the lines for the other mixes were obtained from the individual test values by calculating the ratios $\log(G_i/G_c^o)/(cv_c^o/wv_w)$ and taking the average, one value for each value of M. The lines as drawn have the slopes corresponding to these average ratios. Although each plotted point represents only one test, those for neat cement define a straight line satisfactorily. The lines for the mixtures containing aggregate are less well defined, but they seem to be reasonably good approximations of test results.

Figure 10.29 is to be compared with Fig. 4.1, based on the remolding test, for the mixtures were made with the same kind and grading of sand.

As we saw in Chapter 4, Section 4.1, the water-requirement factor k_2 is a function of M, and it is obtained by plotting $k_2(1 + M)$ versus M, the values being given in Fig. 10.29. The result is given in Fig. 10.30 where it is seen that a straight line is obtained. We recognize at once that all the mixtures tested are in the AB category.

The rclationship is

$$k_2(1 + M) = 2.34 + 0.613M \tag{10.17}$$

or

$$k_2 = 2.34(1 - x) + 0.613x \tag{10.18}$$

In these equations

$$k_c = 2.34$$

$$m_{AB} = 0.613$$

which are the water-requirement factors of the cement and the aggregate for mixtures made with this particular sand and cement, the mixes being in the AB category.

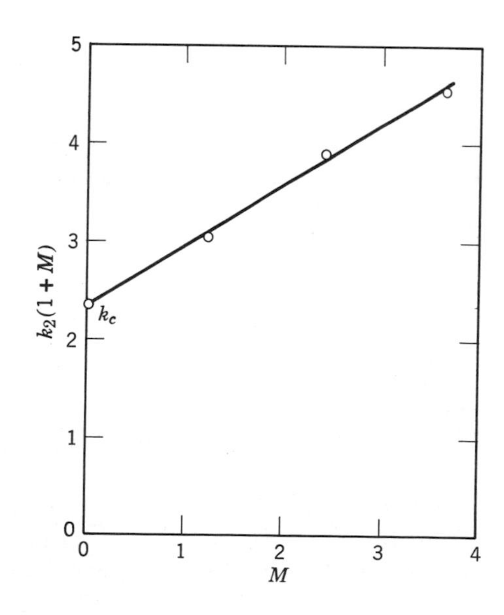

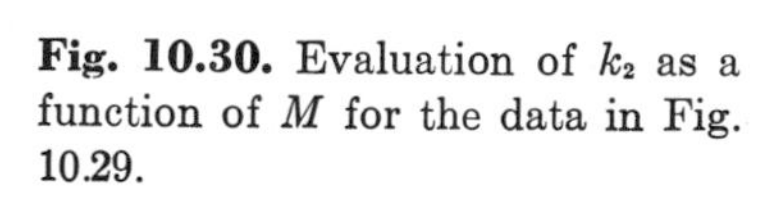

Fig. 10.30. Evaluation of k_2 as a function of M for the data in Fig. 10.29.

On comparing the figures above with those obtained from virtually the same materials by means of the remolding test (Fig. 4.3) we see that the corresponding factors are quantitatively different, as would be expected. However, there is also a qualitative difference: Whereas the data based on actual stiffness gives no indication of point B, that from the remolding test shows it to occur at $M = 3.1$ (see Fig. 4.3). The reason for this difference is not apparent, and there are not enough data on modulus of stiffness to enable us to explore the question. However we shall find below that the modulus of stiffness alone does not determine the molding effort; that is to say, the stiffness indicated by the plastometer does not.

5.6 Modulus of Stiffness versus Slump

Data given in Fig. 10.29 show that different mixtures having the same slump do not necessarily have the same stiffness. The numerals by plotted points indicate slumps. The approximate relationship may be discerned from the accompanying table.

Mix by Weight	M	Slump (in.)	G_i (psi)
1:0	0	3	0.22
1:1	1.20	3	0.5
1:2	2.41	3	0.7
1:3	3.62	3	1.0

It would seem that mixtures having equal stiffness would show equal slumps regardless of differences of composition; but, the figures in the table show that, at least for mixtures in the AB category, the richer the mix, the lower the modulus of stiffness must be for a given slump. Some of the difference might be accounted for in terms of differences in density, for the gravitational force causing the deformation is larger the denser the plastic material. However the actual differences in density are much too small to account for the observed result.

It seems likely that where resistance to deformation is predominantly due to dilatancy, the boundary conditions have more influence than they would if the deformation were resisted only by viscosity. In the plastometer, flow takes place between confining parallel walls, whereas in the slump test the boundary condition permits radial flow which apparently reduces the probability of particle interference.

5.7 *Modulus of Stiffness versus Remolding Number*

Comparisons of the kind described above, but for the remolding test, were made by comparing Fig. 10.29 with Fig. 4.1. The results are given in the accompanying table.

Mix by Weight	M	Remold Number	G_i (psi)
1:0	0	20	0.16
1:1	1.20	20	0.5
1:2	2.41	20	1.0
1:3	3.62	20	0.7

The remolding test is an actual process of molding, and presumably a given remolding number corresponds to practically the same amount of useful work for different mixtures, the differences due to differences in material density being small. Nevertheless we find that for a given remolding number the modulus of initial stiffness is not the same for different mixtures. Although there are quantitative differences between the relationship between G_i and slump and G_i and remolding number, both relationships suggest that the actual stiffness of a mixture is not the same under different boundary conditions.

The observations in this and the preceding section raise the question as to the relevance of conventional rheological measurements to the study of workability of concrete. Conventional rheological procedures were developed from the theory that resistance to deformation is elastic, or viscous, or is a combination of elastic and viscous resistance. At least in its application to plastic concrete, rheology has not much to say about combined thixotropy and the dilatancy ("anti-thixotropy") that was discussed in Section 3.2.1 for the fluidized state, or about visco-dilatant deformation for the plastic state discussed in Section 5.2.1. Numerous references in the foregoing text to "the actual stiffness" are also somewhat questionable. Such usage implies that a visco-dilatant mixture has one intrinsic degree of stiffness dependent only on its composition, but this is probably true only if the mass being deformed is of virtually infinite dimensions. For all finite dimensions, particularly dimensions as small as those of the plastometer, the observed stiffness is probably characteristic not only of the composition of the plastic material but also of sample dimensions. Maximum particle size in relation to sample dimensions must also be considered. It is to be hoped that this area will be explored in future studies of the workability of plastic concrete.

5.8 *Rheological Properties with Different Cements*

Cement pastes having the same water-cement ratio but made with different cements are not likely to have the same modulus of stiffness. The differences are associated with differences in specific surface and chemical composition. Table 10.5 gives data for five cements having the same specific surface as assessed by the permeability method (see Chapter 2), and Table 10.6 gives data for the same cements at equal surface areas as indicated by the Wagner turbidimeter method. The chemical compositions of the cements in Table 10.5 and 10.6 are given in Table 10.7.

On either basis for equalizing specific surface area, the figures show substantial differences between the pastes. In terms of the modulus of softness, cements made from clinkers 15623 and 15670, the last two listed, gave pastes two or three times as soft as those prepared from the other three cements. The difference is somewhat associated with the differences in chemical activity. For the first three cements listed in Table 10.6, the amounts of heat released during the first 5 min of contact

Table 10.5 Stiffness of Pastes Made with Different Cements Having Equal Surface Areas Determined by the Permeability Method[a]

Kind of Cement[b]			Surface Area by Turbidimeter Method (cm^2/g)	Slump (6-in. Cone)[d] (in.)	G_i (psi)	$\frac{1}{G_i}$ (psi^{-1})
Clinker Number	Blend Number	ASTM Type Number[c]				
15367	16186	1	1680	2.8	0.083	12.0
15699	16195	1	1920	2.9	0.105	9.5
15498	16191	3	1620	3.0	0.088	11.4
15623	16189	2	1820	3.8	0.054	18.1
15670	16198	4	2010	3.0	0.049	20.4

[a] The surface area of the cement = 3200 cm^2/g (permeability method); $w/c = 0.28$ by weight. Data are from PCA Series 290.
[b] Different grinds of a given clinker were blended to produce a specific surface area of 3200 cm^2/g as measured by the permeability method.
[c] With respect to chemical composition.
[d] Test made with a cone one-half standard size.

Table 10.6 Stiffness of Pastes Made with Different Cements Having Equal Surface Areas Determined by Wagner Turbidimeter[a]

Kind of Cement[b]			Surface Area by Permeability Method (cm^2/g)	Slump (6-in. Cone)[d] (in.)	G_i (psi)	$\frac{1}{G_i}$ (psi^{-1})
Clinker Number	Blend Number	ASTM Type Number[c]				
15367	15754	1	3516	2.7	0.105	9.5
15699	15761	1	2951	2.8	0.123	8.1
15498	15758	3	3567	2.2	0.135	7.4
15623	15756	2	3060	3.8	0.050	20.0
15670	15763	4	2760	4.2	0.040	25.0

[a] The surface area of the cement = 1800 cm^2/g (Wagner); w/c = 0.28 by weight.
[b] Different grinds of a given clinker were blended to produce a specific surface area of 1800 cm^2/g as measured by the Wagner turbidimeter. Data are from PCA Series 290.
[c] With respect to chemical composition.
[d] The test was made with a cone one-half the standard size.

with water were 3.90, 3.84, and 3.60, cal/g of cement; for the last two, the values were 2.18, and 2.37 cal/g.

Although it is now a matter only for speculation, it seems likely that the effect is not due to any difference in the quantity of reaction products developed during the period of the experiment. Probably the

Table 10.7 Computed Compound Compositions of the Cements Listed in Tables 10.5 and 10.6

Clinker Number	Blend Number	C_3S[a]	C_2S[b]	C_3A[c]	C_4AF[d]	$CaSO_4$[e]
15367	15754	45.03	25.80	13.34	6.69	4.01
15699	15761	45.24	28.94	9.73	7.52	3.09
15498	15758	60.57	11.58	10.26	7.76	3.13
15623	15756	48.51	27.90	4.63	12.60	2.94
15670	15763	28.33	57.52	2.22	5.96	3.20

[a] Tricalcium silicate.
[b] Beta-dicalcium silicate.
[a] Tricalcium aluminate.
[d] Tetracalcium aluminoferrite.
[e] Calcium sulfate.

differences in consistency were due to differences in the electrostatic charge and consequent differences in the depth and position of the potential trough due to the forces between particles, as discussed in Chapter 9.

The practical significance of the indicated differences in consistency at a given water-cement ratio are not immediately apparent. As is discussed in Chapter 3, if the cement is to be used in a mixture that falls to the right of point B in the voids-ratio diagram, a relatively high modulus of stiffness at a given water-cement ratio would be advantageous because the position to the right of point B indicates the paste to have a consistency thinner (softer) than the optimum consistency. If the mixture could be represented to the left of point B, the opposite would be true.

6. VISCOSITY OF CONCRETE DURING VIBRATION

6.1 *The Fluidizing Effect of Vibration*

We visualize freshly mixed cement paste in the quiescent state as a three-dimensional network of particles in water, the particles being held together, and apart, by small interparticle forces acting across very small spaces between the points of near contact between particles. In a paste of ordinary composition, these forces are such as to give the paste the properties of a weak solid. The stiffness of the paste will naturally be some function of the total number of points of virtual contact between particles and of the intensity of forces of interaction at those points. (See Chapter 9.)

In Chapter 3, Section 5.7 the mechanical action of vibration is described; it is considered to be a process of generating compression waves which moves the water molecules much more than it does the solid particles of cement and rock. This amounts to a back-and-forth flow of water between the particles, the relative motion generating hydraulic pressure. The pressure is highest where the restriction to flow is greatest, namely the points of virtual contact between particles. As shown in Chapter 9, any widening of the gaps between particles at points where the particles are closest together, nearly in contact, results in a reduced van der Waals attraction at those points. If the gap width becomes doubled, for example, the attractive force might be decreased to perhaps 1 per cent of what it was before. Since the total attractive force in cement paste is not very high, such a reduction could mean that the effects due to interparticle attraction might be practically wiped out by vibration. Experiments show that cement paste and concrete does become converted from a plastic solid to a fluid, and that state persists as long as the vibration continues, but, the fluid state is thixotropic, not Newtonian.

6.2 Experiments of L'Hermite and Tournon

L'Hermite and Tournon [18,20] conducted experiments on the apparent viscosity of vibrating concrete. Their apparatus is shown in Fig. 10.31. The sample of concrete was placed in a container clamped to a vibrating table, with a burnished metal sphere was embedded in it as shown. Traction was applied to the sphere by the arrangement of pulleys, cord, and weights indicated in the drawing, and any vertical motion of the sphere was automatically traced on a graph. The amplitude of the vibrator was 1 mm, and the frequency 3000 cycles/min. While the concrete was in a state of vibration, the ball would rise or fall through the mixture according to its net weight, which was determined by its buoyancy and the counterweight.

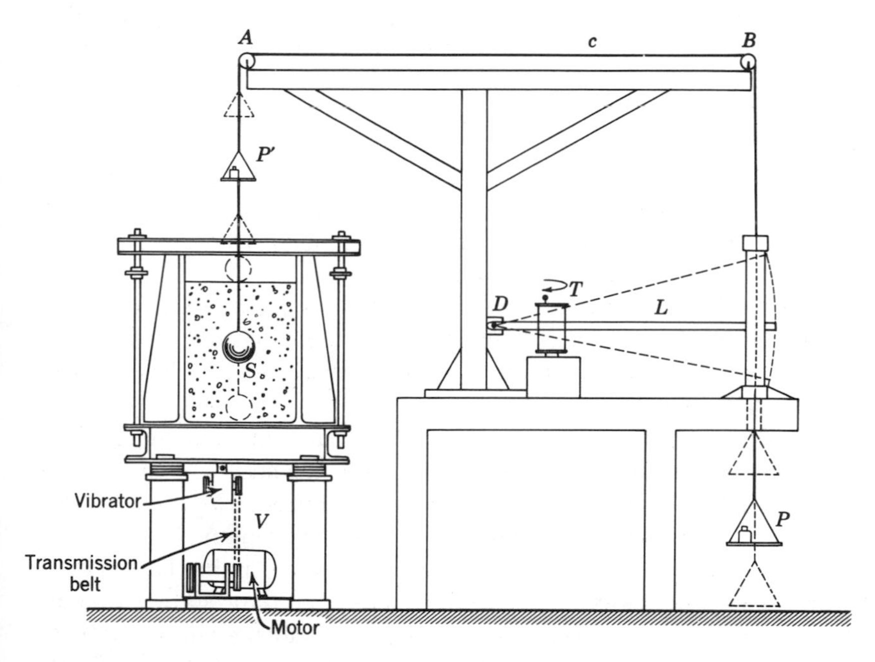

Fig. 10.31. Apparatus designed by L'Hermite and Tournon [18, 20] to measure the viscosity of concrete during vibration. The letters *A* and *B* indicate the pulleys for cord *c*; *D*, *T*, and *L* designate parts of the recording apparatus; *P* and *P'* are counterweights for sphere *S*.

The apparent viscosity of the vibrating mixture through which the ball moved could be calculated from Stokes's law as follows:

$$\eta = \frac{F}{6\pi rv} \tag{10.19}$$

where η is the coefficient of apparent viscosity, F is the driving force due to the net weight of the sphere, r is the radius of the sphere, and v its velocity.

Experiments showed that when the velocity v was held within the limits 0.3 to 0.4 cm/sec while F was varied, the ratio F/v remained constant. Furthermore, when F was held constant and the radius of the ball, r, was changed from 2.5 to 5 cm, the product rv remained constant. This confirmed the application of Stokes's law and in turn demonstrated that concrete is actually in a fluid state while it is being vibrated.

The data given by L'Hermite and Tournon [18] can be represented approximately by the same type of empirical exponential equation used in Section 5.5, as is shown in Fig. 10.32. For the particular mixture used, the data conform approximately to the following equation:

$$\eta = 0.015e^{4.81(c/w)} \tag{10.20}$$

where η is the viscosity in poises of the concrete mixture under vibration, and $0.015 = \eta_o$, which is the indicated viscosity of water. The latter value is too large for pure water at room temperature by perhaps 50 per cent, but it is of the right order of magnitude.

The value 4.81, is determined partly by the circumstance that of the total solids, only the cement is represented in the equation. If the total solids, $c + a$, had been plotted, the value would have been $4.81/(1 + 6.2)$, or 0.668.

The data just considered seem to indicate that vibration of concrete produces Newtonian viscosity. However such a conclusion is not necessarily right because only one amplitude of vibration was used. As we shall see in the next section, the viscosity is non- Newtonian.

Demonstrating that the paste in vibrating concrete is in the fluidized state does not necessarily mean that concrete in that state can flow like a fluid. Although a steel sphere, being one among many particles in the mixture, meets only viscous resistance as it moves through vibrating concrete, it does not follow that all the particles could simultaneously take part in laminar flow. If the shear strains were produced at a sufficiently low rate, such flow might be produced, but at higher rates, particle

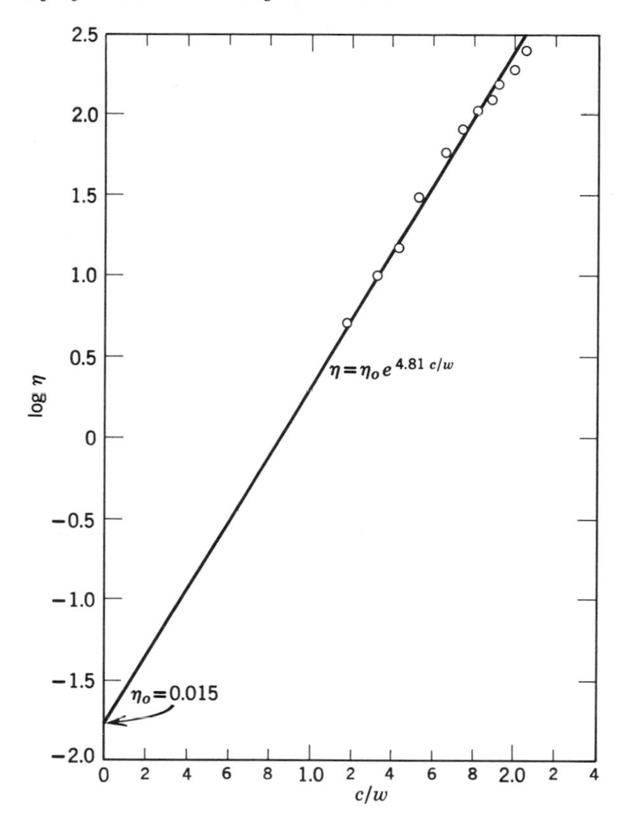

Fig. 10.32. Viscosity of concrete as a function of c/w. A 1:6.2 mix was used. The maximum size of the aggregate was 1 in. (2.5 cm). The vibration frequency was 3000 cpm, the amplitude 1 mm. (Data are from L'Hermite and Tournon [18].)

interference would give rise to extra resistance due to dilatancy, as discussed above.

6.3 *Experiments by Desov*

Using a vibroviscometer of his own design, Desov [19] reached conclusions differing in some respects with those reached by L'Hermite and Tournon. His vibroviscometer was based on the same principle as that described above, but provisions were made for control of temperature, and for varying the amplitude of vibration over a considerable range. Desov found that for either neat cement or mortar the apparent viscosity of a given mixture was different for each different amplitude of vibration. Accordingly, it was concluded that paste or concrete, although fluid

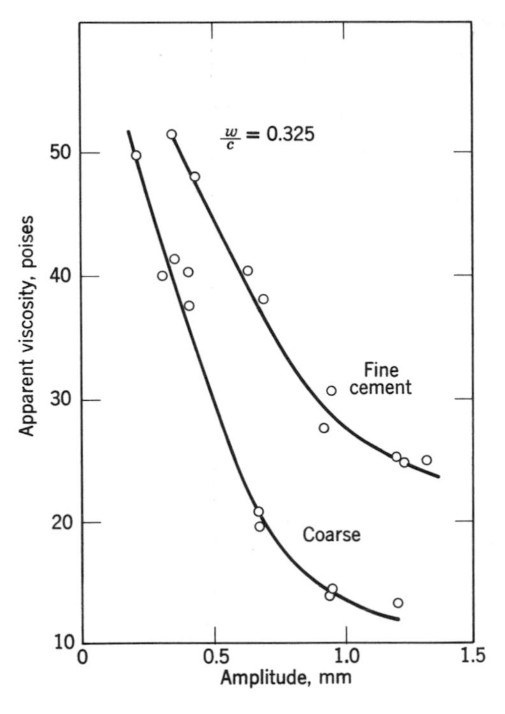

Fig. 10.33. Apparent viscosity of paste as a function of amplitude as measured by Desov [19].

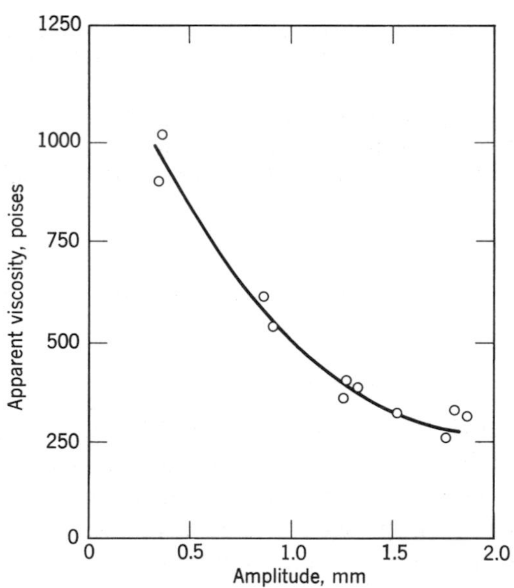

Fig. 10.34. Apparent viscosity of mortar as a function of amplitude of vibration. A 1:2 mortar was used with $w/c = 0.40$ [19].

while vibrating, is not in a Newtonian fluid state. Desov gave the following expression for U_A, the apparent plastic viscosity:

$$U_A = \eta_o + \frac{\alpha}{v} \tag{10.21}$$

where n_o can here be regarded as a numerical constant having dimensions of viscosity, α is a numerical constant called the *coefficient of thixotropy* and v is the velocity of displacement due to vibration. At a given frequency of vibration, the maximum velocity of displacement is proportional to the amplitude of the vibration.

Figures 10.33 and 10.34 are examples of Desov's results, the first for neat cement and the second for 1:2 mortar. The diminution of apparent viscosity with increase in amplitude and thus increase of shearing velocity is quite in agreement, at least qualitatively, with the results obtained with the coaxial cylinder viscometer. (See Section 3.)

7. SHEARING STRENGTH AND COHESIVENESS OF CONCRETE

7.1 Direct Shear Tests

L'Hermite and his coworkers studied the resistance of freshly mixed concrete to shearing stress in terms of shearing strength [18,20]. They used direct-shearing apparatus of the kind used for testing the strength of soils, the essential part of which is shown diagramatically in Fig. 10.35.

During the experiment the upper half of the ring-shaped container was slowly rotated while the lower half was kept stationary, thus subjecting the concrete to maximum shear stress at the horizontal plane between the two halves. The sample of concrete was kept under pressure by a load on the surface which could be varied at will. The angle of rotation and corresponding torque was measured.

Typical data from experiments on nonplastic materials are shown in Fig. 10.36. Each curve represents a test made with a different normal stress, that is, a different load on the surface producing pressure normal to the plane of shear. The shearing strain is given in terms of the angle of rotation, and the corresponding force is given in kilograms per square centimeter of the area of the sample at the nominal shear plane. Since in a nonplastic system, the particles are initially in contact with each other, shear strain gives rise immediately to stress, as is indicated by the rising portion of each curve. Each curve reaches a maximum stress,

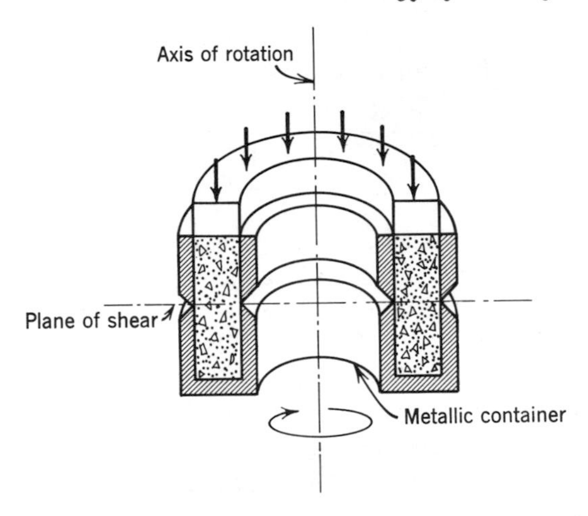

Fig. 10.35. Apparatus of L'Hermite and Tournon for measuring the shearing strength of fresh concrete.

at which point the sample becomes divided into two parts by shearing, and the upper part begins to slide over the lower; this is failure in shear. The decrease of shearing stress and its leveling off indicates that the resistance to continued shearing after failure is less than the stress required to produce failure; the maximum stress corresponds to static friction, the lower stress to sliding friction.

When the shearing strength (static friction) is plotted against the normal stress on the plane of shear the result is a straight line as is shown by the upper straight line in Fig. 10.37; the lower straight line is the corresponding relationship for shearing stresses required to maintain continuous sliding under different normal stresses. Figure 10.37 is

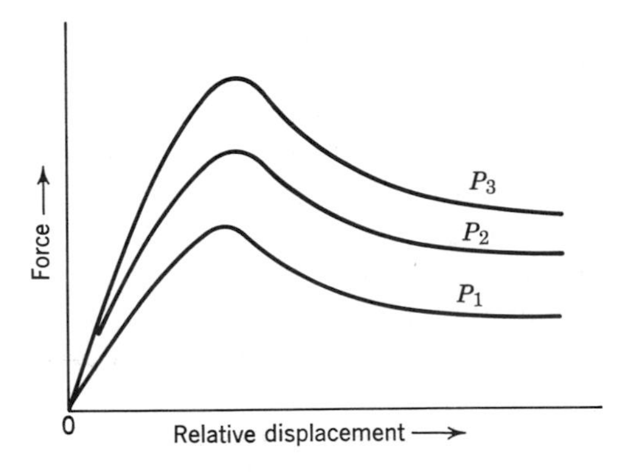

Fig. 10.36. Relationships between force and relative displacement, of the type obtained with the apparatus shown in Fig. 10.35. P stands for normal stress.

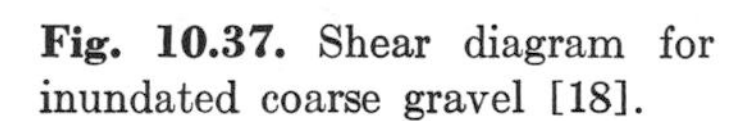
Fig. 10.37. Shear diagram for inundated coarse gravel [18].

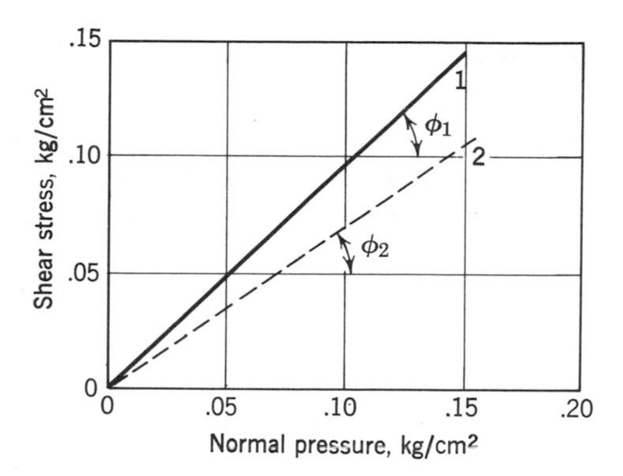

called a *shear diagram*. Similar diagrams were obtained for freshly mixed concrete.

The method the experimenters chose to follow indicates that they considered shearing stress in freshly mixed concrete to be due to internal friction, analogous to the friction between a solid body and a plane solid surface when that body is resting on the surface. For such a simple case, friction is proportional to the pressure acting normal to the surfaces in contact: Coulomb's law.

If f is the coefficient of friction, then, as we see in Fig. 10.38,

$$\tau_r = fP = P \tan \phi \qquad (10.22)$$

where τ_r is shearing stress at failure, P is the normal stress on the sliding plane, and $\tan \phi$ is the tangent of the angle of friction, it being numerically equal to the coefficient of friction f.

When we are considering the shearing strength of a granular material rather than the simple case just described, ϕ *is* the internal angle of

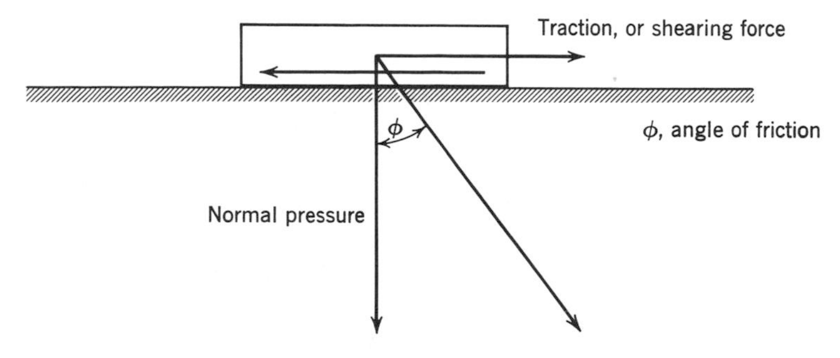

Fig. 10.38. Diagram for Coulomb's law of friction.

friction; it is the angle from the direction normal to the shear plane of the resultant of shear stress and the normal stress. From Eq. 10.22 we see that if shearing strengths are plotted against corresponding effective normal pressures, the latter on the scale of abscissas, the result should be a straight line having a slope equal to the tangent of the angle of friction. Thus the slopes of the lines in Fig. 10.37 indicate the angles of internal friction characteristic of the aggregate from which the data were obtained. The angle ϕ_1 is the angle of friction in the sample just before failure; ϕ_2 is the angle of friction for the sliding of one half over the other.

The shearing strength of a granular system composed entirely of coarse particles is due to the interlocking of particles intersecting the nominal plane of shear. Since the greatest force applied is insufficient to shear the particles themselves, shearing failure involves a displacement of particles away from the nominal plane of shear and against the normal pressure. If there is no normal pressure, such displacement can be produced with virtually no effort, and thus the shearing strength is zero. This is indicated graphically by the 0,0 origin of the straight line and by the statement of Coulomb's law given in Eq. 10.22. The displacement of particles incident to shear failure for a nonplastic granular system requires an increase of volume, and thus identifies the system as a dilatant one.

L'Hermite called the ratio of the angle of friction at failure stress to the angle of friction at sliding stress, the *interlock coefficient*. L'Hermite and Tournon obtained shear diagrams from an inundated sand: tan ϕ_1 was 0.625, and tan ϕ_2 was 0.577, the ratio being 1.08 as compared with 1.38 for coarse, wet gravel. The sand comprised four size groups, whereas the gravel comprised only two. (See Chapter 1.) Thus the indication was that the interlock coefficient and dilatancy are smaller the greater the size range.

For freshly mixed concrete, in which the particles are all slightly separated from one another, and in which the size range of particles is relatively very wide, the difference between ϕ_1 and ϕ_2 is small; L'Hermite and Tournon reported only ϕ_1 for concrete mixtures. This does not mean that concrete mixtures are not dilatant, but it does show that, as compared with that of a compact coarse gravel, the dilatancy of concrete is of a different degree, and as we have already seen, it develops in a different way.

Direct shear tests by L'Hermite and Tournon on concrete mixtures gave the shear diagrams shown in Fig. 10.39. Each shear diagram is a straight line, but its locus is not to the origin. That is to say, when the normal pressure is zero, shearing strength is not zero as it is for

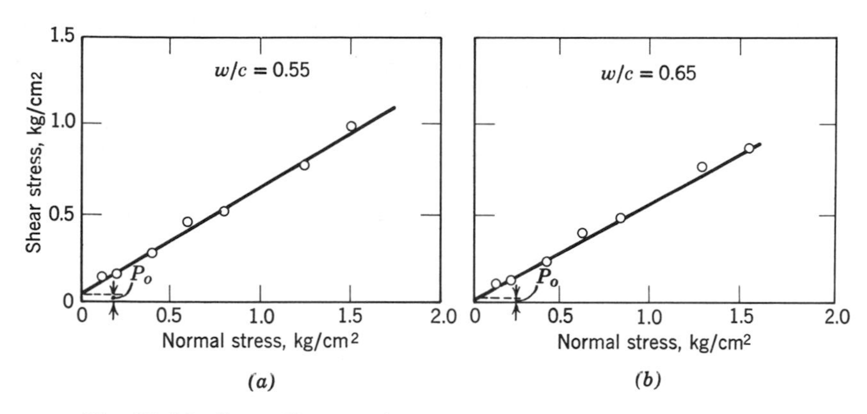

Fig. 10.39. Shear diagrams for concrete. A 1:4.31 concrete was used.

sand or gravel; it has a finite value for each mixture, 40 and 20 g/cm², repectively. If we are to assume that the internal friction of freshly mixed concrete obeys Coulomb's law, we must assume also that the observed normal stress P has been augmented by an unobserved normal stress C such that

$$\tau_r = (P + C) \tan \phi \tag{10.23}$$

By inspection of Fig. 10.39 it is also evident that

$$C = \tau_o \cot \phi \tag{10.24}$$

where τ_o is the shear strength when $P = 0$. Substituting from Eq. 10.24 into Eq. 10.23 we obtain

$$\tau_r = \tau_o + P \tan \phi \tag{10.25}$$

This equation is called the Coulomb law for internal friction in cohesive soils.

Equation 10.25 is required for systems containing very small particles in the flocculent state and thus for those in which interparticle attraction is certainly a prominent attribute. (See Chapter 9.) Associated with interparticle attraction is, of course, a certain cohesive strength, and it is easy to see that in a dilatant system the force required to overcome such strength is mechanically equivalent to the force required to overcome an equivalent normal compressive stress. Thus C in Eqs. 10.23 and 10.24 is the normal compressive stress that would have the same effect as the actual cohesion. Graphically, the tensile stress due to cohesion would be represented in Fig. 10.39 by the negative intercept on the scale of abscissas, to the left of the origin.

7.2 Triaxial Compression Tests

Although the direct shearing test appears to be simple, various results can be obtained from one material depending on the choices the investigator makes concerning several variable test conditions. Some investigators feel that a more accurate assessment of internal friction and cohesion can be obtained by means of triaxial compression tests. Apparatus used by Ritchie [21] is shown in Fig. 10.40. The sample was a cylinder 4 in. in diameter by 8 in. high and was made of freshly mixed concrete compacted in two layers with a standard tamping bar, 25 strokes per layer. The specimen was encircled by a rubber sleeve and subjected to lateral hydraulic pressure giving a radial compressive stress, σ_3. Then,

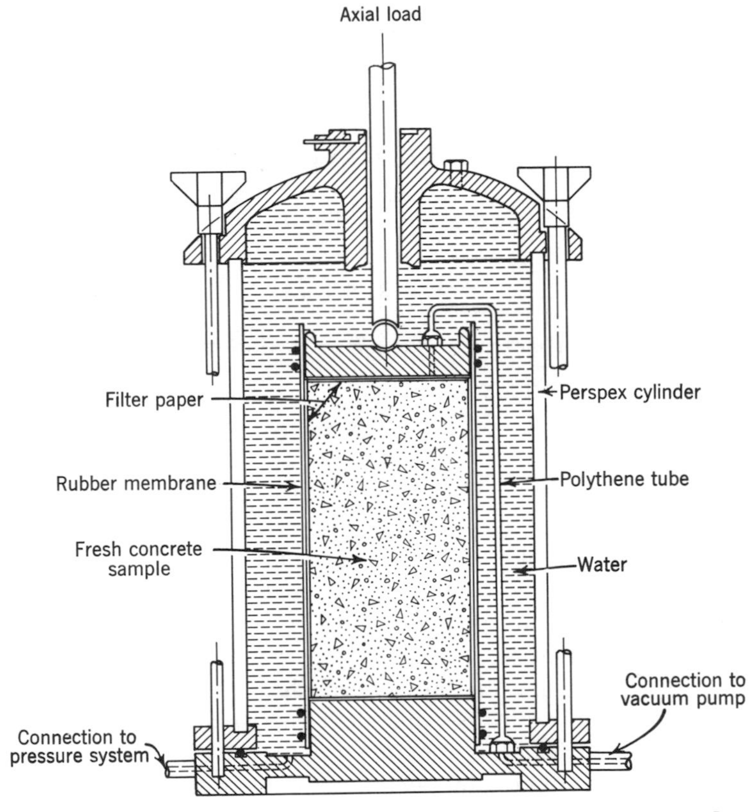

Fig. 10.40. Triaxial compression apparatus used by Ritchie [21].

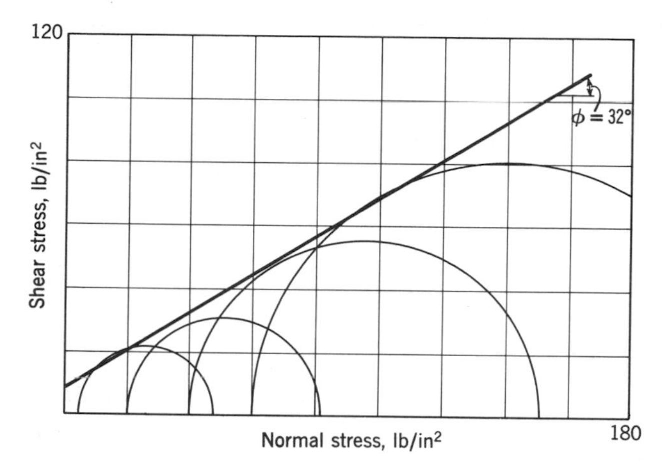

Fig. 10.41. A Mohr diagram.

an increasing axial load was applied, reaching stress σ_1 at which point the deviator stress, $\sigma_1 - \sigma_3$, exceeded the shearing strength of the specimen. Results were analyzed graphically by the Mohr theory of failure.* This analysis requires a graphical construction such as that shown in Fig. 10.41. Each semicircle has a diameter equal to the deviator stress $\sigma_1 - \sigma_3$, the axial stress σ_1 determining the position of the semicircle on the scale of abscissas.

Theoretically, one straight line can be drawn to touch each circle. The point of tangency on a given circle represents the shear stress at which failure occurs, and a vertical dropped from that point to the scale of abscissas indicates the corresponding normal stress, that is, the stress normal to the shearing surface. The line is called the *Mohr envelope;* it is also the Coulomb diagram, its angle with the horizontal axis being ϕ of Eq. 10.25; it could also be called a shear diagram, as in Fig. 10.39. The coordinates of any point on the Mohr envelope gives the shear strength and corresponding normal stress.

Some of Ritchie's results are given in Table 10.8.† The angle of friction was found to be smaller the lower the cement content of the mixture,

* See any textbook on mechanics, especially soil mechanics, for full information on the Mohr theory.

† With each of four different concrete mixtures, Ritchie used three different water-cement ratios and corresponding consistencies. The dryest consistency of each gave results somewhat out of line with the rest of the data, probably because the air contents (not reported) were relatively high. Therefore only the data from the two more workable consistencies were used.

Table 10.8 Data on Shear Strength by the Triaxial-pressure Method[a]

Mix by Weight	w/c by Weight	Slump (in.)	Vebe Time (sec)	ϕ[b] (degrees)	cot ϕ	τ_o[c] (psi)	Cohesive strength[d]	
							psi	g/cm²
1:3	0.477	5	2.0	11	5.1	5.0	26	1830
	0.485	5	1.5	8	7.1	4.0	28	1970
1:4½	0.549	2	6.5	28	1.9	4.0	7.6	530
	0.561	2¾	4.0	25	2.1	7.0	15	1060
1:6	0.665	2¼	4.5	30	1.7	8.0	14	980
1:7.5	0.775	¾	5.0	34	1.5	7.0	10	700

[a] Data are from A. G. B. Ritchie, "The Triaxial Testing of Fresh Concrete," *Magazine of Concrete Research*, **14**(4), 37–41, 1962.
[b] ϕ is the angle of internal friction.
[c] τ_o is the shear strength at zero normal compressive stress.
[d] Cohesive strength is the nominal value obtained by graphical extrapolation.

which is to say the richer mixtures gave lower coefficients of internal friction. Also, cohesive strength was highest for the richest mixture. Both indications seem reasonable, but the results are questionable quantitatively.

Notice that the indicated cohesive strengths range from about 500 to 2000 g/cm², whereas L'Hermite and Tournon found values generally less than 100 g/cm². It seems likely that the values found by L'Hermite and Tournon are of the right magnitude and that Ritchie's results were distorted by the effect of hydraulic pressure within the specimen.

If a specimen were in a perfectly plastic state when it was tested, application of radial stress σ_3 should compress the mixture and raise the pressure of the fluid confined within the boundaries of the specimen. If the specimen were composed entirely of solid material and water, a very slight compression would produce internal hydraulic pressure equal to the external, and the effective radial stress σ_3 would become zero. But, when air is present, as is always the case, the amount of compression required to produce a given internal hydraulic pressure would be greater the higher the air content, owing to the high compressibility of air. Thus, when the specimen contains air, compression tends to force water into air bubbles, allowing solid particles to come closer together and develop mechanical resistance that can support some of the compressive force, and when this state is reached the internal hydraulic pressure is not equal to the outside pressure. It seems, therefore, that in a test on freshly mixed concrete containing air bubbles, the effective radial stress is smaller than the nominal value of σ_3 taken as equal to external hydraulic pressure.

Since shear strength under axial load depends on the effective value of σ_3, it follows that the axial stress σ_1 at the point of failure is also less than it would be if there were no internal hydraulic pressure. The net result is that the deviator stress as plotted is larger than the actual deviator stress. If the Mohr diagram were to be corrected, each semicircle would have to be made smaller, and the Mohr envelope would have a lower position and would indicate smaller values of τ_o and correspondingly smaller values for the indicated cohesive strength.

It is likely also that the nominal values of σ_3 exceed the actual stress by a larger factor at the lowest lateral pressure than they do at the highest lateral pressure, and if this is so, the friction angle reported is smaller than the actual value. Such adjustments would reduce if not eliminate the discrepancy between the results of L'Hermite and Tournon and those of Ritchie.

From the foregoing considerations we may conclude that measurements of strength by application of triaxial pressures cannot be a good way to determine the shear strength and cohesiveness of freshly mixed concrete unless the internal hydraulic pressure is measured and taken into account when the deviator stress [22] is evaluated.

7.3 *Practical Significance of the Shearing Strength*

L'Hermite [20] considered the mechanical work done in producing failure in shear as "determinant from the standpoint of placing a mix." He expressed the work W done in producing shearing failure as follows:

$$W = \frac{f\tau_r}{2} \tag{10.26}$$

where, as before, f is the coefficient of friction and τ_r is the shearing stress at rupture. (Obviously this is an approximation since it ignores the curvature of the stress-strain relationship preceding failure.) L'Hermite suggested that a workability factor could be based on the inverse of the work necessary for shearing failure, thus,

$$\frac{1}{W} = \frac{1}{f\tau_r} \tag{10.27}$$

where $1/W$ is the workability factor, literally the reciprocal of twice the work to produce shearing failure. For a typical concrete mixture he found $(1/W) = 17$, and for the wet aggregate alone, $(1/W) = 12.5$.

The indication that wet aggregate alone is about three-fourths as workable as plastic concrete gives one pause: Surely wet aggregate is not workable at all in the sense that plastic concrete is workable. Indeed, application of the technique for testing shearing strength of soils to freshly mixed concrete seems to be based on the idea that the handling of plastic concrete requires producing deformation and corresponding shear stresses exceeding the shearing strength of the mixture. Possibly such action is involved to some degree, but when it is, the shearing failure is most likely to occur at the surface in contact with the container, such as a chute or the side of a form. Such shear failure does not require the same degree of particle displacement as does a shear failure within the mass of concrete; the effect is similar to that of applying a trowel.

As shown in Sections 3 and 5, shear failure at a boundary is the first step toward converting a plastic solid to a fluid state, mechanically. In practice, such a conversion cannot be carried beyond the initial step because concrete cannot literally become fluid by laminar shearing.

Description of consistency as "wet," or "dry," or "stiff," seems to be based on the observed relative resistance of a material to distortion. Most attempts to measure workability have been based on some means of measuring the relative resistance of a mixture to distortion. The direct shear test described above might be used for that purpose, for applied force and resulting deformation can be observed at all stages of the deformation, and thus a figure closely related to the modulus of stiffness (Section 5.2) might be obtained.

The modulus of stiffness is a major component of consistency, but it is not a sufficient characterization of overall rheological properties; two materials may have the same modulus of stiffness and at the same time different values of cohesive strength. Also two materials may show similar cohesive strength and stiffness at small strains, but one may have a shearing capacity lower than the other, which means that one can be deformed plastically more than the other before shearing failure occurs. L'Hermite reported that shear strain at failure ranged from 0.05 to 0.10 rad, although he did not indicate whether this was true at a given strength. Other data indicate that the shear strain at failure may be considerably larger than that just indicated.

Thus practical considerations of the properties of freshly mixed concrete inclines us to the view that the rheology of freshly mixed plastic concrete mixtures comprises at least three more or less independent factors: the modulus of stiffness, the cohesive strength, and the magnitude of shearing strain at shearing failure. The last named characteristic should be indicative of the amount of strain a given mixture can undergo before it becomes excessively dilatant.

The concept expressed in the foregoing paragraph is the one on which the work of Powers and Wiler [17] was based. Their method gave detailed information about stress-strain relationships, but the direct shear test of L'Hermite and Tournon seems to be the best method for assessing cohesive strength.

We must keep in mind also the evidence discussed in Section 5.7 and Section 5.8 that the coefficient of stiffness as measured in a plastometer may not be indicative of stiffness under different boundary conditions. Owing to the dominant role of dilatancy, stiffness may depend on the manner of deformation as well as on the composition of the mixture. If resistance to deformation is mainly a matter of propagating particle interference, the maximum size of particles in relation to the dimensions of flow-space may be one of the determinants of stiffness.

8. DEVICES FOR ASSESSING RELATIVE WORKABILITY

8.1 *The Nature of Workability*

The term "workability" has been used repeatedly in earlier sections and chapters with the expectation that it would not be misunderstood; yet when pressed for a definition, anyone finds it difficult so say exactly what he understands it to mean. The natural tendency is to formulate a definition as if the term refers to an intrinsic property of fresh concrete, whereas as actually used it usually implies much more than that. It is often used with modifiers such as "excellent," "good," "fair," "poor"; so qualified, it expresses a personal judgment of how well a given concrete mixture responds to the particular method of production and placement being used, under the conditions of the work. For example, it might be observed that a pavement slab is being shaped and finished at a satisfactory rate and with acceptable results; and the observer refers to the concrete as "a workable mix," or he says the workability is "good." But, if he observed the same mixture being deposited in a narrow form containing reinforcement, he might express an opposite judgment. Thus the term "workability" pertains not only to intrinsic qualities of the mixture, but also to extrinsic factors.

Often, appraisal of a given mixture is expressed in somewhat more explicit terms. It may be said that a mixture is too stiff or too soft, or more likely, too dry or too wet. Whether dry or wet, the mixture may also be described as harsh or smooth (plastic) according to the way it looks or feels to the workman trying to change the shape of the mass as deposited, and how it adapts itself to the shapes required. When an unusually rich mixture is being used, or one made with an unusually fine cement, its stickiness is likely to be mentioned. During the handling of a mixture, there may be some separation of coarse aggregate from the mixture, and the observer usually refers to this as segregation, or "ravelling." Since segregation often leads to flaws in the final product it is undesirable, and therefore a mixture prone to segregate is likely to be downgraded with respect to its workability, even though the molding or shaping of it may not require unusual effort.

A mixture that would be judged easily workable immediately after it was discharged from the mixer may be found unworkable at the point of deposit because of segregation of a somewhat different kind from that mentioned above. If the mixture has a relatively wet consistency, some of the solid material, particularly the coarser parts of the aggregate, are liable to settle toward the bottom of the conveyance during transportation to the point of deposit, leaving "soup" at the top. When the conveyance is unloaded by tilting it, the material first discharged is overly wet, while that remaining may be so dry as to be almost immobile.

In common parlance workability is considered to include consistency, but not to be identical with it. This distinction tends to compound the vagueness, for as we found in Section 2.1, consistency also is hard to define. However, if we consider consistency as that rheological quality of a mixture that can be varied by changing its water content, we may say that it is not the same as workability. For example, two mixtures being handled under the same conditions might be said to have the same consistency but at the same time to have different workabilities because of differences in the aggregate grading, or differences in cement content, or differences in both. Speaking in terms of the voids-ratio diagram, we would say that at a given consistency as determined by some accepted criterion, every different mix has a different workability. Mixes in the *AB* category have workabilities different from the workabilities of the *BF* mixtures, and of course different from those of the *FC* mixtures. These differences can be attributed mainly to the differences in degree of aggregate dispersion, which in turn can be related to differences in paste consistency, as discussed in previous chapters.

8.1.1 Capacity for plastic distortion. In earlier sections it was pointed out that concrete mixtures cannot undergo unlimited shear strain plastically: Beyond a certain limit of strain, the mixture becomes excessively dilatant. This is to say, a mass of concrete has a certain capacity for plastic distortion. The existence of such a limit has not been clearly demonstrated in the laboratory, but it is indicated by various evidence, including that from field observations. It is possible that a mixture may show different capacities under different conditions, so that to speak of *the* capacity for plastic shear may be only a first approximation of the truth.

To produce a deformation beyond the capacity for plastic distortion or plastic shear would require volumetric dilatation, and a correspondingly large shearing stress. When the force is transmitted to the mass entirely by friction at two parallel boundaries it may be impossible to develop enough stress in the mass to produce volumetric dilatation because of the zone of slip developed at the boundary. The limit for plastic distortion is frequently seen when handling relatively stiff, cohesive mixtures. When caused to flow over the "lip" of a concrete buggy, such a mixture will often develop visible fractures where the curvature of the flowing (or sliding) mass requires shear strains exceeding the capacity for plastic shear. Also fracturing and perhaps a tendency toward segregation, can be observed where such a mixture is made to flow around a right angle in the form.

Mixtures in the *AB* category have a relatively large capacity for plastic distortion, and at the other extreme, those in the *FC* category have very little. The latter can be deformed very little without developing excessive dilatancy, and with that goes a tendency to segregate during handling, a tendency to leave corners and angles unfilled, and to clog the lines of a concrete-pump system.

Differences in such characteristics at a given consistency are practically significant or not according to the conditions under which the concrete is being handled. If a mixture has a low capacity for plastic distortion, but yet enough for the existing conditions, it might be considered just as workable as a richer mixture having a much larger capacity for plastic distortion. Indeed, from some points of view it might seem preferable because it is less sticky and can be handled with less effort.

Although a lack of stickiness may be desirable under some conditions, under other conditions the cohesiveness that stickiness implies may be essential or at least desirable. This is the case when concrete is being placed under water, or in general when conditions are such that it is not possible to deposit the concrete close to its final position, that is to say, where concrete must be caused to flow or slide into place.

Judgments of workability are greatly modified when vibrators are used for placing concrete. Mixtures that seem too stiff or too harsh when vibration is not used may be preferred, and generally should be.

8.1.2 Subjective factors. In appraising the workability of concrete mixtures under various circumstances we are likely to comment on the wetness, or dryness, on the cohesiveness, or stickiness, and perhaps if an analytical attitude has been developed, on the capacity for plastic distortion; but, we might not use the same terms about the same mixture under different circumstances. It is evident that we cannot deal with workability in quantitative terms; it involves intrinsic rheological properties which perhaps are subject to quantitative measurement, but also it involves various extrinsic and even subjective factors. The subjective factors are evident in matters of opinion, and opinion varies with the point of view. The contractor, or the workman, is likely to base his opinion on the amount of time and effort required to achieve an acceptable result. The concrete technologist is likely to base his opinion on the frequency of visible faults, together with evidence of any sacrifices of quality incidental to reducing the cost of placement.

The term "workability" is associated with experience, general impressions, and personal judgments involving not only the properties of fresh concrete, but also the myriad situations under which it is handled. If the term is applied to something that can be measured in the laboratory under standard conditions, it takes on a different meaning; it tends to become synonymous with the term "rheological characteristics."

To deal with workability in the laboratory, we must confine our attention to intrinsic properties as far as we can, excluding extrinsic factors, or standardizing them; this means that in the laboratory we are pretty much limited to an application of rheological methods. Consciously or not, this has been done by most of the many persons who have invented devices for measuring workablity; they have succeeded to one degree or another in measuring rheological characteristics, almost always in relative terms.

According to Fulton [23] there have been at least 52 more or less different tests described in the literature. Most of them have not found much use. Some have caught on in one country and not in others. We shall review some of those that have proved to be of greatest interest, or usefulness.

8.2 The Remolding Test

The essential features of the apparatus as described by Powers in 1932 are shown in Fig. 10.42 [24]. The main part of the apparatus is a cylindrical container within which a concentric ring is held above the bottom a certain distance equal to or slightly more or less than the final depth of the sample. The assembly as shown rests on a modified standard flow table (ASTM Designation C124) to which it is clamped. In a later development the standard flow table was replaced by one having a diameter only slightly larger than the container, and the cam shaft of the table was driven by a motor.

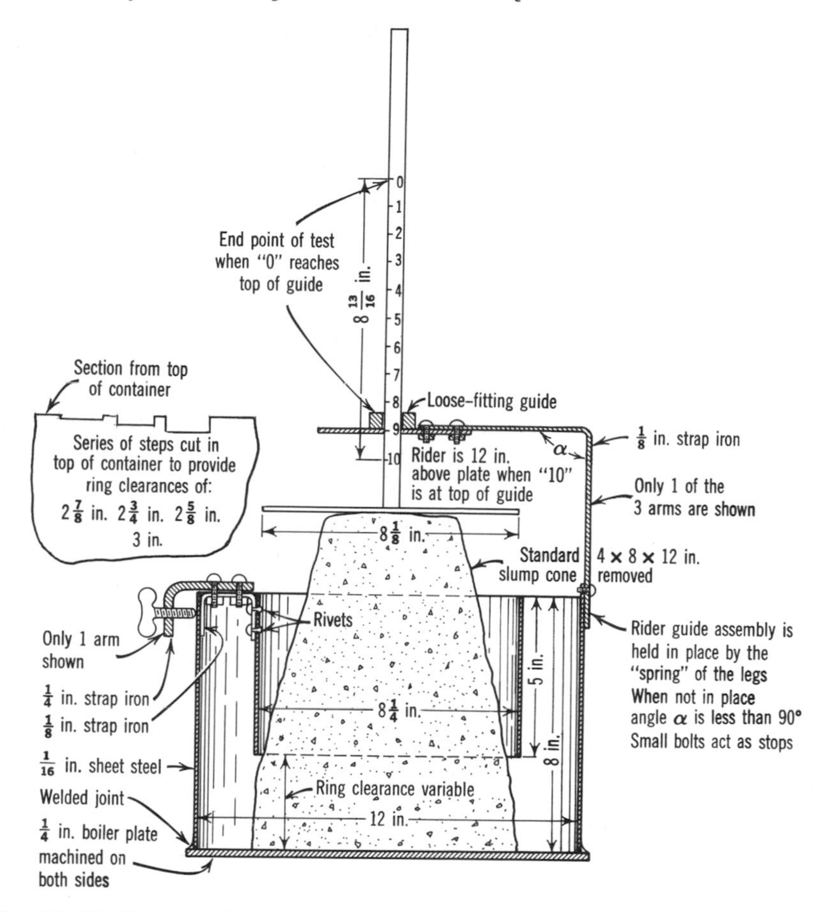

Fig. 10.42. Cross section of the apparatus used for the remolding test [24]. The device shown rests on a flow table set for $\frac{1}{4}$-in. drops. It is held accurately centered by means of blocks clamped onto the table.

The drawing indicates a stiff sample before the remolding process has started. The sample was originally compacted in a standard slump cone set inside the container, each layer being rodded or tamped as much as required to mold it properly. The slump mold is then removed and the threelegged rider assembly partly shown in the drawing is set in place. The flow table is then operated, giving $\frac{1}{4}$-in. drops at the rate of about 1 per sec, until the sample has become remolded into the form of a flat disc, the end point being indicated as shown in the drawing.

The relative amount of effort required to carry out the operation was originally called *remolding effort;* in this book we have called it the *remolding number*.

O. Stern, as reported by Bahrner [25] observed that the volume of the sample after jigging was usually smaller than the volume after it was molded by hand in the slump cone. He proposed that the "workability" be indicated by the remolding number reduced by a factor V/V_o, where V_o is the capacity of the slump cone and V is the volume of the sample after jigging.

The function of the ring within the container is to restrict the movement of the sample during the remolding process; it influences the performance of different types of mixtures differently. A mixture with a relatively large capacity for plastic shear can pass through the space under the ring easily, but one having a low plastic-shear capacity tends to clog the opening, and thus requires extra effort which shows up as a relatively large remolding number.

The remolding test apparatus is a means of assessing the relative amount of effort required to mold concrete under one specific set of conditions, the motive force being developed and applied in a specific way. With the concentric ring set at a certain distance above the bottom, it requires a certain capacity for plastic shear for the change to proceed by a succession of smooth movements of plastic flow. If the capacity of the concrete for plastic shear is insufficient, the process involves extra work to break down the unstable structure that develops from excessive dilatancy (particle interference). However, a mixture that finds difficulty in passing under the ring may adapt itself to the new shape readily if the ring is removed. Not only is less plastic deformation required, but also some of the reshaping might be done without a flow process, that is, by crumbling and reconsolidation. Thus with the ring in place two mixtures might require different amounts of effort, and with the ring removed, or set at a different clearance, the two mixtures might require the same effort.

In making such a test in connection with designing mixtures for a given placing condition the question arises as to whether discriminations between different mixtures truly reflect the differences in effort that would be required under the actual conditions of the work. There is no way to answer the question except on the basis of estimate or personal opinion. This can be said also of any other laboratory device, and it raises a question as to the real utility of such tests as far as the practical designing of concrete mixtures is concerned; it causes us to lean toward the view that mixtures should be developed on the basis of an analytical understanding of the properties of mixtures, together with observation of actual performance under given field conditions.

Nevertheless, meaningful analysis of concrete properties rests on the results of quantitative measurements under controlled conditions, and for this purpose a device such as the remolding test is useful, as we have seen in previous chapters. Although results of such experiments may not be quantitatively applicable to various situations, they lead us to at least a qualitative understanding of the factors involved.

Since the remolding test involves an actual molding process, the utility of other devices can be evaluated on a limited basis in terms of the degree to which the test values obtained from them correlate with the remolding number. When the correlation is *not* good, the question remains open as to how good the correlation might be with the molding effort required under some different placing condition, and of course when the correlation with remolding number is good, the same question arises in the opposite way.

There is enough resemblance between the flow conditions in the remolding apparatus and those that can be seen in concrete forms to warrant the assumption that *deductions* based on the measurement of relative molding effort with this apparatus are relevant to practical considerations. However this can be said without further qualification only with respect to concrete being placed in the plastic state, not to concrete being placed in the fluidized state, that is, by means of vibration.

8.2.1 The Thalow drop table. The Thalow drop table [26] is essentially the remolding test described above, except that it has no ring to restrict the flow during the transformation, no rider, and the diameter of the container is not much larger than the base of the slump cone. A specially constructed portable jig table is provided; it gives four 1-cm drops per revolution of the cam shaft.

8.3 The Vebe Apparatus

About 1937, Wuerpel [27] used a remolding device of essentially the same design as that shown in Fig. 10.42, but it was clamped to a vibrating table instead of to a flow (jig) table. In 1940 Bahrner [25] described a modified but similar apparatus which is manufactured and distributed commercially; it is generally known as the Vebe apparatus, after the developer's initials. (See Fig. 10.43.) This apparatus does not provide a ring to restrict the path of flow during the transformation from the conical shape to the final shape.

Since the Vebe apparatus fluidizes the sample, the time required to complete the transformation no doubt depends on the apparent viscosity, which, as we have seen, is partly a function of the frequency and effective amplitude of the vibrator.

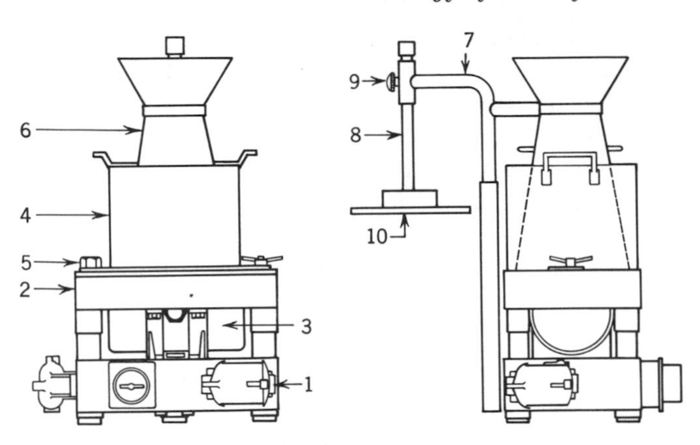

Fig. 10.43. Diagram of the Vebe apparatus. 1. Base resting on three rubber feet. 2. Vibrator table supported on rubber. 3. Counterweight of vibrating motor. 4. Cylindrical pot. 5. Fasteners to hold pot on vibrating table. 6. Standard slump cone. 7. Swivel arm with funnel. 8. Graduated rod. 9. Guide sleeve. 10. Transparent disk fastened to graduated rod.

It seems obvious that this type of apparatus is to be preferred in the laboratory when the concrete under test is to be placed under field conditions by means of vibrators. Its application when the concrete is not to be placed with vibrators is more questionable, although there is no doubt a relationship between stiffness in the plastic state and apparent viscosity in the fluidized state. However, a mixture lacking sufficient capacity for plastic distortion in the plastic state might suffer less from that disadvantage in the fluidized state.

8.3.1 Hallström's apparatus. The Hallström apparatus, described in a paper by Eriksson [28], is shown in Fig. 10.44. It consists of two main parts: a tublike vessel, and a cylinder that can be held concentrically as shown, and a third part, a "rider" resembling that of the remolding apparatus. The vessel was clamped to a vibrating table.

The method of operation was described by the author as follows:

With the upper parts detached, the lower container was filled with concrete by vibration to a designated standard level below the rim. Then the cylinder was set in place, extending 2.5 cm into the concrete, and filled to its brim with concrete. The rider was added, the vibrator started, and the time required for the rider to descend 13 cm was measured.

This test involves flow of concrete from the cylinder into fluidized concrete in the tub under conditions of restricted flow. Eriksson's experimental data indicated that those mixtures that have high particle interference (as discussed in Chapter 6, Section 3) and low capacity for plastic shear also show relatively high resistance to flow in the fluidized state under the condition of this test. The same mixtures did not show up unfavorably in the Vebe apparatus.

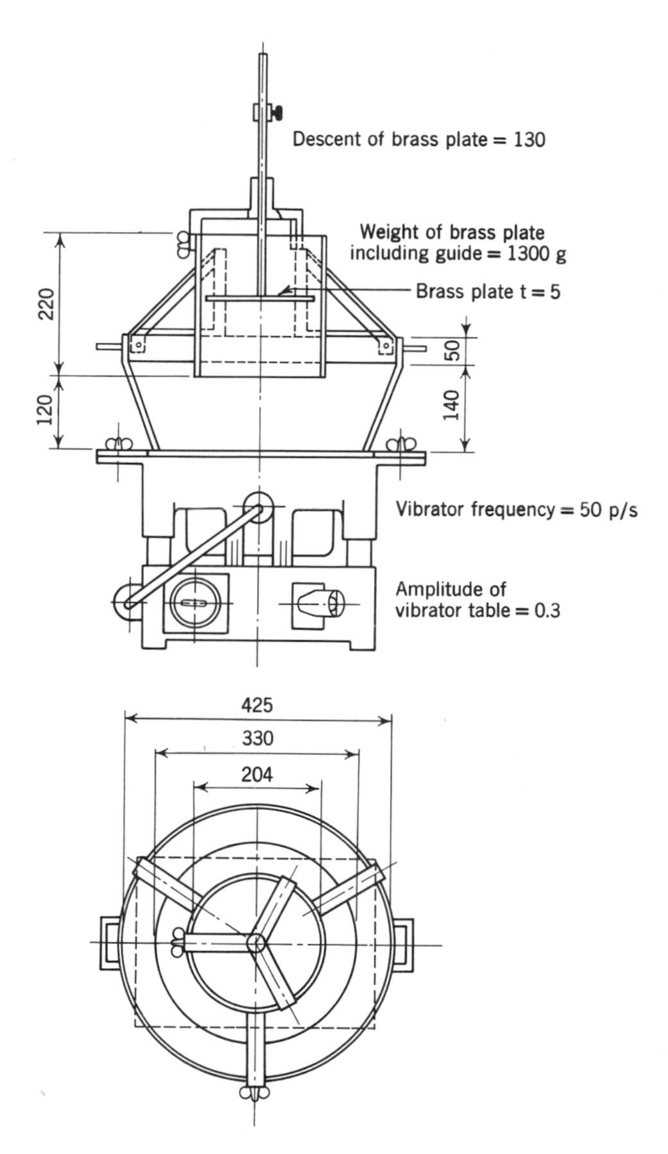

Fig. 10.44. The Hallström mobility meter. (Dimensions are in millimeters [28].)

Thus it appears that the conditions under which flow takes place has an important effect on the rating of different mixtures with respect to relative consistency, or mobility, in the fluidized state as well as in the plastic state. Probably the differences that appear when the mixtures are in the fluidized state are not as large as those that appear when in the plastic state.

8.4 The Flow Table: ASTM Designation C124

The ASTM standard flow test for concrete employs a circular metal table top 30 in. in diameter that can be repeatedly raised and dropped with a cam. The apparatus is mounted on a concrete base weighing not less than 300 lb.

At the start of the test, the concrete is molded in the form of a frustrum of a cone, with the mold held down over the center of the flow table. The mold has a base diameter of 10 in., top diameter of $6\frac{3}{4}$ in., and a height of 5 in.

As soon as the mold is removed, the table is raised and dropped 15 times in about 15 sec by revolving the cam shaft at a steady rate. The "flow" is the increase in the diameter expressed as a per cent of the original diameter, 10 in.

A mixture can have relatively high particle interference and low capacity for plastic shear without being penalized by this test. When there is a tendency for particle interference to become excessive, some of the particles are free to move into a more favorable position, some of them becoming only half embedded and thus increasing the freedom for relative movement of those remaining within the mass. It is common for mixtures having little plasticity and cohesiveness to give relatively high "per cent flow" because of separation of "soup" from the main mass during jigging. Some mixtures may show segregation of coarse aggregate, the segregated particles rolling ahead of the spreading mass during jigging. Attempts have been made to take segregation into account when interpreting the per cent flow given by the standard test [29,30].

Differences in remolding number for a given flow are shown in Fig. 10.45. The curves show that at a given aggregate grading, and with a given paste composition, the relative effort required to mold the material is about inversely proportional to the flow; but, at a given flow the remolding number is different for different gradings. There probably would be less difference if the ring in the remolding apparatus had been set higher and still less if the ring had been removed.

8.4.1 Flow table for standard mortar tests: ASTM Designation C230. The apparatus used for mortar tests is a 10-in. diameter flow table specified primarily for regulating the consistency of standard mor-

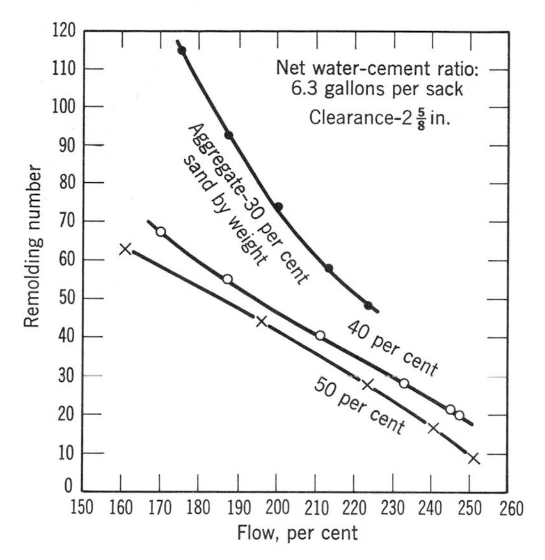

Fig. 10.45. The relation of remolding number to flow. (ASTM Designation C124.)

tar mixtures specified for tests of cement. When used for other purposes it may show the same limitations as discussed above for the large flow table.

8.5 The Slump Test: ASTM Designation C143

By means of a bottomless sheet metal mold resting on a plane rigid surface, the fresh concrete is molded into the form of the frustrum of a cone with a base diameter of 8 in., a top diameter of 4 in., and a height of 12 in. The mold is filled in a standardized manner, and then it is lifted vertically, leaving the molded concrete unsupported laterally. Under the force of gravity, the mass slumps symmetrically (more or less), increasing its average diameter as its height diminishes; or there may be a limited amount of slump with some of the upper part of the cone shearing off; or in extreme cases, the mass may crumble. The value reported is the difference between 12 in. and the height after the mold is removed.

A note in the ASTM Standard Method suggests that the slump test is properly applicable only to mixtures that slump without shearing off or crumbling. In terms of earlier discussion this may be interpreted to mean that the test should be applied only to mixtures having a sufficient capacity for plastic deformation together with a certain minimum of cohesiveness.

Without cohesiveness, an unsupported granular material will be unstable if its surface presents a slope exceeding its internal angle of friction ϕ (Section 7.1). For a dry, granular material, and presumably for an inundated one too, the angle ϕ is likely to be about 32°, and thus if its initial slope is greater than that the material will slide from the top until the proper angle of repose is established. (Actually the angle of repose may be slightly smaller than the internal angle of friction owing to special conditions at the surface.) On this basis, it can be shown that the slump of a noncohesive granular material should be about 5.6 in. Slumps of concrete greater or less than this value are possible because of plasticity and cohesion.

We have already made much use of slump data and have shown how the slump is related to water content (Chapter 4). We have also seen that the relative stiffness as indicated by slump does not seem to be closely related to the modulus of initial stiffness as measured in the plastometer. Which of the two indications of stiffness is the more closely related to the effort required for placing concrete under field conditions is, as with all other test methods, impossible to say. However, the effort required under the conditions of the remolding apparatus is closely related to slump under limited conditions, as is shown in Fig. 10.46. It seems that for very stiff mixtures, slump 1–2 in., the slump is a good

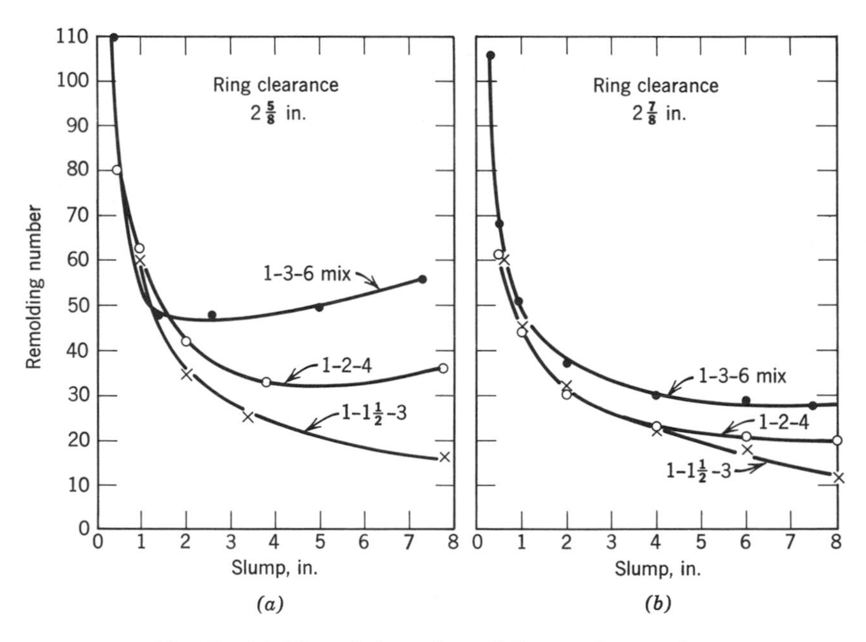

Fig. 10.46. The relation of remolding number to slump.

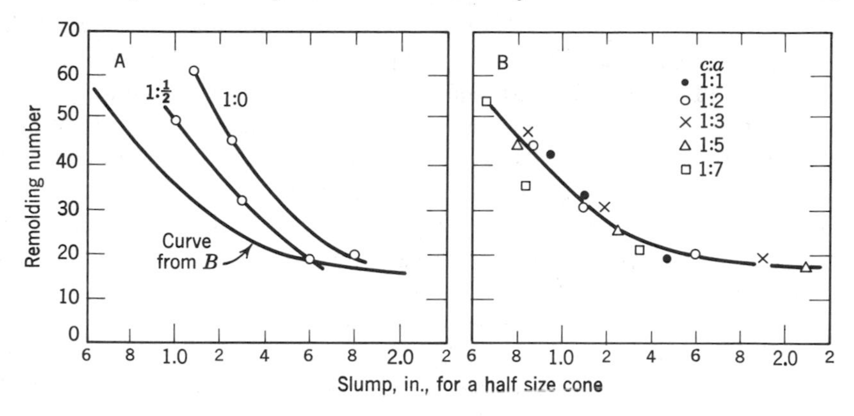

Fig. 10.47. The relation of remolding number to slump. Aggregate No. 265-11 was used. (See Table 3.1.)

indication of the amount of effort required for molding. But at higher slumps, the molding effort may vary widely or not, depending on the degree of flow restriction. It would appear that if the ring were removed completely, slump could be a good indication of the relative molding effort for these three mixes.

Another indication of the relationship between slump and work of molding is given in Fig. 10.47. These data were obtained with a half-size remolding apparatus, and a half-size slump cone, using aggregate with No. 4 maximum size; slump values are the differences between 6 in. and the height after slumping. Thus a slump of 2 in. is about equal to a standard slump of 4 in. These data show a fair relationship between slump and remolding number except for mixes 1:0, 1:½ and 1:7, weight proportions. In this connection we may recall Chapter 4, Section 3.2, where it was pointed out that the relationship between slump and water content is not the same for neat cement and very rich concrete mixtures (maximum size ¾ in.) as for ordinary mixtures, the values of S_o being different.

8.6 *Penetration Tests*

8.6.1 Rod penetration. Various devices have been developed that are based on the relative amount of energy required to cause a rod to penetrate a mass of fresh concrete. Perhaps the first of this kind was that made by Pearson and Hitchcock [31], described in 1923. The apparatus consisted essentially of a metal rod, ¾ in. in diameter and 20 in. long, held vertically by means of a spider and sleeve over the center of a

6 by 12 in. cylinder mold that was mounted on a small flow table. The concrete was placed in the mold and compacted by 30 drops of the flow table; then the rod was inserted in the sleeve and allowed to penetrate the concrete as much as it would under its own weight. Then the flow table was operated, and the number of jigs required to produce a total penetration of 10 in. was determined.

In 1928 a modification of the apparatus devised by Pearson and Hitchcock was described by Smith and Conahey [32]. They used an $8\frac{1}{2}$ by 12 in. cylinder as a sample container, and they consolidated the sample by jigging, as did Pearson and Hitchcock. Instead of a spider, sleeve, and rod assembly, a miniature pile driver was mounted on the cylinder, and the amount of mechanical work required to drive three $\frac{1}{2}$-in. rods simultaneously to a total depth of 11 in. was taken as the measure of workability.

8.6.2 Ball-penetration. A ball-penetration test consists of placing a heavy steel ball or plunger in contact with the surface of a mass of fresh concrete, releasing it, and measuring the depth of its penetration. A test of this kind was developed by Kelly [33,34] and his associates and is now known as the Kelly ball test. In 1963 it was adopted as a standard method of test. (ASTM Designation C360.)

The "Kelly ball" is a short cylinder with a hemispherical bottom end and a handle attached to the flat end; it weighs 30 lb. The distance of penetration is measured by means of a scale on the handle and a frame that rests on the surface of the concrete being tested.

Data published by Kelly and Polivka [34] indicate that ball-penetration is related to slump by an equation of the following form:

$$P = P_0 + kS$$

where P is the ball-penetration reading; P_0 is the penetration at zero slump, by extrapolation; S is slump; and k is an empirical constant. For concrete containing 5 bags of cement per cubic yard, laboratory experiments gave the following relationships:

for $\frac{3}{4}$ in. maximum size

$$P = 0.36 + 0.48S \text{ in.}$$

for $1\frac{1}{4}$ in. maximum size

$$P = 0.57 + 0.36S \text{ in.}$$

There are indications that different relationships may be found with different materials.

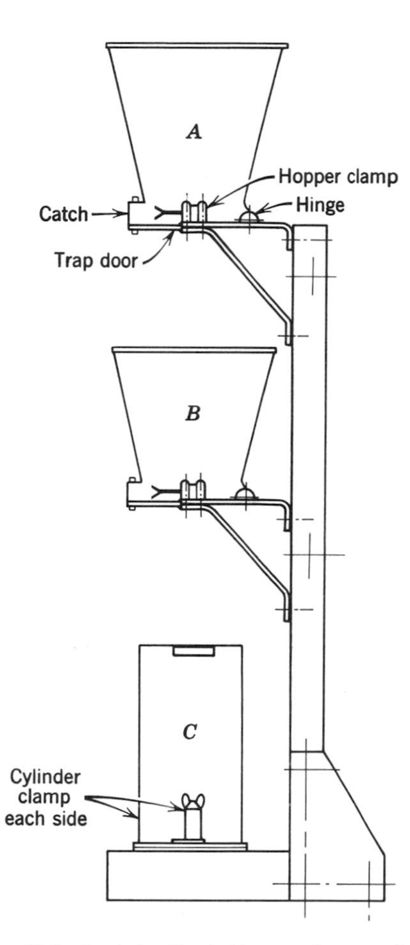

Fig. 10.48. Diagram of the compacting factor apparatus [35,36].

The principal advantage claimed for this test is that it can be used on the job, and used frequently to check uniformity of production, with relatively little expenditure of time and effort.

8.7 The Compacting Factor Test

The compacting factor test involves determining the relative density of concrete in a mold after the concrete has dropped from a standard height into the mold. The relative density is expressed as the ratio of the observed weight of the filled mold to the weight calculated from the composition of the mixture without air voids. The apparatus is illustrated in Fig. 10.48. It was developed by Glanville and his associates, and was first described in 1937 [35,36]. It has been adopted as a British Standard Test. Glanville described the procedure as follows:

"The hopper A is filled with concrete by hand from a scoop. The release of a hinged door at the bottom of A then deposits the concrete into B, a hopper with a smaller volume. Finally, the concrete is released from B into C, which is a 12-inch high $\times$ 6 inch internal diameter cylinder. C is filled to overflowing and the surplus struck off by simultaneously working two steel floats from the outside to the center" [35].

The purpose of the middle hopper is to bring the concrete to a standard state before its final fall. The final fall represents the expenditure of a certain amount of work.

According to Neville [37] the calculation of the compacting factor is sometimes based on the actual density obtained by tamping or vibrating rather than on the computed density for zero air content. As shown in Chapter 3, a state of "normal compactness" is not found at zero air content, but at an air content amounting to about $\frac{1}{7}$ the water content, more or less, depending on consistency. The same is true of concrete placed by ordinary vibration, although the air-water ratio might not be the same as for other methods. In any case a direct determination of density in the completely compacted state obviously does not give the same basis for computing compacting factor as the calculated density at zero air content.

In this connection it is worth noting that the common practice of using the actual specific gravity of cement instead of its displacement factor in water, (discussed in Chapter 2) leads to a small overestimate of volume and thus an underestimate of calculated density for an air-free mixture. The difference is not large, but it could be significant when the compacting factor is not far below unity.

Glanville pointed out that if his apparatus is to be considered a measure of workability, it must be assumed that the lower the compacting factor obtained by a fixed amount of work, the greater the amount of additional work required to complete the compaction.

As with other arbitrary tests, a given test value obtained from the Glanville apparatus does not have exactly the same significance with respect to different materials and different conditions of placement. Different mixtures having the same slump may give different compacting factors. Data for comparisons with remolding number do not appear to be available.

REFERENCES

[1] Powers, T. C., "The Bleeding of Portland Cement Paste, Mortar and Concrete," Portland Cement Association Research Bulletin No. 2, 1939.

[2] Steinour, H. H., Portland Cement Association Laboratory Report No. 288-11-D4, 1941 (unpublished).

[3] "Glossary of Terms on Cement and Concrete Technology—Increments 2, 3, and 4," *Proc. ACI,* **61,** 487–507, 1964, Report by ACI Committee 116, Walter H. Price, Chairman.

[4] *Webster's Third New International Dictionary,* G. & C. Merriam Company, Springfield, Mass., 1961.

[5] Reiner, M., *Deformation Strain and Flow,* Interscience, New York, 1960, p. 4.

[6] Bingham, E. C., *Fluidity and Plasticity,* McGraw-Hill, New York, 1922, Chapter 8.

[7] Dellyes, R., "The Rheology of Cement Slurries." *Revue des Materiaux de Construction,* Publication Technique No. 68. Centre d'Études et des Recherches de l'Industrie des Liants Hydrauliques, Paris, 1954. Translated by L. M. L. Booth, Department of Scientific and Industrial Research, Building Research Station, Library Communication No. 715, September 1955.

[8] de la Peña, C., "The Workability of Concrete and Its Measurement," RILEM Bulletin No. 8, 1952, Chapter 2.

[9] Greenberg, S. A., R. Jarnutowski, and T. N. Chang, "The Rheology of Silica Suspensions," Portland Cement Association Serial No. 1045, 1963.

[10] Papadakis, M., "Rheology of Cement Suspensions." Tech. Publ. No. 72. Centre d'Études et de Recherches de l'Industrie des Liants Hydrauliques, Paris, 1955.

[11] Ish-Shalom, M., and S. E. Greenberg, "The Rheology of Fresh Portland Cement Pastes." "Proc. Fourth International Symposium on the Chemistry of Cement," *Nat. Bur. Std. (U.S.) Monograph 43,* **2,** 731–743, 1962.

[12] Gaskin, A. J., "Discussion of 'The Rheology of Fresh Portland Cement Pastes,' by M. Ish-Shalom and S. E. Greenberg," "Proc. Fourth International Symposium on the Chemistry of Cement," *Nat. Bur. Std. (U.S.) Monograph 43,* 744–748, 1962.

[13] Green, H., *Industrial Rheology and Rheological Structures,* Wiley, New York, 1949, p. 164.

[14] Bruere, G. M., "Mechanisms by Which Air-Entraining Agents Affect Viscosities and Bleeding Properties of Cement Pastes." *Australian J. Appl. Sci.,* **9**(4), 349–359, 1958.

[15] Bruere, G. M., "Air Entrainment in Cement and Silica Pastes," *Proc. ACI,* **51,** 905–919, 1955.

[16] Stolnikov, V. V., "Investigation of the Elasto-Plasto-Viscous Properties of Cement-Water Pastes," *Izv. Vses. Nauchn.-Issled. Inst. Gidrotekhn.* (*Bull. Joint Sci. Res. Inst. Hydroelectric Eng.*), **41,** 98–109, 1949.

[17] Powers, T. C., and E. M. Wiler, "A Device for Studying the Workability of Concrete," *Proc. ASTM,* **41,** 1003–1015, 1941.

[18] L'Hermite, R., and G. Tournon, "Vibration of Fresh Concrete," Technical Publication No. 2. Centre d'Etudes et de Recherches de L'Industrie des Liants Hydraulique, Paris, 1948.

[19] Desov, A. E., "Structural Viscosity of Cement and Concrete Mixes," *Kolloid. Zhur.,* **13**(5), 346–356, 1951.

[20] L'Hermite, R., "Vibration and Rheology of Fresh Concrete," *Rev. Mater. Construc.* (405), Centre d'Études de Recherches de l'Industrie des Liants Hydraulique, Technical Publication No. 14, 1949.

[21] Ritchie, A. G. B., "The Triaxial Testing of Fresh Concrete," *Magazine of Concrete Research,* **14**(4), 37–41, 1962.

[22] Bishop, A. W., and D. J. Henkel, *The Traixial Test,* Edward Arnold, London 1957.

[23] Fulton, F. S., *Concrete Technology, A South African Hand Book,* Portland Cement Institute, Kew Road, Richmond, Johannesburg, 1961.

[24] Powers, T. C., "Studies of Workability of Concrete," *Proc. ACI,* **28,** 419–448, 1932.

[25] Bahrner, V., "New Swedish Consistency Test Apparatus and Method," *Betong,* **1,** 27–38, 1940.

[26] Thaulow, S., *Field Testing of Concrete,* Norsk Cementforening, Oslo, 1952.

[27] Wuerpel, C. E., "Vibratory Remolding Test as a Measure of Concrete Workability," *Proc. ACI,* **40,** 70–75, 1943.

[28] Eriksson, A. G., "Development of Fluidity and Mobility Meters for Concrete Consistency Tests," *Proceedings No. 12,* Swedish Cement and Concrete Institute, Stockholm, 1949.

[29] Smith, G. A., and G. Conahey, "A Study of Some Methods of Measuring Workability of Concrete," *Proc. ACI,* **24,** 24–42, 1928.

[30] Williams, G. M., "Admixtures and Workability of Concrete," *Proc. ACI,* **27,** 647–653, 1931.

[31] Pearson, J. C., and F. A. Hitchcock, "A Penetration Test for Workability of Concrete," *Proc. ASTM,* **23,** 276, 1923.

[32] Smith, G. A. and G. Conahey, "A Study of Some Methods of Measuring Workability of Concrete," *Proc. ACI,* **24,** 24–42, 1928.

[33] Kelly, J. W., and N. E. Haavik, "A Simple Field Test for Consistency of Concrete," *ASTM Bull.* (163), 70, 1950.

[34] Kelly, J. W., and M. Polivka, "Ball Test for Field Control of Concrete Consistency," *Proc. ACI,* **51,** 881–888, 1955.

[35] Glanville, W. H., "Grading and Workability," *Proc. ACI,* **33,** 319–326, 1937.

[36] Glanville, W. H., A. R. Collins, and D. D. Matthews, "The Grading of Aggregate and Workability of Concrete," *Road Research Technical Paper No. 5,* Department of Scientific and Industrial Research and Ministry of Transport, London, Her Majesty's Stationary Office, 1938.

[37] Neville, A., *Properties of Concrete,* Wiley, New York, 1963, p. 179.

[38] Bruere, G. M., and J. K. McGowan, "Synthetic Polyelectrolytes as Concrete Admixtures," *Australian J. Appl. Sci.,* **9**(2), 127–140, 1958.

11

Settlement, Bleeding, and Shrinkage

1. INTRODUCTION

In this chapter we consider changes that occur in concrete after mixing and placing or molding, and before the cement paste begins to set. As we have seen in earlier chapters, the aggregate in plastic concrete is dispersed by the paste to one degree or another; and the particles in the paste are dispersed in the water, their spatial distribution being determined to some degree by the forces of interparticle attraction and repulsion that are discussed in Chapter 9. This dispersed and suspended state of all the particles in plastic concrete is created during the process of mixing.

The suspension in its initial state is not a stable one. Being denser than water, the particles experience a downward force that tends to make them settle, and, in certain zones yet to be discussed, to reduce the interparticle distance, principally in the vertical direction. Wherever the interparticle distance is being reduced, a point will be reached where the combined forces of electrostatic repulsion and increased disjoining pressure of adsorbed water will become equal to the unbuoyed weight of the particles, and settlement ceases. At this stage the particles are virtually in contact but not actually so. Eventually, the small interparticle gaps become bridged with the products of chemical reaction which are called cement gel; indeed, under suitable conditions, the original cement grains become almost if not entirely replaced by cement gel, but we are not concerned with that transformation here.

Because of the settlement just mentioned, the final volume of hardened concrete is usually less than the volume immediately after placement. Consequently, settlement is sometimes called shrinkage. However, it is better to restrict the term shrinkage and another term, *volume change,* to changes of volume that are not due to gravitational forces. That is to say, it is better to restrict these terms to three-dimensional changes that come about as a result of temperature change or of certain internal changes that affect the overall volume, particularly, changes in hydrostatic tension or film tension. We shall see further on that a mass of fresh concrete does undergo expansion or contraction or both, in addition to settlement.

If freshly placed concrete is in a mold or form, settlement of the solids leaves a layer of water over the surface. Indeed, the settlement is usually so slight that it would hardly be noticed were it not for the appearance of water. The water seems to have exuded from the plastic mass, and thus the phenomenon came to be called *bleeding.*

Bleeding may take place gradually by uniform seepage over the whole surface. In addition to general seepage, a number of localized "pipes" or channels from the interior to the surface may sometimes develop, from each of which water flows with sufficient velocity sometimes to transport small solid particles and build up miniature craters around the mouth of each of the channels. The formation of channels and craters is characteristic of lean mixtures and wet consistency. It signifies a need for corrective measures. On the other hand, when bleeding occurs by uniform seepage only, the composition of the mix is thereby indicated to be within reasonable bounds, and the bleeding itself is not necessarily undesirable. Such bleeding is called *normal bleeding.*

Casual observation sometimes gives the impression that during bleeding water rises above the level of the original surface. For example, in the laboratory a mold full of concrete will appear to overflow. Such a phenomenon is not a direct result of the bleeding process, however, for bleeding is due to settlement of solids, as described above, and that cannot account for any elevation of the surface above the original level. When water runs off the top of a level surface the phenomenon may be due to either or both of two causes: Rising temperature may cause air bubbles in the mixture to expand and thus raise the fluid level, or if the original screeded or trowelled surface was left slightly higher than the level of the mold, a common occurrence, settlement of solids thickens the water film at the surface so much that surface tension can no longer prevent the water from running off.

It is necessary to make a distinction between settlement and bleeding because settlement does not always result in an accumulation of water at the surface. The rate of bleeding is ordinarily greater than the rate of evaporation from the surface, but, on warm, windy days it may not be. The rate of bleeding from a paving mixture, for example, may become smaller than the rate of evaporation; and when it does so, bleeding can not be seen, although, as will be seen further on, the rate and amount of settlement may be greater than normal. Besides loss of water by evaporation there also may be an internal loss through water-absorption by unsaturated aggregate particles. Moreover, if the concrete is on the ground, some water may be absorbed from the mixture by the subgrade, although normally the subgrade is dampened beforehand to avoid this. (Under laboratory conditions, settlement is almost always accompanied by visible bleeding, unless the rate of evaporation is artifically increased.)

There are two aspects of bleeding to be considered: the rate at which it occurs, and the total amount. For cement pastes, the rate of settlement and the amount are about proportional, but for concrete mixtures they may be quite independent. As we shall see below, the rate of settlement is governed by the coefficient of permeability of the mixture, whereas the amount of settlement is governed mainly by the degree of dispersion of the aggregate.

The rate of bleeding under conditions where capillary tension cannot develop can be conveniently divided into two stages: a stage at which bleeding occurs at a constant rate, followed by a period of diminishing rate. The conditions which determine the length of the period of constant rate are various, and are brought out further on.

The total amount of settlement tends to be proportional to the depth of the freshly placed mass, and the amount of settlement per unit of depth tends to be constant for a given mixture. The amount of unhindered settlement divided by the depth of the mass is called the *bleeding capacity*.

The amount of bleeding is sometimes expressed as the total loss per unit amount of cement in the mixture, or per unit of water originally in the mixture. Such figures are of doubtful significance. The nature of the phenomenon is such that the loss of water is not the same from the upper and lower parts of the mass. Under some circumstances, the water content of the upper part may remain unchanged, all the loss being from the bottom part. The various factors determining these relationships are discussed further on.

The termination of settlement and bleeding occurs when the process is arrested mechanically, as described above. Or it occurs at the end of the dormant period as discussed in Chapter 10, Section 1.1.1, which is to say, settlement can be arrested by a bridging of the narrowest interparticle gaps with hydration products. However, if the frail structure developed this way is destroyed by remixing or by vibration soon after it is formed, settlement and bleeding will recur.

We should note in passing that the destruction of structure in the early stages of hardening does no harm to the ultimate quality of the concrete. On the contrary, concrete can be improved by such a process as long as the mixture can be completely reconsolidated.

The practical consequences of settlement and bleeding are discussed in Section 6. A related phenomenon, the so-called "plastic shrinkage," is also discussed. Finally in Section 6 we discuss post-bleeding expansion.

1.1 Reasons for the Detailed Study of Bleeding

The systematic experimental study of settlement and bleeding was carried considerably farther than would have been needed only to establish the phenomenological aspects of the subject. Such study yielded important contributions to our understanding of properties of concrete not only in the freshly mixed state, but also in the hardened state. It originally gave evidence of the peculiarities of the flocculent state, particularly the ability of cement paste to show a degree of firmness, and at the same time show evidence of the freedom of individual particles.

Studies of settlement and bleeding revealed the nature of relatively gross flaws that occur in concrete, some visible, some not. They led also to a concept of the microscopic structure of cement paste, which structure is the pattern from which the structure of hardened paste develops. For example, it has been found that the permeability of hardened cement paste to the flow of water under pressure may differ for different hardened pastes even though their densities and the average hydraulic radii are the same. Specifically, it was found that with pastes having a water cement ratio of 0.71, one made with a cement having an ordinary specific surface area had a coefficient of permeability of 1100×10^{-12} cm/sec, whereas another paste having exactly the same density, but made with a cement having a much higher specific surface area, had a coefficient of permeability of 18×10^{-12}, only 1.6 per cent of the first value [1]. At the same time, it was established that the amount and specific surface area of cement gel were the same in each of the two specimens. It was concluded that the permeabilities were different because the distributions of cement gel within the same volume of space were different. The finer cement contained a relatively larger number of cement particles, cement particles being the sources of cement gel; consequently the average distance between cement particles was smaller in the finer cement. Thus, of the two subsequently developed gel structures that formed from the finer cement had the higher degree of continuity and lower degree of permeability. This is believed to be only an example of how the fresh-paste structure influences the hardened-paste structure, although it is probably a somewhat extreme example from a relatively unexplored area.

Because of such connections with the practical as well as the theoretical aspects of concrete technology, we shall consider the results of studies of settlement and bleeding in detail.

2. MEASUREMENT OF SETTLEMENT AND BLEEDING

Early attempts to measure bleeding led to enigmatic results because the methods of measurement used then influenced the amount and rate of settlement. Finally, a method was developed that was practically free of this defect (2).

2.1 *The Float Method*

The float method of measuring settlement and bleeding involves measuring the subsidence of a small central area of the top of a sample. The float is a disk of bakelite on which a straight glass fiber is mounted like a mast. The disk is about $\frac{1}{2}$ in. in diameter and $\frac{1}{8}$ in. thick.

The sample to be measured is contained in a cylindrical flat-bottom vessel. If the sample is cement paste, the paste must be vigorously stirred with a mechanical mixer before pouring it into the test vessel, the mixing schedule being such as to prevent the occurrence of false set. (See Chapter 10, Section 1.2.2.)

A satisfactory setup for measuring the bleeding of cement paste is shown in Figs. 11.1 and 11.2. The lower photograph shows the top of the sample with a float installed at the center. Because of the thixotropic structure of cement paste (see Chapter 9), the float remains in a fixed position relative to the top of the paste during settlement. As soon as the float is installed, the sample is covered with a layer of water to prevent possible development of capillary forces if the rate of evaporation should exceed the rate of bleeding, as it may under some conditions.

After the sample is prepared and the float installed as indicated above, the movement of the float is measured with a micrometer microscope, as is indicated in the upper photograph of Fig. 11.1. Time is measured with a stopwatch.

For concrete, the method described above is used, except that the sample is much larger. A cylindrical steel tub or pot with vertical sides and a diameter of at least 20 in. is satisfactory for obtaining intrinsic bleeding characteristics.

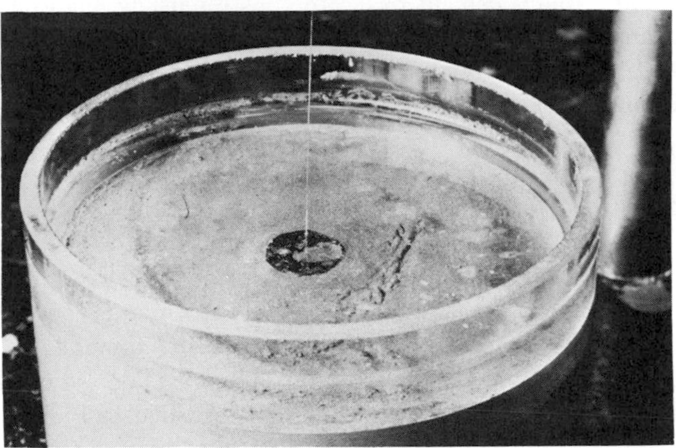

Fig. 11.1. Apparatus for measuring the bleeding of paste. Before beginning measurements, the sample is covered with a layer of water about $\frac{1}{4}$ in. deep.

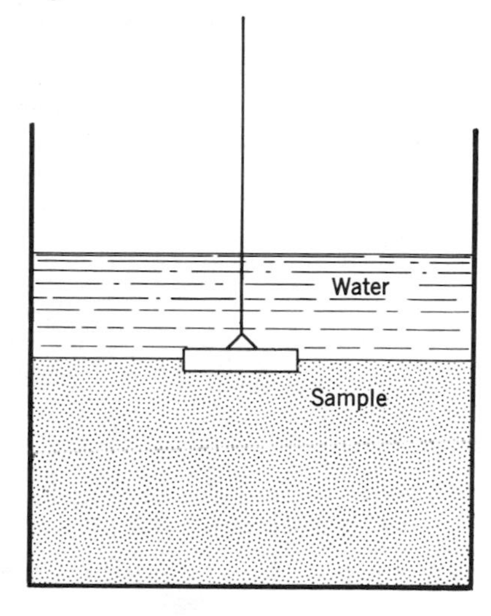

Fig. 11.2. Diagram of bleeding-test float seated in the surface of a flocculent sample.

2.2 *Bleeding Curves*

Figure 11.3 is a bleeding curve obtained from a sample of cement paste that was in a vessel 4.25 cm deep and 10 cm (4 in.) in diameter. The curve is made up of two sections: a straight line that indicates a period of constant rate of settlement and bleeding, followed by a period of diminishing rate, a zero rate being reached within 60 min in this case.

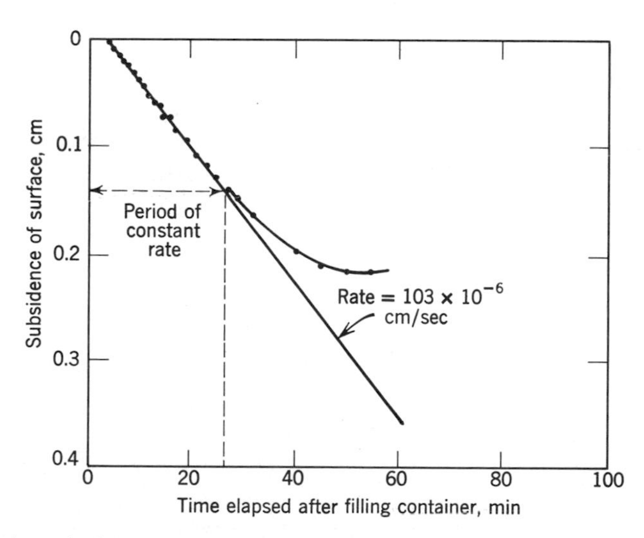

Fig. 11.3. A typical bleeding curve unmodified by wall effects or setting. Cement No. 14502 was used with $w/c = 0.41$. The depth of the sample was 4.25 cm.

The curve shown in Fig. 11.3 is typical of all curves representing normal bleeding that is not modified by some faulty feature of measuring technique, or by the influence of the wall effect that is discussed further on.

Figure 11.4 gives bleeding curves obtained from concrete mixtures. The characteristics of these curves are generally the same as those of the curve for cement paste, although there are certain variants that are discussed further on. Generally, the period of constant rate is relatively short as compared with that for paste. Sometimes the period of constant rate is so short that it is nearly finished before the first reading can be obtained.

For example, in one of the diagrams of Fig. 11.4, there are two curves, *A* and *B*. Although the curves represent the same mixture, they seem to indicate different initial bleeding rates. This is an example of how experimental results can be deceiving. Notice that curve *B* represents a sample that was 23.7 cm (9.3 in.) deep and that it showed a constant bleeding rate of 76 millionths cm/sec for about 16 min; thereafter the rate diminished. Since the length of the constant rate period is proportional to the depth of the sample, the length of the constant rate period for sample *A*, 9 cm deep ($3\frac{1}{2}$ in.) would be $(9/23.7) \times 16$ min = 6.1 min. Looking at curve *A*, we see that the first reading was obtained after about 6 min of bleeding, which means that the period of constant rate had already ended.

2.2.1 The wall effect. The rate of settlement at the periphery of a sample is less than the rate at the center. This was demonstrated experimentally by means of the setup shown in Fig. 11.5. A hydrometer-like float was seated in the center of the top of the paste, mostly submerged in water that was overlayed with kerosine to prevent evaporation and chemical reaction of the lime water with carbon dioxide. The downward movement of the float was measured with a micrometer microscope, and it was indicated also by the short stem which was withdrawn below the surface of the kerosine as the float descended and the layer of water became deeper. By opening the cock on the siphon periodically, just enough water would be withdrawn from the layer of water to restore its original depth, that is, thickness. In this way the rate of subsidence at the center of the surface was compared with the average rate of subsidence of the whole surface, which is $cm^3/cm^2/sec$, or cm/sec.

Results showed that the average rate of descent for the whole surface was at first always less than the rate of descent of the float at the center. For example, a sample that showed a rate of 50×10^{6} cm/sec at the center had an average rate for the whole surface of 42×10^{-6}.

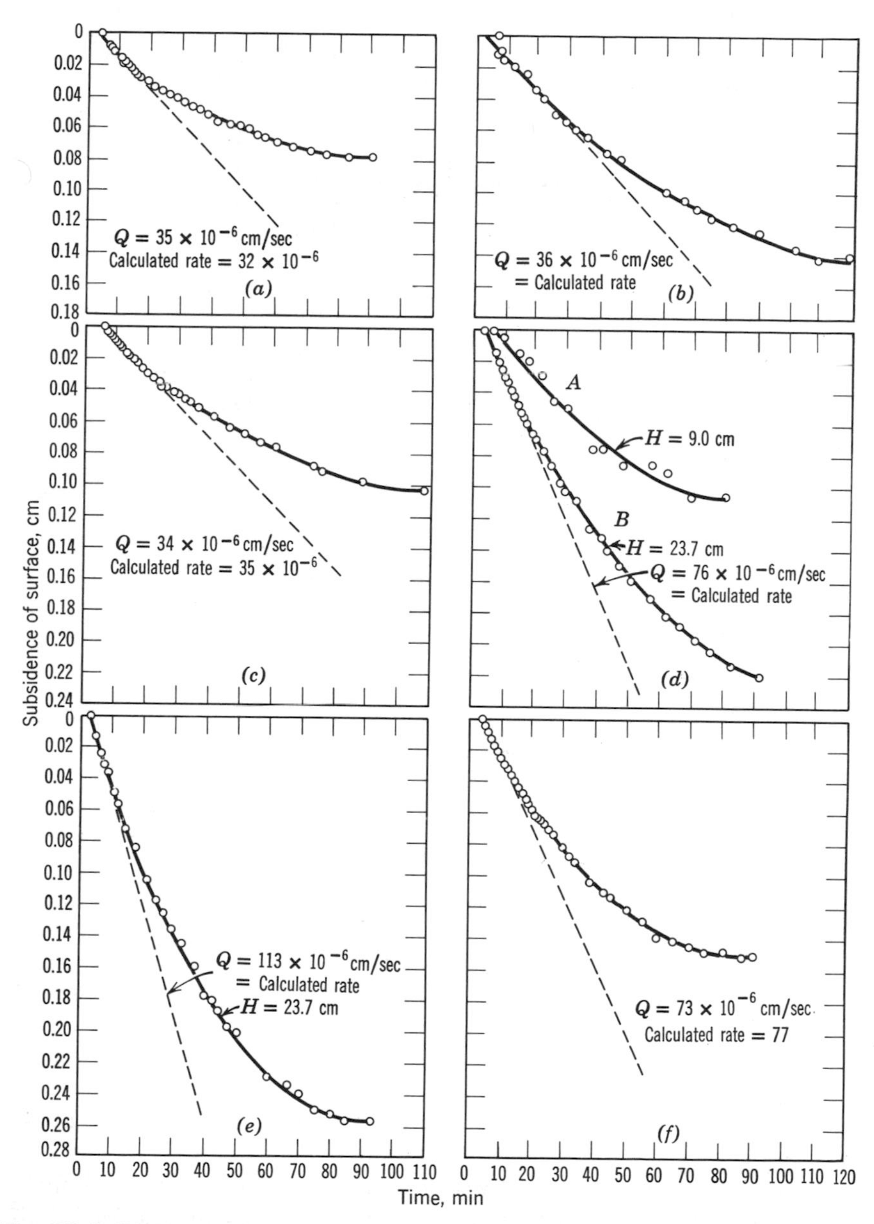

Fig. 11.4. Bleeding curves for concrete mixtures. Q is the initial rate of bleeding. (a) The mix by weight was 1:1.08:1.2; w/c was 0.31. (b) The mix by weight was 1:1.12:1.8; w/c was 0.38. (c) The mix by weight was 1:1.6:2.4; w/c was 0.40. (d) The mix by weight was 1:1.9:2.85; w/c was 0.49. (e) The mix by weight was 1:2.4:3.6; w/c was 0.57. (f) The mix by weight was 1:2.4:3.6; w/c was 0.52.

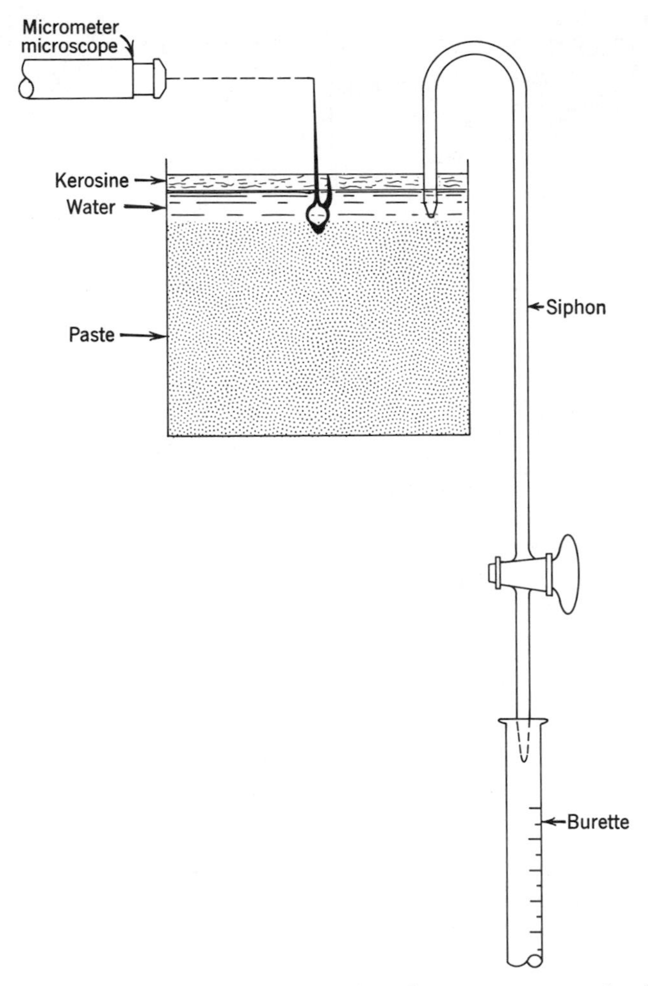

Fig. 11.5. Experimental setup for comparing the average rate of subsidence of the whole surface with the rate of subsidence at the center.

This particular difference, of course, depended on the diameter of the vessel.

In one experiment, the vessel was coated with collodion, and in another with paraffin, to modify forces of adhesion. No effect was noticed, indicating that lagging at the periphery is not due to adhesion.

To understand why the peripheral rate of subsidence is lower than the rate at the center, or lower than the average rate for the whole surface, we should consider first conditions in the interior of the mass.

We shall see further on that as the particles settle through the water, the water flows around the particles individually, rather than around clusters of particles. Hence during the downward settlement we may imagine a pattern of flow lines determined by the positions and sizes of the solid particles. The pattern of flow around an individual particle as it moves downward at a certain rate is determined in part by the circumstance that all the neighboring particles are moving at the same rate; there is as a consequence a certain symmetry of flow lines around individual particles. But, the same symmetry of flow lines cannot exist around a particle that is next to a wall, partly because the wall is smooth and vertical, but mainly because it does not move downward with the particle. A peripheral particle, therefore, experiences a greater viscous drag than does a similar particle in the interior, and it therefore tends to fall at a lower rate.

Since the fall of a peripheral particle is retarded, the next particle inward must also be retarded; the symmetry of the flow pattern mentioned above becomes altered, and the viscous drag thereby increases; but since the peripheral particle still moves, the retardation of the inner particle is not immediately as much as that of the peripheral one. Probably, the retardation of a given vertical layer of particles is due not only to the lower rate of fall of the adjacent layer but also to a reduction of interparticle distance. The difference between the rates of fall produces strain in the paste structure, and such a strain in a structure composed of randomly arrayed particles tends to produce virtual contact between particles. The same process extends from particle to particle (or layer to layer) inward until all the particles have acquired the peripheral rate.

The phenomenon described above manifests itself when paste is allowed to settle in a container that is narrow and deep. The rate of settlement at the center is at first the normal rate for the composition of the mixture, but after a certain time the rate diminishes to the peripheral rate, and that rate is maintained until the end of the dormant period if the sample is sufficiently deep. Figure 11.6 illustrates the phenomenon. The slope of the straight line OA is the rate of settlement at the wall, and that of OB is the normal initial rate at the center. At the time t_1 the slope abruptly changes to that of line BC, which is the same as the slope OA. Finally, the period of gradually diminishing rate begins at time t_2.

It is not always easy to distinguish between point B of Fig. 11.6 and the normal end of the constant rate period as seen in Fig. 11.3. However the occurrence of the break at B can be forestalled by using a sample diameter greater than the depth.

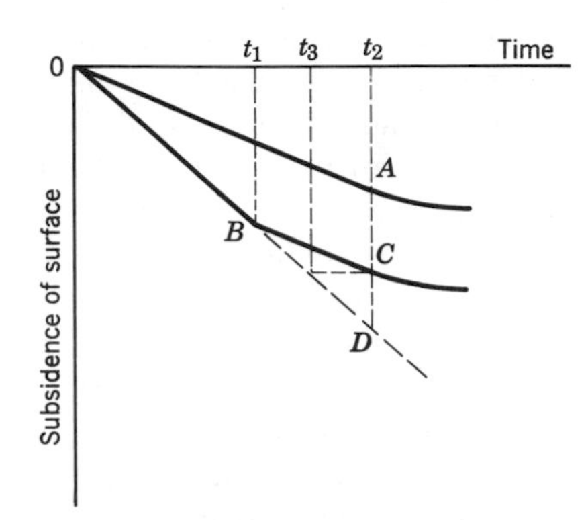

Fig. 11.6. Diagram of the wall effect on a bleeding curve.

2.3 *The Valore, Bowling, and Blaine Method (ASTM Designation C 243)*

In 1949 an apparatus (Fig. 11.7) was described for automatically collecting the water that appears at the surface of a sample of paste or mortar [3]. (The method was adopted as an ASTM Standard Method of Test in 1965.)

Figure 11.7 is a diagram of the apparatus. The cylindrical part of the container is filled with the sample. Then the collecting ring is put in place, supported so that it penetrates the sample as shown. The upper part of the container ("outer ring") is nearly filled with carbon tetrachloride, a liquid denser than water and immiscible with it. The burette and collecting funnel is then supported in the position shown, and filled with carbon tetrachloride by means of the aspirator.

As the surface of the sample settles, it releases water that rises through the carbon tetrachloride to the top of the burette where it can be measured. Such measurements are made at regular intervals until settlement stops.

The water is collected only from the area defined by the collecting ring, and therefore the rate observed is the average rate of settlement for that area. Presumably, the result is influenced by the ring, through its wall-effect, as discussed above.

2.4 *ASTM Method for Concrete (Designation C 232)*

The ASTM method of testing for bleeding of concrete requires a flat-bottom cylindrical container having an inside diameter of 10 in. and a height of 11 in. The container is filled to a depth of 10 in. in a specified manner, and the rate and amount of bleeding is then measured as follows. The container is allowed to set on a level surface for 10 min, reckoning from the time filling was complete. Then it is gently

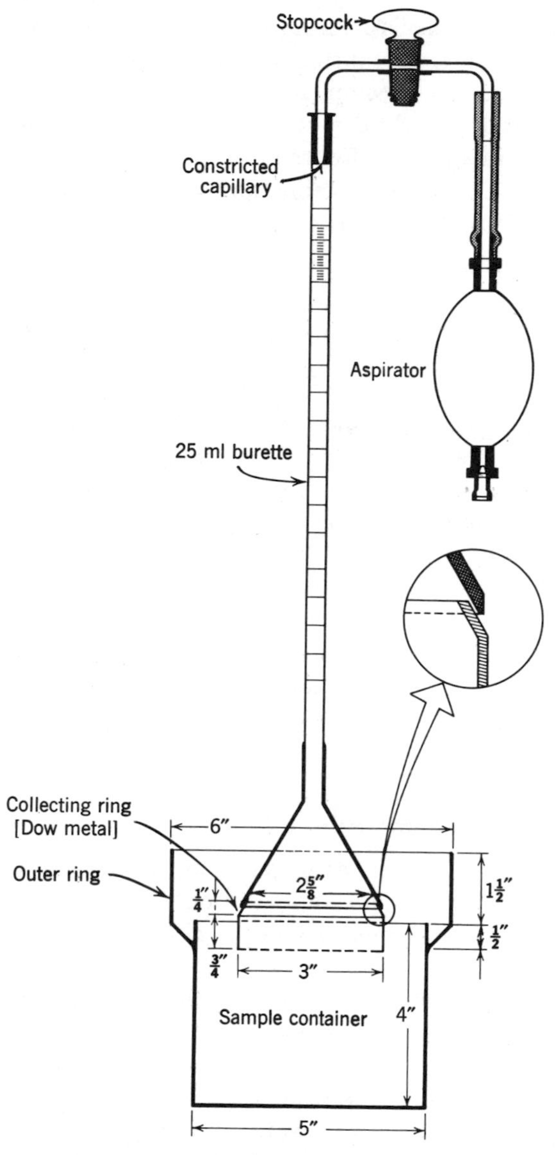

Fig. 11.7. The Valore, Bowling, and Blaine apparatus for measurement of bleeding. Reprinted by permission of the American Society for Testing and Materials.

tilted and held by a 2 in. block while the accumulated water is drawn off and transferred to a 100 ml graduate by means of a pipette. The container is then returned to its horizontal position. This procedure is repeated at 10-min intervals during the first 40 min, and then at 30-min intervals until settlement is complete.

By following the procedure described above, it is possible to obtain the average rate of bleeding for each 10-min interval and the average total bleeding. The test is suitable for comparing different mixtures, but it is not sufficiently accurate for making quantitative studies of the phenomenon.

3. SEDIMENTATION ZONES (COMPRESSION ZONES)

Settlement indicates that the granular material is becoming more consolidated in some manner and degree. Consolidation begins at the bottom, the consolidated part becoming progressively deeper as settlement proceeds. The consolidated portion of a settled mass is called the *sedimentation zone* or the *compression zone.*

The sedimentation zone may build up in a simple way, or the process may be complicated by fixed objects within the mass, by wall effects, or by other variations in boundary conditions. Such variations are perhaps the most important aspect of settlement and bleeding from the practical point of view, as is indicated in the introduction. However, we shall first consider the development of sedimentation zones under the simplest conditions.

3.1 Settlement of a Semi-infinite Body

Observation of the settlement of paste or concrete at a point remote from the nearest boundary gives information applicable to a body having finite thickness but extending laterally without limit—a semi-infinite body. A curve such as that shown in Fig. 11.3, although obtained from a small sample, does give the intrinsic characteristics of the mixture; a curve such as *OBC* in Fig. 11.6 does not, unless corrections are applied. The prerequisite is that the settlement at the center of the sample can be completed before wall effects reach the center.

Data virtually free from the wall effect show that the initial rate of settlement of a given paste is independent of the depth of the sample, H. This must mean that the amount of settlement at a constant rate is directly proportional to the depth of the sample, as is shown in Fig. 11.8*a*. In this case, the amount of settlement at a constant rate was proportional to depth for depths not exceeding 12 cm. For greater depths

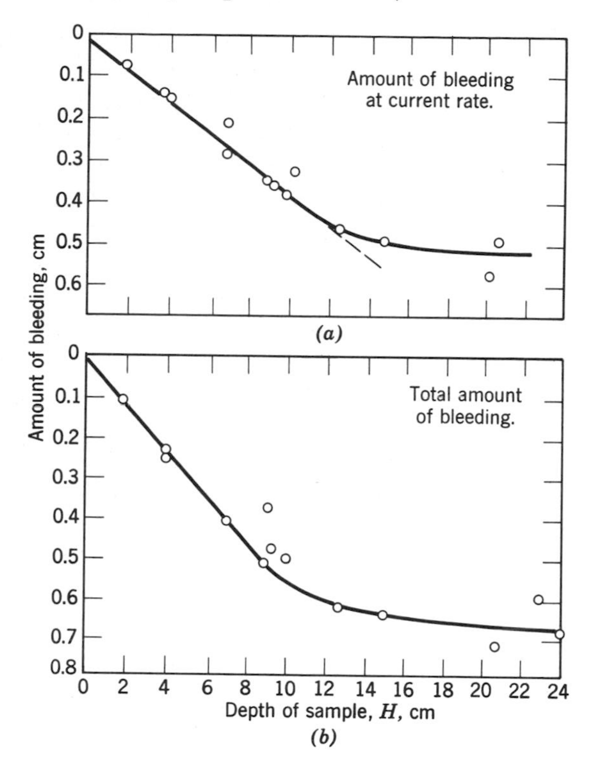

Fig. 11.8. Amount of bleeding versus the depth of sample. The value of w/c for the mixture used is 0.42. (*a*) Amount of bleeding at a constant rate. (*b*) Total amount of bleeding.

the settlement was not proportional because the amount of time required exceeded the period during which the paste remained free from stiffening due to the beginning of setting; that is to say, the time required exceeded the dormant period of the paste. For the paste in Fig. 11.8 the dormant period was about 1 hr. It might be different for different cements or different ambient conditions.

In Fig. 11.8*b* we see that the total amount of settlement is also proportional to the depth of the sample. In this case the greatest depth for proportional settlement is about 9 cm, indicating that the average rate for the whole process was about $\frac{9}{12}$ of the initial rate.

The observation that samples having different depths settle at the same rate indicates that in a sample 10 cm deep, for example, the settlement at the 5 cm level, or any other lower level, is for a limited time going on at the same rate as that observed at the top. It shows also that the particles at a given level do not bear any of the weight of the particles above them during the constant rate period.

We can imagine a mass of cement paste to be composed of identical layers of particles, the particles being held in a net of interparticle forces of attraction and repulsion already described (Chapter 9). Just as the particles in a given layer are separated from each other by water, the imaginary layers are separated from each other. For our present purposes, we consider the layers to be horizontal, although not necessarily bounded by planes. The network of particles in each layer constitutes a permeable body denser than water, and thus it is one that tends to fall through the water at a characteristic rate. If the mass is semi-infinite, this can occur without distortion of the layer. Because all layers are identical, they settle at the same rate, and thus at first a given layer cannot overtake the one below it. From what has already been said in connection with Fig. 11.8, it is apparent that such a relationship between layers can exist only for a limited time, unless the mass has unlimited depth and the settlement is not arrested by setting.*

If we could measure the rates of settlement at lower levels simultaneously with the rate of settlement at the top, we could obtain a diagram such as that shown in Fig. 11.9. The column at the left represents a section of a virtually semi-infinite sample having a depth of 16 cm and that at the right the same section 140 min after the beginning of settlement, the change in depth being plotted to the right-hand scale, which is 10 times as large as the scale at the left.

The line $0A$, in Fig. 11.9, passes through the points marking the termination of the period of constant rate of settlement. As shown, the duration of the period of constant rate for any "layer" within the specimen is proportional to the distance of the layer from the bottom of the specimen. The slope of the line gives the rate of rise of the sedimentation zone, or compression zone, within the specimen, in this case about 1 mm/min. The amount of settlement at a constant rate, per unit of initial depth is of course the same at all points along the line, in this case 0.48 cm/12 cm = 0.04 cm per cm of depth, which is *the bleeding capacity at constant rate* for this specimen. The settlement process after the end of the period of constant rate is somewhat more complicated.

For a given sample, the end of the constant rate period is the time

* In this connection, we should not confuse the beginning of setting with the standard time of initial set. The latter is the time at which a stiff paste (normal consistency) reaches an arbitrarily selected degree of firmness; it is reached normally after about the third hour. The actual process of setting begins much earlier than that.

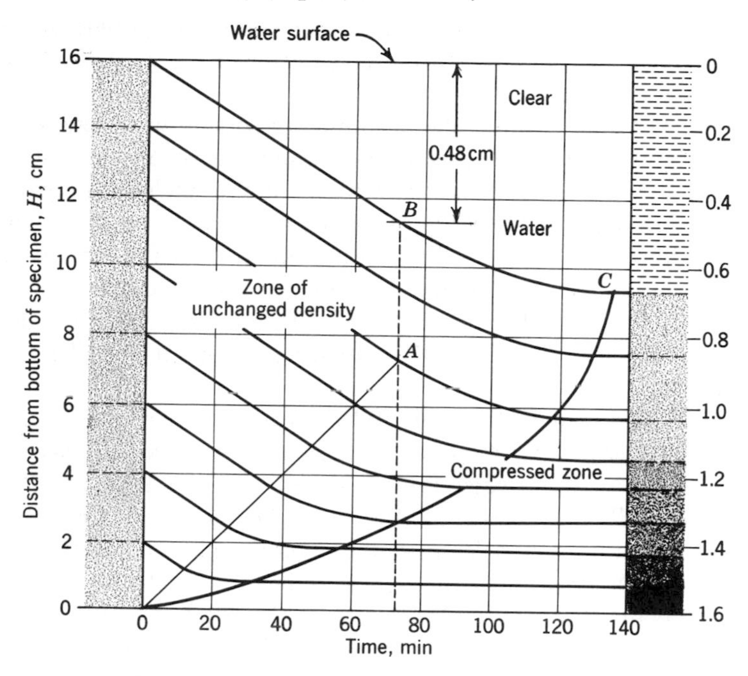

Fig. 11.9. Development of sedimentation (compression) zones in a semi-infinite sample. Settlement curves are plotted to the right-hand scale. The bleeding rate of the paste was 108×10^{-6} cm/sec.

when the top layer of particles begins to overtake the layer below it, or it is the time when the beginning of the setting process begins to retard the settlement. For the sample represented in Fig. 11.9, line $0A$ terminates about 72 min after the start, which means that point A marks the end of the dormant period. The diagram thus shows that only the layers not over 12 cm from the bottom were able to complete the constant-rate portion of total settlement before the beginning of setting; the bleeding capacity at constant rate for the sample as a whole was $0.48/16 = 0.03$ cm per cm of depth, as compared with 0.04 for points below the 12 cm level.

The curve $0C$, in Fig. 11.9, indicates approximately the total duration of the settlement process, and thus points along it indicate the total amount of settlement for the respective levels within the specimen. This line shows that the total settlement was completed within the dormant period only at levels not more than 6 cm above the bottom. At levels between 6 and 12 cm, some of the post-constant-rate settlement occurred unhindered by setting, and the rest at a rate controlled in part by setting; at levels above 12 cm the setting process presumably controlled the rate of settlement.

It might be supposed that the beginning of setting would arrest settlement suddenly so that the settlement would cease at the end of the dormant period. Experimental observations make it clear, however, that setting is a gradual process giving bleeding curves such as those indicated to the right of line *AB* in Fig. 11.9. But the curves drawn in this region are only estimates, available data being insufficient to show how the shape differs from that of a curve that terminates within the dormant period. The locus of the curve 0*C* beyond the dormant period is likewise uncertain.

As mentioned above, the end of the period of constant rate is the time when the top layer of particles begins to overtake the layer below it. It is also the time when the weight of the top layer of particles, originally totally supported by the fluid, begins to be transferred to the paste structure. On reflection, it becomes apparent that at the time when the weight transfer of the top layer begins, the transfer is already at more advanced stages at lower levels, the lower the level the more nearly complete the transfer. The combined effect of the densification of structure and the transfer of weight from fluid to the paste structure is to decelerate the rate of fall of the particles; hence the graphs become curved to the right of line 0*A*; finally they become horizontal indicating a state of rest.

In the final state of rest, the unbuoyed weight of the solid material is transferred from particle to particle to the bottom of the container. However, we have seen in Chapter 9 that such forces cannot be transmitted by direct solid-to-solid contacts because of electrostatic repulsion backed up by the disjoining pressure of adsorbed water that develops between solid surfaces when they are brought near to each other. Thus the proximity of surfaces almost in contact depends on the magnitude of the force tending to bring the surfaces together, and the magnitude of that force in the final state of rest depends on the depth of the overlying material.

From the discussion above we would expect that the average density of a sediment depends on the depth of the sediment and that there should be a density gradient from the bottom to the top of the sediment. Direct measurements of the densities of slices cut from hardened specimens show that such is indeed the case; they show, moreover, that the density gradient is linear or nearly so, indicating that the degree of compression below a given level is proportional to the weight of the material above that level. Of course, the compressibility may be due to partial collapse of structure as well as to the diminution of gaps between points of near contact.

On the basis of considerations given in the foregoing paragraphs, we are able to distinguish the following zones in a sufficiently deep sample of paste after settlement:

1. A zone of clear water at the top.
2. A zone of uniform density equal to the original density.
3. A compressed zone in which there is a gradient of density, increasing toward the bottom.

It was once thought that there would be a fourth zone of maximum compression having uniform density [2]. However, later experiments showed that such a zone does not develop, at least not under compressive forces of the magnitude developed from the weight of the sediment.

Zone 1 is always present if the surface water is not allowed to evaporate, with the following possible exception: if the settling rate is exceedingly low, of the order of 2 or 3 millionths cm/sec, the chemical processes going on during the dormant period can cause an absorption of water at the same rate, thus preventing the appearance of water at the surface.

The clearness of the water that appears at the top is significant. It shows that the finest cement particles have not been able to separate themselves from the coarser ones even though they are subjected to a relatively high viscous drag; they are held in the mass by the interparticle forces discussed in Chapter 9.

If the depth of the specimen is not too great, only zones 1 and 3 will be found at the end of the period of settlement. But before settlement is complete, all three zones are present, zone 2 diminishing while 1 and 3 increase.

Sedimentation zone 3 can be called the *bottom compression zone.* Under the conditions described above, the top of zone 3 is horizontal. However if the sample has such dimensions that wall effects are present, a peripheral zone can develop simultaneously with the bottom zone.

3.2 The Peripheral Sedimentation Zone

In Section 2.2.1 the retarding effect of the wall of the container on the rate of settlement was described and it was pointed out that one consequence was shear strain between perpendicular "layers." Whether or not such strain produces compression as it does in the bottom zone, it does produce a degree of mechanical stability. With cement paste this can be demonstrated when the settlement is about two-thirds complete by gently pouring the paste out of the container. A stabilized sediment will be found sticking to the bottom and extending up the walls. It seems clear that this sediment is composed of a combined peripheral and bottom sedimentation zone.

Considering Fig. 11.6 again, we see that the curve $0C$ indicates that the peripheral sedimentation zone reached the point of measurement before the bottom zone reached the top, time t_1 being the time when the peripheral zone reached the float. Time t_3 is the time when the bottom zone would have reached the top if the peripheral zone had not arrived first.

The top of the bottom sedimentation zone can be made to reach the top of the mixture before the peripheral zone reaches the center, in the example of Fig. 11.6, if we increase the diameter of the container sufficiently, or if we leave the diameter unchanged and decrease the depth of the sample in proportion to the fractional decrease in time t_3 desired. In either case the measurement at the center gives virtually the same result we would obtain from a semi-infinite body of the same depth. However, it is evident that such values pertain to a certain fraction of the sample under observation, and not to the rest. Different parts of the sample will have settled at different rates and to different degrees, giving rise to a pattern of nonuniform density corresponding to the pattern of the sedimentation zones. Although the differences in density are not large, they are sufficient to account for significant variations in properties of concrete specimens.

An added degree of variation is possible when the depth and lateral dimensions of the settling mass are both relatively large. In such cases the dormant period may terminate before either the bottom of peripheral sedimentation zones are fully developed. The resulting pattern of densities can be imagined with the aid of Fig. 11.9.

Considering the interstitial spaces in concrete, we can visualize paste-filled spaces under aggregate particles where the paste layer is wide relative to its depth. In such places we should expect the settlement of the paste to be relatively unhindered, and, where particle interference arrests the settlement of the aggregate, the development of fissures under the aggregate particles is easy to explain. In other places the situation of the paste may resemble a vertical sheet or column that is narrow relative to the depth; indeed, as we learned in Chapter 6 the average width of such places is of the same order as the average size of the cement particles, and the average width is considerably narrower than the diameter of the largest cement particles. Settlement of the cement paste in such places is therefore dominated by the wall effect and particle sizes as well, so much so that there is probably very little settlement relative to the aggregate particles except in the more or less horizontal spaces under the aggregate particles.

The wall effect and the resulting peripheral, or lateral, sedimentation zones are dominant also in concrete in deep forms such as walls or columns. Under such conditions the rate and amount of settlement may be governed more by the width than by the depth of the form.

4. THE BLEEDING RATE AND BLEEDING CAPACITY OF CEMENT PASTE

We shall now consider the principal determinants of the initial rate of settlement and of the total amount of settlement per unit of initial depth—the *bleeding capacity*. We have already seen that the rate and amount of settlement of a given mixture may vary according to boundary conditions, but the intrinsic characteristics of a mixture appear in an uncomplicated way when the measurements are made so that wall effects are eliminated. In what follows we deal only with intrinsic characteristics, insofar as that is possible.

4.1 Bleeding and the Flocculent State

Settlement and bleeding as already described are phenomena peculiar to the moderately flocculent state, which is to say, a state in which the interparticle attraction is neither too strong nor too weak. That this is true may be seen by comparing a mixture of cement and alcohol with one of cement and kerosine, both liquids being virtually water free. A mixture of cement and alcohol does not bleed at all in the proper sense of that term; the smaller particles remain in suspension long after the coarse material has settled out. A sharp demarkation between the top of the sediment and the clear liquid is a long time developing, if it ever does develop. There is abundant evidence that in alcohol the particles of cement are in a nonflocculent state. For example, the float method of measuring settlement cannot be used because there is no thixotropic structure to keep the float from drifting about, and since there are no forces to hold the large and small particles together during settlement, the float is further unstabilized by variation of buoyancy with time.

In contrast, in a similar mixture of cement and toluene or any other nonpolar organic liquid, the cement particles stick together strongly, so strongly indeed that they form a rigid structure that settles relatively little.

The sediments formed by the two types of mixtures just mentioned are quite different: The sediment formed in alcohol is relatively dense, and the density is practically the same for different ratios of solid to liquid in the suspension. There is a tendency toward segregation of particle sizes, the top strata of the sediment being relatively fine. However cement settling out of a nonpolar liquid produces relatively bulky sediment in which there is no tendency for size segregation. The two extremes of behavior just described correspond to very weak, zero, or negative interparticle attraction on the one hand, and strong attraction on the other.

Either of these two kinds of mixture can be made to settle and bleed in the manner described above as normal bleeding by changing the interparticle forces in the proper way and to the right degree. We can reduce the interparticle attraction in a strongly flocculent suspension of cement in kerosine by adding a surfactant such as oleic acid. As we add the agent to a batch drop by drop while the batch is being stirred, the rheological characteristics of the mixture gradually change until the mixture becomes a true paste having properties much like those of cement-water paste. In this state kerosine paste settles and bleeds normally, forming a bulky sediment which is less dense the lower the initial proportion of solids in the mixture. However beyond this point an increase in the dosage of the surfactant transforms the plastic paste to the condition exemplified by cement in alcohol, that is, a nonflocculent mixture which forms a dense sediment more or less stratified with respect to particle size.

As we shall see further on, in paste and concrete the density of the sediment is lower the higher the initial water content of the mixture, a characteristic normal to the flocculent state. Since, as mentioned above, a nonflocculent mixture produces a relatively dense sediment, it might seem advantageous to make concrete with nonflocculent paste, for in that case it would not matter how much excess water was used; the cement would always form the same dense sediment and therefore the same strong cementing medium. Unfortunately such a possible advantage cannot be realized. As already mentioned, in most concrete mixtures the mixture as a whole does not settle as much as the paste it contains, and thus it does not become equally densified, because of particle interference as discussed in Chapter 6, Section 3.4. Even with a flocculent paste, flaws develop at the undersides of aggregate particles, and to use a nonflocculent paste would only increase the development of flaws.

4.2 The Bleeding Rate of Cement Paste

In Section 3.1 the process of bleeding during the period of constant rate is described as one involving the flow of fluid through permeable structures composed of particles held in a net of interparticle forces. It follows that the rate of bleeding should be controlled by Darcy's law, which may be stated as

$$Q = K'\rho_f g \frac{\Delta h}{L} \tag{11.1}$$

where Q is the rate of flow at a given temperature cm/sec; $K'\rho_f g$ stands for K, the coefficient of permeability, ρ_f being the density of the liquid g/cm^3 and g the gravitational constant, cm/sec^2; $\Delta h/L$ is the dimensionless hydraulic gradient.

In Chapter 12 it is shown that the hydraulic gradient depends on the unbuoyed weight of the solid material, as follows:

$$\rho_f g \frac{\Delta h}{L} = (\rho_s - \rho_f)g \frac{cv_c^o}{V}(1 + M) \tag{11.2}$$

where ρ_s is density of the solid material g/cm^3; $(cv_c^o/V)(1 + M)$ is the solids content of the mixture. Thus the bleeding equation is

$$Q = K'(\rho_s - \rho_f)g \frac{cv_c^o}{V}(1 + M) \tag{11.3}$$

In the present case $M = 0$.

The coefficient K is discussed at length in Chapter 12. Here we may note that it is a function of the specific surface area of the solid material and the solids content of the mixture. The observations reviewed below are direct manifestations of the major factors controlling K, that is, $K'\rho_f g$.

4.2.1 Effect of water content and specific surface area. Table 11.1 is an example of the relationship between the specific surface area of the cement (as indicated by the Wagner turbidimeter method), water content, and bleeding rate. The data pertain to four cements prepared from the same clinker.

Table 11.1 shows that the bleeding rates depend on the specific surface area of the cement and the water content of the paste. It shows also that normal bleeding is obtained only when the bleeding rate is less than perhaps 220×10^{-6} cm/sec; that is to say, at higher rates channels develop and surface craters appear, as described in the introduction.

During settlement, water flows through the paste structure, upward relative to the grains. Each grain is pulled downward with a force equal to its unbuoyed weight and upward by the viscous drag of the liquid. The viscous drag is proportional to the effective radius of the particle; the downward force is proportional to the third power of the radius; hence the viscous drag per unit weight is greater the smaller the particle.

Calculation based on Stokes's law shows that if a particle is forced to fall at the same rate as one 10 times its size, it will experience a viscous drag equal to 100 times its own unbuoyed weight. Thus it is clear that during settlement of cement paste, in which the smallest particles are perhaps a hundredth the size of the largest, hydraulic forces are generated that tend to destroy the paste structure.

Table 11.1 Effect of Surface Area and Water Content on the Bleeding Rate of Cement Paste[a]

Water-cement Ratio w/c	Bleeding Rate Q (cm/sec × 10^6) — Specific Surface Area (cm^2/g): 1085	1540	2045	2550
0.26	103	39	17	9
0.32	210	80	38	20
0.39	[b]	150	73	40
0.48	[b]	270	133	75
0.59	[b]	[b]	223	128
0.74	[b]	[b]	[b]	213
0.95	[b]	[b]	[b]	[b]

[a] Data are from T. C. Powers, "The Bleeding of Portland Cement Paste, Mortar, and Concrete," Portland Cement Association Bulletin No. 2, 1939. Surface area was determined with a Wagner turbidimeter.

[b] Channeling observed.

When normal bleeding is observed, we can take it as evidence that the interparticle forces are sufficiently strong to hold the structure together. Conversely, when fine particles are seen to be transported upward to the surface through channels, it is clear that the structure has been partially disrupted.

The lower the water content of the paste, the smaller the average distance between particles, the greater the number of points of near-contact, and the stronger the cohesive force. It is to be expected, therefore, that for any given cement there is a limiting water content beyond which the cohesive force is unable to withstand the disruptive hydraulic force, as is indicated in Table 11.1. Since the average distance between particles is smaller at a given water content the higher the specific surface area, the limiting water content for normal bleeding is greater the higher the specific surface area.

There is experimental evidence that at the critical water content for the development of channels, the paste structure may remain intact for a short time, after which channels appear and small particles begin to be transported to the surface. If the water content is considerably higher than the critical one, it seems that the paste structure lacks continuity from the beginning. If the water content is very high, this is obviously so, for during settlement the falling bodies are seen as discrete floccules that settle at a relatively high rate.

At a given water content the bleeding rate of a paste is not determined by water content and specific surface area alone, at least not by the specific surface as it is indicated by any of the methods now used for measuring it. There is an effect of temperature which within limits can be ascribed to the effect of temperature on the viscosity of water, as will be shown further on; but, there is a considerable variation among different cements, independent of temperature, which may be ascribed to differences in chemical reactivity and perhaps to certain physical characteristics. It is shown in Chapter 12 that correlation with chemical differences may only reflect differences in the effective specific surface area that remains after the changes produced by the initial reactions.

The range in the bleeding rates of cement pastes with a given water content and specific surface area is shown for room temperature in Fig. 11.10.

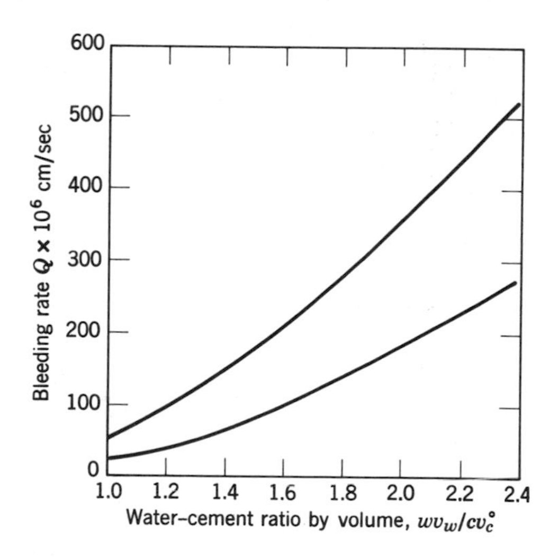

Fig. 11.10. Range of bleeding rates found among various cements having about the same specific surface area. Note by Steinour: "Most cements with Wagner specific surfaces of 1850 ± 100 cm^2/g will give pastes whose bleeding rates fall within the band shown."

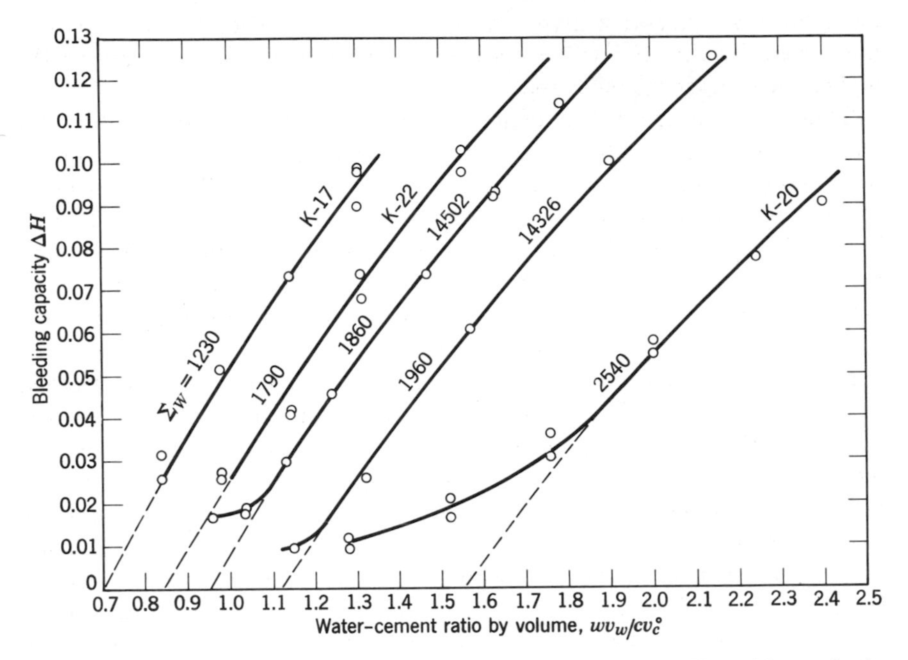

Fig. 11.11. Bleeding capacities of cement pastes in relation to Eq. 11.5. Σ_w is the Wagner specific surface area in cm²/g. Each point represents one test.

4.3 The Bleeding Capacity of Cement Paste

4.3.1 Effect of water content and specific surface area. As is shown in Fig. 11.11, the bleeding capacity of cement paste is greater the higher the water content, and at a given water content it is greater the lower the specific surface area of the cement. The water content of the sediment is greater the greater the initial water content of the paste.

If it is assumed that the settlement occurs without lateral shifting of the particles, and without an appreciable change in the degree of "nesting" of the settled particles when the water content is changed, the amount of settlement is related to the amount of water in the mixture in excess of a certain base amount as follows:

$$\frac{\Delta H}{cv_c^o/V} = \frac{1}{3}\left[\frac{wv_w}{cv_c^o} - \left(\frac{wv_w}{cv_c^o}\right)_B\right] \tag{11.4}$$

Since

$$\frac{cv_c^o}{V_o} = \frac{1}{1 + (wv_w/cv_c^o)}$$

the expression for settlement per unit original height is

$$\Delta H = \frac{\frac{1}{3}[(wv_w/cv_c^o) - (wv_w/cv_c^o)_B]}{1 + (wv_w/cv_c^o)} \tag{11.5}$$

The curves drawn in Fig. 11.11 represent Eq. 11.5 with the value of $(wv_w/cv_c^o)_B$ being established empirically for each curve as indicated by the dash-line extensions of the curves. The conformity of the data to these lines is good except for the lowest water contents in some cases, and except for the finest cement, K-20.

The theory represented in Eq. 11.5 is that the forces of flocculation give each paste a "base" structure, and that pastes are able to bleed because of the presence of water in excess of that required by the base structure. Moreover, whatever the amount of "dilation" of the base volume by excess water may be, the maximum settlement is limited to about one-third of the excess water unless particles can shift laterally as they settle, or unless some of the particles "nest" more compactly into "pockets" presumably created by the excess water. The forces of flocculation are expected to prevent lateral shifting during settlement. The smallness of the difference between the initial volume and the "base" volume of any given sample is considered to nullify the "nesting" effect.

Figure 11.11 indicates that the expectations just mentioned were fulfilled to a considerable degree. For the cases of noncomformity, particularly those pastes made with the finest cement, it may be assumed that different base structures are produced at different water contents when the original water content is below a certain minimum value characteristic of the cement. Thus, for pastes made with cement K-20, a base water-cement ratio of 1.55 is indicated for pastes having water-cement ratios above about 1.9 ($w/c = 0.6$ by weight), but for water-cement ratios below 1.9, the base water-cement ratio is lowered proportionately; that is to say, the nonconforming points are thought to belong on different curves paralleling the dashed portion of the curve shown.

The principal point to notice about the theory above is that it gives a plausible explanation of the variation in the amount of bleeding with the initial water content, or, what amounts to the same thing, it accounts for the variation in the density of the sediment as a function of the original water content. Approximately one-third the "excess" water in cement paste cannot be lost by normal bleeding.

Steinour [4] extended the studies of settlement of paste described above by greatly increasing the range of water contents of the samples as well as the number of cements. He found several different empirical

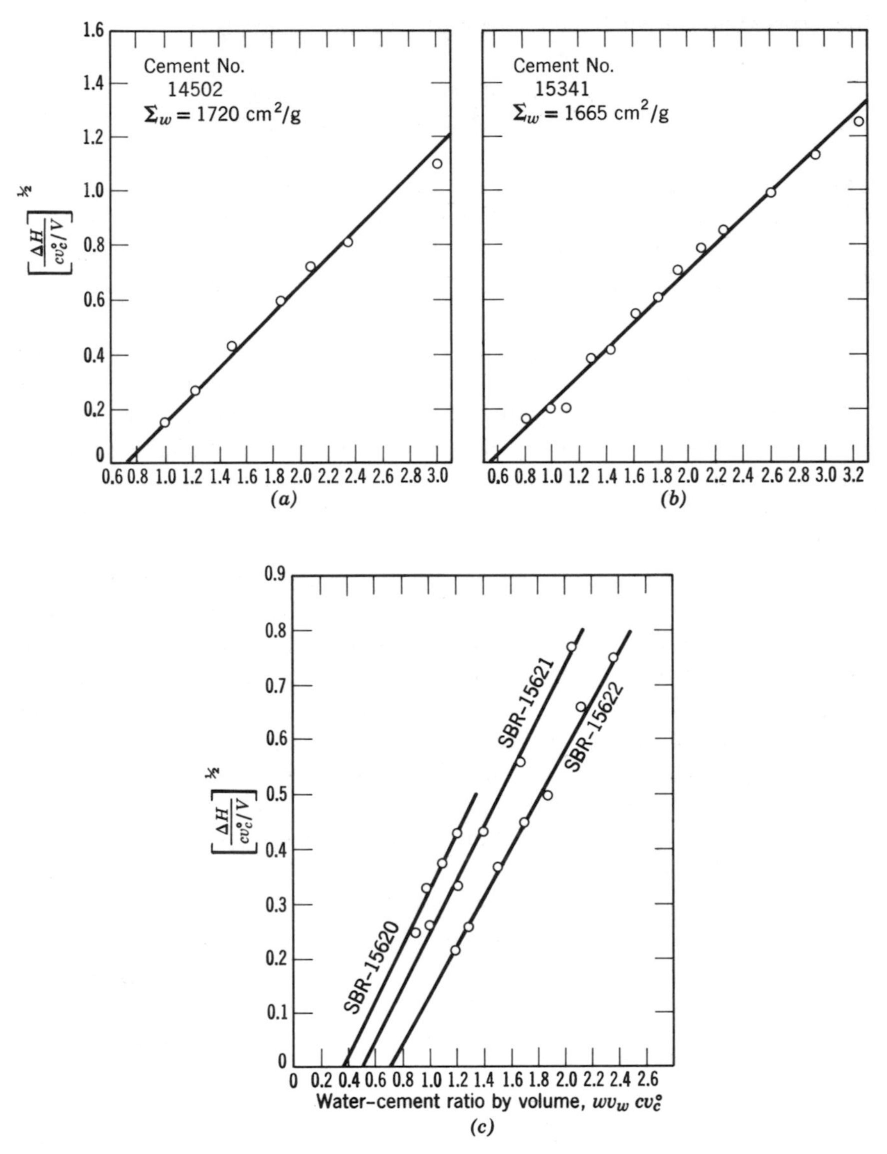

Fig. 11.12. Steinour's function for the bleeding capacity of cement pastes. For the three cements represented in (*c*), the values of Σ_w were 1375, 1820, and 2200 cm²/g.

relationships that would fit the experimental data over a wide range of water contents, and recommended the following as the most practical one:

$$\frac{\Delta H}{cv_c^o} = k^2 \left[\frac{wv_w}{cv_c^o} - \left(\frac{wv_w}{cv_c^o} \right)_m \right]^2 \tag{11.6}$$

where k is an empirical constant, about 0.5 for most cements, and $(wv_w/cv_c^o)_m$ plays somewhat the same role as the base water-cement ratio in Eq. 11.5. Examples of the application of Eq. 11.6 are given in Fig. 11.12.

Cement 14502 is represented in Fig. 11.11 as well as in Fig. 11.12. Its base water cement ratio is 0.95 in Fig. 11.11 and 0.70 in Fig. 11.12, corresponding to weight ratios of 0.29 and 0.215 respectively.

Equation 11.6 has the virtue of fitting experimental data very well, but its physical significance is not as easy to understand as is that of Eq. 11.5.

4.4 *Various Determinants of Bleeding Rate and Capacity*

On the basis of Fig. 11.12, it might appear that the bleeding capacity of cement paste at a given water content depends only on the specific surface area. However there are actually significant variations among different cements, as we see in Fig. 11.13. Also, Table 11.2 shows the bleeding capacities along with bleeding rates for 21 different cements comprising five ASTM types. In general there is a fairly high coefficient of correlation between bleeding rate and bleeding capacity, indicating that the factors that influence the bleeding rate at a given water content and specific surface area are the same as those that control the bleeding capacity, but there are exceptions. Increasing the intensity of interparti-

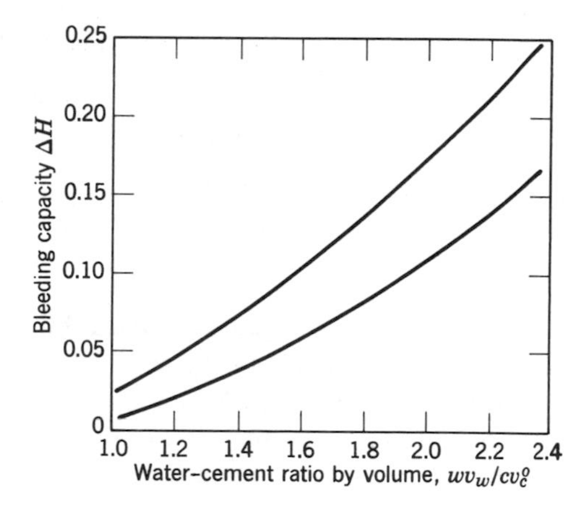

Fig. 11.13. Range of bleeding capacities found among various cements having similar specific surface areas. Note by Steinour: "Most cements with Wagner specific surfaces of 1850 ± 100 cm²/g will give pastes whose bleeding capacities fall within the band shown.

Table 11.2 Bleeding Rates and Capacities of Different Cements[a]

Cement[b] Number	Bleeding Rate Q (cm per sec $\times 10^6$)	Bleeding Capacity (ΔH)
LTS-42	163	0.123
LTS-23	156	0.096
LTS-51	144	0.097
LTS-15	140	0.084
LTS-21	140	0.099
LTS-16	129	0.097
LTS-11	117	0.078
LTS-12	116	0.081
LTS-18	115	0.068
LTS-22	111	0.046
LTS-25	107	0.084
LTS-13	104	0.048
LTS-43A	98	0.061
LTS-43	88	0.067
LTS-41	85	0.069
LTS-24	80	0.062
LTS-14	79	0.054
LTS-17	78	0.050

[a] Data are from H. H. Steinour, "Further Studies of the Bleeding of Portland Cement Paste," Portland Cement Association Research Bulletin No. 4, 1945. Values have been adjusted to correspond to a specific surface area by the Wagner turbidimeter of 1840 cm^2/g. The water-cement ratio was 0.466 by weight. The temperature was 23.5°C. The coefficient of correlation between the bleeding rate and bleeding capacity is 0.84.

[b] The first digit of each cement number indicates the ASTM type.

cle attraction by changing the electrolytic environment of the particles reduces bleeding capacity while usually having relatively little effect on the bleeding rate.

4.4.1 Effect of calcium chloride. The effect of calcium chloride is of special interest in view of the commercial use of this material. As is shown in Table 11.3, it reduced both the bleeding rate and bleeding capacity. The reduction of rate of bleeding can be accounted for partly on the basis of the increase in density and viscosity of the liquid. It was noted that the constant-rate periods were relatively brief, perhaps signifying a shortening of the dormant period, and that the pastes were relatively susceptible to channeling, which could indicate a reduction in interparticle attraction.

Table 11.3 Effect of Calcium Chloride on Bleeding Rate and Bleeding Capacity[a]

			Bleeding Rate ($10Q^6$ cm/sec)		
Cement Number and ASTM Type	Addition	Bleeding Time (min)	Actual	Calculated for Same Density and Viscosity as Pure Water	Bleeding Capacity (ΔH)
15365	None	71	163		0.085
I	$CaCl_2$	38	146	156	0.040
15496	None	46	129		0.049
III	$CaCl_2$	26	89	95	0.020
15621	None	71	139		0.072
II	$CaCl_2$	46	103	110	0.039
15697	None	81	122		0.077
I	$CaCl_2$	58	94	100	0.047
15668	None	87	156		0.093
IV	$CaCl_2$	71	122	131	0.068

[a] Amount of $CaCl_2$ was 1 percent of the cement weight and about 2.1 percent of water. The water-cement ratio was 0.466 by weight, except for 15668 for which it was 0.42.

4.4.2 Effect of temperature. Considering bleeding to be a phenomenon of fluid flow through a permeable body (as discussed above, and particularly as discussed in Chapter 12), we should suppose the rate of bleeding to vary inversely with the viscosity of the fluid. Since viscosity of water varies inversely with temperature, we should expect the rate of bleeding to increase as temperature increases. Powers [2] found the relationship between bleeding rate and the reciprocal of viscosity (the fluidity) of pure water to be as it is shown in Fig. 11.14. The relationship for this particular cement seems to be approximately as supposed above except at the highest and lowest temperatures. At tem-

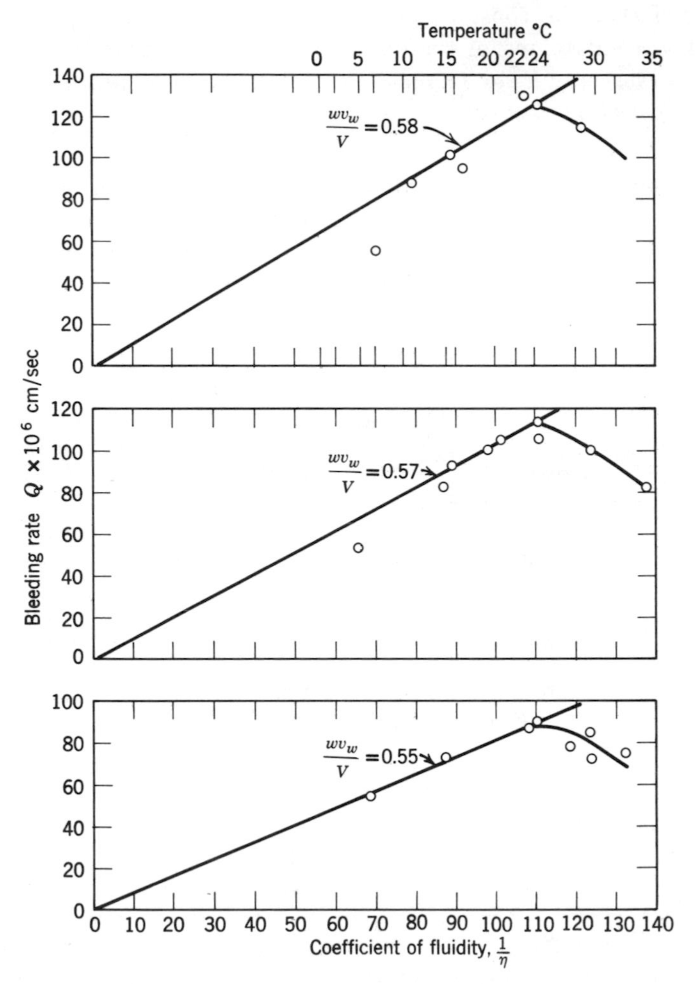

Fig. 11.14. Relation of bleeding rate to the fluidity of water, estimated from water temperature, for a particular cement.

peratures above about 24°C (75°F), increasing the temperature had but little effect, and the effect was to decrease rather than increase the rate. Evidently some factor other than viscosity is involved but not yet understood.

Steinour [3] made further tests on various cements in pastes having the same water content, for the purpose of observing the effect of temperature above the range where, as indicated in Fig. 11.14, viscosity governs the rate of bleeding for a given cement at a given water-cement ratio. According to Fig. 11.14, at a given water content,

$$\frac{Q_1}{Q_2} = \frac{\eta_2}{\eta_1}$$

For Steinour's experiments the temperatures were 74°F (23.5°C) and 90°F (32°C). The corresponding viscosities, of necessity neglecting the effects of solutes, were assumed to be 0.00925 and 0.00767 poises. From these figures for viscosity, we should expect the bleeding rate at the higher temperature to be 1.21 times that at the lower temperature.

Of the 25 cements listed in Table 11.4, three show a lower bleeding rate at 90°F than that at 75°F. (Fig. 11.14 is not typical.) The average ratio for the rest of the cements is 1.24, as compared with the calculated value, 1.21, but the deviation from the average is considerable.

In general it seems that the rate of bleeding decreases as the temperature falls below normal room temperature, but above room temperature the rate of bleeding may or may not be higher depending on characteristics of cement not now understood. It must be kept in mind, however, that data for temperatures below room temperature are available for only one cement and that the supposition that those data are typical might not be justified.

Concerning the relation of bleeding capacity to temperature, the data in Table 11.4 show that the average for the higher temperature is the same as that at the lower; 60 per cent of the cements show a ratio within 10 per cent of 1.00 This would indicate that the settlement was completed within the dormant period at the higher temperature as well as at the lower, and that the degrees of flocculation, that is, the strengths of interparticle attraction, were the same at both temperatures. However among the various cements there were pronounced deviations from the average. Four of the cements showed ratios exceeding 1.00 by more than 10 per cent, and for these it seems that the interparticle attraction was relatively weak at the higher temperature. For five of the cements, the ratio was more than 10 per cent below 1.00, and for these it seems that the interparticle attraction was stronger at the higher temperature, or that the dormant period was shortened too much to permit completion of settlement, or that both factors were involved.

4.4.3 Influence of cement composition. The rate and amount of bleeding of various pastes having the same cement content was found by Steinour to be smaller the greater the initial chemical reactivity of the cement. There was good evidence that these are the cements that produce a relatively large amount of calcium sulphoaluminate during the initial reaction period. Steinour wrote,

Table 11.4 Bleeding Data for Pastes at 23.5°C and 32°C[a]

Cement Number (LTS Group)[b]	Bleeding Rate (10^6Q cm/sec)			Bleeding Capacity (ΔH)		
	23.5°C	32°C	$\frac{32°}{23.5°}$	23.5°C	32°C	$\frac{32°}{23.5°}$
11 - 616	120	134	1.12	0.080	0.065	0.81
12 - 625	120	160	1.33	0.084	0.081	0.96
12T- 726	127	80	0.63	0.038	0.046	1.21
13 - 621	134	159	1.19	0.062	0.053	0.85
14 - 423	75	66	0.88	0.050	0.042	0.84
15 - 526	130	170	1.31	0.079	0.080	1.01
16 - 834	139	167	1.20	0.086	0.083	0.96
16T- 835	115	131	1.14	0.069	0.067	0.97
17 - 640	86	120	1.40	0.059	0.067	1.13
18T-3025	105	115	1.10	0.044	0.023	0.52
21 - 223	190	217	1.14	0.120	0.120	1.00
21T- 310	120	151	1.26	0.085	0.079	0.93
22 - 618	141	176	1.25	0.053	0.076	1.43
23 - 521	148	173	1.17	0.091	0.090	0.99
24 - 631	71	60	0.84	0.053	0.042	0.79
25 - 655	109	130	1.19	0.086	0.085	0.99
31 - 628	35	46	1.31	0.015	0.012	0.80
33 - 430	54	57	1.06	0.025	0.023	0.92
33T- 511	36	53	1.47	0.015	0.017	1.13
34 - 621	60	71	1.18	0.029	0.029	1.00
41 - 660	76	86	1.13	0.061	0.056	0.92
42 - 630	146	162	1.11	0.114	0.112	0.98
43 - 722	75	93	1.24	0.053	0.056	1.06
43A- 322	89	112	1.26	0.053	0.050	0.94
51 - 532	112	183	1.63	0.077	0.111	1.44

[a] Data are from H. H. Steinour, "Further Studies of the Bleeding of Portland Cement Paste," Portland Cement Association Research Bulletin No. 4, 1945. The water-cement ratio was 0.466 by weight.

[b] A "T" in the cement number indicates that the cement contained Vinsol resin and was therefore an air-entraining cement. The first digit of the cement number indicates the ASTM type.

"Apparently the amount of reaction produced is related to the amount of sulfate that goes readily into solution, and to a lesser degree, the amount of alkali in the cement (which often supplies a considerable part of the soluble sulfate). The amount of reaction produced appears to increase with the amount of tricalcium aluminate in the cement when (the calculated amount) of this is above 7 per cent" [4].

Thus it appears that cements high in tricalcium aluminate are likely to show relatively low rates and amounts of bleeding.

4.4.4 Effect of entrained air. Entrained air bubbles in cement paste reduce the bleeding rate and bleeding capacity. Examples are shown by the data in Table 11.5 and in Fig. 11.15. For the cement pastes

Table 11.5 Effect of Entrained Air on the Bleeding Rate of Cement Paste[a]

Air Content by Volume	Bleeding Rate (10^6Q cm/sec)
0.00	106
0.14	75
0.24	53
0.31	28

[a] Data are from T. C. Powers, "The Bleeding of Portland Cement Paste, Mortar, and Concrete," Portland Cement Association Research Bulletin No. 2, 1939. The water-cement ratio was 0.40 by weight.

represented in Table 11.5 the different amounts of air were produced by using different amounts of Vinsol resin as an air entraining agent, the bleeding rate for zero air content being estimated by graphical extrapolation. For the pastes represented in Fig. 11.15, the pastes were mixed in a closed system in the presence of a fixed amount of air-entraining agent, the amount of entrained air being regulated by controlling the quantity of air admitted to the system. Where comparable, the results from the two investigations seem to be alike.

Bleeding rate and capacity are reduced by a greater fraction than the fraction of paste displaced by air bubbles. The reduction is due not only to the displacement of paste, but also to the buoyancy of the bubbles and to their surface area.

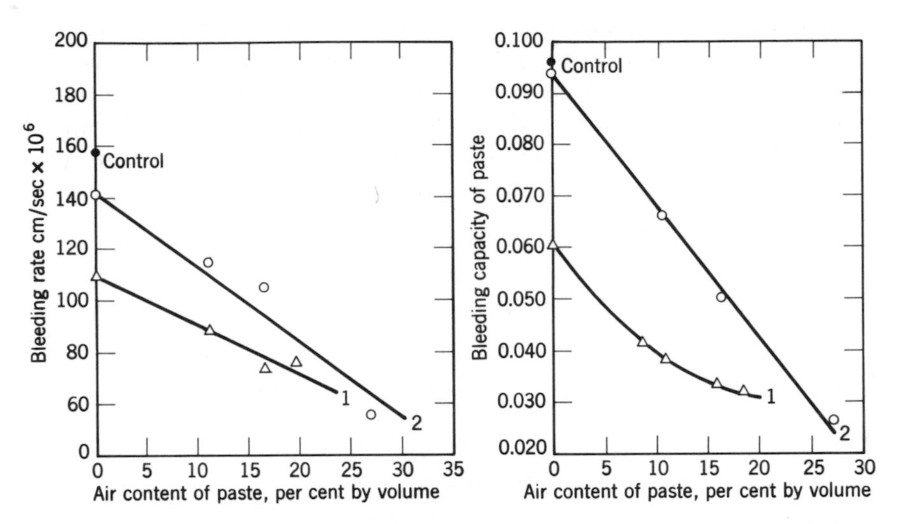

Fig. 11.15. Bleeding rates of pastes with various amount of entrained air but a fixed amount of air-entraining agent. The water cement ratio was 0.48 by weight. The air-entraining agents used are indicated: 1, sodium abietate, 0.05 per cent of weight of cement; 2, saponin, 0.10 per cent of weight of cement. The temperature was 20 ± 1°C.

Bruere [5] found that for two different air entraining agents, bleeding was reduced not only by the air bubbles, but also by the agent itself. The effect may be seen in Fig. 11.15 by comparing the left-hand terminal points of the curves with the points marked control.*

4.4.5 Effect of varying interparticle attraction. It was pointed out in Section 4.1 that normal bleeding is a characteristic of a flocculent state in which the interparticle attraction is neither too strong nor too weak. Silica in pure water does not bleed normally because it lacks the thixotropic structure characteristic of paste. It can be made to bleed normally by creating interparticle attraction. For the reasons given in Chapter 9, small additions of calcium hydroxide will transform a silica-water suspension to a paste resembling cement paste. Moreover, the mixture can be given different bleeding capacities by varying the concentration of the calcium ion, as is shown in Table 11.6. Notice that the

* In Chapter 10, Section 3.2.3, it was brought out that the lack of effect of sodium abietate on bleeding capacity is evidence that that agent does not act as a surface-active material in cement paste. The bleeding capacity appears to be slightly reduced (if at all) by the insoluble "calcium resinate" produced by the reaction between sodium abietate and calcium hydroxide. That could account, too, for the reduction in the bleeding rate.

bleeding capacity was reduced by a factor of 10 by increasing the calcium hydroxide concentration from one-fourth saturated to fully saturated. It is significant also that such a change of bleeding capacity was produced without much change in the bleeding rate, if any.

Some of the differences in bleeding characteristics of different cements with the same water content is probably due to differences in electrolytic environment and corresponding differences in interparticle attraction. There are indications that interparticle attraction is higher the higher the soluble alkali content of the cement.

Table 11.6 Effect of Calcium Hydroxide on the Bleeding Rate and Bleeding Capacity of Silica Paste[a]

Concentration of Solution	Bleeding Rate ($10^6 Q$ cm/sec)	Bleeding Capacity (ΔH)
One-quarter saturated	58	0.10
One-half saturated	53	0.05
Saturated	53	0.01

[a] Data are from T. C. Powers, "The Bleeding of Portland Cement Paste, Mortar, and Concrete," Portland Cement Association Research Bulletin No. 2, 1939. Sixty ml of calcium hydroxide solution were used per 100 g of pulverized silica.

In connection with Fig. 11.15 we have already seen that certain air-entraining agents are able to reduce the bleeding rate and capacity in the absence of the effect of air bubbles. Bruere attributed the effect to *increases* in interparticle attraction. Although some surfactants reduce interparticle attraction in cement paste, and that seems to be the effect generally expected, other agents increase it.

A particularly interesting example of that kind of agent is the synthetic polyelectrolyte called Krilium [6]. This material (as well as others of the same type) increased the stiffness of the paste markedly, reduced the bleeding capacity, and increased the bleeding rate, as is shown in Table 11.7. Also, it prevented the development of channels. At the higher water-cement ratios, the mixtures without Krilium bled rapidly and developed channels; with it they bled even more rapidly but did not develop channels, or at least not the craters of fine particles ordinarily associated with channeled bleeding. This observation suggests that the smallest particles were held so tightly to the larger ones that small

Table 11.7 Effect of Krilium on the Bleeding Rate and Bleeding Capacity of Cement Paste[a]

Krilium (per cent by weight of cement)	Bleeding Rate ($10^6 Q$ cm/sec)	Bleeding Capacity (ΔH)	Bleeding Rate ($10^6 Q$ cm/sec)	Bleeding Capacity (ΔH)
	w/c = 0.50		w/c = 0.60	
0.0	130	0.080	240	0.160
0.05	135	0.055	265	0.116
0.10	145	0.040	290	0.080
0.15	180	0.035	310	0.070
	w/c = 0.70		w/c = 0.80	
0.0	380[b]	0.233[b]	460[b]	0.240[b]
0.05	410	0.167	530	0.200
0.10	440	0.132	580	0.157
0.15	430	0.100	650	0.139

[a] Data are from G. M. Bruere and J. K. McGowan, "Synthetic Polyelectrolytes as Concrete Admixtures," *Australian J. Appl. Sci.*, **9**, 127–140, 1968.
[b] Channeled bleeding was observed.

floccules rather than individual particles became the unit of paste structure. (There is further discussion of this in Chapter 12, Section 3.5.2.)

Krilium produces an increase in cohesiveness as well as stiffness. This has been found advantageous under certain circumstances, as for example when placing concrete by the Shotcrete method [7].

In Chapter 12 it is shown that in some pastes made with organic liquids both the rate and amount of bleeding are affected by strong interparticle attraction.

4.4.6 Sedimentation volume and bleeding capacity. It is pointed out in Chapter 9, Section 9.2 that a comparison of sedimentation volumes shows the relative intensities of interparticle attractions. The volume of the sediment per unit weight or solid volume of material is observed, sedimentation occurring from a dilute suspension.

Table 11.8 gives data on the sedimentation volumes for five different cements having approximately the same specific surface area, each under three different conditions. One condition is that of strong interparticle attraction (suspension in toluene), one is that of virtually no interparticle attraction (suspension in dry alcohol), and the third is that of intermediate attraction (suspension in water). Column 4 of Table 11.8 gives the results in relative terms.

In the third and fourth columns the figures are as would be expected: the densest sediment was that formed in alcohol where the attraction would be zero, if not negative; the bulkiest sediment was that formed in toulene, interparticle attraction being strong; the sedimentation volume in water is intermediate.

In the fifth column corresponding figures for the bleeding capacities of cement-water pastes are given for comparison with the corresponding sedimentation volumes. The outstanding feature is that there is no correlation between the sedimentation volumes and bleeding capacities. In view of other assertions and evidence that interparticle attraction is

Table 11.8 Data on Sedimentation Volumes of Cement in Different Liquids

1	2	3	4	5	6
Number of Cement ASTM Type	Suspending Medium	Sedimentation Volume (cm^3/cm^3 of cement)	Relative Sediment Volume	Bleeding Capacity[a]	Material in Suspension at 24th hr
15754	Water	3.30[b]	0.86	0.062	None
I	Alcohol	2.15	0.56		Little
	Toluene	3.86	1.00		None
15756	Water	3.30[b]	0.84	0.095	None
II	Alcohol	2.24	0.57		Much
	Toluene	3.92	1.00		None
15758	Water	3.12[b]	0.83	0.059	None
III	Alcohol	2.12	0.56		Much
	Toluene	3.77	1.00		None
15761	Water	3.50[b]	0.95	0.090	None
I	Alcohol	2.13	0.58		Little
	Toluene	3.66	1.00		None
15763	Water	3.30[b]	0.83	0.113	None
IV	Alcohol	2.28	0.57		Much
	Toluene	3.98	1.00		None

[a] Bleeding capacity for $w/c = 0.466$; $wv_w/cv_c^o = 1.50$.
[b] Based on volume of sediment after three hr; others after 24 hr.

a principal determinant of bleeding capacity, as well as of sedimentation volume, this lack of correlation might seem surprising. However it is not difficult to reconcile the apparently conflicting observations.

When interparticle attraction is strong enough to bind the particles of a dilute suspension into a number of discrete floccules, during sedimentation each floccule falls as a unit, and the structure of the sediment becomes composed of such units. The interstitial space in the structure is determined primarily by the packing pattern of the floccules and secondarily by the density of each floccule. After sedimentation, the water-cement ratio of the sediment for the first item was $(wv_w/cv_c^o) = (3.30 - 1.0)/1.0 = 2.3$. This figure is to be compared with the water-cement ratio of the paste for which the bleeding capacity was determined: 1.50 cm^3/cm^3. Thus we see that before as well as after settlement the paste is much denser than the sediment formed from the dilute suspension.

If the floccules are like equidimensional solid particles, they should form a sediment comprising about one-half floccules and one-half interfloccule space. Thus of the 4.3 cm^3 of total volume about 2.15 cm^3 would be interfloccule, leaving $3.30 - 2.15 = 1.25$ cm^3 as the estimated water-cement ratio of the individual floccules. This indicates that the cement paste on which the bleeding test was made had a density comparable with that of an individual floccule. It thus becomes clear that the structure formed by floccules during sedimentation from a dilute suspension cannot exist in ordinary cement paste.

The volume of the sediment formed out of a dilute suspension is apparently controlled by the geometry of the floccules, which in view of the similarities of the figures, seems to be about the same for different cements; any small variation in the density of the floccules themselves due to differences in interparticle forces would make very little difference in the relatively large over-all volume. But the bleeding capacity of cement paste is controlled in part by the interparticle force and not at all by conditions pertaining to floccules, for individual floccules cannot exist in cement paste.

It is interesting to note also that the sediment formed in alcohol has, for the first item, a density corresponding to a water-cement ratio of $2.15 - 1.0 = 1.15$, or $w/c = 0.36$ by weight. Flocculent pastes having this water-cement ratio are able to bleed a small amount and thus form a slightly denser sediment than that formed in alcohol.

4.5 *Effect of Remixing*

A settled sample of cement paste is capable of settling and bleeding again after being remixed, with little change in bleeding characteristics.

Table 11.9 Bleeding Tests on Remixed Pastes[a]

Rest[b] Period, (min)	Bleeding Rate ($10^6 Q$ cm/sec)	Bleeding Capacity (ΔH)	Duration of Constant Rate Period, (min) (Approximate)	Duration of Bleeding (min)	Time between Initial Mix and End of Bleeding
0	194	0.122	20	55	55
15	189	0.113	18	55	70
30	196	0.125	15	57	87
45	192	0.106	17	52	97
60	185	0.103	18	48	108
90	172	0.090	20	47	137
120	167	0.075	12	45	165

[a] Data are from H. H. Steinour, unpublished report on PCA Series 290, April, 1941. The water cement ratio was 0.469 by weight. The schedule for the initial mixing was: 2 min mix, 2 min wait, followed by 2 min mix. The final remixing was $\frac{1}{2}$ min continuously. The depth of the paste was 36 mm. The temperature was 23.5°C. The results for remixed pastes are averages of two or three tests.
[b] This is the period after the initial mixing, at the end of which the final remixing was done.

The data in Table 11.9 pertain to seven identical samples that were allowed to stand for different periods before being remixed. They show that, with the particular cement used, samples remixed within the first hour were able to bleed for about the same length of time, with only slightly reduced bleeding rates and capacities, as compared with the corresponding values for the first settlement.

Other experiments indicated that the paste discussed above was able to settle in a deep container for about 1 hr before the rate of bleeding began to diminish because of the beginning of setting. Thus we see that a paste remixed within its dormant period will show almost normal bleeding characteristics, even though the settlement cannot be completed within the normal dormant period. Even for a sample that had rested 2 hr before being remixed, bleeding could continue for 45 min, although showing a considerably reduced rate and bleeding capacity.

Such experiments indicate that the cement grains in a quiescent body of paste may begin to establish interparticle connections composed of hydration products after the first hour, but if that initial structure is destroyed mechanically, the dispersed grains may show very little physical alteration from the state attained during the initial mixing period.

Another period of quiescence is required for new interparticle connections to develop.

5. THE BLEEDING RATE AND BLEEDING CAPACITY OF CONCRETE

5.1 Bleeding Rate

The initial rate of bleeding and settlement of concrete can be represented by Eq. 11.3 by using material constants corresponding to the composition of the concrete mixture; that is to say, it depends on the coefficient of permeability of the mixture and on the unbuoyed weight of the material in the mixture. The rate of settlement of concrete depends approximately on the coefficient of permeability of the paste it contains and on the paste content. However this is true only while the settlement is unhindered by particle interference. As we shall see further on, particle interference is a prominent factor under some circumstances.

In general the initial bleeding rate of a concrete mixture is lower the greater the ratio of the surface area of solids to the volume of water in the mixture. Most of the surface area is that presented by the cement, but that presented by the aggregate counts also. The surface area of air bubbles is also significant when an airentraining agent is used to increase the air content above normal.

5.1.1 Relation of bleeding rate of concrete to that of paste. It was shown by Powers [2], on the basis of an empirical bleeding rate equation discussed in Chapter 12, that the rate of bleeding of concrete relative to that of the paste alone can be calculated from the following relationship:

$$\frac{Q}{Q_p} = \left(\frac{p}{V}\right)^2 \cdot (1 + M) \cdot \frac{(\bar{\rho}_s - \rho_f)}{(\rho_c^o - \rho_f)} \cdot \frac{1}{[1 + M(\sigma_{ab}/\sigma_c)]} \tag{11.7}$$

where (p/V) is the paste content of the concrete; M is the mix, solid volume basis; $\bar{\rho}_s$ is the average density of the solids in the concrete mixture; ρ_c^o is the apparent density of the cement in water; σ_{ab} is the specific surface area of the aggregate, and σ_c that of the cement, in cm^2/cm^3.*

Below is an example of the application of Eq. 11.7 to data obtained

* The various figures that may be obtained for the specific surface area of cement are discussed in Chapter 12. It appears that the value that should be used here is σ_0, the Stokes equivalent-sphere value as determined from the bleeding rate. However, in the original use of Eq. 11.7 the Wagner value was used, and in the following examples also it is used.

from an actual concrete mixture. The bleeding rate of the concrete as well as that of the paste had been measured.

$$\text{Mix by weight: } 1:1.6:2.4$$

$$M = 4.97$$

$$\frac{p}{V} = 0.314; \left(\frac{p}{V}\right)^2 = 0.0986$$

$$\sigma_{ab} = 43 \text{ cm}^2/\text{cm}^3$$

$$\sigma_W = 5850 \text{ cm}^2/\text{cm}^3$$

$$\bar{\rho}_s = 2.72 \text{ g/cm}^3$$

$$\rho_c^o = 3.252$$

$$\left(\frac{\sigma_{ab}}{\sigma_W}\right) = 0.0073$$

$$\bar{\rho}_s - \rho_f = 1.72 \text{ g/cm}^3$$

$$\rho_c^o - \rho_f = 2.25 \text{ g/cm}^3$$

Using these values, we obtain $Q/Q_p = 0.413$.

For the paste alone, the measured bleeding rate was 85×10^{-6} cm/sec; hence the initial bleeding rate of the concrete mixture above should be $0.413 \times 85 \times 10^{-6} = 35 \times 10^{-6}$. The observed rate was 34×10^{-6}, as is shown in Fig. 11.4. Other examples of agreement between calculated bleeding rates are shown in Fig. 11.4.

It will be noted that adding aggregate to the paste did not reduce the bleeding rate as much as it did the paste content, the paste content being 0.314 as compared with the relative bleeding rate of 0.413. It is thus clear that although the rate of bleeding of the concrete is less than that of a cement paste with the same water-cement ratio, the velocity of water movement through the interstitial spaces in concrete is greater than it would be in the paste alone, because of the greater hydraulic head corresponding to the greater unit weight of the concrete. Therefore the tendency of hydraulic forces to disrupt the paste structure is somewhat higher in concrete than it is in paste alone. It has been observed that channeled bleeding tends to develop in concrete at a lower water-cement ratio than it does in cement paste alone. Of course, this tendency would be greatly modified by appreciable quantities of silt or clay in the aggregate.

An uncompleted although rather extensive series of experiments indicated that the rate of bleeding of concrete mixtures is controlled by many variable factors some of which seem to be related to the relative skills of different laboratory technicians. In the laboratory, specimens of concrete prepared for a bleeding test by a standard rodding procedure tend to show relatively low rates and amounts of bleeding because of premature particle interference. It was found that if the container holding the sample is placed on a vibrating table for 2 or 3 sec, the period during which the rate of settlement is constant is prolonged, making it easier to measure accurately. Also the bleeding capacity is increased. Apparently the standard rodding procedure disturbs the pattern the aggregate would normally establish in a plastic mass that is subject only to the vibrations and distortions involved in transporting and placing concrete. The disturbed pattern seems to be one having reduced minimum clearances between particles, minimum rather than average clearance being the determinant not only of the duration of constant rate bleeding but also of bleeding capacity. The data shown in Fig. 11.4 and the like were obtained from samples that had been subjected to brief vibration.

In the uncompleted, more extensive tests referred to above, carried out by different technicians, concretes were made with 18 different cements and seven of them showed bleeding rates in agreement with Eq. 11.7; but the rest of them showed rates only about half as high as that which the bleeding rates of the respective pastes would call for. In view of the remarks above, it would seem that the samples were not prepared with uniform skill, or that the different samples did not respond alike to the same "homogenizing" treatment. It seems significant that where the theoretically expected result was not obtained, the observed result was always too low, indicating that particle interference prevented free settlement almost from the beginning.

However there is also evidence that the departures from the theoretical relationship may not be due entirely to matters of technique. Among the 11 nonconforming mixtures, all the eight cements relatively high in soluble alkali were included. Moreover, all but three of the 11 cements were of ASTM Type II or Type IV. In Chapter 10, Section 5.8, it was shown that the rheological properties of these types are distinctly different from those of the other types. This is to say that differences in physical and chemical characteristics of the cement paste have to be considered when we are accounting for the relatively low bleeding rates of some concretes in relation to the bleeding rates of the paste they contain. It would seem, however, that such characteristics are involved only indirectly, for the interference with free settlement must be mechanical, there being no basis for assuming that the presence of aggregate could shorten the dormant period of the paste enough to account for the relatively low bleeding rates.

If such variation in bleeding rate as that described above is found in the laboratory, it would seem that also in the field the various bleeding characteristics observed cannot be accounted for entirely in terms of the differences in bleeding characteristics of the different cement pastes. Even where only one brand and type of cement is involved, variation in the hindrance of free settlement by particle interference from place to place and from time to time may be expected according to chance variations in the dispersion of aggregate and the effects of shocks and vibrations incidental to construction. Such variation of behavior should be more prominent the leaner the mix. Mixtures near point B should be relatively free from it.

5.1.2 Effect of aggregate grading. Defining aggregate as that material retained on the No. 200 sieve, we find that ordinary variations

Table 11.10 Effect of the Specific Surface Area of Sand on the Bleeding Rate of Mortars

Reference Number	Mix by Solid Volume (M)	Water-cement Ratio (w/c)	Bleeding Rate (10^6Q cm/sec)		Specific Surface Area of Sand (cm^2/cm^3)
			Observed	Corrected[a]	
47	1.90	0.384	30	30	86
51	1.90	0.393	36	33	99
55	1.90	0.402	37	31	113
59	1.90	0.393	32	29	126
48	2.38	0.431	43	43	86
52	2.38	0.443	50	44	99
56	2.38	0.452	47	38	113
60	2.38	0.439	48	44	126
49	3.18	0.477	52	52	86
53	3.18	0.490	63	57	99
57	3.18	0.508	60	47	113
61	3.18	0.490	60	54	126
50	4.80	0.646	114	114	86
54	4.80	0.668	121	106	99
58	4.80	0.693	113	95	113
62	4.80	0.668	112	98	126

[a] The observed rates of bleeding are corrected so that all values in a group of four are comparable as to the effect of water content, the first of each group being taken as the base.

of aggregate grading have practically no effect on the rate of bleeding of concrete. Test results in Table 11.10 show the effect of varying the proportions of the two smallest size groups and producing the different specific surface areas of the aggregate shown in the last column. In this case, the specific surface area of the aggregate was increased as much as 45 per cent, and there was some reduction in the bleeding rate, but the effect was so small that it was somewhat masked by the normal experimental variation, which amounted to 7 or 8 per cent. The smallness of the effect is due to the smallness of the contribution by the aggregate to the total surface area in the mixture.

Imperfectly washed aggregate may contain silt or clay or both. When it contains such fine material, the aggregate is likely to have an appreciable effect on the bleeding rate. This circumstance probably accounts for the widely held impression that bleeding can be controlled by regulating the grading of the sand. It can, if the definition of aggregate and aggregate grading is made sufficiently flexible.

Reducing excessive bleeding rate by leaving very fine material in the sand may sometimes be undesirable. Certain kinds of clay cause substantial increases in drying shrinkage of the mature concrete. In general, the amount and kind of paste-making material in concrete should always be known and kept under close control.

5.2 The Bleeding Capacity of Concrete

In Table 11.11 the bleeding capacities of several concrete and mortar mixtures are given. The remarkable feature of these data is that the bleeding capacities of the different mortar mixes were all alike (as were the consistencies); also, among the concretes, mixtures having about the same slump also had about the same bleeding capacity. These results depended partly on the circumstance that the same percentage of sand was used in each concrete mixture and that the same lot of cement was used in all mixtures. In the series of mixtures mentioned in Section 5.1.1, involving 18 different cements, concretes of the same slump and cement content showed various bleeding capacities. In general the higher the bleeding capacity of the paste, the higher the bleeding capacity of the concrete, but there were exceptions. Some particulars are given below.

5.2.1 The relationship between the bleeding capacity of concrete and that of paste. We might suppose that when aggregate is added to cement paste without changing the composition of the paste the bleeding capacity is reduced in proportion to the volume of paste displaced by the aggregate particles, except perhaps for an increment that might

Table 11.11 Bleeding Capacities of Concretes and Mortars[a]

Mix by Weight	Mix by Solid Volume (M)	Water Cement Ratio (w/c)	Approximate Slump (in.)	Bleeding Capacity (ΔH)	Paste Content (p/V)	$\frac{\Delta H}{p/V}$
		No. 200 – $\frac{3}{4}$ in. Concrete				
1:0.8:1.2	2.38	0.31	4.0	0.009	0.446	0.020
1:1.2:1.8	3.60	0.38	8.4	0.011	0.374	0.028
1:1.6:2.4	4.80	0.43	8.0	0.013	0.323	0.042
1:1.6:2.4	4.80	0.40	4.7	0.009	0.314	0.028
1:1.9:2.85	5.68	0.49	8.0	0.012	0.304	0.041
1:2.4:3.60	7.15	0.53	9.0	0.013	0.266	0.041
		No. 200 – No. 4 Mortar				
1:0.8	0.955	0.34	—	0.018	0.688	0.028
1:1.2	1.40	0.38	—	0.019	0.614	0.034
1:1.6	1.75	0.41	—	0.019	0.570	0.034
1:2.0	2.40	0.45	—	0.019	0.506	0.037

[a] Data are from T. C. Powers, "The Bleeding of Portland Cement Paste, Mortar and Concrete," Portland Cement Association Research Bulletin No. 2, 1939.

develop because of the weight of the aggregate. Thus it might be expected that

$$\frac{\Delta H}{p/V} \geq \Delta H_p \tag{11.8}$$

where ΔH is the bleeding capacity of the concrete and ΔH_p is that of the paste. But the observed relationship for most concrete mixtures is not that, as we may see in Fig. 11.16 where data from the last column of Table 11.11 are plotted. Only the richest of the mixtures conformed to Eq. 11.8, those mixtures being well within the *AB* category. For all the other mixes, the concrete was unable to settle as much as the paste itself would permit. This behavior is another manifestation of the particle interference discussed in Chapter 6.

As already indicated in Section 5.1.1, various factors involving manipulative techniques and chemical compositions of cements influence the bleeding capacity as well as the bleeding rate. Generally those cements among the 18 that deviated from the theoretical relationship between the bleeding rate of paste and that of concrete showed deviations also

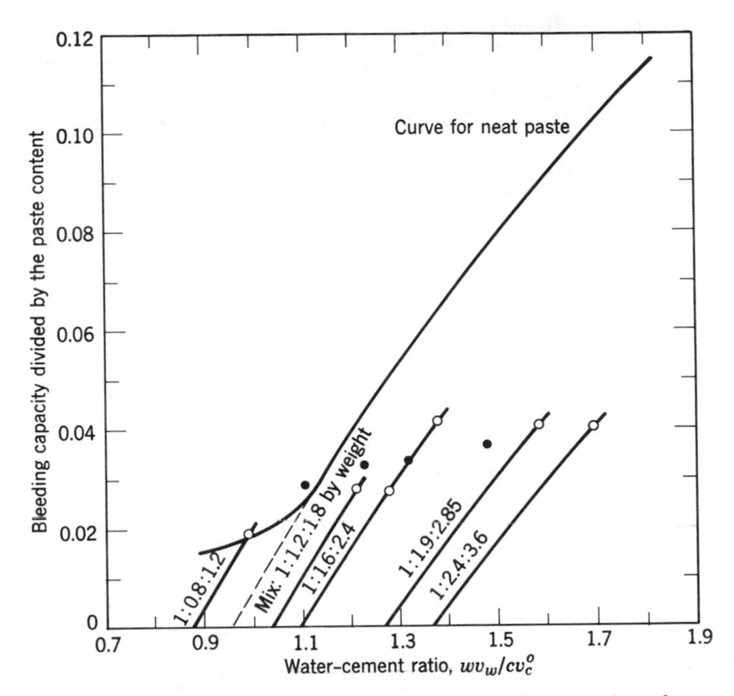

Fig. 11.16. Bleeding capacities of mortars and concretes per unit volume of paste in the mixture, compared wtih bleeding capacities of pastes having the same water-cement ratios. The solid dot indicates mortar, the circle, concrete. (Data are from [2].)

from the expected bleeding capacity. The concretes made with the 18 different cements were in the *BF* category and therefore conformity with Eq. 11.8 was not expected.

The degree of the inequality was not the same for the different cements. All the cements having high soluble alkali contents fell into one group, along with some other cements, and in this group the average bleeding capacity was about 13 per cent of $\Delta H/p/V$. For the rest of the cements, six out of 18, the bleeding capacity of the concrete was about 37 per cent of $\Delta H/p/V$.

As to observations in the field, the remarks made above relative to bleeding rate should be applicable also to bleeding capacity. Probably the amount of bleeding is more sensitive to shocks and vibrations incidental to construction than is the rate of bleeding.

5.2.2 Effect of settlement differential on strength. When the composition is such that a concrete mixture cannot settle as much as the paste it contains, the mixture does not develop as much strength as it otherwise would. This effect of differential settlement is shown in Fig. 11.17. The upper diagram is like Fig. 11.16; it shows for a series of mortars the bleeding capacities divided by the respective paste contents. The curve for cement paste is shown also. The lower diagram shows the results of strength tests made on the same mixtures, along with results from another series in which pulverized silica was added to each mixture leaner than 1:1 (be weight) in such quantity as to keep the paste content (cement plus silica plus water) the same as that of the 1:1 mix. Because the specific surface area of the silica was about the same as that of the cement, the bleeding characteristics remained almost independent of cement content.

Notice that the points for the very rich mixtures conform to the straight line *A*, defined by the points representing mixtures containing pulverized silica. The points for mixtures leaner than 1:1.3 fall on a different line, *B*, showing progressively larger losses of strength as the cement content diminishes, relative to line *A*.

Thus loss of strength was observed if the settlement was limited by particle interference rather than by the bleeding capacity of the paste. If a mixture made without pulverized silica settled without particle interference, adding silica had virtually no effect on the strength.

The water-cement ratios given in the lower diagram of Fig. 11.17 are in every case based on the amount of water left in the specimen after settlement; they are accurate figures for the water content of each specimen. Therefore the differences in strength at a given water-cement ratio are due to factors other than differences in composition of the cement paste. Unfortunately, the air contents of these mixtures were not reported, but other data leave no reason to doubt that the air content of the mixtures represented on line *A* was practically the same at every point, and relatively low, whereas the air contents at points along line *B* increased with a decrease in cement content. Thus some of the strength loss can be accounted for in terms of differences in air content, the higher air content being an effect of particle interference.

Strength loss may also be due to lack of any bond, or bond weakness, at the undersides of aggregate particles. Settlement of the aggregate is evidently arrested by the development of point contacts, and thus a structure stable enough to withstand the forces developed by settlement is formed, the coarser cement particles probably being a part of this structure. After this structure is formed, the cement particles in the interstices continue to settle, leaving microscopic fissures under the aggregate particles.

The effect of differential settlement on the properties of hardened

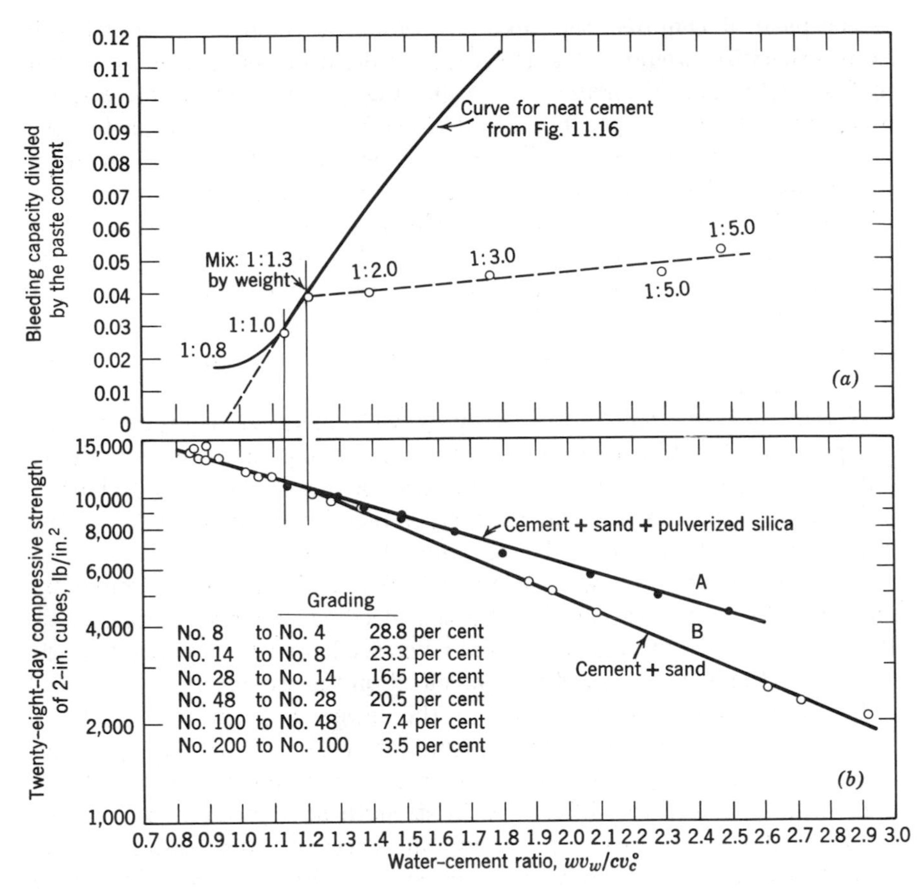

Fig. 11.17. Effect of settlement-differential on strength. (a) Each point represents the average of two or more tests. (Data are from [2].)

concrete is easy to observe. When a wall made of ordinary concrete is destroyed, the undersides of fragments present surfaces composed largely of half-embedded aggregate particles; the top sides of fragments show the cavities from which aggregate particles were pulled, thus revealing the bond-weakness at the undersides of the particles in such concrete. Microscopical examination of sections of relatively lean concrete reveal open fissures at the undersides of the larger aggregate particles.

The effect of bond weakness under small particles is less conspicuous than it is under the larger ones, but it seems to be present, as is shown in Fig. 11.18. These data pertain to mixtures using the same aggregate represented in Fig. 11.17, but the base mix for supplementary silica was 1:2 instead of 1:1. Strength is shown in relation to the ratio of cement content to voids content, the voids being the sum of the water volume after settlement and the air content. Thus in this diagram the effect on strength of the extra air in the leaner mixtures without supplementary silica is at least partially accounted for; still there is a marked difference in strength. Some of the difference in strength could be due to puzzolanic activity of the silica, but this seems unlikely. The silica used was pulverized quartz, and whatever the puzzolanic action, it develops slowly. It is significant that tests at the ages of three and seven days showed as much difference in strength as was shown at the later age. It thus seems probable that the loss was due to differential settlement, as discussed above.

In those mixtures in which the aggregate remains free to move with the paste throughout the settling process (mixtures without particle inter-

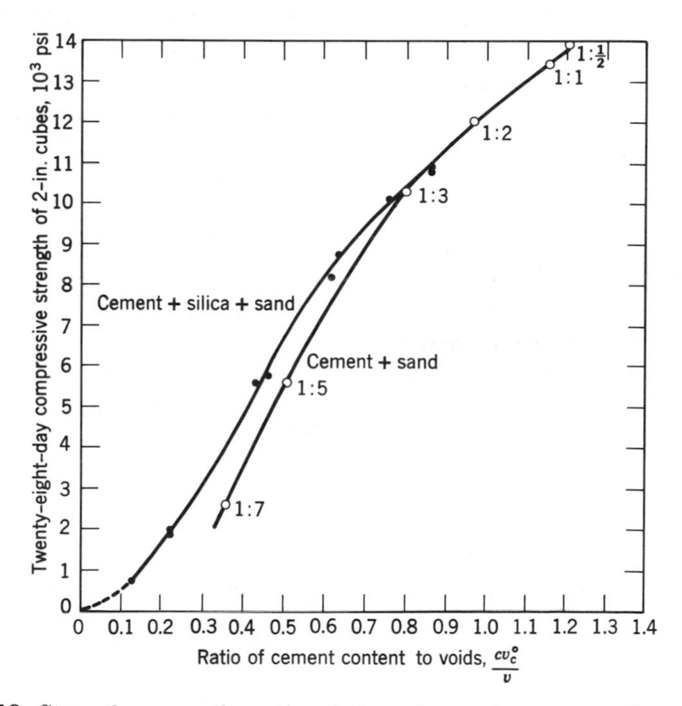

Fig. 11.18. Strength versus the ratio of the volume of cement to the volume of voids (water and air). The sand grade used was #200–#4; *FM* = 3.258; *SM* = 17.4; the voids content was 0.294. Pulverized silica used had a specific surface area of 2500 cm^2/g or 6680 cm^2/cm^3. (Data from Portland Cement Association Research Series No. 265 G.)

ference), uniform bond around the particles is possible, but such mixtures are far richer in cement than those ordinarily used; therefore differential settlement is probably the rule, although for mixtures in the *BF* category not too far from point *B* there is little evidence of nonuniform bond. In such concrete, well cured, the bond strength is such that most if not all the aggregate particles in a plane of fracture are themselves broken. For example, with a certain sand and gravel aggregate, it was found that at compressive strengths (using 6 × 12-in. cylinders) upwards of 5500 psi, fracture surfaces intersected the aggregate particles.*

5.2.3 Effect of prolonged vibration. Higginson [8] reported the effect of using an internal vibrator for periods up to 10 min on the bleeding capacity of concrete. Bleeding capacity was determined by essentially the ASTM method for concrete. The sample was vibrated in the container used for the bleeding test. For periods of vibration exceeding about 20 sec the effect was to reduce the bleeding capacity.

The result just mentioned seems at first to be in conflict with the observation in Section 5.1.1 that vibration prolongs the constant rate period and increases the bleeding capacity. However the period of vibration referred to before was so brief that it could effect the orientation of particles without changing the average degree of dispersion of the aggregate, whereas in Higginson's experiment it is known that a considerable fraction of air was dispelled, thus reducing the volume of the matrix and decreasing the degree of dispersion of the aggregate correspondingly. Also, the fluidizing effect of vibration may allow the aggregate to settle through the paste, thus further decreasing the dispersion and corresponding freedom for downward movement during the bleeding period following the cessation of vibration.

Under ordinary field conditions it does not seem likely that vibration could reduce the amount of bleeding perceptibly, owing to the relatively low average vibration intensity and relatively short duration.

5.2.4 Effect of revibration. When concrete is revibrated after it has completed its settlement it becomes able to settle and bleed again as much or more than it did the first time. Experimental results obtained by Vollick [9] are given in Table 11.12. The measurements was made in a container 8 in. in diameter and 10 in. high, filled to a depth of 9 in. The mixture was vibrated for 20 sec with an internal vibrator,

* In the discussion above, the data pertained to aggregate material having much greater strength than that of ordinary cement paste. However there are aggregates, particularly those used for lightweight concrete, which are made of material that is not as strong as the matrix material, and in such cases some of the remarks made above do not apply.

and then again for 20 sec after an interval of 1, 2, 3, or 4 hr. After the vibration, the container rested in a tilted position so that the water released by settlement could be removed with a pipette and measured. Although the results were doubtless influenced by the tilted position as well as by the wall effect, the indications given are probably reliable.

Table 11.12 Effect of Revibration on Bleeding and Strength[a]

Interval before Revibration	Bleeding (per cent of mix water)	Compressive Strength (psi)
0	2.9	4000
1 hour	3.5	4650
2 hours	3.4	5000
3 hours	3.2	4500
4 hours	2.8	4200

[a] Data are from C. A. Vollick, "Effects of Revibrating Concrete," *Proc. ACI*, **54**, 721–732, 1958. The cement was an ASTM Type I. The aggregate was natural sand and crushed trap rock with a maximum size of 1-in. The cement content was $5\frac{1}{2}$ bags per cubic yard. The slump of the concrete was 3 in. Compressive strength was estimated from resilience as measured by an impact hammer.

Concrete 4 hr old normally appears rather firm. Nevertheless it can be returned to the plastic state by using a vibrator, as was demonstrated by Sawyer and Lee [10]. Even for concrete 4 hr old, Vollick found the bleeding capacity of revibrated concrete to be as large as it was when the concrete was fresh.

It is significant that revibration does not impair strength; indeed the 28-day strength increases about as much as should be expected from the increase in density resulting from the second settlement.

5.3 *Effect of Evaporation on Settlement*

Table 11.13 shows that the rate of evaporation from a flat surface of water may vary from zero to as much as 100×10^{-6} cm³/cm²/sec (or cm/sec), depending on the temperature of the water, the temperature and humidity of the air, and on wind velocity. Data such as those

Table 11.13 Rates of Water Evaporation under Various Conditions[a]

Concrete Temperature (°F)	Air Temperature (°F)	Relative Humidity (per cent)	Dew Point (°F)	Wind Speed (mph)	Rate of Evaporation (cm/sec)
(1) Increase in Wind Speed					
70	70	70	59	0	2
70	70	70	59	5	5
70	70	70	59	10	8
70	70	70	59	15	12
70	70	70	59	20	15
70	70	70	59	25	18
(2) Decrease in Relative Humidity					
70	70	90	67	10	3
70	70	70	59	10	8
70	70	50	50	10	14
70	70	30	37	10	18
70	70	10	13	10	24
(3) Increase in Concrete and Air Temperatures					
50	50	70	41	10	4
60	60	70	50	10	6
70	70	70	59	10	8
80	80	70	70	10	10
90	90	70	79	10	15
100	100	70	88	10	24
(4) Concrete at 70°F, Decrease in Air Temperature					
70	80	70	70	10	0
70	70	70	59	10	8
70	50	70	41	10	17
70	30	70	21	10	22
(5) Concrete at High Temperature, Air at 40°F and 100 per cent Relative Humidity					
80	40	100	40	10	28
70	40	100	40	10	18
60	40	100	40	10	10
(6) Concrete at High Temperature, Air at 40°F, Variable Wind					
70	40	50	23	0	5
70	40	50	23	10	22
70	40	50	23	25	48

Table 11.13 ***Continued***

Concrete Temperature (°F)	Air Temperature (°F)	Relative Humidity (per cent)	Dew Point (°F)	Wind Speed (mph)	Rate of Evaporation (cm/sec)
(7) Decrease in Concrete Temperature, Air at 70°F					
80	70	50	50	10	24
70	70	50	50	10	14
60	70	50	50	10	6
(8) Concrete and Air at High Temperature, 10 per cent Relative Humidity, Variable Wind					
90	90	10	26	0	10
90	90	10	26	10	46
90	90	10	26	25	100

[a] This table is a modification of a table compiled by C. A. Menzel, "Causes and Prevention of Crack Development in Plastic Concretes," Portland Cement Association Office Memorandum, November 1954.

shown in Fig. 11.4 show that the initial rates of bleeding among different mixes made with the same materials may range from 20×10^{-6} to over 100×10^{-6} cm/sec depending on the cement content. In all cases the period of constant rate lasts 15 to 30 min at most, and thereafter the rate diminishes, reaching zero within an hour and a half. Thus we see that there are many atmospheric conditions in which the rate of evaporation may equal or exceed the rate of bleeding, and that there are very few atmospheric conditions under which the rate of evaporation may not exceed the rate of bleeding before the settlement is completed.

When the rate of evaporation exceeds the rate of bleeding during the initial constant rate stage, there is a measurable increase in the rate of settlement, and if that condition prevails throughout the settling process, the total settlement will be greater than it otherwise would have been; in other words, the bleeding capacity will exceed the normal value. The increase in bleeding rate and capacity is due to the development of capillary force, which is discussed further on.

5.3.1 Laboratory data. Using an adaptation of the float method, Klieger [12] measured the settlement of small specimens of concrete under controlled laboratory conditions. A typical effect on rate of settlement produced by varying the wind velocity is shown in Fig. 11.19.

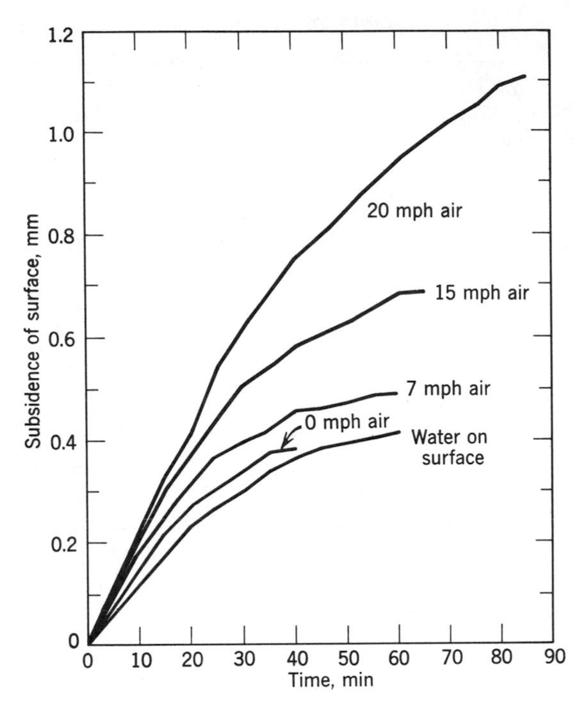

Fig. 11.19. Effect of capillary force on rate of settlement of concrete. The concrete was made without an air-entraining agent. The temperature of the experiment was 73°F. (Data are from [12].) By permission of the American Concrete Institute.

With the surface of the specimen covered with water, no capillary force could develop, and the lowest rate of settlement was observed. The graph shows that the initial rate was 17.5×10^{-6} cm/sec and that it lasted for 20 min. With no water on the surface and in still air at humidities ranging from 50 to 57 per cent, the rate was slightly higher. The rate of settlement increased with wind velocity, and at 20 mph it was 37×10^{-6} cm/sec, which is about double the rate in the absence of capillary force.

The effect of capillary force on the total amount of settlement is shown in Table 11.14. Under the conditions of Klieger's experiments the settlement was more than doubled by a wind of 20 mph, as was the initial bleeding rate. Since the rate of bleeding is proportional to the motive force, it appears that the total settlement is also proportional to that force, a least for forces of the magnitude considered here.

5.3.2 Capillary force. When the rate of evaporation is less than the rate of bleeding, the surface of the concrete becomes covered with

Table 11.14 Effect of Capillary Force on the Amount of Settlement[a]

Surface Exposure Condition	Total Subsidence, (mm)							
	Non-Air Entrained				Air Entrained			
	50°F	73°F	90°F	Average	50°F	73°F	90°F	Average
Under $\frac{1}{4}$ in. of water	0.555	0.482	0.454	0.497	0.475	0.533	0.488	0.499
0 mph air	0.502	0.413	0.454	0.456	0.516	0.507	0.457	0.493
7 mph air	0.698	0.511	0.698	0.636	0.559	0.639	0.782	0.660
15 mph air	0.679	0.590	0.899	0.723	0.751	0.910	0.911	0.857
20 mph air	0.840	1.010	0.825	0.892	0.777	0.965	0.895	0.879

[a] Data are from P. Klieger, "The Effect of Atmospheric Conditions during the Bleeding Period and Time of Finishing on the Scale Resistance of Concrete," *Proc. ACI*, **52,** 309–326, 1955. Each figure is the average of four tests on different specimens of the same kind. The relative humidity during the tests ranged from 50 to 57 per cent. To convert the results to bleeding capacity ΔH, divide by 76.

a layer of water, plane and horizontal or nearly so. But when the rate of evaporation exceeds the rate of bleeding the liquid surface loses its planeness, becoming converted to myriad curved surfaces (meniscuses), which are concave between the particles. Since the fluid pressure on the convex side of a meniscus is less than it is on the concave side, that is, less than the pressure of the atmosphere, the difference constitutes a motive force in addition to gravity driving the topmost particles downward.

If we let P_c stand for that part of the total pressure on the topmost particles which is due to capillary force, it is equal in magnitude and opposite in sign to the hydrostatic pressure difference across the meniscuses. That is

$$P_c = -\gamma \left[\frac{1}{r_1} + \frac{1}{r_2}\right] \tag{11.9}$$

Since the r's are negative, P_c comes out to be a positive downward pressure; it may be called the pressure on the topmost particles that is due to capillary tension.

The curvature of the water surface is limited by the dimensions of the interstitial spaces among the particles at the surface, analogous to the relationship between the curvature of a meniscus in a cylindrical capillary and the radius of the capillary. Carman [13] has shown that the maximum possible curvature and hence the maximum pressure due to hydrostatic tension in a water-filled granular bed can be calculated from the following relationship, which is an adaptation of the original:

$$P_{\text{max}} = \frac{\gamma S}{wv_w} = \frac{\gamma S/cv_c^o}{wv_w/cv_c^o} \tag{11.10}$$

S is the total surface area in the mix; wv_w is the volume of water, and cv_c^o is the of volume of cement. For the present purpose we may assume the surface area of the aggregate and air bubbles to be negligible, and thus we may write

$$S = cv_c^o\sigma_c \tag{11.11}$$

where σ_c is the specific surface area of the cement in cm^2/cm^3. Strictly speaking, the effective specific surface as the cement exists in fresh paste should be used, which, as is shown in Chapter 12, is usually considerably smaller than the air-permeability value; however we shall not be concerned with accuracy of computation of maximum possible force, for during bleeding it turns out that only a small fraction of the maximum is developed, as is shown below.

Eliminating S from Eq. 11.10, we obtain

$$P_{\text{max}} = \frac{\gamma\sigma_c}{wv_w/cv_c^o} \tag{11.12}$$

For a typical cement, and for mixtures at an ordinary temperature, we find the maximum capillary tension as follows:

$$\begin{aligned} \gamma &= 73 \text{ dyn/cm} \\ \sigma_c &= 10^4 \text{ cm}^2/\text{cm}^3 \\ v_c^o &= 0.31 \\ v_w &= 1.002 \end{aligned}$$

Then, for $w/c = 0.448$, $wv_w/cv_c^o = 1.45$, the value for Klieger's experiment, we have

$$P_{\text{max}} = \frac{73 \times 10^4}{1.45} = 0.5 \times 10^6 \text{ dyn/cm}^2$$

This is about 0.49 atm or about 7 psi.

There is no way at hand to tell how accurate this estimate of the maximum possible downward pressure due to capillary tension may be, but it is evident that no such force actually developed in Klieger's experiments. As we noted above, the highest initial rate observed was about double that observed when no capillary tension could develop. Therefore it may be assumed as a first approximation that the motive force produced by evaporation was about equal to that produced by the unbuoyed weight of the solid material. The latter force is given by the following expression (Eq. 11.2):

$$(\rho_s - \rho_f) g \frac{cv_c^o}{V} (1 + M)$$

For the concrete used in Klieger's experiments, $\rho_s = 2.74$, and 1.0 g/cm^3 may be taken as the estimated density of the aqueous solution: the cement content was 0.1043 (6.05 sacks per cubic yard) and $M = 6.98$; $g = 980.6$ cm/sec^2. Using these figures we find the force developed by evaporation to be 1419 dyn/cm^2, or 0.0014 atm.

The result just given would indicate that the force due to capillary tension was only $\frac{1}{350}$ of the indicated maximum. This ratio indicates that the curvature of the water surfaces was only about $\frac{1}{20}$ of the maximum possible curvature, as may be seen from Eq. 11.9, assuming that the meniscii are spherical. It shows, of course, that the maximum possible curvature was not needed to bring water to the surface as fast as it could evaporate. Yet further consideration shows that a somewhat higher fraction of maximum capillary force than that indicated by the above calculation must have been developed.

During the initial period of constant-rate settling, the top layer of particles descends faster than the underlying particles because of capillary force. Consequently, the top particles overtake the ones below, and a progressively thickening compression zone develops from the top downward. As this zone thickens, the resistance to the capillary force increases, not only because of the increasing thickness, but also because compression reduces the coefficient of permeability. If the capillary force remained constant, the initial settlement would be nonlinear, and the rate would soon drop to the normal bleeding rate without capillary tension. But since the normal bleeding rate was less than the rate of evaporation, this could not happen. Instead the curvature of the water surface increased at the rate required just to offset the increase in resistance, and the rate of bleeding remained equal to the rate of evaporation as long as this was physically possible. The development of water curvature can go only as far as it is necessary for it to go to increase the hyrostatic tension to the level required to balance the increase of resistance.

Thus it appears that at a wind velocity of 20 mph the *initial* capillary tension was perhaps $\frac{1}{350}$ of the maximum possible, and that it increased steadily during the experiment to some unknown larger fraction of the maximum.

6. SOME CONSEQUENCES OF SETTLEMENT AND BLEEDING

6.1 *Water Gain*

Accumulation of water at the top surface of a mass of concrete is often undesirable. For example, when we are placing concrete continuously in a deep form, water may accumulate during the process if the mixture is such that it bleeds copiously; and then as added concrete plunges into the mass it tends to carry the surface water back in. Such action, together with whatever puddling there may be in the placing process, amounts to an imperfect remixing process which increases the water content of the material being added; and it causes the upper part of the lift to become progressively lower in quality relative to the quality of the bottom part as the filling of the form progresses. This phenomenon has been called "water-gain."

However, accumulation of water at the surface is not always undesirable. Particularly, some surface water is required to prevent capillary tension and to lubricate the tools used for finishing the surface of a floor slab. However, if the amount of water is excessive, or is too easily "brought to the surface" mechanically, the finisher tends to produce a thin layer of slurry and consequently a surface of poor quality. Competent finishers do not begin finishing until the bleeding period is over, or nearly so. We shall see further on that premature finishing not only may produce poor results because of localized water gain, but also it may produce a condition conducive to the later loss of a top layer by abrasion or frost action.

6.2 *Effect of Fixed Objects in the Form*

Fixed objects in the forms interfere with settlement and may give rise to structural flaws. Horizontal reinforcing steel is the most common example. Since the bars are unable to follow the downward movement of the cement and aggregate particles, settlement leaves a layer of water under each bar. Later, the chemical process of hardening causes the water to be absorbed by the hardening paste, and leaves a permanent space under the bar. Since the amount of settlement is usually greater the greater the depth, the higher a horizontal bar is above the bottom of the form, the larger the space under the bar after settlement. In this way up to half the area of contact between the steel and the concrete is lost, with corresponding loss of bond strength. This loss is reduced to some degree by using deformed bars. It can be greatly reduced by revibrating the concrete, but this is seldom done.

Flaws or zones of weakness may develop in the concrete at places other than under bars. For example, a rectangular grid of horizontal bars such as is used in bridge decks may cause the development of more or less stable arches of fresh concrete bridging or partially bridging the spaces between bars. The concrete under such arches settles, and even though the arches are unstable, their collapse involves a loosening of structure and the development of a zone of weakness.

A fixed object may be nothing more than an inconspicuous ledge resulting from uneven thicknesses of form boards or misalignment of panels of a wall form. An arch may develop across the section sufficiently stable to form a stationary "bottom" for the material above that level, thus allowing a flaw to develop. The evidence of the flaw may only be an irregular fissure just below the obstruction, the fissure extending inward some unobservable distance.

6.3 *Effect of Inequalities of Permeability*

Flaws may develop because of differences between the permeabilities of different parts of the same mass. For example, if a layer of freshly placed concrete is covered immediately with a layer of less permeable concrete, the bottom layer can settle faster than the upper, and thus a plane of weakness may develop between the two layers, particularly if the lower layer is relatively deep. Such flaws are not common, for usually the lower layer is allowed to set before a second layer of different concrete is added. An old method of making concrete walkways was to cover a lean base mixture with a rich topping, but the base mixture was usually a nonplastic one, and thus any weaknesses of bond between the two was not likely to be due to a difference in rate of settlement. However, it is clear that if a plastic base mixture is used, the rich topping should not be added until after the base has become firm.

6.3.1 Premature finishing. Present-day practice favors monolithic floors and walkways rather than the two-course construction mentioned above. This involves producing a smooth surface by floating and troweling the concrete before it becomes too hard. The processes involve application of pressure to the topmost particles, which tends to compress the top layer and thus reduce its permeability. If this process is completed before the end of the bleeding period, the troweled surface tends to settle more slowly than the underlying material, and a horizontal zone of weakness may develop. Under traffic, and especially under the action of frost, the compressed top layer is prone to scale off.

The zone of weakness developed in the way just indicated may be visualized as the formation of a layer of water between the compressed layer and the rest, with the upper compressed layer tending to disintegrate, fragments of its underside falling through the water layer to form a relatively loose aggregation; but it is perhaps more realistic to picture a sort of loosening of the structure rather than a falling down of fragments. It is not always easy to find a plane of demarcation, even when other evidence indicates the presence of flaws due to premature finishing.

6.3.2 Evaporation. It is shown in Section 5.3 that when the rate of evaporation exceeds the normal rate of bleeding, the rate of settlement becomes equal to the rate of evaporation and remains at that level as long as it is physically possible. Consequently, a top compression zone develops. Usually, this should not have undesirable results because the final state is simply a somewhat higher degree of compression of the whole mass than would otherwise have occurred.

Klieger's experiments [12] indicated that the effect of relatively rapid evaporation was to improve the resistance to frost action. However he found also that if the concrete was subjected to a wood-float treatment either after the settlement was half-finished, or completely finished, the effect was to induce surface scaling under frost action. Presumably this was due to the further compression of the already compressed top layer and development of a plane of weakness owing to retarded settlement of the top layer.

However, the explanation given above will not stand up unless another condition can be assumed, as follows. As discussed in Section 5.3.2, the initial constant rate of settlement when the rate of evaporation exceeds the normal rate of bleeding may be accounted for in terms of progressively increasing curvature of the menisci as required to maintain the rate of subsidence equal to the rate of evaporation. As long as this is possible, the top compressed layer moves downward faster than the underlying material and becomes progressively thicker. But if the curvature reaches the maximum corresponding to Eq. 11.12, the rate of settlement of the top compression zone must begin to diminish; and if the delayed finishing operation disturbed the underlying structure enough to increase its bleeding capacity or to cause it to resume bleeding after bleeding had stopped, the development of a plane of weakness becomes possible. Klieger's experiments, together with others, show that delayed finishing should be repeated at intervals until the additional bleeding induced by the finishing operation becomes negligible.

Klieger concluded that if the concrete surface is finished immediately after casting, before bleeding, it is better not to perform a second finishing. However the finish referred to was a simple strike-off with a wood float which produces relatively little surface compression and does not produce a smooth surface. The kind of surface discussed in Section 6.1.1 cannot be produced until the concrete is near the end of its dormant period.

The paragraphs above indicate that a compression zone at the top induced by evaporation and capillary force is undesirable only when conditions are complicated by effects of mechanical finishing, and that those undesirable effects are due to the mechanical factors rather than to the effects of evaporation. However, the discussion pertained to a situation where temperature, humidity, and wind velocity are constant throughout the period of settlement. If ambient conditions are variable, we can foresee some undesirable consequences. Particularly, if during the early stages of settlement the wind velocity is high and the humidity low, a top compression zone will develop rapidly. If the process is interrupted, perhaps by a rain shower, or by the erection of some protection against wind and sun, or by the spraying on of a curing compound, the capillary force will suddenly diminish or disappear, leaving the top compression zone to settle under the force of its own weight. If this happens before the bottom compression zone reaches the top, that is to say, while the underlying material is able to settle at the normal rate, the compressed top layer may lag, and thus a zone of weakness may develop beneath it.

A set of circumstances leading to very undesirable results was described by Ryell [14]. In placing and finishing a concrete bridge deck, six finishing operations were performed during a period of 90 min, and immediately afterward a sprayed-on curing compound was applied. When the weather was hot and windy, the rate of evaporation exceeded the normal rate of bleeding, and evidently the combined effect of evaporation and mechanical working of the surface during the 90-min period was to produce a top compression zone, that is, a "crust" of compressed material, which subsequently was found to be up to $\frac{3}{16}$ in. thick. Immediate application of the curing compound, which presumably formed a nearly impervious film, reduced the rate of evaporation so much that the capillary force became insufficient to drive the crust downward at the normal bleeding rate of the material below it. Consequently, a layer of water developed under the top layer, and in some areas rivulets of water could be observed coursing down the sloping deck under the compressed top layer. When the slab became dry, inspection revealed "a general flaking over large areas of the slab."

The effects described above were unusually severe because the concrete contained an organic material that prolonged the dormant period and

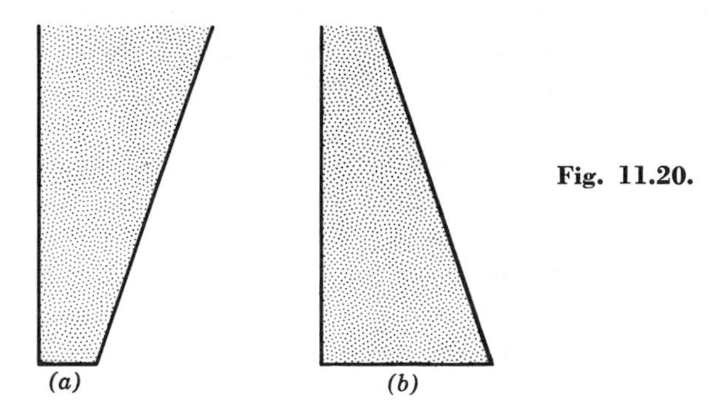

Fig. 11.20.

increased the bleeding capacity. Apparently the latter characteristic was the significant one, for a retarder that did not increase the bleeding capacity did not contribute to the undesirable behavior.*

6.4 Flaws Induced by the Shape of the Form

Since fresh concrete has some shear strength, and shear stress is about proportional to strain, and the amount of settlement tends to be proportional to depth, the shape of the forms can give rise to strains, stresses, and sometimes cracks (fissures), or perhaps zones of weakness. Some of the possible phenomena can be seen in exaggerated degree using a laboratory filtering funnel as a form, the outlet being plugged, and filling it with cement paste at, say, $w/c = 0.45$. We can think of the sample as being made up of concentric vertical cylindrical shells, each shell tending to settle to an extent proportional to its depth, and each one tending to settle more and for a longer time than the one encircling it. Thus settlement produces shear stress between the shells, which can become greater than the shearing strength of the paste. Evidence of this is that the top surface becomes sloped downward from the periphery, but the slope becomes broken by a series of steps, that is, faults.

Figure 11.20 illustrates some types of situations for the settlement of concrete mixtures in relatively deep and narrow places where the walls are not parallel. The situation shown in Fig. 11.20*a* is similar

* It has been suggested that the formation of a crust at the surface in the case discussed above was due to a delayed false set induced by the organic admixture [15]. Laboratory experiments on the cement in question indicated that false set could be so induced. However, when paste sets it ceases to settle, and therefore it is not clear how false set, delayed or not, could account for the rivulets of water flowing under the top crust or for that matter the separation of the crust from the mass below, since it would seem that false set should occur through the mass. Also, it does not seem to explain the observed influence of weather conditions.

to that of the funnel experiment, but it is somewhat complicated by the wall effect on the vertical side. The rate of the settlement of particles next to the sloping form may be taken as zero. The left side consequently tends to settle more than the right, shear strain develops, and faults may or may not develop depending on the angle of the slope, the distance between walls, the depth, rate of filling, and bleeding capacity.

Figure 11.20*b* shows the opposite situation. On the right-hand side, the particles settle *away* from the wall leaving a layer of water that tends to be thicker the farther the point of observation from the bottom; however such settlement leaves the particle network unsupported laterally, and it responds to the force of gravity by flowing laterally, displacing the water layer at the wall upwards. Often such displacement develops localized channels of flow, which, after the sample has become hard, and demolded, are visible as variously branched surface markings, sometimes called "sand streaks."

It is apparent that the total amount of settlement is greater in the situation shown in Fig. 11.20*b* than in Fig. 11.20*a*, and that the patterns of density variation in the final state will be different.

If a section has a shape of the kind shown in Fig. 11.21, and if it is filled in one continuous operation, a fissure or zone of weakness is

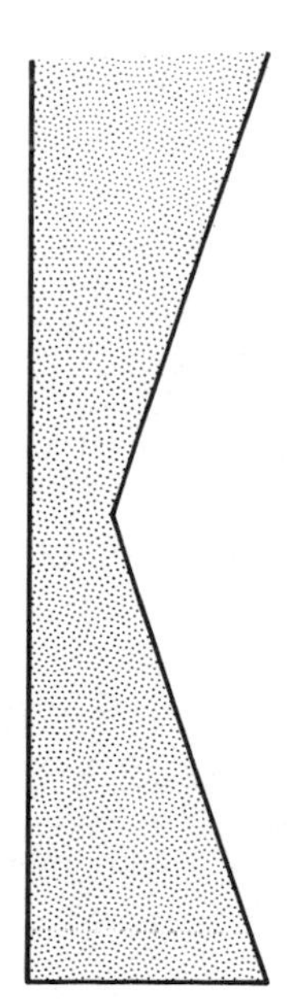

Fig. 11.21.

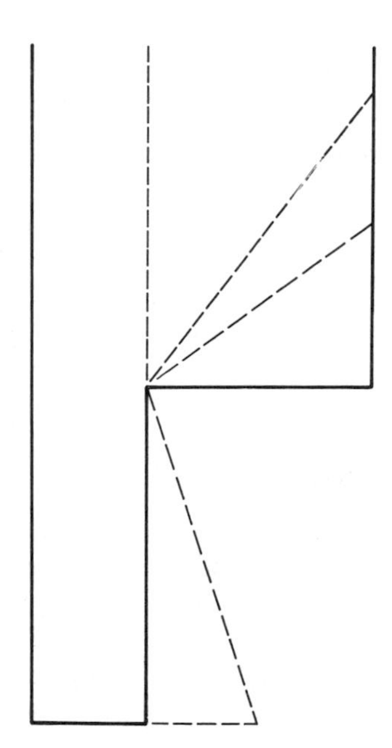

Fig. 11.22.

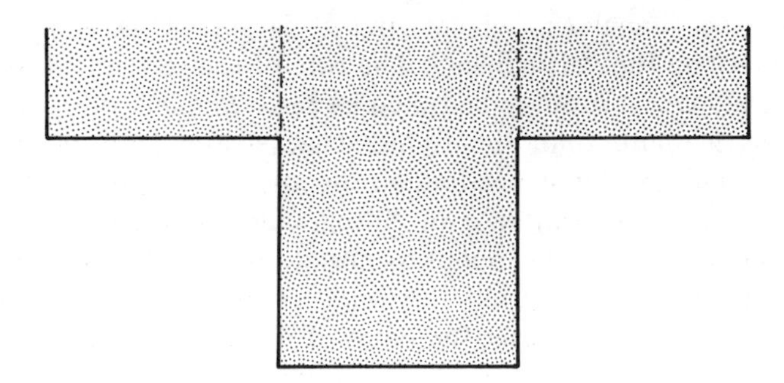

Fig. 11.23.

liable to appear at the narrowest part, as is apparent from the discussion of the shapes in Fig. 11.20. A shape like that indicated by the solid outline in Fig. 11.22 would be especially prone to develop such a flaw. Various degrees of this effect would be shown by section outlines given by different combinations of the upper and lower dash lines.

A common T section is represented in Fig. 11.23. If such a form is filled rapidly in one operation, it is obvious that faulting might develop. If the deep part of the section is relatively narrow, an arch may form, allowing a fissure or zone of weakness to develop below it. Of course the position of reinforcing steel can have a considerable influence on the development of flaws, as discussed above.

If the deep parts of a section are filled first and allowed to settle before the rest of the concrete is placed, there is little or no tendency for faults or arches to form.

Of course, the effects discussed above vary in degree according to the differences in the characteristics of the mixture. The greater the bleeding capacity the greater the tendency to form faults or arches. Flaws can be eliminated by revibration, but under most circumstances it is probably more practical to minimize them by proper design and control of mixtures, and by management of placement schedules.

6.5 Densification

In the other paragraphs of this section the undesirable consequences of settlement and bleeding are emphasized. It must not be overlooked, however, that settlement and bleeding does result in a reduction of water content and that if not offset by one or more of the phenomena discussed above, the effect is beneficial to the principal properties of concrete, particularly strength, impermeability, and volume stability. It was pointed out, in connection with Fig. 11.19, that uninterrupted evaporation could produce slab surfaces more resistant to scaling than slabs having the extra compaction due to capillary force induced by evaporation at rates exceeding the normal bleeding rate.

In a 1942 paper Swayze [16] pointed to the advantage of withholding curing water from paving concrete during the bleeding period, and while the temperature is rising, and then applying curing water while the temperature is falling, as a means of minimizing volume changes. In a later paper [17], anticipating Klieger's experiments, he advocated utilizing the capillary force as a means of producing dense and durable surfaces; water should be allowed to evaporate from the fresh concrete "at any reasonable rate," using an awning to prevent excessive rates when the weather is too hot and windy.

7. PLASTIC SHRINKAGE

7.1 Cause

When the rate of evaporation exceeds the rate of bleeding, the surface may be driven downward at a rate equaling the rate of evaporation, as discussed above. But this phenomenon can prevail for only a limited time, for as the resistance to compression increases it must finally exceed the maximum downward force that can be developed by evaporation (see Eq. 11.12). During the period of free movement of the top compression zone no hydrostatic tension can develop in the liquid below the top; the capillary force is balanced by the resistance to downward movement. However evaporation beyond this stage results in little further compression, and hydrostatic tension begins to develop throughout the mass. Since hydrostatic tension produces lateral compressive forces as well as vertical, lateral shrinkage begins as soon as hydrostatic tension can develop. In the laboratory, lateral shrinkage is manifested as a separation of the margin of the mass from the mold. In the field it may be manifested on a flat slab as the development of a pattern of cracks. Since this phenomenon occurs while the concrete is plastic, or potentially so, it is called *plastic shrinkage,* or *plastic shrinkage cracking* [18].

The crack pattern is determined by the nature of the restraints against lateral contraction. If the concrete is homogeneous, of uniform depth, and does not contain fixed objects, a three-branch pattern develops, the restraint being from the bottom which in turn is restrained by the subgrade or bottom form. When the concrete is not homogeneous, the pattern of cracking may be influenced to some degree by the pattern of the nonuniform dispersion of the aggregate. When the concrete contains fixed objects, for example a rectangular grid of reinforcement, the restraint of lateral movement by the steel bars may be reflected in the crack pattern. Also, the pattern may be influenced by flaws developed during settlement as discussed above. It may be influenced by a previously induced strain pattern even when no flaws actually were produced.

7.2 *Prevention*

Prevention of plastic shrinkage is obviously a matter of preventing the development of hydrostatic tension after free settlement is over and before the concrete has become strong enough to withstand the rather low maximum stresses that can be developed in green concrete. High shrinkage stress is not possible until the hydration of the cement is well advanced, say 10 or 12 hr after the start under normal conditions. Remedial measures may involve sun shades and windbreaks; prewetting of the aggregate if it is unusually dry and absorptive; cooling of the materials, particularly the mixing water, to reduce the difference between the temperature of the air and that of the concrete; application of water sprays, often impractical, or application of a curing compound to arrest evaporation.

In any case the remedy should not be such as to induce flaws of the kind discussed in Section 6, particularly those due to premature termination of evaporation. Sometimes it is feasible to eliminate cracks already formed by revibration, or by retroweling, provided the latter is not merely a smearing over of the cracks.

8. POSTBLEEDING EXPANSION

The period of bleeding is followed by one of expansion. This was learned by Steinour [19] when he continued observations of the position of the float (Fig. 11.1) beyond the end of the usual bleeding experiment, the specimen remaining submerged. An example is shown in Table 11.15: the rate of expansion was high at first, and then gradually diminished.

The amount of expansion was considerably different for different cements; even with the same cement it was not always the same in check tests, as is shown in Table 11.16 which contains data by Brownyard and Dannis [20] as well as by Steinour. The variability of the phenomenon is probably inherent in its nature. The postbleeding expansion is apparently a concomitant of the physical and chemical reactions that occur during the initial stages of setting and may involve the disruption of the gel coating on the cement grains [21]. It is certainly a movement that occurs while the structure is delicate and unstable, and, as observed, it occurs under conditions of restraint that allow upward movement only. Under this restraint the structure probably tends to fluctuate be-

Table 11.15 Post-bleeding Expansion of Cement Paste[a]

Age Interval (hr)	Observed Rate of Expansion (per cent/hr)	Total Expansion (per cent)
$0\text{–}1\frac{1}{2}$	0	0
$1\frac{1}{2}\text{–}2\frac{1}{2}$	0.10	0.10
$2\frac{1}{2}\text{–}3\frac{1}{2}$	0.04	0.14
$3\frac{1}{2}\text{–}4\frac{1}{2}$	0.04	0.18
$4\frac{1}{2}\text{–}5\frac{1}{2}$	0.02	0.20
$5\frac{1}{2}\text{–}6\frac{1}{2}$	0.02	0.22
$6\frac{1}{2}\text{–}23\frac{1}{2}$	0.01	0.38

[a] Type I cement, No. 15754, was used with a water-cement ratio of 0.38 by weight. The age at the end of the bleeding period of the cement was 1 hr 12 min. The age at the beginning of the expansion was 1 hr 30 min. Expansion is expressed as the per cent of the depth of the sample.

tween behaving elastically and plastically, and thus appears to be acting erratically. Some experiments indicated the results to be influenced by vibrations not considered significant when we are measuring bleeding.

The data discussed above indicate that the amount of expansion that may occur during the first day is considerably greater than the amount

Table 11.16 Post-bleeding Expansion of Cement Paste[a]

Cement Number	ASTM Type[b]	Clinker Number	Wagner Surface Area (cm^2/g)	Age at End of Bleeding (min)	Post-bleeding Expansion after Approximately 20 hrs (per cent)	
					Steinour	Brownyard and Dannis
15761	I	5	1800	68	2.1	0.86
15754	I	1	1800	78	0.38	
15496	III	2	1700	53	0.91	
15758	III	2	1800			0.64
15621	II	3	1820	88	0.09	
15668	IV	4	1830	83	0.83	
15763	IV	4	1800			0.05

[a] The water-cement ratio was 0.38 by weight.
[b] With respect to composition.

that occurs during a prolonged period of wet curing after the first day. During the first day an expansion of at least 0.09 is to be expected, and expansions up to 10 times that amount are probable, whereas after the first day additional expansion is not likely to exceed 0.045. Thus the expansion during the first day may be 10 times as great as later expansion, the later expansion being that which is normally observed and reported.

The one cement in Table 11.16 that showed expansion as much as 2 per cent is probably exceptional. In fact, it was the unusual behavior of mortar specimens made with this cement that called attention to the phenomenon. The specimens expanded so much that the protrusion above the top of the mold was easily visible. The high alkali content of this cement may be significant; the Na_2O content was 1.13 per cent.

REFERENCES

[1] Powers, T. C., L. E. Copeland, and H. M. Mann, "Capillary Continuity or Discontinuity in Cement Pastes," *J. PCA Research and Development Laboratories,* **1**(2), 2–12, 1959.
[2] Powers, T. C., "The Bleeding of Portland Cement Paste, Mortar and Concrete," Portland Cement Association Research Bulletin No. 2, 1939.
[3] Valore, Jr., R. C., J. E. Bowling, and R. L. Blaine, "The Direct and Continuous Measurement of Bleeding in Portland Cement Mixtures," *Proc. ASTM,* **49,** 891, 1949.
[4] Steinour, H. H., "Further Studies of the Bleeding of Portland Cement Paste," Portland Cement Association Research Bulletin No. 4, 1945.
[5] Bruere, G. M., "Mechanism by which Air-Entraining Agents Affect Viscosities and Bleeding Properties of Cement Pastes," *Australian J. Appl. Sci.,* **9,** 349–359, 1958.
[6] Bruere, G. M., and J. K. McGowan, "Synthetic Polyelectrolytes as Concrete Admixtures," *Australian J. Appl. Sci.,* **9,** 127–140, 1958.
[7] Bruere, G. M., "Further Studies of a Synthetic Polyelectrolyte Used as a Concrete Admixture," *Australian J. Appl. Sci.,* **11,** 205–208, 1960.
[8] Higginson, E. C., "Some Effects of Vibration and Handling on Concrete Containing Entrained Air," *Proc. ACI,* **49,** 1–12, 1952.
[9] Vollick, C. A., "Effects of Revibrating Concrete," *Proc. ACI,* **54,** 721–732, 1958.
[10] Sawyer, D. H., and S. F. Lee, "The Effects of Revibration on the Properties of Portland Cement Concrete," *Proc. ASTM,* **56,** 1215–1227, 1956.
[11] Menzel, C. A., "Causes and Prevention of Crack Development in Plastic Concrete," Portland Cement Association Office Memorandum, November 1954.
[12] Klieger, P., "The Effect of Atmospheric Conditions during the Bleeding Period and Time of Finishing on the Scale Resistance of Concrete," *Proc. ACI,* **52,** 309–326, 1955.
[13] Carman, P. C., *Soil Science,* **52,** 1–14, 1941.
[14] Ryell, J., "An Unusual Case of Surface Deterioration on a Concrete Bridge Deck," *Proc. ACI,* **62,** 421–440, 1965.

[15] Goetz, Hans. W., "False Set of Cement as Influenced by Hydroxylated Carboxylic-Acid-Type Admixture," *Mater. Res. Std.,* **7,** 246–250, 1967.
[16] Swayze, M. A., "Early Concrete Volume Changes and Their Control," *Proc. ACI,* **38,** 425–440, 1942.
[17] Swayze, M. A., "Finishing and Curing: A Key to Durable Concrete Surfaces," *Proc. ACI,* **47,** 317–331, 1950.
[18] Lerch, W., "Plastic Shrinkage," *Proc. ACI,* **53,** 797–802, 1957.
[19] Steinour, H. H., Unpublished report on PCA Series 290, April, 1941.
[20] Brownyard, T. L., and M. L. Dannis, Unfinished and unpublished PCA Project Series J354, 1941.
[21] Powers, T. C., "Some Physical Aspects of the Hydration of Portland Cement," *J. PCA Research and Development Laboratories,* **3**(1), 47–56, 1961.

12

Theoretical Aspects of Bleeding Rate

1. INTRODUCTION

In Chapter 11 we found that paste or plastic concrete behaves as if it is at first composed of independent permeable layers, each layer having the same coefficient of permeability. In this chapter we try to find a quantitative relationship between the composition of a mixture and its coefficient of permeability. The attempt will perhaps be found instructive regarding the nature of plastic concrete, and more successful than might be expected.

The starting point of the analysis is Darcy's law, which is that the velocity of flow of a liquid through a porous medium due to a difference in pressure is proportional to the pressure gradient in the direction of flow. Thus when the macroscopic flow is unidirectional, the law may be stated as follows:

$$Q = K\frac{\Delta h}{L} \tag{12.1}$$

where Q is the volume of liquid passing through a unit area of the porous medium in a unit of time, or it is the apparent or approach velocity of flow in cm/sec; K is the proportionality constant in cm/sec and is called the *coefficient of permeability* of the porous medium to a specified liquid at a given temperature; Δh is the drop in hydraulic head across the thickness of the permeable body, L being the thickness.

The problem is to find the value of K in terms of the characteristics of the solids and the liquid, and their relative proportions. The problem has been studied by many, particularly those interested in the movement of fluids in soils and earth structures. The most successful development is the Kozeny-Carman equation, the history of which has been reviewed by Wyllie [1]. There has been much discussion of the details of the development and the physical significance of the parameters of the equation. We shall find reason to question previous interpretations, particularly with respect to the concept of stagnant water, and the degree to which cement becomes physically altered by the initial chemical reactions.

Because it has been adapted to the phenomenon of bleeding, we shall discuss the Kozeny-Carman equation in its simplest aspects, and then discuss an equation based on a different model, essentially one developed by Steinour [2], although modified and simplified.

2. ANALYSIS BASED ON POISSEUILLE'S LAW

We may imagine the flow through a permeable body composed of particles and interstices to be analogous to flow through a solid body which is pierced by a number of identical, parallel, cylindrical capillaries. The coefficient of permeability of such a body to flow in the direction of the capillaries can readily be calculated in terms of Poisseuille's law, as follows.

2.1 *The Poisseuille Model*

Consider the arrangement shown in Fig. 12.1. It represents a setup for measuring the coefficient of permeability of a permeable body installed as an obstruction, or plug, in a conduit served by inlet and outlet pipes. The plug is shown as a granular body, but for our present purpose we will imagine it to be a solid body pierced by parallel, straight, cylindrical capillaries. Means of observing the rate of flow through the conduit and the corresponding pressure drop through the plug are provided. We assume that the velocity of flow is always below the critical Reynolds number, which is to say, we shall assume steady, streamline flow.

Under the conditions assumed, it is obvious that the total flow in the conduit is equal to the sum of the amounts flowing through individual capillaries. Also, the resistance to flow is made up of the resistances generated in the individual capillaries.

According to Poisseuille's law, the rate of flow of a liquid through a tube is directly proportional to the difference in liquid pressures at the ends of the tube, is proportional to the fourth power of the radius of the tube, and is inversely proportional to the viscosity of the liquid.

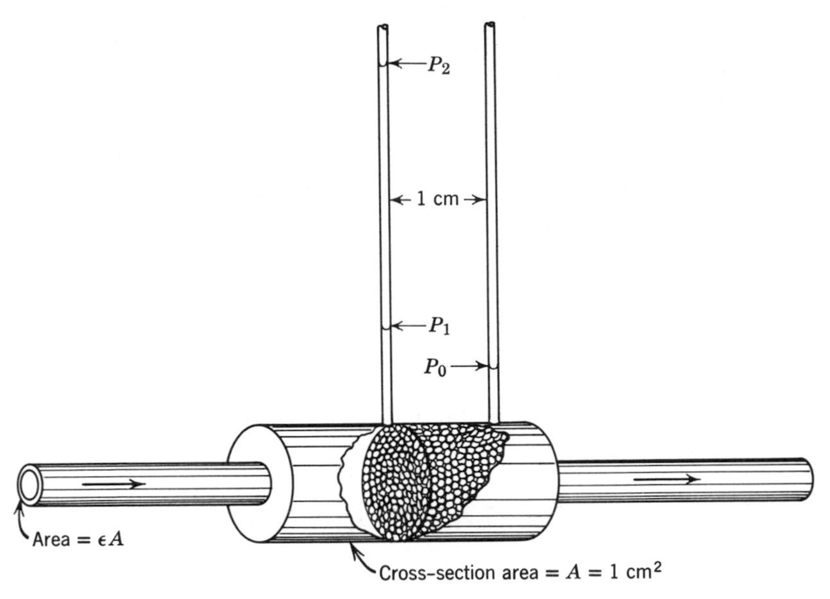

Fig. 12.1.

The quantitative relationship between rate of flow and pressure drop may be expressed as follows:

$$q = \frac{\pi R^4}{8\eta} \rho_f g \frac{\Delta h}{L_e} \tag{12.2}$$

where q is the volume efflux from a single capillary, in cm³/sec; R is the radius of the capillary; L_e is the length of the capillary; ρ_f is the density of the fluid and η its viscosity; g is the gravitational constant; Δh is the hydraulic head, as in Eq. 12.1. Since the total flow per unit area of the plug is the flow from N capillary openings in area A, it follows that

$$Q = \frac{N}{A} q \tag{12.3}$$

and

$$\frac{N}{A} = \frac{v}{A} \cdot \frac{1}{\pi R^2} = \frac{\epsilon}{\pi R^2} \tag{12.4}$$

where Q is the macroscopic approach velocity, v is the total volume of capillary space in a unit length of the plug, and πR^2 is the capacity of a single capillary 1 cm long. Since Δh is the same for the plug as a whole as it is for a single capillary, the expression for flow through the plug can be written

$$Q = \frac{\rho_f g}{8\eta} \cdot \frac{\pi R^4}{\pi R^2} \cdot \epsilon \cdot \frac{\Delta h}{L_e} = \frac{\rho_f g R^2}{8\eta} \epsilon \frac{\Delta h}{L_e} \tag{12.5}$$

Comparing Eq. 12.1 with Eq. 12.5, we see that

$$K = \frac{R^2 \epsilon \rho_f g}{8\eta} \tag{12.6}$$

which gives the coefficient of permeability of the plug in terms of its porosity, pore size, and the viscosity of the liquid. Since viscosity is a function of temperature, constant temperature is implied.

The model used above hardly needs verification, for it is as valid as Poisseuille's law, which has long since been experimentally proved.

2.2 The Kozeny-Carman Model

It is obvious that Eq. 12.6 is not directly applicable to a permeable body composed of particles and interstices. However, there is a certain similarity between a granular bed and a system of parallel capillaries with respect to permeability. Specifically, if the porous medium is uniform and isotropic, the area for flow in the direction of the pressure gradient may be assumed to be the fraction ϵ of unit volume, as in Eq. 12.6. However, although the area for flow in any plane normal to the direction of flow may be ϵ, the openings in that plane will not be circular, nor all of the same diameter. Also it is apparent that the flow paths through the interstitial spaces will not be parallel straight lines in the direction of flow. Nevertheless, Kozeny [3] showed that the Poisseuille model can be adapted to it.

The derivation of Kozeny's equation, about as presented by Carman [4], is as follows. For a noncircular conduit, the radius of the conduit is indicated by the mean hydraulic radius m, defined as the volume (capacity) of the conduit divided by the wetted area. For a circular section

$$m = \frac{\pi R^2 L}{2\pi R L} = \frac{R}{2} \tag{12.7}$$

Substituting $2m$ for R in Eq. 12.5, we obtain

$$\frac{Q}{\epsilon} = \frac{\rho_f g}{2\eta} m^2 \frac{\Delta h}{L_e} \tag{12.8}$$

For a section other than circular, the factor 2 is replaced by a shape factor k_o. For elipses, its value is between 2 and 2.5, depending on the ratio of major to minor axis lengths; for a square section, it is 1.78, and for various rectangles having different ratios of width to height it ranges upward from that value to 3.0 for infinite width. Carman found that the value for the openings in granular beds may be taken as 2.5.

Since the pressure drop in the conduit depends on the actual velocity and on the effective length of the flow path, we must find an expression that takes these factors into account. For the Poisseuille model, the effective microscopic velocity was simply Q/ϵ, but in the present case we must apply also the factor of L_e/L where L_e is the actual distance of flow in traversing thickness L. Hence Eq. 12.8 must be modified as follows:

$$\frac{Q}{\epsilon} \cdot \frac{L_e}{L} = \frac{\rho_f \eta}{k_o g} \cdot m^2 \cdot \frac{\Delta h}{L_e}$$

which means that

$$Q = \frac{\rho_f g}{k_o (L_e/L)^2 \eta} \cdot \epsilon m^2 \frac{\Delta h}{L} = K \frac{\Delta h}{L} \tag{12.9}$$

Carman observed colored streams passing through beds of particles and concluded that on the average their direction was inclined 45° from the general direction of flow. Hence the ratio $(L_e/L) = 2^{1/2}$, or $(L_e/L)^2 = 2$, this being called the *tortuosity factor*. It follows that $k_o(L_e/L)^2 = k_c = 2.5 \times 2 = 5$, which was Carman's evaluation of that which we shall call the *Carman constant*, but which is also called the *Kozeny constant*.

Studies by Wyllie and others [5] have shown that k_c is not the same for all systems and that in a given system it may take on different values if ϵ is varied. Nevertheless, on the basis of many experiments by Carman and others, $k_c = 5$ has become widely used. At this point we shall leave the question open as to its value as far as application to cement pastes is concerned.

If we assume that the space available for flow is ϵ, the hydraulic radius may be expressed as follows:

$$m = \frac{\epsilon}{(1 - \epsilon)\sigma} \tag{12.10}$$

where σ is the specific surface area of the wetted material, cm^2/cm^3.

Finally, we obtain the Kozeny-Carman equation:

$$Q = \frac{\rho_f g}{k_c \eta \sigma^2} \cdot \frac{\epsilon^3}{(1 - \epsilon)^2} \cdot \frac{\Delta h}{L} \tag{12.11}$$

2.2.1 Application to a nonflocculent suspension of spheres. To apply Eq. 12.11 to the settlement of a thick suspension of solid material in a liquid, we must express the hydraulic gradient in terms of the effective weight of the solid material. A particle falling through a liquid at constant rate (acceleration equal to zero) is supported entirely by the liquid. Consequently, the hydraulic pressure 1 cm below the top of the suspension exceeds the normal hydrostatic pressure at that depth by an amount equal to the unbuoyed weight of the solids in 1 cm^3 of the mixture. Hence

$$\rho_f g \frac{\Delta h}{L} = (\rho_s - \rho_f)(1 - \epsilon)g \tag{12.12}$$

where ρ_s is the density of the solid material, and ρ_f that of the fluid in g/cm^3; g is the gravitational constant, the pressure being in dyn/cm^2. Combining this equation with Eq. 12.11, we obtain the bleeding-rate equation according to the Kozeny-Carman model:

$$Q = \left[\frac{1}{k_c \eta \sigma^2} \frac{\epsilon^3}{(1 - \epsilon)^2}\right] (\rho_s - \rho_f)(1 - \epsilon)g$$

or

$$Q = \left[\frac{(\rho_s - \rho_f)g}{k_c \eta \sigma^2}\right] \frac{\epsilon^3}{(1 - \epsilon)} \tag{12.13}$$

It will be convenient to let C stand for some of the quantity in brackets, that is,

$$C = \frac{(\rho_s - \rho_f)g}{\eta} \tag{12.14}$$

and to change Eq. 12.13 to a linear form as follows:

$$[Q(1 - \epsilon)]^{1/3} = \left(\frac{C}{k_c \sigma^2}\right)^{1/3} \epsilon \tag{12.15}$$

Steinour [2] found that Eq. 12.15 represented experimental results accurately, provided that the settling mixture is nonflocculent and composed of spheres. For example, Fig. 12.2 shows the data he obtained from measurements of the settling rate of tapioca spheres in lubricating oil. The particles were about 0.174 cm in diameter, with little deviation from the average among them. Steinour estimated the actual size closely by measuring the rates of fall of 150 individual tapioca particles in a 1-liter oil-filled cylinder having a diameter of 62 mm, and calculating the specific surface area from Stokes's law. The average velocity, after a correction for the wall effect by Francis' formula [6] was

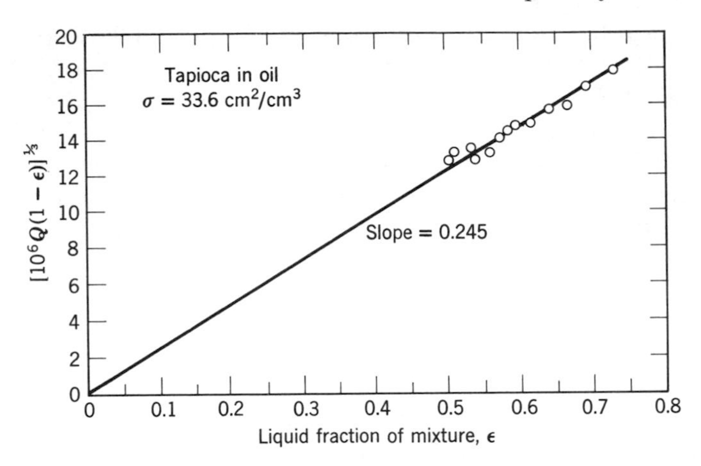

Fig. 12.2. Settlement of nonflocculent spheres plotted according to the Kozeny-Carman equation, Eq. 12.15. (Data are from [2].)

0.1194 ± 0.0095 cm/sec, and the calculated specific surface area, 33.6 cm^2/cm^3.

Essential data were as follows:

$$\rho_s = 1.38 \text{ g/cm}^3$$

$$\rho_f = 0.89 \text{ g/cm}^3$$

$$\eta = 7.13 \text{ poises at } 25°\text{C}$$

$$g = 980.6 \text{ cm/sec}^2$$

$$\sigma = 33.6 \text{ cm}^2/\text{cm}^3$$

Using these figures in Eq. 12.13, we find

$$C = 67.39$$

Then from Eq. 12.15, we see that the slope of the line in Fig. 12.2 is related to the quantity in brackets as follows:

$$\frac{C}{k_c\sigma^2} = (\text{slope})^3$$

As we see from the notation on the graph, the observed slope was 0.245, and thus $(0.245)^3 = 14.71 \times 10^{-3}$. Also, $\sigma^2 = 33.6^2 = 1129$. Using these values in the expression above, we find

$$k_c = \frac{67.39}{1.129 \times 14.71} = 4.06$$

Thus Steinour found that the Carman constant for this material was 4.06, which is to be compared with 4 obtained by Carman for experiments on fluidized beds of spheres.

2.2.2 Application to cement paste. A typical result of applying Eq. 12.15 to the bleeding of cement paste is shown in Fig. 12.3. Within limits, the relationship is linear, as required by the Kozeny-Carman equation, but the line does not pass through the origin. At $y = 0$, the intercept in this case is at $\epsilon = 0.28$; values ranging from 0.24 to 0.32 were found among pastes made with a wide variety of portland cements.

Powers [7] considered the intercept to be a quantity of water in the capillaries that does not take part in the flow, and called it *immobile water*. The Kozeny-Carman equation was accordingly altered as follows:

$$Q = \frac{C}{k_c \eta \sigma^2} \frac{(\epsilon - w_i)^3}{(1 - \epsilon)} \qquad (12.16)$$

where w_i is an empirical constant, interpreted as representing immobile water.

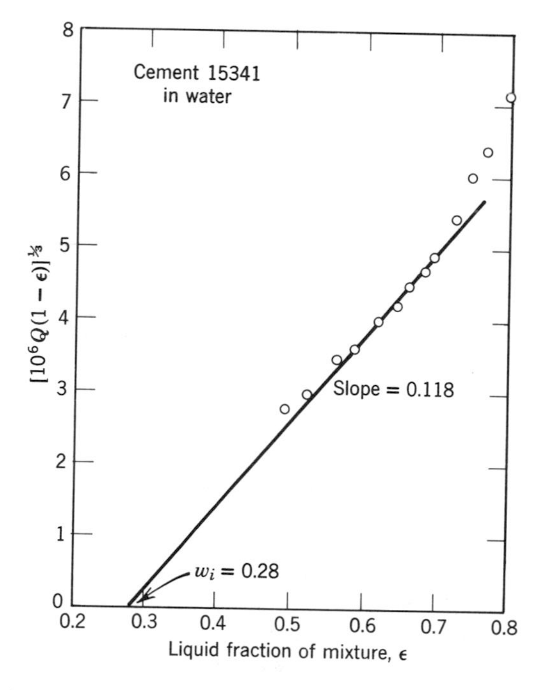

Fig. 12.3. Application of Eq. 12.15 to cement paste. Each point is the average of four tests. (Data are from Portland Cement Association Series 288-5-Cl.)

The linear form of this equation is

$$y = [Q(1 - \epsilon)]^{1/3} = \left(\frac{C}{k_c \eta \sigma^2}\right)^{1/3} (\epsilon - w_i) \qquad (12.17)$$

It turned out that Eq. 12.17 fits experimental data very well if $\sigma = \sigma_W$, where the latter is the value given by the Wagner turbidimeter, and if $k_c = 5$, w_i being in general different for different cements. For example, in tests made on 21 different cements comprising ASTM Types I, II, III, and IV, 19 of them gave satisfactory linear relationships between y and ϵ (Eq. 12.17) and 18 of the 19 gave slopes corresponding to values calculated from the bracketed quantities in Eq. 12.13 or Eq. 12.15, with $k_c = 5$ and $\sigma = \sigma_W$. This means that if the Wagner specific surface is known, only the constant w_i has to be established empirically.

The relationship discussed above achieved a satisfactory fit of experimental data, but it did not enable us to predict the bleeding rate of a given paste from its water content and the specific surface area of the cement. Moreover, it was something of an enigma: the physical meaning of w_i was not certain, and use of the Wagner value for specific surface area seemed questionable, for as is pointed out in Chapter 2, it represents only about half the specific surface area of the cement.

2.2.3 Steinour's studies. The enigma mentioned above was studied extensively by Steinour [2]. Systematic experiments were carried out leading to conclusions given below.

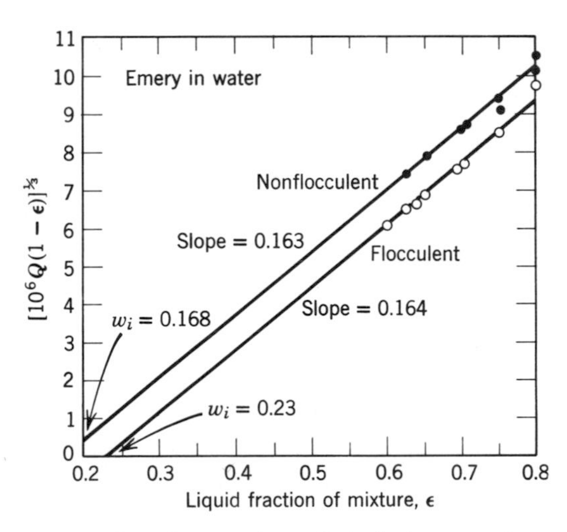

Fig. 12.4. Settlement of flocculent and nonflocculent suspensions of uniform-size angular particles. Data are from [2].

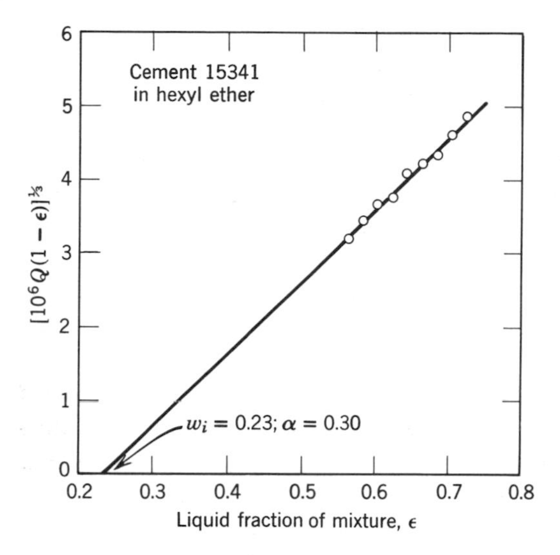

Fig. 12.5. Bleeding data for a flocculent paste made of portland cement and hexyl ether. (Data are from the Portland Cement Association Series 288-11-D4.)

We have already seen evidence in Fig. 12.2 that the original Kozeny-Carman equation is satisfactory for thick suspensions of uniform spheres. But, it is not satisfactory for suspensions of irregular particles, as is shown by the upper line in Fig. 12.4. These are data for angular particles of emery. Seventy-three per cent of the material had Stokes diameters between 10 and 14 μ (Chapter 1, Section 3.2) and a specific surface area, based on average Stokes diameters, of 4930 cm^2/cm^3. The plotted points conform to a straight line fairly well, but the extended line does not pass through the origin as did the line for spherical particles. In this case the intercept is at $\epsilon = 0.168$, $y = 0$, that is, $w_i = 0.168$.

The lower line in Fig. 12.4 represents the material just described, but the mixtures were in the flocculent state, the mix water being an 0.12 per cent solution of zinc sulfate. The points conform to a straight line very well, but this line also does not pass through the origin. It is evident that changing from the nonflocculent to the flocculent state moved the intercept to the right, in this case from $w_i = 0.168$ to $w_i = 0.230$. Other data showed that changing suspensions of spheres from the nonflocculent state increased the intercept from $w_i = 0$ to $w_i = 0.10$, approximately.

Figure 12.5 shows bleeding data for flocculent pastes made with cement and hexyl ether, the cement being the same as that represented in Fig. 12.3. Comparing the intercepts, we see that the effect of changing from water to nonreactive organic liquid was, in this case, to reduce w_i from 0.28 to 0.23.

Thus Steinour found that w_i appears when the particles are angular rather than spherical; it increases when a given kind of suspension is changed from the nonflocculent to the flocculent state; it has a larger value for cement in water than for cement in a nonreactive liquid, presumably because of an effect of chemical reaction. Other data, some of which is shown further on, show that the value of w_i is usually independent of particle size.

From the observations just mentioned and many other collateral investigations Steinour came to the conclusion that w_i does represent water that stays with the particles as they settle, and that in any case where chemical reaction is not involved the quantity of this stagnant water is a characteristic fraction of the amount of solid material present. The stagnant water seems to have the effect of streamlining irregular particles, so that the interface between the stream and the particle is defined by the surfaces of such stagnant bodies of water. Thus each irregular particle is supposed to function as a composite body composed of solid material and an envelope of stagnant water. The specific surface area of the composite body is its surface area divided by its volume, the latter being, of course, the sum of the material and stagnant water. In cement pastes it seems that the particles may be augmented also by a coating of hydration product (essentially cement gel) formed by the initial chemical reactions which were discussed in Chapter 10. Indeed, the supposed layers of stagnant water must be held outside this thin layer of cement gel.

The increase in w_i that Steinour observed when a mixture is changing from the nonflocculent to the flocculent state is not so easily explained. A plausible explanation is that it represents water in isolated pockets, excluded from the floc structure and thus is not to be included when we are calculating the hydraulic radius of the floc structure within which practically all the resistance to settlement is developed. But if the amount of such water is greater the higher the water content of the paste, which would seem to be a reasonable expectation, it does not seem that the amount so excluded should be proportional to the amount of solid material. Alternatively, we may assume that flocculation increases the amount of stagnant water by producing points of virtual contact between particles around which the water is relatively stagnant. On this basis it would be reasonable to expect the quantity to be proportional to the quantity of solid particulate material.

However that may be, it proved to be feasible to assume that w_i is made up as shown in the following equation:

$$w_i = \frac{\mathbf{a} + \mathbf{b} + \mathbf{c}}{1 + (\mathbf{a} + \mathbf{b} + \mathbf{c})} = \frac{\alpha}{1 + \alpha} \tag{12.18}$$

where $\mathbf{a}(1 - \epsilon)$ is the quantity required to streamline irregular particles; $\mathbf{b}(1 - \epsilon)$ is an additional quantity due to the flocculent state of the particles; $\mathbf{c}(1 - \epsilon)$ is the volume added to the grains by chemical reaction and is a quantity depending on the surface area of the material. $\sigma(1 - \epsilon)$. Parameter α is defined by the equation: It is the factor by which the solid volume must be multiplied to obtain the total particle augmentation from all causes.

2.3 *Modification of the Kozeny-Carman Equation*

When w_i is interpreted as above, it follows that not only must the space for flow be considered to be reduced, but also the volume of solid material must be considered to be augmented by a factor $1 + \alpha$. That is to say, each particle is supposed to be a composite body consisting of the particle, its thin coating of hydration products, and its streamlining envelope of stagnant water. The surface of such a composite particle is supposed to be that presented by the streamlined envelope, or mostly so, and it may therefore be called the hydrodynamic surface.

The equation for permeability, Eq. 12.9, is restated in terms of effective quantities, as follows:

$$K = \frac{m_e^2 \epsilon_e}{k_c \eta} \rho_f g \tag{12.19}$$

where the subscript e indicates the effective values of the indicated parameters. The effective porosity is the total porosity reduced by a factor proportional to the volume of solid material, as discussed above. That is

$$\epsilon_e = \epsilon - \alpha(1 - \epsilon)$$

Also, the effective hydraulic radius is

$$m_e = \frac{\epsilon_e}{S_e}$$

where S_e is the effective surface area per unit volume of mixture. It may be expressed as the product of the volume of the composite particles described above, multiplied by the surface area of the composite material. On that basis, we may write

$$m_e = \frac{\epsilon - \alpha(1-\epsilon)}{(1-\epsilon)(1+\alpha)\sigma_h} \tag{12.20}$$

where σ_h is the *hydrodynamic* specific surface area.

If we now define the *apparent* specific area as the hydrodynamic surface area divided by the volume of solid material, we find that

$$\sigma_{ap} = \frac{S_e}{1-\epsilon} = (1+\alpha)\sigma_h = \frac{\sigma_h}{1-w_i} \tag{12.21}$$

where σ_{ap} is the apparent specific surface area.

The effective hydraulic radius in terms of apparent specific surface area is

$$m_e = \frac{\epsilon - \alpha(1-\epsilon)}{(1-\epsilon)\sigma_{ap}} = \frac{[\epsilon/(1-\epsilon)] - \alpha}{\sigma_{ap}} \tag{12.22}$$

Similarly, the effective porosity may be written

$$\epsilon_e = \left[\frac{\epsilon}{1-\epsilon} - \alpha\right](1-\alpha) \tag{12.23}$$

Substituting the relationships above in Eq. 12.19, we obtain the following expression for the coefficient of permeability:

$$K = \frac{\rho_f g}{k_c \eta \sigma_{ap}^2}\left[\frac{\epsilon}{1-\epsilon} - \alpha\right]^3 (1-\epsilon) \tag{12.24}$$

The bleeding rate equation corresponding to Eq. 12.13 is

$$Q = \frac{C}{k_c \sigma_{ap}^2}\left[\frac{\epsilon}{1-\epsilon} - \alpha\right]^3 (1-\epsilon)^2 \tag{12.25}$$

The linear form of Eq. 12.25 is

$$\left[\frac{Q}{(1-\epsilon)^2}\right]^{1/3} = \left[\frac{C}{k_c \sigma_{ap}^2}\right]^{1/3}\left[\frac{\epsilon}{1-\epsilon} - \alpha\right] \tag{12.26}$$

or, in terms of the hydrodynamic specific surface area,

$$\left[\frac{Q}{(1-\epsilon)^2}\right]^{1/3} = \left[\frac{C}{k_c}\frac{1}{(1-\alpha)^2\sigma_h^2}\right]^{1/3}\left[\frac{\epsilon}{1-\epsilon} - \alpha\right] \tag{12.27}$$

It is useful also to obtain the above relationships in terms of w_i, using the relationship $\alpha = w_i/(1-w_i)$. The effective porosity is

$$\epsilon_e = \frac{\epsilon - w_i}{1-w_i} \tag{12.28}$$

For the effective hydraulic radius we have

$$m_e = \frac{\epsilon - w_i}{(1 - \epsilon)\sigma_h} \tag{12.29}$$

These relationships give the following for coefficient of permeability:

$$K = \frac{\rho_f g}{k_c \eta} \cdot \frac{1}{(1 - w_i)\sigma_h^2} \cdot \frac{(\epsilon - w_i)^3}{(1 - \epsilon)^2} \tag{12.30}$$

For the apparent specific surface area we have, by Eq. 12.21,

$$K = \frac{\rho_f g}{k_c} \frac{1}{(1 - w_i)^3 \sigma_{ap}^2} \frac{(\epsilon - w_i)^3}{(1 - \epsilon)^2} \tag{12.31}$$

The bleeding equation in terms of hydrodynamic specific surface area is

$$Q = \frac{C}{k_c(1 - w_i)\sigma_h^2} \frac{(\epsilon - w_i)^3}{(1 - \epsilon)} \tag{12.32}$$

and in terms of apparent specific surface area it is

$$Q = \left[\frac{C}{k_c(1 - w_i)^3 \sigma_{ap}^2}\right] \frac{(\epsilon - w_i)^3}{(1 - \epsilon)} \tag{12.33}$$

2.3.1 Indicated specific surface areas. According to the foregoing considerations, the specific surface area of cement as it exists in freshly mixed cement paste is obtainable from bleeding data. According to Eq. 12.26

$$(\text{slope}_{26})^3 = \frac{C}{k_c \sigma_{ap}^2} = \frac{C}{k_c} \frac{1}{(1 - \alpha)^2 \sigma_h^2}$$

where $(\text{slope}_{26})^3$ is the experimentally observed value, as for example that of the line shown in Fig. 12.6. Hence the apparent specific surface is given by

$$\sigma_{ap}^2 = \frac{C}{k_c} \frac{1}{(\text{slope}_{26})^3} \tag{12.34}$$

and the hydrodynamic specific surface area, which is supposed to be the actual surface area per unit volume of solid plus stagnant liquid, is given by

$$\sigma_h^2 = \frac{C}{k_c} \frac{1}{(1 - \alpha)^2 (\text{slope}_{26})^3} \tag{12.35}$$

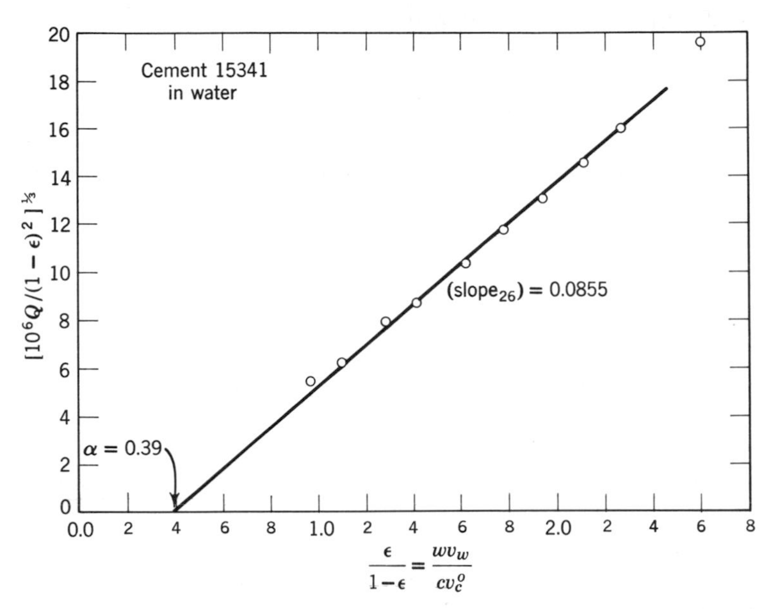

Fig. 12.6. Bleeding data for cement paste plotted according to the modified Kozeny-Carman equation, Eq. 12.26.

The corresponding relationships on the basis of Eq. 12.33 are

$$\sigma_{ap}^2 = \frac{C}{k_c} \frac{1}{(1 - w_i)^3(\text{slope}_{33})^3} \tag{12.36}$$

and

$$\sigma_h^2 = \frac{C}{k_c} \frac{1}{(1 - w_i)(\text{slope}_{33})^3} \tag{12.37}$$

To illustrate the method of calculating specific surface area, we use bleeding data for Cement 15341, which are plotted according to Eq. 12.26 in Fig. 12.6. The necessary data are the following:

Density of the cement:	$\rho_s = 3.147$ g/cm^3
Density of fluid:	$\rho_f = 1.00$
Viscosity of fluid:	$\eta = 0.0094$ poises
Specific surface area by the permeability method:	$\sigma_p = 10600$ cm^2/cm^3
Slope from Fig. 12.6:	$(\text{slope})_{26} = 0.0855$
Parameter α:	$\alpha = 10.39$
Gravitational constant:	$g = 980.6$
Carman constant:	$k_c = 4.06$

These data give (Eq. 12.14),

$$\frac{C}{k_c} = 5.52 \times 10^4$$

Then by Eq. 12.33

$$\sigma_{ap}^2 = \frac{5.52 \times 10^4}{(\text{slope}_{26})^3} = \frac{5.52 \times 10^4}{0.665 \times 10^{-3}} = 83.0 \times 10^{-6}$$

$$\sigma_{ap} = 9110 \text{ cm}^2/\text{cm}^3$$

By Eq. 12.21, we have

$$\sigma_h = \frac{9110}{1.39} = 6560 \text{ cm}^2/\text{cm}^3$$

The relationships between the various figures for specific surface are discussed further on.

There is some uncertainty as to how best to compute the unbuoyed weight of the solid material in the mixture. We need to know the actual density of the solid material after it has gone through the initial reaction, and the density and viscosity of the fluid also at that time, the latter being functions of time. Such data not being available, we must be content with approximate figures knowing that no large error is involved. The question is, How best do we minimize the error? Whether we should use the true density of the cement or its apparent density in water is debatable. The procedure actually followed was to use the true density when calculating the unbuoyed weight, and the apparent density when computing the volume composition of the mixture.

In the above calculation the Carman constant $k_c = 4.06$ was used in accordance with the findings of Steinour. However in what follows the calculations are based on $k_c = 5.0$, the value established by Carman as suitable for many materials and which is used in the standard formula for measuring the specific surface area by the air-permeability method. This choice was made on the ground that it gave the better agreement with results obtained from the viscous-drag equation presented in the last part of this chapter.

Figure 12.7 gives data for three cements prepared from the same clinker, the data being treated according to Eq. 12.26. The features of this diagram are typical of those found in five such groups prepared from different clinkers, with a few exceptions. The outstanding exception is the finest of the three cements represented in Fig. 12.8; it shows a value of α, and hence of w_i, distinctly different from that of the

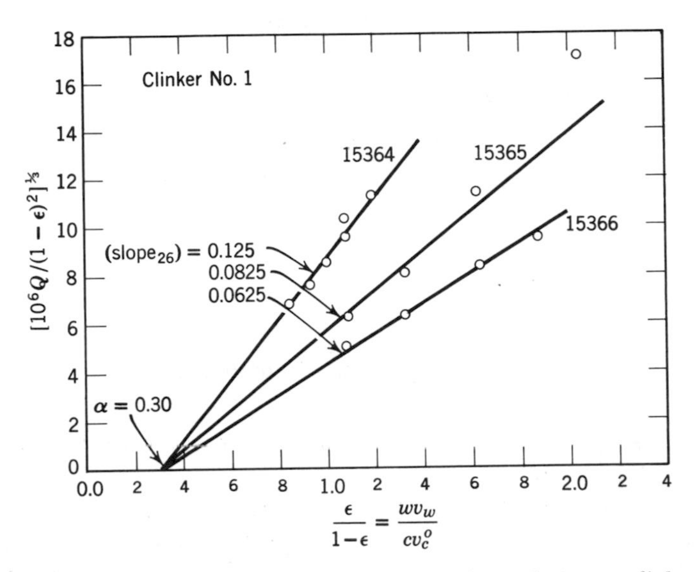

Fig. 12.7. Bleeding data for three cements made from the same clinker. ASTM Type-I composition. (Data are from [15].)

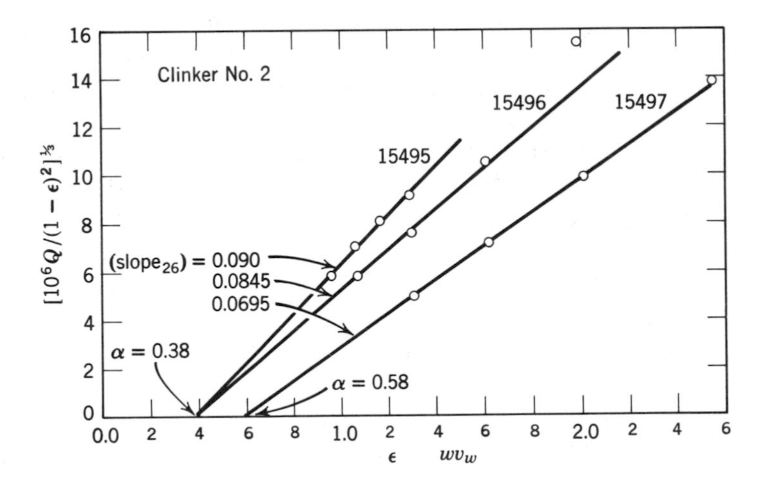

Fig. 12.8. Bleeding data for three cements made from the same clinker. ASTM Type-III composition. (Data are from [15].)

other two. The calculated specific surface falls out of line, as will be seen later.

From the initial report published by Powers [7], it was concluded that w_i is generally higher the finer the cement. From later experiments, Steinour found some instances where the value of w_i seemed to increase with fineness, and others where it did not. Usually the data were such that a more or less arbitrary choice could be made whether the values were or were not independent of the specific surface area. In the early papers the data were always plotted as in Fig. 12.3, and often the value of w_i was less clearly indicated than it is there. When the data are plotted according to Eq. 12.26, that is, as in Fig. 12.7, there appears to be little reason to doubt that the same value holds for all cements made from the same clinker, with the exception noted.

A third example is shown in Fig. 12.9. The finest and coarsest cements seem to conform closely to the lines converging to the same point, but the middle series would permit a number of choices, and probably the statistically best value of α indicated by the four data points would not give the intercept used. However on the basis of five such patterns together, it seems clear that a single value for α would serve as well

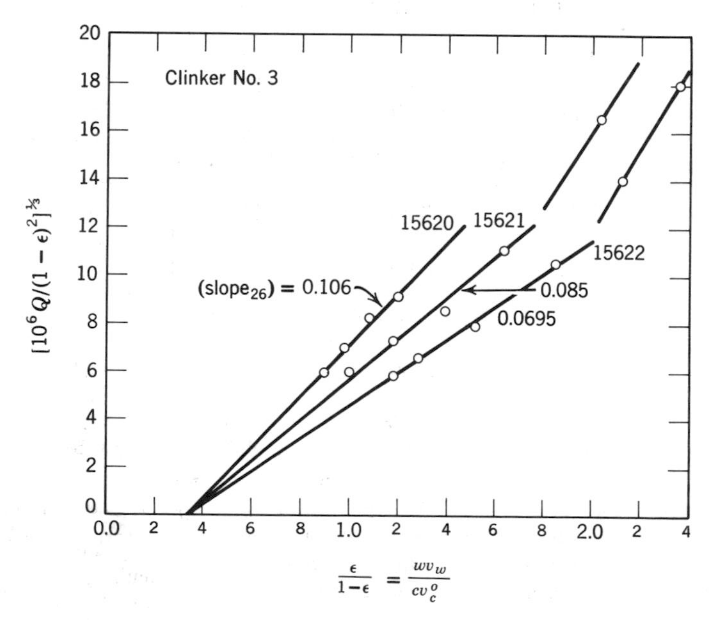

Fig. 12.9. Bleeding data for three cements made from the same clinker. ASTM Type-II composition. (Data are from [15].)

Table 12.1 Specific Surface Areas and Stagnant Water Factors from the Modified Kozeny-Carman Equation[a]

	Data from Cement Paste			Data from Dry Cement	
Cement Number	Stagnant Water Factor (α)	Apparent Specific Surface Area (σ_{ap})	Hydrodynamic Specific Surface Area (σ_h)	By Air Perme- ability (σ_p)	By Wagner Turbidime- eter (σ_W)
Clinker No. 1. ASTM Type-I Composition					
15364	0.30	4800	3770	5980	3310
15365	0.30	8900	6850	9780	5310
15366	0.30	13500	10400	14060	7060
Clinker No. 5 (High Na_2O content) Type-I					
15696	0.39	5600	4020	6800	4160
15760	0.39	7200	5180	7760	4780
15761	0.39	9200	6600	9310	5740
15698	0.39	11900	8500	12000	7280
Clinker No. 2. ASTM Type-III Composition					
15495	0.38	6450	4670	8630	4610
15496	0.38	8550	6200	10540	5480
15497	0.58	11450	7240	16020	7940
Clinker No. 3. ASTM Type-II Composition					
15620	0.33	6250	4700	7460	4400
15621	0.33	8600	6470	9930	5820
15622	0.33	11700	8800	12920	7100
Clinker No. 4. ASTM Type-IV Composition					
15667	0.26	6950	5510	6660	4070
15762	0.26	7590	6010	7680	4880
15763	0.26	9060	7190	8870	5860
15669	0.26	12100	9600	12250	7460

[a] All specific surface areas are in cm^2/cm^3. Commercial clinkers were ground to three degrees of fineness at the respective mills. In some cases intermediate degrees of fineness were obtained by blending in the laboratory. The calculation by the modified Kozeny-Carman equation were based on $k_c = 5.0$. (See Table 12.2 for compositions.)

Table 12.2 Chemical Constitutions of Clinkers from which Cements of Tables 12.1 and 12.5 Were Prepared[a]

Clinker Number	ASTM Type	Calculated compounds				MgO	Free CaO	SO_3	Alkalies		Loss on Ign.
		C_3S	C_2S	C_3A	C_4AF				Na_2O	K_2O	
1	I	46.5	27.6	14.3	6.8	2.6	1.0	0.4	0.17	0.16	0.2
2	III	62.9	12.6	11.1	7.7	1.4	2.7	0.2	0.30	0.40	0.9
3	II	47.6	30.9	4.8	13.2	1.4	0.7	0.0	0.05	0.22	1.1
4	IV	37.1	52.2	2.3	5.7	1.8	0.2	0.2	1.13	0.44	0.3
5	I	48.4	28.7	10.1	7.6	3.4	0.0	0.2			0.4

[a] C_3S stands for tricalcium silicate. C_2S stands for beta-dicalcium silicate. C_3A stands for tricalcium aluminate. C_4AF stands for tetracalcium aluminoferrite. All quantities are given as the per cent by weight.

or better than would allowing it to be determined by the data for each cement individually. Anyhow, as we shall see in the final part of this chapter, there is reason to doubt the reality of α or w_i as a measure of the augmentation of the grains, and therefore the considerations that lead us to expect the parameter in question to vary with fineness seem less cogent than they once did.

Table 12.1 gives the values of specific surface area of the cements discussed above as computed from the bleeding data, and as determined on the dry cements by two standard ASTM methods. The chemical constitutions of the clinkers from which the cements were prepared are given in Table 12.2.

We see that the *apparent* specific surface area in water is generally similar to the figure given by the air permeability method; the hydrodynamic specific surface area is comparable with, although generally larger than, the value given by the ASTM turbidimeter method. The theoretical relationship between the various specific surface areas obtained from a given cement are discussed next. Further discussion of the surface areas of the different cements is deferred to the latter part of the chapter.

2.4 *Relationships between Hydrodynamic, Apparent, and Equivalent-Sphere Specific Surface Areas from Sedimentation Analysis*

To interpret the bleeding equations, it is helpful to compare the various specific surface areas with the specific surface areas of the same materials as determined by sedimentation analysis performed on very dilute suspensions. Before doing so it is necessary to show to what degree the different values might be expected to agree.

When it is observed that a certain particle having a known density ρ_s and volume v settles through a liquid having density ρ_f at velocity V_s the surface area per unit volume can be calculated from the following form of Stokes's law:

$$\sigma_o^2 = \frac{2(\rho_s - \rho_f)g}{\eta} \cdot \frac{1}{V_s} \tag{12.38}$$

where σ_o is the specific surface area if the particle is spherical, or it is the equivalent-sphere specific surface area if it is not spherical; it is the value obtained by sedimentation analysis in a very dilute suspension.

If the particle is not spherical, we may choose to assume that it becomes streamlined by stagnant liquid. The velocity of fall is then determined by the effective particle volume, the effective surface area, and the average density of the composite particle. Thus for effective volume v_e we have

$$v_e = v(1 + \alpha)$$

and for effective surface area

$$S_e = v(1 + \alpha)\sigma_h$$

For the average density, we have

$$(\rho_s)_e = \frac{\rho_s + \alpha\rho_f}{1 + \alpha}$$

where $(\rho_s)_e$ is the density of the composite particle, and $\alpha\rho_f$ is the weight of the stagnant water associated with 1 cm³ of solid material. Then

$$(\rho_s)_e - \rho_f = \frac{(\rho_s - \rho_f)}{1 + \alpha}$$

Using the above expressions in Stokes's law, we obtain

$$\sigma_h = \frac{S_e}{v_e} = \frac{2(\rho_s - \rho_f)g}{\eta(1 - \alpha)} \cdot \frac{1}{V_s} \tag{12.39}$$

Thus, if the factor α is known, the hydrodynamic specific surface area of an irregular particle can be computed from its Stokes velocity. The hydrodynamic surface area per unit of solid material, that is to say, the apparent specific surface area, is S_e/v. Hence, as before (Eq. 12.21),

$$\sigma_h = \frac{\sigma_{ap}}{(1 + \alpha)}$$

and

$$\sigma_{ap} = \frac{2(\rho_s - \rho_f)g}{\eta V_s} = \sigma_o \qquad (12.40)$$

Thus we see that if the particles in a sedimentation analysis actually carry an envelope of water equal to α times its own volume, the value reported is the apparent specific surface area and is $(1 + \alpha)$ times the hydrodynamic surface area of the particle-plus-water complex: σ_0 and σ_{ap} are identical.

In connection with Eq. 12.17 it was pointed out that the slope of the linear plot is found, in nearly all cases, to satisfy the relationship

$$(\text{slope})^3 = \frac{C}{5} \cdot \frac{1}{\sigma_W^2}$$

To the degree that this is true, it means that the Wagner specific surface has the same value as other expressions used in its place in Eq. 12.32 and 12.33. Thus

$$5\sigma_W^2 = k_c(1 - w_i)\sigma_h^2$$

and

$$\sigma_h = \left(\frac{5}{k_c}\right)^{1/2} \frac{\sigma_W}{(1 - w_i)^{1/2}} = \left(\frac{5}{k_c}\right)^{1/2} (1 + \alpha)^{1/2} \sigma_W \qquad (12.41)$$

Similarly,

$$\sigma_{ap} = \left(\frac{5}{k_c}\right)^{1/2} \frac{\sigma_W}{(1 - w_i)^{3/2}} = \left(\frac{5}{k_c}\right)^{1/2} (1 + \alpha)^{3/2} \sigma_W \qquad (12.42)$$

For example, the data in Fig. 12.6, which represents four tests per point, give $\sigma_h = 6560$, using $k_c = 4.06$; $(1 + \alpha)^{1/2} = 1.39^{1/2} = 1.18$. The Wagner surface factor is $\sigma_w = 5240$. Then from Eq. 12.41 we have

$$\sigma_h = \left[\frac{5}{4.06}\right]^{1/2} 1.18 \times 5240 = 1.31 \times 5240 = 6860 \text{ cm}^2/\text{cm}^3$$

which is in this case 4 per cent larger than that computed directly from the observed slope.

It is not required by the assumptions made in deriving the modified

Kozeny-Carman equation that the amount of stagnant water in cement paste be the same as the amount of stagnant water associated with the same particles in a dilute suspension such as is used in a sedimentation analysis. However if we assume that they are the same, it would follow that the apparent specific surface area given by the bleeding test data and Eq. 12.34 should be the same, or approximately the same, as the equivalent-sphere value σ_o, as shown in Eq. 12.41. Values of σ_o from sedimentation analyses are not available. However the value obtained from the permeability method was found by Lea and Nurse [8] to be, on the average, 1.22 σ_o. Therefore for a general comparison we can use that estimate of σ_o.

The average value of σ_p in Table 12.1 is 9440 cm^2/cm^3, which gives $\sigma_o = 9440/1.22 = 7738$. The average value σ_{ap} for the data in Table 12.1 is 8616, or 1.14 times the estimated σ_o.

If this result is taken to indicate that the apparent specific surface area of cement in paste is at least as much as the apparent specific surface area of the same cement as determined by the sedimentation method, we are confronted with the rather incredible proposition that the stagnant water around the individual particles in a thick cement paste presents approximately the same surface area per particle as it does in a very dilute suspension.

Moreover, it appears that the specific surface area of cements after reaction with water is just as high as it was before; indeed, it would appear to have been increased on the average about 14 per cent. The Type-III cements of Table 12.1 give $\sigma_{ap} = 1.3\ \sigma_p$, whereas for the Type-IV group the corresponding ratio is 0.99. Thus it appears that the least chemically active cement has the same specific surface area after the initial reaction as it had before, and the most active cement has a 30 per cent *greater* specific surface area than before. These particular comparisons are of special interest in connection with the analysis in the next part of this chapter. That analysis leads to the conclusion that the specific surface area of this particular group of Type-IV cements remains unchanged, but the Type-III group loses about 2500 cm^2/cm^3 during the initial reaction with water. Most cements lose surface area during the initial reaction.

3. AN EMPIRICAL BLEEDING EQUATION BASED ON STOKES'S LAW

As far as representation of experimental data is concerned, the modified Kozeny-Carman equation serves the purpose well enough. But we have already seen reasons to doubt that the irregular cement grains are actually enveloped in stagnant water. As long as we think of relatively isolated particles, as in the sedimentation analysis, a streamlining envelope amounting to about 30 per cent of the volume of the particle is not difficult to visualize, although it is not self-evident that cement particles are actually streamlined to any such degree.

The principal difficulty arises when we are considering thick, flocculent pastes. The average distance between grains is usually smaller than the average size of the grains. Moreover, owing to interparticle attraction, the space around a particle tends to be occupied by smaller particles. It is therefore not clear how the envelopes of the small and large grains can coexist if the envelopes amount to as much as 30 per cent of the volume of each particle. In other words, the experimental evidence given above indicates that the small grains do contribute to the effective surface area, yet it would seem that many if not most of them would have to be within the stagnant envelopes of the larger grains.

However that may be, we shall see in the following discussion that it is possible to develop a bleeding equation that involves no assumption about stagnant water. We shall see, also, that it gives the same specific surface for portland cement in cement paste as is found by sedimentation analysis, provided that the surface area has not been reduced by chemical reaction; and, as is mentioned above, this is the case only for certain Type-IV cements.

3.1 *Viscous Drag of Particles*

The following very simple development is based on a statement of what the resistance to flow would be if each particle making up the permeable body developed the same amount of viscous drag that it would were it alone in a virtually infinite volume of liquid having viscosity η and moving at velocity V relative to the position of the particle.

According to Stokes's law, for a particle of radius r,

$$f = 6\pi\eta r V \tag{12.43}$$

where f is the drag in dynes.

By way of rationalization of Stokes's law, it may be helpful to rewrite it as follows:

$$\begin{pmatrix}\text{Viscous drag of one}\\ \text{sphere moving at}\\ \text{unit velocity in a}\\ \text{liquid of unit viscosity}\end{pmatrix} = 2\left(\frac{\text{surface area}}{\text{volume}}\right)(\text{projected area})$$

$$= 2 \cdot \frac{4\pi r^2}{(4/3)\pi r^3} \cdot \pi r^2 = 6\pi r$$

The drag force per unit volume of material in the particle is

$$\frac{f}{\frac{4}{3}\pi r^3} = F = \frac{9}{2}\frac{\eta}{r^2}V = \frac{\eta\sigma_o^2}{2}V \tag{12.44}$$

In terms of the conditions shown in Fig. 12.1, the total force represented by Eq. 12.44 accounts for pressure drop through the plug if the viscous drag per unit volume of solid material is multiplied by the total volume of solid material $(1 - \epsilon)$. Thus the pressure drop in dyn/cm² is given by

$$F(1 - \epsilon) = \frac{\Delta P}{L} = \frac{\eta\sigma_o^2}{2}(1 - \epsilon)V \tag{12.45}$$

V should be the velocity at which the liquid approaches the spheres. If the apparent rate of flow through the porous body is Q cm³/cm², the apparent velocity within the body is Q/ϵ, which can be substituted for V.*

Also, remembering that the development above accounts only for Stokes resistance when there is an infinite dispersion of particles, we must insert a factor to account for additional viscous resistance due to hydrodynamic interaction. We shall indicate this to be a function of the particle concentration, that is, $\phi(1 - \epsilon)$.

Now we can write the expression for rate of flow through the porous body as follows:

$$Q = \frac{2}{\eta\sigma_o^2} \cdot \frac{\epsilon}{1 - \epsilon} \cdot \phi(1 - \epsilon)\frac{\Delta P}{L} \tag{12.46}$$

or

$$Q = \left[\frac{2}{\eta\sigma_o^2} \cdot \frac{\epsilon}{1 - \epsilon} \cdot \phi(1 - \epsilon)\rho_f g\right]\frac{\Delta h}{L} \tag{12.47}$$

in which the expression in brackets is the permeability coefficient K of Eq. 12.1.

To obtain a bleeding equation, we may specify that the hydraulic gradient shall be that corresponding to the unbuoyed weight of the submerged solid material. Thus, as in Eq. 12.12,

$$\rho_f g\frac{\Delta h}{L} = (\rho_s - \rho_f)g(1 - \epsilon)$$

and

$$Q = \left[\frac{2(\rho_s - \rho_f)g}{\eta\sigma_o^2}\right]\epsilon \cdot \phi(1 - \epsilon) \tag{12.48}$$

* It should be noticed that the tortuosity need not be explicitly represented because the flow around each particle is implicit in Stokes's law. Q/ϵ corresponds to the Stokes velocity of a falling particle, although the relationship here is not restricted to that velocity.

The quantity in brackets may be recognized as Stokes's law for the velocity of fall of a single sphere in an infinite body of liquid.

If the equation is to be applied to a mixture of particles of many sizes, as it will be, we must assume that if there were an infinite dispersion of particles all particles somehow are forced to fall at the same rate as one particle having the same specific surface area as that of all the particles together. This assumption is defensible on the ground that results from cement are obtained that agree with the results of sedimentation analysis, which is made under conditions that allow each size to settle almost uninfluenced by other particles.

3.1.1 Application to settlement of uniform spheres. Experiments show that the rate of settlement is a semilogarithmic function of the particle concentration, as is shown in Fig. 12.10. (These data are from the same source as the data in Fig. 12.2.) The data conform fairly well to a line drawn so that it terminates at the measured Stokes velocity of a single particle. (As mentioned in Section 2.2.1. the Stokes velocity

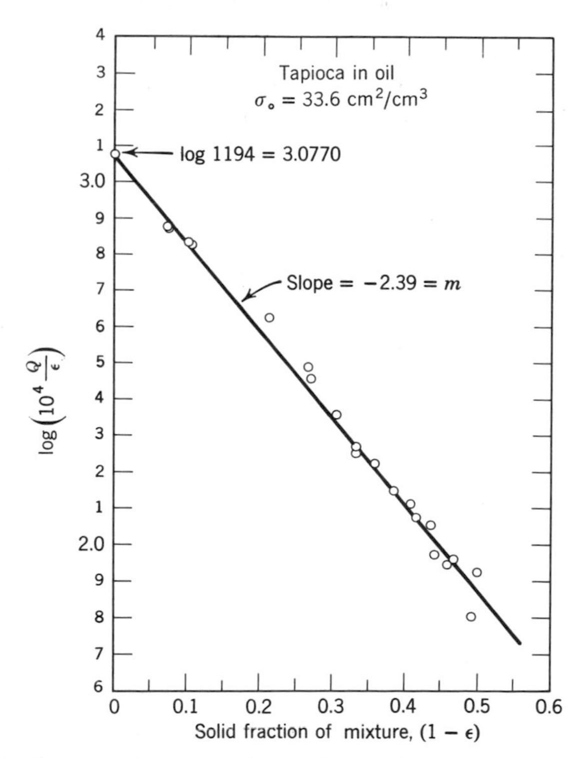

Fig. 12.10. Settlement of nonflocculent spheres plotted according to the viscous-drag equation. Eq. 12.50.

was evaluated by measuring the rates of fall of 150 individual tapioca particles.) The average velocity was 0.1120 ± 0.009 cm/sec. Application of a correction for a wall effect gave 0.1194 ± 0.0095 cm/sec.

Thus the equation for settlement has the form

$$Q = Q_o \epsilon e^{-k(1-\epsilon)} \tag{12.49}$$

where Q_0 is the Stokes velocity (see Eq. 12.48) and k is an empirical constant. In the form plotted in Figure 12.10, Eq. 12.49 appears as follows:

$$\log \frac{Q}{\epsilon} = \log Q_o - m(1 - \epsilon) \tag{12.50}$$

where

$$m = \log k \log e = 0.4343k \tag{12.51}$$

The slope of the line is the same for different sizes of uniform spheres. Figure 12.11 represents microscopic, uniform glass spheres with a specific surface area of 4440 cm^2/cm^3 as compared with 33.6 cm^2/cm^3 for the tapioca. The slopes are the same at $-m = 2.39$.

3.1.2 Graded spheres. The settlement of graded spheres also complies with Eq. 12.50, but with a different value of m. Figure 12.12 represents materials made by spherizing the particles of pulverized glass, and

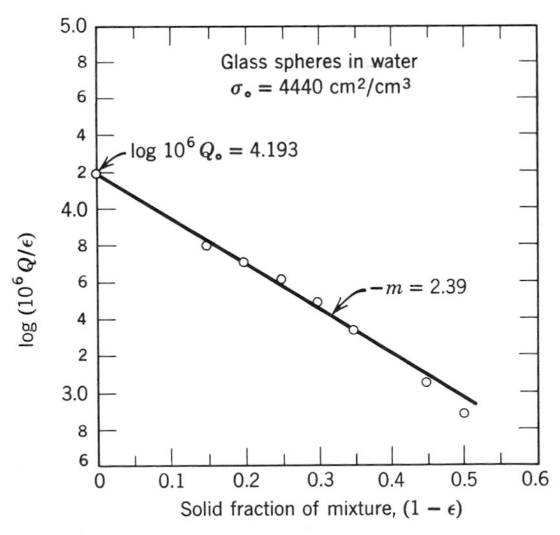

Fig. 12.11. Settlement of nonflocculent microscopic glass spheres in water plotted according to Eq. 12.50. (Data are from [2].)

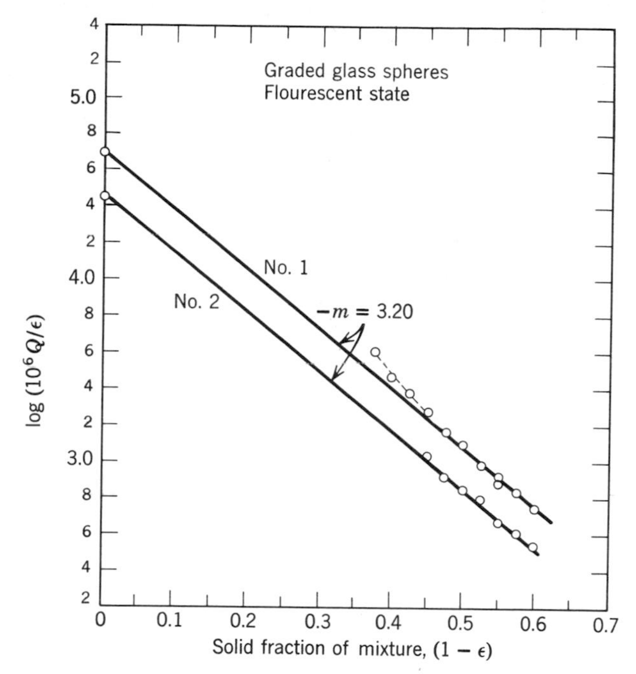

Fig. 12.12. Settlement of flocculent pastes of graded glass spheres. For paste No. 1, $\sigma_o = 2880\ cm^2/cm^3$. For paste No. 2, $\sigma_o = 3190\ cm^2/cm^3$. Values of σ_o were cal-calculated from σ_p using the relationship $\sigma_o = 0.8\ \sigma_p$.

therefore each comprises a wide range of particle size. The terminal points correspond to 0.8 σ_p, for it has been found that the factor usually brings the specific surface area as determined by the permeability method in line with that found by sedimentation analysis. Some of the points for the coarser material do not conform, apparently because of channeled bleeding.

As before, the value of m is seen to be independent of particle size, but it is larger than it is for a uniform, nonflocculent suspensions of spheres: 3.20 as compared with 2.39. The increase is due to flocculation, as will be seen below, but some of it is probably due to the range of sizes.

3.1.3 Settlement of uniform angular particles. Equation 12.50 also holds for the settlement of angular particles whether they are in the flocculent or nonflocculent state, as is shown in Fig. 12.13, which is plotted from Table 12.3. Emery particles of nearly uniform size are represented. From a sedimentation analysis, Steinour found the average

Table 12.3 Data on the Settlement of Emery Particles[a]

Fluid Content (ϵ)	Solid Content ($1-\epsilon$)	Settling Rate ($10^4 Q$ cm/sec)	$\frac{10^4 Q}{\epsilon}$	Log $\frac{10^4 Q}{\epsilon}$
Non-flocculent State—in water				
0.65	0.35	13.9	21.4	1.33
0.70	0.30	21.6	30.9	1.49
0.75	0.25	33.8	45.1	1.65
0.80	0.20	54.1	67.6	1.83
Non-flocculent State—Hexametaphosphate Added				
0.62	0.38	11.1	17.7	1.25
0.65	0.35	13.7	21.1	1.32
0.70	0.30	21.9	31.2	1.49
0.75	0.25	32.5	43.4	1.63
0.80	0.20	51.3	64.1	1.81
0.85	0.15	75.5	88.8	1.95
Flocculent State				
0.60	0.40	5.66	9.43	0.98
0.62	0.38	7.36	11.8	1.07
0.64	0.36	8.06	12.6	1.10
0.65	0.35	9.32	14.3	1.16
0.70	0.30	14.5	20.8	1.32
0.70	0.30	15.2	21.7	1.34
0.75	0.25	25.0	33.3	1.52

[a] Data are from H. H. Steinour, "Rate of Sedimentation," *Ind. Eng. Chem.*, **36**, 840–847, 1944.

$$\eta = 0.916 \times 10^{-3} \text{ poises}$$

$$\rho_f = 0.997 \text{ g/cm}^3$$

$$\rho_s = 3.79 \text{ g/cm}^3$$

$$\rho_s - \rho_f = 2.79 \text{ g/cm}^3$$

Stokes velocity to be $Q_o = 250 \times 10^{-4}$ cm/sec, and $\log(10^4 Q_o) = 2.398$. Accordingly, the lines in Fig. 12.13 are drawn to a terminal point of 2.40. Again we find good agreement.

For the emery particles in the nonflocculent state we find $-m = 3.00$ as compared with 2.39 for nonflocculent spheres. Therefore we conclude

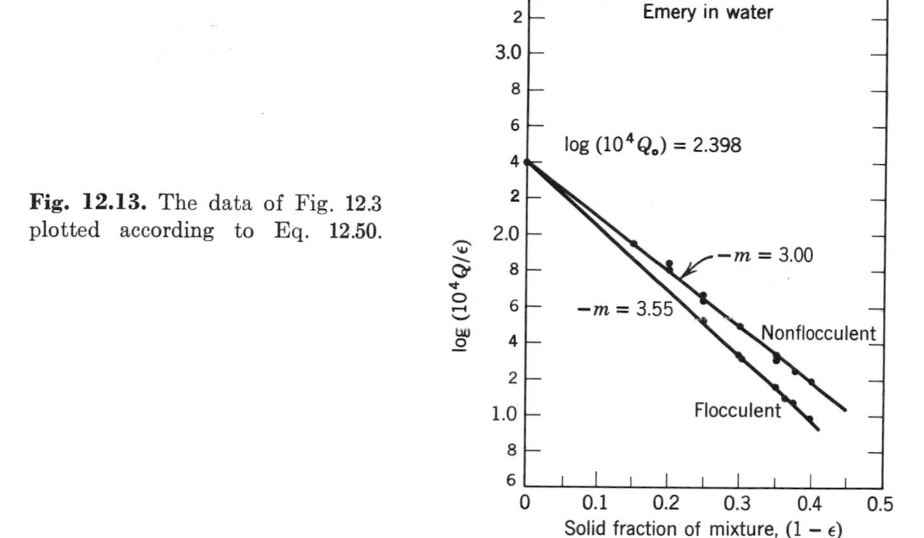

Fig. 12.13. The data of Fig. 12.3 plotted according to Eq. 12.50.

that m is greater for angular particles than it is for spheres, and for a given particle shape it is greater for the flocculent state than for the nonflocculent.

3.1.4 Significance of differences in *m* or *k*. Some of the relationships brought out in the discussion above are summarized in Table 12.4. Among the examples given, the value of σ_o was in every case computed from the indicated Stokes velocity Q_o, together with the densities and the viscosity. This means that any difference in bleeding rate not accounted for in terms of the specific surface area as determined by sedimentation analysis is expressed in terms of the differences in m. For example, if spheres and angular particles have the same specific surface area, σ_o, at a given solids content, the paste made with the angular particles will show the lower bleeding rate, its value of m being the higher. Since only flocculent cement pastes bleed normally, we are not especially concerned with the influence on m of the change from the nonflocculent to the flocculent state.

Thus m in the viscous-drag equation seems to play somewhat the same role as does α or w_i in the modified Kozeny-Carman equation. There is a significant difference however: In the Kozeny-Carman equation the indicated specific surface areas generally seem to be the same as, or greater than, that of the original dry cement, and thus all the observed differences between cements having the same fineness seem

Table 12.4 Factors Influencing the Hydrodynamic Interaction Factor in Suspensions

Material	Specific Surface Area σ_o (cm^2/cm^3)	State of the Suspension[a]	Interaction Factor	
			$-m$	$-k$
Uniform spheres	33.6	N-F	2.39	5.50
Uniform spheres	4440	N-F	2.39	5.50
Graded spheres	2480	F	3.38	7.78
Graded spheres	3190	F	3.38	7.78
Uniform emery	4880	N-F	3.00	6.91
Uniform emery	4880	F	3.55	8.18

[a] The abbreviation N-F stands for the nonflocculent state and F stands for flocculent state.

to be expressed in the differences in α or w_i; but as we shall see later, the viscous-drag equation shows that in most cases the specific surface area of the cement after it has been mixed with water is distinctly lower than it was originally. The main differences in the bleeding rate are due to these changes in effective specific surface area, differences in m or k being of relatively minor practical importance.

We may note in passing that m seems to be larger for packed beds than for suspensions. The data by Loudon [9] gives the results for uniform glass beads presented in the accompanying table. The average, 2.65, is to be compared with 2.39 for suspensions.

Specific Surface Area (cm^2/cm^3)	$-m$
72.2	2.624
110.9	2.664
198.0	2.644
365.9	2.656
Average	2.647

Considering the steps that were taken in developing Eqs. 12.49 and 12.50, we can consider k or m to be a hydrodynamic interaction factor. It seems to depend on the pattern of flow as determined by the particle concentration, particle shape, and by the pattern of particle distribution in the water-filled space. Within whatever limits of $(1 - \epsilon)$ the value of k or m remains constant, changing the particle concentration, that is $(1 - \epsilon)$, does not seem to change the flow pattern, and thus the simple exponential term serves to evaluate the effect of changing the particle concentration on the coefficient of permeability, or bleeding rate. For cement paste this range of concentration seems to be identical with that which we have previously defined as the range of normal bleeding. Some examples are given in the following section.

3.2 *Application of the Viscous-drag Equation to Cement Pastes*

Within certain limits of solids content, the bleeding rates of cement pastes conform to Eq. 12.49. This is illustrated in Fig. 12.14 where the same data that are shown in Fig. 12.3 are plotted according to Eq. 12.50. It is evident that for cement contents greater than 0.3 the points fall on a straight line having the slope $m = -4.12$ and having an intercept indicating $10^6 Q_o = 10^4$.

The points for the lower cement concentrations do not fall on that line; rather they seem to describe a straight line having a much greater slope. (Compare the line with that of the coarser of the glass spheres in Fig. 12.12. Also see Fig. 12.3.) However, considering the composition

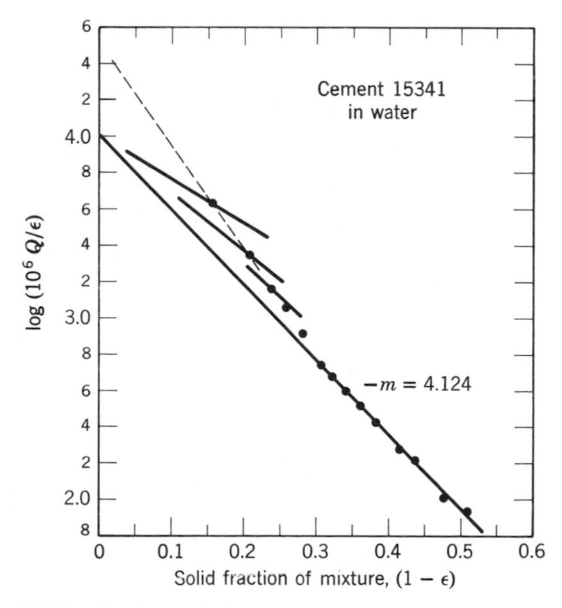

Fig. 12.14. The data of Fig. 12.3 plotted according to Eq. 12.50.

of parameter Q_o, particularly that it is an inverse function of the specific surface area, and that each point represents the same cement, we must regard each of these points as falling on a different line, as indicated.

In line with the discussion above, we may regard all those mixtures that fall on one line as having geometrically similar flow patterns, whereas the points that do not conform represent mixtures that have patterns geometrically dissimilar from that of the pastes that conform. The dissimilarity varies in degree with differences in the cement content within that range. The difference is probably the difference between the pattern of normal bleeding and that of channeled bleeding, as discussed before.

The method of computing specific surface area may be illustrated in terms of this example. From the data for this cement, which are given in Section 2.3.1, we find

$$\frac{2(\rho_s - \rho_f)g}{\eta} = 4.47 \times 10^5$$

Hence

$$Q_o = \frac{4.47 \times 10^5}{\sigma_o^2}$$

From Fig. 12.14, we see that $10^6 Q_o = 10^4$ cm/sec, and thus $Q_o = 10^{-2}$. Therefore

$$\sigma_o^2 = \frac{4.47 \times 10^5}{10^{-2}} = 44.7 \times 10^6$$

$$\sigma_o = 6690 \text{ cm}^2/\text{cm}^5$$

This result may be compared with those given in Section 2.3.1: $\sigma_{ap} = 9110$, and $\sigma_h = 6560$. (We make detailed comparisons further on.)*

Plots for other cements are shown in Figs. 12.15, 12.16, and 12.17, which correspond to Figs. 12.7, 12.8, and 12.9. Looking first at Fig. 12.15, we see that the three straight lines are drawn parallel, somewhat

* Incidentally it is of some interest to observe that if a straight line through the nonconforming points in Fig. 12.14 is extrapolated, the indicated value of $10^6 Q_o$ is about 4×10^4, and the calculated specific surface area is about 3400 cm^2/cm^3. It is perhaps of greater interest to note from the same extrapolation that the value of m based on $10^6 Q_o = 10^4$ becomes zero when $(1 - \epsilon)$ is about 0.09. This would seem to be the point where the extra viscous drag disappears, and the rate of fall cannot rise any more as the content of solids is reduced. However we must remember that the conditions for flocculence exist, and as the suspension is made more and more dilute it finally becomes a suspension of discrete floccules which individually are able to settle faster than the individual particles. Hence the value of Q could continue to rise as indicated.

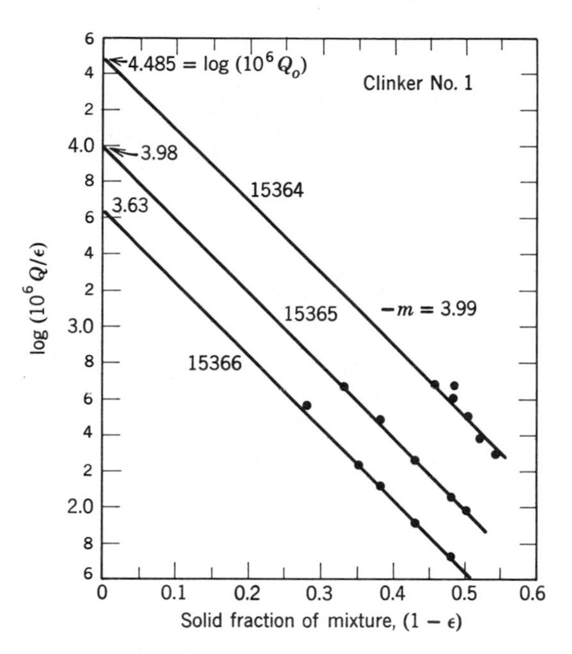

Fig. 12.15. The data of Fig. 12.7 plotted according to Eq. 12.50.

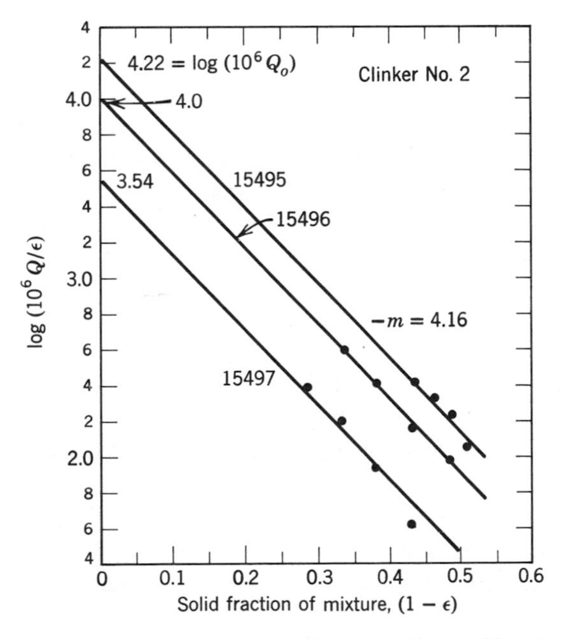

Fig. 12.16. The data of Fig. 12.8 plotted according to Eq. 12.50.

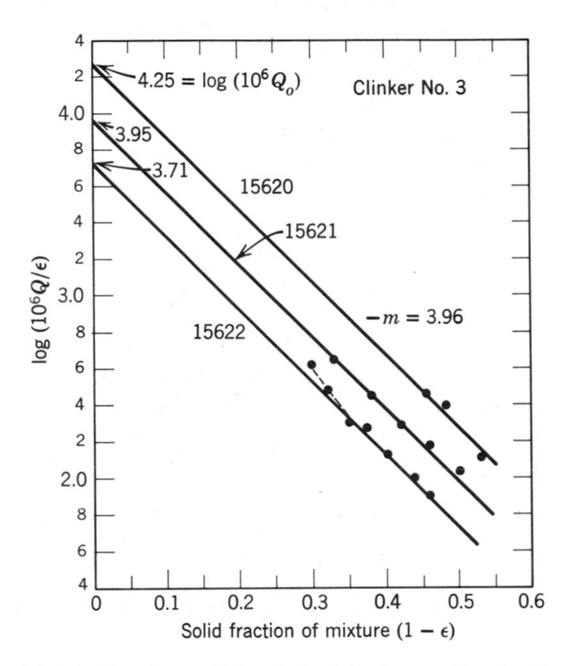

Fig. 12.17. The data of Fig. 12.9 plotted according to Eq. 12.50.

arbitrarily, to give the best fit (approximately) to all points. That procedure amounts to assuming that the empirical constant m is characteristic of a given kind of material and given pattern of particle distribution, and is independent of specific surface area, in line with other data such as those shown in Fig. 12.12 for glass spheres.

It is apparent that there are some cases for which keeping the lines parallel does not give the best fit—an obviously different trend seems to have been ignored. However it turns out in each case that if the indicated trend is used, an absurd value of Q_o is obtained. The corresponding value for the specific surface area is completely out of line with the specific surfaces indicated for the other members of the group; usually it is absurdly low. Therefore it was concluded on the basis of five such groups that such departures represent experimental vagaries.

The specific surface areas indicated by the values of Q_o for the five groups of cements are given in Table 12.5. Along with them, the values of σ_h as calculated from the modified Kozeny-Carman equation and the air-permeability value σ_p are shown for comparison.

Table 12.5 Specific Surface Areas and Hydrodynamic Interaction Factors from the Viscous-drag Equation[a]

1	2	3[b]	4[c]	5[d]
Cement Number	Interaction Factor $-m$	σ_o	σ_h	σ_p
Clinker No. 1 ASTM Type-I				
15364	3.99	3800	3770	5980
15365	3.99	6800	6850	9780
15366	3.99	10200	10400	14060
Clinker No. 5 ASTM Type-I, High Na_2O Content				
15696	4.36	3980	4020	6800
15760	4.36	5010	5180	7760
15761	4.36	6680	6600	9310
15698	4.36	8720	8550	12000
Clinker No. 2 ASTM Type-III				
15495	4.16	5140	4670	8630
15491	4.16	6600	6200	10540
15497	4.16	11200	7240	16020
Clinker No. 3 ASTM Type-II				
15620	3.96	5130	4700	7460
15621	3.96	7120	6470	9930
15622	3.96	9400	8800	12920
Clinker No. 4 ASTM Type-IV				
15667	3.70	5540	5510	6660
15762	3.70	6200	6010	7680
15763	3.70	7100	7190	8870
15669	3.70	9600	9600	12250

[a] All surface data are given as cm^2/cm^3. For the chemical constitution of the clinkers see Table 12.2.

[b] This is the specific surface area determined from the bleeding rate by Eq. 12.50.

[c] This is the "hydrodynamic" specific surface area determined from the bleeding rate by Eq. 12.35.

[d] This is the specific surface area by the permeability method.

3.2.1 Comparison with Kozeny-Carman results. Figure 12.18 shows the relationship between the hydrodynamic specific surface area, as computed from the modified Kozeny-Carman equation, and the surface area given by the equation based on Stokes's law. For three of the five materials there is practically a perfect correlation at all values of σ_o, the points describe one straight line. For the other two groups the points fall on regression lines parallel to the line just mentioned. In those cases the discrepancies can easily be eliminated by making small changes in the original estimate of the best value of α or w_i. One cement is an exception: It is the finest of the three in Fig. 12.8, for which the value of α appropriate for the other two does not seem suitable.

It seems that usually σ_h as obtained from cement paste by the modified Kozeny-Carman equation is, within experimental uncertainty, the equivalent of σ_o as computed from cement paste by the equation based on Stokes's law. It is obvious, also, that this value is smaller than the air-permeability value σ_p and, in most cases, smaller than the estimated specific surface area that would be given by sedimentation analysis of very dilute suspensions. It will be shown in the ensuing discussion that the lowness of the specific surface area values found in cement paste is probably due to the actual loss of surface area resulting from the using up of grains of gypsum and small grains of cement during the initial chemical reactions. First we shall examine cases where no loss of surface area is indicated.

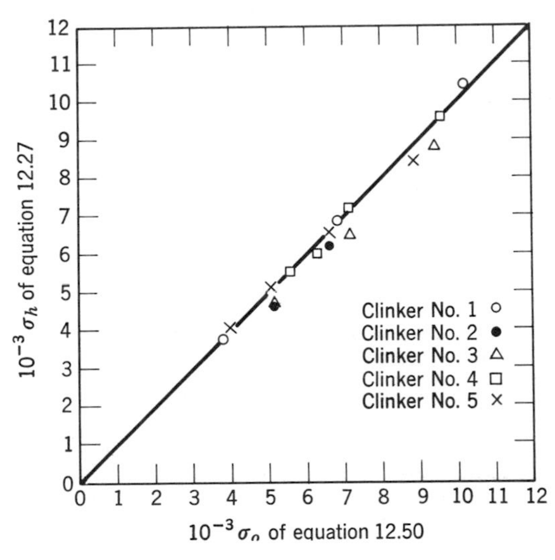

Fig. 12.18. Plot showing equivalence of σ_h in Eq. 12.27 and σ_o in Eq. 12.50.

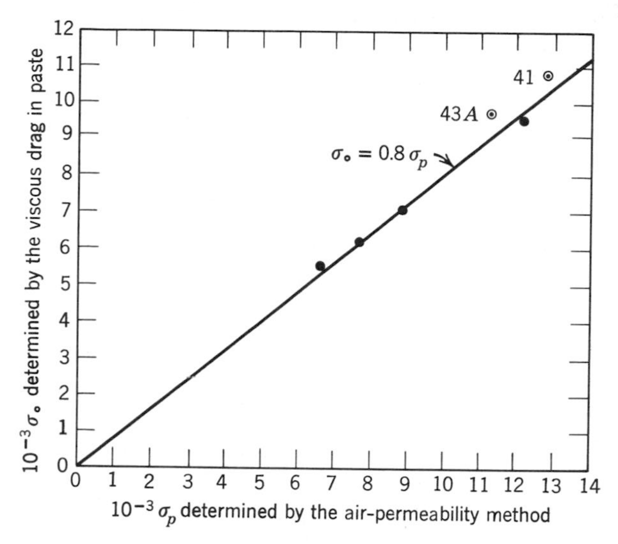

Fig. 12.19. The relationship between σ_o in Eq. 12.50 and the air-permeability specific surface area for certain Type-IV cements. The solid dots represent cement made with clinker No. 4. The other points are for the cements indicated by number.

3.3 *Agreement with Sedimentation Analysis*

We could test the validity of Eq. 12.49 or 12.50 in a direct way by comparing the value of specific surface area obtained from it with the value obtained from the same material by a sedimentation analysis of a very dilute suspension in a nonreactive liquid, but, unfortunately, sedimentation data (by the Andreasen method, for example), are not available for the same cements for which bleeding data are available. However, the value that would be obtained from a sedimentation analysis, σ_o, can be estimated from the air-permeability value, σ_p; Lea and Nurse [8] reported that the specific surface area obtained from the Andreasen sedimentation analysis is, on the average, 82 per cent of the value given by the air-permeability method. That is to say, we assume that $\sigma_o = 0.8\sigma_p$, and by plotting the values obtained from the bleeding rate against σ_p we obtain a reasonable approximation of the desired comparison.

Application of the method of comparison just described showed that with certain cements the specific surface area derived from the bleeding rate is the same as that obtained from sedimentation analysis, σ_o. This is shown in Fig. 12.19. Four of the points represent cements prepared from clinker No. 4. (See Table 12.2.) The two other points are for cements prepared from other clinkers. Chemical compound compositions are given in Table 12.6.

It thus seems that the specific surface area calculated from the bleeding rate is actually σ_o, the specific surface area corresponding to the Stokes equivalent-sphere, and that is why the symbol was so marked in Eq. 12.48.

When the value of σ_o is equal to 0.8 σ_p it signifies that the specific surface area of the cement after the cement has been made into paste is the same as it was before the cement made contact with water, or virtually so. However cements for which this is true are exceptional, as is shown below.

3.4 Loss of Surface Area in Cement Paste

Most of the data in Table 12.5 are shown in Fig. 12.20. The straight line to the origin is the same as that shown in Fig. 12.19 and thus represents cements made with certain Type IV clinkers. We see that points representing other types of clinker also conform to straight lines having the same slope, but these lines do not pass through the origin. That is to say, in general,

$$\sigma_o = 0.8(\sigma_p - \sigma_l) \tag{12.52}$$

and

$$\sigma_l = \sigma_p - 1.25\sigma_o \tag{12.53}$$

where σ_l is the loss of specific surface area a cement suffers when it is mixed with water to form paste.

For the group of clinkers represented in Table 12.5 the losses are related to type of composition as is indicated by the data in the accompanying table. The figures in the table should be taken as being only indicative of the role played by chemical composition, for there are various exceptions among cements of the same type, as will be discussed further on.

Clinker Number	ASTM Composition Type	Surface Loss (cm^2/cm^3)
4	IV	0
1 and 3	I and II	1200
2	III	2250

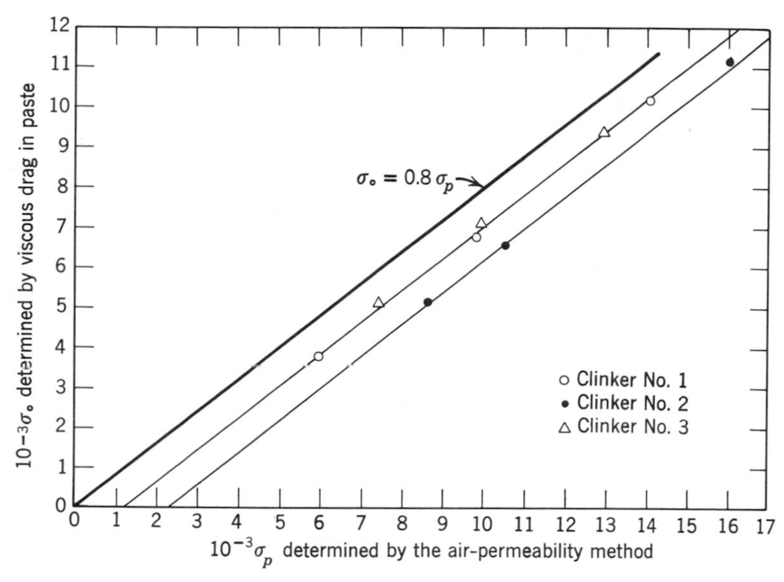

Fig. 12.20. Data showing loss of surface area by certain cements when they are made into pastes.

It may seem anomalous that the loss of specific surface area is indicated to be independent of the fineness of the cement. However, within the range of finenesses with which we are concerned even the coarsest of the cements contain much more of the finest material than can possi-

Table 12.6 Compositions of Certain Type-IV Cements Represented in Figure 12.19[a]

Reference Number	Specific Surface Area (cm^2/cm^3)	Calculated Compounds (per cent by weight)					Alkalies	
		C_3S	C_2S	C_3A	C_4AF	$CaSO_4$	Na_2O	K_2O
LTS 41	12850	20.0	51.0	4.5	15.2	3.4	0.06	1.19
LTS 43A	11710	29.0	52.0	5.3	9.3	3.2	0.33	0.01
Clinker No. 1	Various	34.0	54.0	2.3	6.0	3.2	0.05	0.22

[a] Data are from Lerch, W. and C. L. Ford, "Long-Time Study of Cement Performance in Concrete, Chapter 3—Chemical and Physical Tests of the Cement," *Proc. ACI*, **44**, 761, 1948. The average analysis of cements prepared from Clinker No. 4 shown here is not exactly the same as that shown in Table 12.2.

bly be used up by solution and the initial chemical reactions. As is shown in Chapter 2, Fig. 2.1, upward of 10 per cent of a cement has Stokes's diameters between about 0.3 and 7.5 μ with a specific surface area of about 40,000 cm^2/cm^3. If only 6 per cent of this material, or not more than $\frac{6}{10}$ per cent of the whole cement, is consumed by the initial reaction, the maximum amount of surface loss indicated above could be accounted for.

Probably much of the loss of surface area is due to the consumption of gypsum as it dissolves and reacts with tricalcium aluminate. We know also that an appreciable amount of tricalcium silicate is used up in the same period, and it may be that some of the flour-fine crystals of this material are consumed.

The products of the reactions just mentioned do not remain as separate particles; if they did, they would probably greatly reduce the bleeding rate, for the particles produced are much smaller than those consumed. The product forms a thin gel coating on the remaining grains and since there is still a large amount of surface remaining, the coating is on the average very thin. Thus it seems possible to account for a substantial loss of surface area with only a very slight augmentation of the size of the remaining grains, which probably constitutes well over 99 per cent of the original cement.

3.5 Exceptions

3.5.1 Nonconforming cements. Among cements nominally of the same type, not all show the same loss of surface area. For example, the cements prepared from Clinker No. 5 (see Table 12.2) deviate somewhat from the other Type-I cement, Clinker No. 1. They are unusually high in Na_2O.

Among Type-IV cements one showed a surface loss of about 2100 cm^2/cm^3. Like Clinker No. 5, it had a high Na_2O content, 1 per cent. However Cement No. 41 (see Table 12.6), which showed no loss of surface area, had a K_2O content of 1.19 per cent. This result seems compatible with Lerch's [10] finding: "With cements of similar tricalciumaluminate content, the alkalis influence the rate of reaction of gypsum with the cements, those of high alkali content react with gypsum more rapidly than those low in alkalies." As far as loss of surface area during the initial reaction is concerned, it seems that a distinction must be made between the effects of Na_2O and K_2O.

3.5.2 Abnormal states of flocculence. The orderly relationships between bleeding rate, specific surface area, and mixture composition that we have dealt with in the foregoing pages is apparently dependent on the peculiar state of flocculence that most cement pastes have. Experiments with pastes made with other materials sometimes give perplexing results. For example, when the same cement as that represented in Fig. 12.14 was mixed with hexyl ether it produced a paste that seemed to bleed normally at a constant rate, but the indicated specific surface area in comparison with σ_p and the value found in water paste was as follows:

Specific surface area by air permeability:	10600 cm^2/cm^3
Specific surface area by bleeding in water:	6700 cm^2/cm^3
Specific surface area by bleeding in hexyl ether:	7350 cm^2/cm^3

On the basis of Equation 12.53, this cement seems to have lost 2560 cm^2/cm^3 in water, but on the same basis it seems to have lost 1780 cm^2/cm^3 in hexyl ether. Since chemical reaction with hexyl ether seems impossible, the question is, why is the loss indicated?

The question seems to involve the degree of interparticle attraction as indicated by the relative bleeding capacity; the bleeding capacity of hexyl ether paste was much less than that of water paste at the same bleeding rate. Support for this interpretation is found in data published by Bruere and McGowan [11] who made water pastes with and without an admixture of a polyelectrolyte (Krilium), which acted as a strong flocculating agent. Their data gave the following figures:

Specific surface area by air permeability:	11500 cm^2/cm^3
Specific surface area in water paste:	6650 cm^2/cm^3
Specific surface area with 0.12 per cent Krilium:	4080 cm^2/cm^3

It is evident that the flocculating agent greatly reduces the effective surface area, in line with the interpretation of results from hexyl ether. There is no evidence that Krilium accelerates chemical action.

The evidence of difference in interparticle attraction is given in Fig. 12.21 where the relationships between bleeding capacity and bleeding rate for the mixtures just discussed are shown. It is evident that those losses of specific surface area which were *not* attributable to chemical reaction are associated with relatively large increases in interparticle attraction as indicated by large reductions in bleeding capacity at a given bleeding rate.

It seems that the strength of interparticle attraction when Krilium is present is such that the paste structure is composed of a network of small floccules, the floccules playing the same role as do the individual grains in normal paste structure. That is to say, some of the finest material that remains after the initial chemical reaction is bound so tightly to larger grains that not all of its surface area is presented

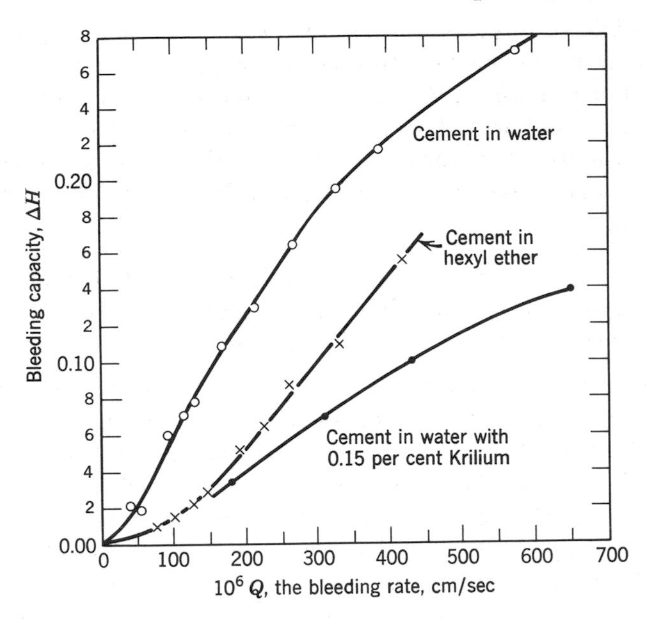

Fig. 12.21. Relationships between the bleeding rate and bleeding capacity showing relatively strong interparticle attraction in hexyl ether or in water containing Krilium.

to flowing water.* Presumably, the same state exists with cement in hexyl ether, but to a lesser degree.

The experimental indications discussed above naturally raise the question as to whether the same considerations apply to normal cement-water pastes, in which case the evidence of loss of surface during the initial chemical reaction would have to be discounted. It turns out that although there are differences in the interparticle attraction among the cements, the differences are not of the magnitude indicated in Fig. 12.21. In this connection we may recall the experiments with silica powder which were discussed in Chapter 11, Section 4.4.5: Changing the interparticle attraction by changing the concentration of calcium hydroxide had little or no effect on the bleeding rate and therefore no effect on the effective specific surface area. Among the cements of Table 12.4

* An alternative interpretation to that given in the text is that when Krilium is present the value of m is not the same for different mixtures, and thus the whole graph is like the nonconforming part of Fig. 12.14, giving a spurious value for specific surface area. The explanation given in the text seems to hold together, although it is only a presumption that strong flocculation can reduce effective surface area of the particles in a thick suspension.

there is no correlation between the indicated differences in interparticle attraction and the loss of surface.

As we have seen, it is not impossible for all the surface area in a flocculent mixture to be effective, and we know also that it is chemically possible for some of the cement to be used up. It is to be expected that if any kind of cement shows no loss of surface, that kind should be Type IV because of its relatively low degree of initial chemical activity [12].

3.6 *Rationale of Eq. 12.49*

The coefficient of viscosity of a fluid that is recorded in a handbook represents an intrinsic property of the fluid. The viscosity can be determined from experiments made with perfectly cylindrical capillaries in which the liquid is caused to flow telescopically (streamline flow) provided that energy losses associated with the entry and exit of the tube have been corrected for or eliminated; or it can be based on the rate of fall of a sphere in the liquid, provided that the body of liquid is very large or that suitable corrections are applied for extra resistance due to the reflections of the disturbance created by the sphere from the walls of the container; or it can be evaluated from the rate of flow between perfectly flat, parallel planes, provided that corrections are applied for edge conditions if the parallel planes do not constitute a semi-infiinite system.

Under nonideal conditions of flow, the intrinsic viscosity of a liquid can hardly be derived from measurements of rate of flow, pressure gradient, and dimensions of the flow space. Neither can the rate of flow under nonideal conditions be accurately calculated on the basis of the viscosity coefficient. For example, if a pipe is lined with small sand particles cemented on the metal surface, the resistance to flow through the pipe is greater than can be accounted for from the clear diameter of the pipe and the viscosity of the liquid. The rough boundary of the stream creates disturbances in the normally steady streamlines and extra energy is required to maintain a given rate of flow.

When we discussed the Poisseuille model (in Section 2.1), we encountered no difficulty, for the conditions necessary for measuring the coefficient of viscosity were stipulated: viscous flow through straight, cylindrical, smooth capillaries. Only the effects of end conditions were overlooked. But when we used the Kozeny-Carman equation to describe flow through porous bodies, or thick, flocculent suspensions, and still assumed that the proportionality constant between the rate of flow and the pressure gradient for a given average pore size is the same coefficient as it is for the Poisseuille model, the step was a long one. The difficulty has been recognized, of course, and many have attempted to cope with it with varying degrees of success. In the Kozeny-Carman model the problem seemed to be disposed of by a tortuosity and shape factor which could be evaluated empirically, the average pore size being taken care of by using the hydraulic radius. Within certain limits this has been made to work successfully, but in the present case we seem to have gone beyond those limits, for the assumption of stagnant liquid must be introduced, which is subject to question as we have already seen.

Even with corrections for pore tortuosity and shape, it seems very doubtful that flow through a granular bed, or a thick suspension is strictly laminar, however low the velocity of flow. Scheidegger [13] expressed it as follows:

"The fluctuations of the velocities of the [liquid] particles can be claimed as being analogous to the fluctuations of velocity during eddy motion in turbulent flow. Thus, the flow path of a particle through a porous medium can be regarded as analogous to the trajectory of a particle in turbulent hydraulic motion."

In other words, the flow patterns around individual particles interfere with each other, and the effect of the interaction is more severe the more concentrated the particles. It has proved extremely difficult if not impossible to deal with the effects of such hydrodynamic interaction in terms of detailed considerations of the flow patterns.

In the development leading to Eq. 12.49 we simply assumed that the least possible viscous resistance is the combined resistances of the individual particles according to Stokes's law, plus an added resistance due to hydrodynamic interaction. This added resistance can be expressed as a correction factor applied to the normal coefficient of viscosity, regardless of the actual geometry of the flow pattern. This gives the *effective* coefficient of viscous resistance, η_e, which is

$$\eta_e = \eta e^{k(1-\epsilon)} \tag{12.54}$$

Instead of attempting to apply corrections to the porosity, or to the effective surface area, we correct only the coefficient of viscous resistance, letting the characteristics of the flow pattern be represented by the concentration $(1 - \epsilon)$ and by k. Surprising as it may seem, the constant k does hold for a considerable range of concentrations, making it possible to evaluate Q_o analytically or graphically. This has proved to be true not only for fresh paste, but also for paste in the fully mature, hardened state.*

* The development referred to here and in Section 3.6.1 has not been published. It is a simplification and improvement of the interpretation of experimental data

3.6.1 Upper limit for specific surface area. As we have seen, the value of k or m in Eqs. 12.49 and 12.50 seems to depend on the flow pattern produced by particle shape and arrangement. It is smaller for spheres than for angular particles; it seems to be different for a packed bed than for a suspension; it is smaller for a nonflocculent than for a flocculent one; and apparently it is independent of particle size, that is, of specific surface area. With regard to size independence, that can be true only for material having a relatively low specific surface area as compared with that of colloidal material. In porous bodies made of colloidal material a considerable part of the water that is in the pores is within the range of the forces of adsorption. The effective viscosity has been found to be greatly increased by those forces, as is shown by elevated values of the activation energy for flow [14]. When adsorbed water is a significant fraction of the mobile water in a permeable body of colloidal material, as it is in *hardened* cement paste, the equation for effective viscosity is like Eq. 12.54 except that the exponential term has the form shown here:

$$\eta_e = \eta \exp \frac{\theta}{T} \left[\frac{1 - \epsilon}{\epsilon} \right] \tag{12.55}$$

where θ/T corresponds to k in Eq. 12.54 and is the interaction coefficient, θ being related to the extra activation energy for flow due to the effects of adsorption on the energy content of the flowing water, and T being the absolute temperature. The form of $f(\epsilon)$ is determined by the dominant role played by adsorbed water and thus by the surface area of the particles, which is proportional to $(1 - \epsilon)$; the ratio $(1 - \epsilon)/\epsilon$ shows that the effect depends inversely on the amount of water per unit of surface area. Data on hardened cement paste analyzed on this basis gives a value for specific surface of cement gel the same as that obtained by water vapor adsorption.

The data given in this chapter together with that referred to just above shows that Eq. 12.47 with

$$\frac{\phi(1 - \epsilon)}{\eta} \text{ replaced by } \frac{1}{\eta e^{k(1-\epsilon)}}$$

---

given by T. C. Powers, H. M. Mann, and L. E. Copeland, Highway Research Special Report No. 40, 1959.

It seems almost certain that the constant k is a function of extra activation for flow, the extra activation energy, over that for normal viscous flow, being due to hydrodynamic interaction; if so, its value is inversely proportional to the absolute temperature. However, the point is not brought out in the text, for as yet no experiments have been made to test it.

is suitable for relatively coarse materials such as portland cement. Also it is suitable for colloidal material if Eq. 12.55 is used.

There is probably an intermediate range of specific surface area in which neither expression for η_e is suitable. It seems likely that the effects of adsorption cannot be neglected if the specific surface area is upwards of perhaps 5×10^4 cm²/cm³.

3.7 Final Remarks on the Kozeny-Carman Equation

The Kozeny-Carman equation, as it is applied to the bleeding rate of cement paste, seems to give accurate figures for the specific surface area of cement as it exists in a paste, provided that the form based on the "hydrodynamic" specific surface area is used. (See Eqs. 12.27 or 12.32.) That is to say, the data indicate that $\sigma_o = \sigma_h$. Since it is physically impossible for independent streamlining envelopes to exist in a thick, flocculent paste and since the indicated surface area in a thick paste is the same as that in a very dilute suspension where such envelopes might exist, it follows that such envelopes do not exist under either condition, at least not in anywhere near the amount indicated by a typical value of α or w_i.

It seems therefore that α or w_i is an empirical constant which together with functions of ϵ gives the factor by which the normal, or intrinsic, coefficient of viscosity must be multiplied to obtain the effective coefficient of viscous resistance.

It follows that over the range of data where the two types of equations (Eqs. 12.32 and 12.49) are found to apply, there must be a constant value of α or w_i corresponding to that of k or m. Thus, letting the Carman constant be $k_c = 5.0$, and writing σ_o, for σ_h, Eq. 12.27 may be written as follows:

$$Q = \frac{2(\rho_s - \rho_f)g}{10\eta(1 + \alpha)^2\sigma_o^2}\left[\frac{\epsilon}{1 - \epsilon} - \alpha\right]^3 (1 - \epsilon)^2$$

The equivalent expression on the other basis is

$$Q = \left[\frac{2(\rho_s - \rho_f)g}{\eta\sigma_o^2}\right] \epsilon e^{-k(1-\epsilon)}$$

The quantity in brackets is the Stokes velocity for a dilute suspension. Eliminating this quantity from both equations, we have remaining the expressions for the extra viscous resistance, which should be equal to each other. Thus

$$\frac{(1 - \epsilon)^2}{10(1 + \alpha)^2}\left[\frac{\epsilon}{1 - \epsilon} - \alpha\right]^3 = \epsilon e^{-k(1-\epsilon)} = \epsilon e^{-2.203m(1-\epsilon)} \qquad (12.56)$$

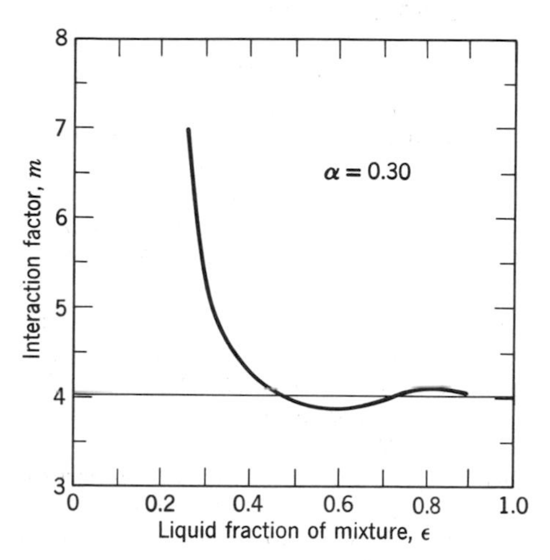

Fig. 12.22. Solution of Eq. 12.56 for m with $\alpha = 0.30$.

By assigning a value to α, Eq. 12.56 can readily be solved for k at various values of ϵ. Results for $\alpha = 0.3$ are shown in Fig. 12.22. We see that for $\epsilon = 0.45$ to $\epsilon = 0.9$, m oscillates around an average value of 4.0, which should agree, and does, with the figures shown for the first group in Tables 12.1 and 12.4.

Since ϵ for pastes is generally upwards of 0.5 we can say that the modified Kozeny-Carman model is as accurate as the other, or nearly so, but the constant α is needed in connection with $f(\epsilon)$ to account for extra viscous resistance due to interference between the flow patterns around adjacent grains in a thick suspension or packed bed.

REFERENCES

[1] Wylie, M. R. J., "The Historical Development of the Kozeny-Carman Equation," together with a translation by W. F. Striedeck and C. M. Davis of the original paper, "Concerning Capillary Conduction of Water in the Soil," by Joseph Kozeny, *Proc. Royal Acad. Sci., Class 1,* **136,** 217–306, Vienna, 1927. Published by The Petroleum Branch of the American Institute of Mining and Metallurgical Engineers. (Not dated)

[2] Steinour, H. H, "Rate of Sedimentation," *Ind. Eng. Chem.,* **36,** 618–624, 840–847, 901–907, 1944.

[3] Kozeny, J., "Concerning Capillary Conduction of Water in the Soil," *Proc. Royal Acad. Sci., Class 1,* **136,** 271–306, Vienna, 1927.

[4] Carman, P. C., "Fluid Flow Through Granular Beds," *Trans. Inst. Chem. Engrs.,* **15,** 150, 1937.

[5] Wyllie, M. R. J., and M. B. Spangler, "Application of Electrical Resistivity Measurements to Problem of Fluid Flow in Porous Media," *Bull. Am. Petrol. Geologists,* **36**(2), 359–403, 1952.

[6] Francis, A. W., "Wall Effect in Falling Ball Method for Viscosity," *Physics,* **4,** 403–406 (1933).

[7] Powers, T. C., "The Bleeding of Portland Cement Paste, Mortar and Concrete," Portland Cement Association Research Bulletin, No. 2, 1939.

[8] Lea, F. M., and R. W. Nurse, *J. Soc. Chem. Ind.,* **58,** 277–83T, 1939.

[9] Loudon, A. G., "The Computation of Permeability from Simple Soil Tests," *Geotechnique,* **3,** 165, 1952–53.

[10] Lerch, W., "The Influence of Gýpsum on the Hdyration and Properties of Portland Cement Pastes," *Proc. ASTM,* **46,** 1252–1293, 1946.

[11] Bruere, G. M., and J. K. McGowan, "Synthetic Polyelectrolytes as Concrete Admixtures," *Australian J. Appl. Sci.,* **9,** 127–140, 1958.

[12] Copeland, L. E., and D. L. Kantro, "Chemistry of Hydration of Portland Cement at Ordinary Temperature," in H. F. W. Taylor, ed., *The Chemistry of Cements,* Academic, New York, 1964.

[13] Scheidegger, A. E., *The Physics of Flow Through Porous Media,* Macmillan, New York, 1957.

[14] Powers, T. C., "Physical Properties of Cement Paste," *Proceedings of the Fourth International Symposium on the Chemistry of Cement, Monograph No. 43,* National Bureau of Standards, U.S. Department of Commerce, Washington, D.C., 1960, Vol. II, Paper V-1, pp. 577–604.

[15] Steinour, H. H., "Further Studies of the Bleeding of Portland Cement Paste," Portland Cement Association Research Bulletin No. 4, 1945.

Index

Page numbers in italics refer to entries in chapter references.